# Artificial Intelligence

V. W. Marek M. Truszczyński

# Nonmonotonic Logic

## Context-Dependent Reasoning

Foreword by Ray Reiter

Springer-Verlag
Berlin Heidelberg GmbH

V. Wiktor Marek
Mirosław Truszczyński

Department of Computer Science, University of Kentucky
Lexington, KY 40506-0027, USA

CR Subject Classification (1991): I.2.0, I.2.3-4, F.4.0-1

With 14 Figures

ISBN 978-3-662-02908-4

Library of Congress Cataloging-in-Publication Data
Marek, V. Wiktor, 1943- . Nonmonotonic logic: context-dependent reasoning / V. Wiktor Marek, M. Truszczyński. p. cm. – (Artificial intelligence) (Berlin, Germany)
ISBN 978-3-662-02908-4 ISBN 978-3-662-02906-0 (eBook)
DOI 10.1007/978-3-662-02906-0

1. Artificial intelligence. 2. Reasoning. 3. Optoelectronics. I. Truszczyński, Mirosław. II. Title. III. Series. Q335.M37 1993 006.3–dc20 93-31084


Originally published by Springer-Verlag Berlin Heidelberg New York in 1993
Softcover reprint of the hardcover 1st edition 1993



Typesetting: Camera-ready by authors
45/3140 - 5 4 3 2 1 0 - Printed on acid-free paper

To Ela, Małgosia and Natalia
W.M.

To Lena, Ania and Natalka
M.T.

# Foreword

When I first participated in exploring theories of nonmonotonic reasoning in the late 1970s, I had no idea of the wealth of conceptual and mathematical results that would emerge from those halting first steps. This book by Wiktor Marek and Mirosław Truszczyński is an elegant treatment of a large body of these results. It provides the first comprehensive treatment of two influential nonmonotonic logics – autoepistemic and default logic – and describes a number of surprising and deep unifying relationships between them. It also relates them to various modal logics studied in the philosophical logic literature, and provides a thorough treatment of their applications as foundations for logic programming semantics and for truth maintenance systems.

It is particularly appropriate that Marek and Truszczyński should have authored this book, since so much of the research that went into these results is due to them. Both authors were trained in the Polish school of logic and they bring to their research and writing the logical insights and sophisticated mathematics that one would expect from such a background. I believe that this book is a splendid example of the intellectual maturity of the field of artificial intelligence, and that it will provide a model of scholarship for us all for many years to come.

Ray Reiter
Department of Computer Science
University of Toronto
Toronto, Canada M5S 1A4
and
The Canadian Institute for
Advanced Research

# Table of Contents

# List of Figures

# 1. Introduction

## 1.1 The subject of this book

In recent years a significant effort in logical foundations of artificial intelligence has been devoted to formalizations of commonsense reasoning. An inherent property of commonsense reasoning is its nonmonotonicity. That is, a larger set of "initial assumptions" does not necessarily imply a larger set of "consequences". To model this phenomenon several nonstandard formal systems have been proposed. They are commonly referred to as *nonmonotonic logics*. Our goal in this book is to present a thorough study of some of the most important nonmonotonic logics, including default logic and modal nonmonotonic logics.

The reasons why nonmonotonicity arises in the context of commonsense reasoning are quite natural. We will discuss them briefly below. Artificial intelligence, in its attempts to built systems for simulating human commonsense reasoning, employs the language of classical logic. However, it does not consider all realities modeling an agent's assumptions but only those which are "intended" ("preferred", "more probable"). Such an approach reflects a common feature of human reasoning agents who usually distinguish between the models and do not regard all of them as equally possible. For example, when a mechanic tries to find an explanation for a problem with a car then he or she excludes some models knowing they are possible but unlikely. Sometimes the remaining models (explanations) are arranged according to their likelihood. An immediate consequence of restricting the class of models is nonmonotonicity. Assign to a given set $S$ of formulas the set $c(S)$ of formulas true in all the "intended" realities which satisfy $S$. Since not all models are considered but only the "intended" ones, the set $c(S)$ may be larger than the set of first-order consequences of $S$, $Cn(S)$. Moreover, there is no reason why intended models of a smaller set of formulas should be intended models of a larger set. Therefore there is no reason for $S_1 \subseteq S_2$ to imply $c(S_1) \subseteq c(S_2)$.

The presence of nonmonotonicity in commonsense reasoning can also be explained by analyzing *syntactic* aspects of the reasoning process. In commonsense reasoning we often include into our belief set statements which have no justification in our initial assumptions beyond the fact that we have no evidence in our belief set to contradict them. We often use the term "jumping to conclusions" to denote the following inference technique: if there is no evidence that would contradict $\varphi$, conclude $\varphi$. It should be quite clear that "jumping to conclusions" is not monotonic. After all, if $\varphi$ was included into our belief set by jumping to conclusions, then after the set of initial assumptions is expanded by $\neg\varphi$, $\varphi$ can no longer be derived.

Both aspects of nonmonotonicity, restrictions on classes of models and syntactic "jumping to conclusions", are discussed in our book. However, our main intuitions and the way we think about nonmonotonicity are rooted in syntax. Reasoning systems which formalize and generalize the process of jumping to conclusions are the main subject of our book. We investigate proof systems that extend the classical propositional logic proof system and we study modal nonmonotonic logics. We discuss connections between nonmodal and modal nonmonotonic logics. Our syntactic considerations lead to several algorithms for computing belief sets. We also discuss ensuing general complexity issues. We try not to neglect the "semantic side" of the theory of nonmonotonic logic. We exhibit several semantics for some of our nonmonotonic proof systems.

In the first part of this book we investigate nonmonotonic rules of proof. This mode of reasoning is based on the concept of *context-dependent proof* or *derivation*. The idea is to *relativize* the concept of a proof using a *context* to control the applicability of rules. Speaking more precisely, a context determines what is and what is not a valid derivation. Hence, we deal with a parameterized notion of proof. Fixpoint conditions lead then to structures that one can assign to a (nonmonotonic) proof system. Such proof systems are usually called *default theories*, and the most commonly studied structures associated with a default theory are its *Reiter extensions*. It turns out that similar structures appear in other proof systems such as logic programming and truth maintenance and are also present in McDermott and Doyle's modal nonmonotonic logics.

We study McDermott and Doyle's family of modal nonmonotonic logics in detail in the second part of this book. Default logic and modal nonmonotonic systems are seemingly different but, in fact, they are closely connected (although not in a simple fashion). We study these connections at the end of this book.

Proof systems going beyond classical logic (also beyond classical modal logic) require reevaluation of semantics. We will construct several semantics properly modeling the "works" of default logic and logic programming.

The current investigations in nonmonotonic logics concentrate, roughly speaking, in three areas. First, there is Reiter's default logic [Rei80] and its modifications. Second, there are modal nonmonotonic logics with their origins in the work of McDermott and Doyle [MD80, McD82]. Finally, there is McCarthy's circumscription [McC80]. Even though this is by no means an exhaustive classification, it corresponds well to the first-order, modal and second-order logics in the classical setting. Only the first two areas are treated in our book. The main reason why we do not study circumscription here is that circumscription is a second-order system. This fact has an important consequence. It has been shown by Schlipf [Sch87] that, in its full generality, that is, admitting arbitrary recursive first-order theories, circumscription is so expressive that it allows one to define very complex sets of natural numbers. Specifically, Schlipf proved that all $\Delta^1_2$ sets of natural numbers are definable by means of circumscription. Thus, in particular, circumscription is more expressive than dynamic logic. Without entering into details, let us express the conviction that this goes beyond currently accepted notions of reasonable computability. However, it needs to be mentioned that if we restrict the system to the propositional case then the situation changes and the circumscription becomes computable (Przymusinski [Prz89]).

Circumscription is associated with minimality of models, and minimizing models leads to the nonmonotonicity of circumscription. The syntactic paradigm of jumping to conclusions seems to be absent from circumscription. It so happens that default logic and some McDermott-Doyle logics also associate belief sets with minimal models. But for these systems, it is only a consequence of the "jumping to conclusions" process on which they are based and not the defining property. We do not claim that there is no connection between circumscription and the syntactically defined nonmonotonic systems discussed in this book. In fact, there are results connecting both systems in particular cases (for instance Imielinski [Imi87], Lifschitz [Lif88]), but there is a difference in perspective in both approaches.

The roots of our investigations can be traced to the pioneering work of Doyle, McCarthy, McDermott, Moore and Reiter mentioned above. Their classic papers [Doy79, McC80, MD80, McD82, Moo85, Rei80] looked at various aspects of commonsense reasoning and proposed seemingly different nonmonotonic logics to formalize them. Those papers went against the accepted opinions of logicians and philosophers who often claimed that classical logic in its various manifestations is the ideal logic of reasoning. There is a 2000-year tradition of classical logic which seems to support such contention. One can attempt to interpret the Gödel completeness theorem in this fashion. This, however, is misleading. In fact, Gödel, in his another classic paper on the incompleteness theorem deals with *intended* models (specifically, $\omega$-models). Similarly, Henkin's proof of the completeness theorem deals with *intended* models built on "prescribed" constants.

In hindsight it is easy to see that the ideas that underlie nonmonotonic logics are firmly grounded in the solid foundations created by past generations of logicians. One needs to realize, though, that the ideas of Reiter, and McDermott and Doyle, about the possibility of a logical theory of educated and justifiable guesses were novel and went against the tradition of classical logic.

Reading current issues of journals publishing papers on mathematical logic, one often wonders about the connections between specialized issues raised there and the analysis of human reasoning, which is, after all, the subject of logic. In fact one may get the impression that the logic community is no longer interested in issues related to the correctness of reasoning processes. The papers related to these issues have been relegated mainly to the philosophical and artificial intelligence journals. The idea that the task of analyzing the rules of "correct arguments" is essentially completed can often be detected in conversations with logicians. We can only hope that this book will be a reminder that there are still modes of reasoning that deserve thorough investigation and that there are vast and undeveloped areas in the field.

## 1.2 What do we do in this book?

Dedekind, in his fundamental study [Ded88, Ded48] on the foundations of real number systems says that "In science nothing capable of proof ought to be accepted without proof. Though this demand seems so reasonable yet I cannot regard it as having been met even in the most recent methods of laying the foundations of the simplest science; viz. that part of logic which deals with the theory of numbers". When we look at the current literature of logical foundations of artificial intelligence we often see a hypertrophy of ideas relative to substance. That is, the authors often do not subject their ideas to the rigor prescribed by Frege, Schröder, Hilbert, Gödel and Tarski and many others that followed them. This may lead some, especially in the more formal areas of Computer Science, to ask if the area of nonmonotonic logics can afford a rigorous presentation at all. We hope that our work addresses these issues. In particular we hope that this work shows that problems stemming out of logical foundations of artificial intelligence can be given a rigorous presentation. There is one additional aspect of our project. We believe that further exploration of the area of nonmonotonicity requires a firm foundation. Only with such a foundation it will be possible to subject the domain to the usual investigations, variations, and extensions.

We have no doubt that a rigorous presentation of the ideas of nonmonotonic logics is possible. In fact we try to give such presentation in this book.

We do not cover all of the aspects of the domain in our book, but we sincerely hope that it shows that the domain is fascinating and still vastly unexplored.

Our approach is to study the logics of Reiter, McDermott and Doyle, and Moore according to the standards of formal logic, and using the methods stemming from the tradition of proof theory and model theory. In short, we attempt to present the domain of nonmonotonic logic as a mature part of formal logic.

There are other monographs devoted to the subject, and one can ask if yet another one is indeed necessary. The difference between our work and other books on the subject is that we are less concerned with motivations. After all the best motivations are given in the papers by the founders. For instance the collected works of McCarthy [McC90] or the survey paper by Reiter [Rei87] contain much better arguments for the domain that we would be able to present (without extensive quoting). Hence we concentrate on the "technology" of nonmonotonic logics — an aspect of the domain which is rather inadequately treated in other monographs.

## 1.3 To whom do we address this book?

This book is principally addressed to:

- Researchers in the area of knowledge representation who apply the methods of logic and nonmonotonic logic to various representation schemes. In particular, we hope that the theory presented in the book will be applied by people trying to model commonsense aspects of knowledge and belief. The methods studied here try specifically to address these issues.

- Researchers in practical artificial intelligence who want to design their systems on the basis of sound theory. We hope that the rigorous treatment of the area of nonmonotonic logics presented here, including algorithms, may serve at least as a guide for their work.

- Researchers in logic programming who may want to address the issues of nonmonotonicity arising in the theory and practice of their domain.

- Computer scientists, logicians and philosophers studying logics presented in this book as well as graduate students in these fields.

The book can be used as a textbook. In fact we used earlier drafts of this text in our "special topics" courses offered at the University of Kentucky.

We have tried to write this book so that it is self-contained. Most of the basic material of mathematical logic needed to understand the book can be found in two chapters: "Rudiments" and "Modal Logic". Yet, in a sense, the book requires from the reader some maturity and familiarity with formal, rigorous mathematical proofs. Although the demonstrations do not go beyond standard proof techniques of logic there will be a lot of them and the user must be willing to take such a bitter pill.

## 1.4 The contents of this book

This book contains (besides this introduction) 12 chapters divided into three parts. The first part, dealing with the propositional proof systems and related nonmonotonic logics starts with an introductory chapter which sets up the technical terminology of logic and set theory used in the book. For completeness' sake we briefly treat propositional and predicate logics and provide some basic facts on partial orderings, well-orderings and ordinals. We also study properties of set operators. This chapter serves only as a review and provides a common language with the reader.

The next chapter contains a rigorous treatment of proof systems. These proof systems extend propositional logic. We introduce monotonic rules of proof and the associated consequence operators. Next, we introduce nonmonotonic rules of proof (defaults), default theories and Reiter's extensions for default theories. We investigate extensions from both syntactical and semantical viewpoints. We look at the relationship between well-orderings of the default rules and extensions. We also study algorithms implied by the theory developed in this chapter.

The following chapter is devoted to a special case of default theories, namely normal default theories. These theories have the desired property of always possessing at least one extension. We find that the precursor of all known forms of nonmonotonic logics, the Closed World Assumption of Reiter [Rei78] can be faithfully represented and characterized within normal default logic.

The next chapter gives several "normal form theorems" for default logic. We introduce the notion of equivalence between default theories. We prove that every default theory is equivalent to a clausal one. We also prove that by introducing new constants one can reduce an arbitrary default theory to a theory consisting of semi-normal defaults and justification-free defaults only.

The final chapter of the first part is devoted to these aspects of logic programming that are related to nonmonotonic reasoning. This chapter is not meant as a comprehensive study of logic programming but rather as a study of issues on the borderline of nonmonotonic logic and logic programming. We discuss (locally) stratified logic programs and their perfect models. We investigate stable models of logic programs. We deal with the generalizations of logic programming, discussing the Gelfond-Lifschitz theory of logic programming with classical negation and the Yang-Blair-Brown result on logic programming with clauses. Both these generalizations turn out to be very closely related to Reiter's default logic.

The second part of the book also contains five chapters and is concerned with modal nonmonotonic logics. As in the first part, we devote the first chapter to the rudiments, in this case of modal logic. In contrast to the first part, this chapter does more for us than just setting up the terminology. The point here is that we use the variant of provability in modal logic where necessitation can be applied not only to axioms but to other formulas as well. Hence the basic theory of modal logic needs to be modified somewhat.

The second chapter of this part deals with stable theories, that is, theories in modal language which consist of sentences true in some fixed Kripke model with a universal accessibility relation. These theories are basic for further development and we study them in detail.

The next chapter presents the theory of $\mathcal{S}$-expansions for logics $\mathcal{S} \subseteq \mathbf{S5}$. We investigate their properties and various characterizations. We study both syntactical and semantical characterizations of $\mathcal{S}$-expansions.

This theory is applied to several special cases of modal logic $\mathcal{S}$. The most important of these are three modal logics: the logic of pure necessitation, **N**, the logic **kd45** and the logic **Sw5** (more commonly known under the name **S4.4**). The nonmonotonic logic **kd45** corresponds to the well-known autoepistemic logic of Moore. The nonmonotonic logic **Sw5** corresponds to the logic of reflexive knowledge of Schwarz.

The final chapter of this part is devoted to various topics in modal nonmonotonic logics. Besides of the discussion of bounding introspection and the issues of groundedness we address the algorithmic aspects of $\mathcal{S}$-expansions. We present algorithms due to Schwarz, Niemelä and ourselves.

The third part of the book consists of two chapters. The first one presents connections between various reasoning systems discussed in the previous two parts. We show how extensions for default theories can be reconstructed within the McDermott scheme and how the fragments of McDermott's scheme can be reconstructed in default logic. We present embeddings of logic programming into the logic of reflexive knowledge and into autoepistemic logic. We also discuss briefly truth maintenance systems.

Finally, in the last chapter we present results on the complexity of default reasoning, autoepistemic logic and reasoning with propositional logic programs. We also briefly discuss the complexity of stable semantics of finite predicate logic programs.

## 1.5 Acknowledgments

This book would not have been written without the assistance of a great many colleagues and friends.

There are three people to whom we are most indebted. First of all, we are indebted to Raymond Reiter, who initiated many of the ideas underlying this book. In fact only after we understood the gothic simplicity of his idea for an extension of a default theory did this book become possible.

The whole theory studied in the part of this book devoted to modal nonmonotonic logics owes its presentation to Grigori Schwarz. It so happened that he joined us in Lexington precisely when we needed his assistance most. For this we should probably thank Michail Gorbachev and the KGB of the former Soviet Union for delivering his passport in time. Schwarz's presence and his ideas changed our way of thinking about modal logic.

The perspective of logic as presented here we owe to Anil Nerode and his way of looking at the connections of logic to computation.

We are indebted to those colleagues whose ideas were explicitly used throughout this book: Melvin Fitting, Michael Gelfond, Georg Gottlob, Vladimir Lifschitz and Jeffrey Remmel.

Finally, we gratefully acknowledge conversations with and helpful assistance of Krzysztof Apt, Howard Blair, Allen Brown, David Etherington, Audrey Ferry, Joe Halpern, Jack Minker, Ilkka Niemelä, Halina Przymusinska, Teodor Przymusinski, Helena Rasiowa, Cecylia Rauszer, John Schlipf and V.S. Subrahmanian. We also immensely appreciate the effort of our students Paweł Cholewiński and Artur Mikitiuk, and J. Andrew Ross of Springer-Verlag, who very carefully read the manuscript and helped to eliminate countless mistakes.

The first author gratefully acknowledges the support of the Mathematical Sciences Institute of Cornell University during 1989/90 and later.

# 2. Rudiments of logic and set theory

The material we present in this book requires familiarity with classical propositional logic and some elements of predicate logic and set theory. For the sake of convenience several definitions and facts from these disciplines are gathered below. Our selection is by no means comprehensive. For a detailed treatment of propositional and predicate logics the reader is referred to the books by Mendelson [Men64], Rasiowa and Sikorski [RS70], Fitting [Fit90] and Nerode and Shore [NS93]. An in-depth presentation of set theory can be found in the monographs by Kuratowski and Mostowski [KM76] and Jech [Jec78].

## 2.1 Syntax of propositional logic

We first define the language of propositional logic. The elements of the language, referred to as *well-formed formulas* or simply *formulas*, are built of special elements called *atoms*, of *logical connectives* $\neg, \vee, \wedge, \supset$, and of *brackets* ( and ). The connectives $\neg$, $\vee$, $\wedge$ and $\supset$ are referred to as *negation, disjunction, conjunction* and *implication*, respectively. Other selections of logical connectives are also possible. The set of atoms is denoted by *At*. We will assume (unless explicitly stated otherwise) that *At* is countably infinite and that it contains two distinguished elements $\top$ and $\bot$. In addition, for technical reasons, we assume that the set *At* does not contain any string of connectives or brackets. The language of propositional logic based on the set of atoms *At* will be denoted by $\mathcal{L}_{At}$ and, when the set *At* is fixed, simply by $\mathcal{L}$. A formal definition of the language $\mathcal{L}_{At}$ follows.

**Definition 2.1** Given a set of atoms *At*, by the language of propositional logic, denoted $\mathcal{L}_{At}$, we mean the least set $U$ of strings over the alphabet $At \cup \{\neg, \vee, \wedge, \supset, (, )\}$ such that:

1. $At \subseteq U$.

2. If $\varphi \in U$ then $(\neg\varphi) \in U$.
3. If $\varphi, \psi \in U$ then $(\varphi \vee \psi) \in U$.
4. If $\varphi, \psi \in U$ then $(\varphi \wedge \psi) \in U$.
5. If $\varphi, \psi \in U$ then $(\varphi \supset \psi) \in U$.

The elements of $\mathcal{L}_{At}$ are called *well-formed formulas*, or simply *formulas*. A *theory* is any set of formulas from $\mathcal{L}_{At}$. ▲

One can show that the least set of strings satisfying requirements (1) – (5) indeed exists, that is, the language $\mathcal{L}_{At}$ is well-defined.

Following the standard convention, when writing formulas we will omit the outermost pair of brackets. A formula $(\ldots(l_1 \vee l_2)\vee\ldots)\vee l_k$ will be written as $l_1 \vee \ldots \vee l_k$ and, similarly, a formula $(\ldots(l_1 \wedge l_2)\wedge\ldots)\wedge l_k$ will be written as $l_1 \wedge \ldots \wedge l_k$. We will also write $\varphi \equiv \psi$ to denote the formula $(\varphi \supset \psi)\wedge(\psi \supset \varphi)$.

The definition of the language $\mathcal{L}_{At}$ directly implies the following result, which provides a useful technique for proving properties of formulas.

**Theorem 2.2** (Principle of Induction for Formulas) *Let $Q$ be any property of strings over the alphabet $At \cup \{\neg, \wedge, \vee, \supset, (, )\}$. Assume that:*

1. *For every $p \in At$, the string $p$ has the property $Q$.*
2. *Whenever a string $s$ possesses the property $Q$ then also the string $(\neg s)$ possesses the property $Q$.*
3. *Whenever strings $s_1, s_2$ possess the property $Q$ then the strings $(s_1 \wedge s_2)$, $(s_1 \vee s_2)$, and $(s_1 \supset s_2)$ possess the property $Q$.*

*Then every well-formed formula possesses the property $Q$.* ▲

It turns out that rather than consider arbitrary formulas it is often convenient to restrict attention to classes of formulas of particularly simple structure: conjunctions of *clauses* and disjunctions of *elementary conjunctions*. Clauses and elementary conjunctions as well as some other related classes of formulas are defined below.

**Definition 2.3** (Literals, clauses, elementary conjunctions)

1. A *literal* is a formula of the form $a$ or of the form $\neg a$, where $a$ is an atom from $At$. A literal of the form $\neg a$ is called *negative*. The set of all literals will be denoted by $Lit_{At}$.

2. A *clause* is a formula of the form $l_1 \vee \ldots \vee l_k$ where $l_1, \ldots, l_k$ are literals. The set of all clauses in $\mathcal{L}_{At}$ will be denoted by $Clause_{At}$.

3. A *Horn clause* is a clause $l_1 \vee \ldots \vee l_k$ such that at most one of $l_1, \ldots, l_k$ is an atom.

4. A *program Horn clause* is a Horn clause in which *exactly* one of $l_1, \ldots, l_k$ is an atom.

5. An *elementary conjunction* is a formula of the form $l_1 \wedge \ldots \wedge l_k$ where $l_1, \ldots, l_k$ are literals. ▲

Different sets of atoms generate different languages and, in particular, different sets of literals and clauses. However, it turns out that when we extend the set of atoms of a language, the resulting language and the classes of atoms, literals and clauses conservatively extend the corresponding classes of formulas in the initial language. Formally, this observation can be stated as follows.

**Theorem 2.4** *Let $At_1 \subseteq At_2$ be two sets of atoms. Then:*

1. $\mathcal{L}_{At_1} \subseteq \mathcal{L}_{At_2}$.

2. $\mathcal{L}_{At_1} \cap At_2 = At_1$.

3. $\mathcal{L}_{At_1} \cap Lit_{At_2} = Lit_{At_1}$.

4. $\mathcal{L}_{At_1} \cap Clause_{At_2} = Clause_{At_1}$. ▲

## 2.2 Semantics of propositional logic

We will now introduce a semantics for the language $\mathcal{L}$. To this end, we distinguish two elements **t** and **f**, later referred to as the *truth values*. They represent truth and falsehood, respectively.

There are several natural operations on the truth values. They reflect the intuitive meaning of the logical connectives and are denoted by the same symbols.

**Definition 2.5** (Operations on truth values) We specify the operations $\neg, \vee, \wedge, \supset$ in a table. The operation $\neg$ is unary, the operations $\vee, \wedge, \supset$ are binary.

| $p$ | $q$ | $\neg p$ | $p \wedge q$ | $p \vee q$ | $p \supset q$ |
|---|---|---|---|---|---|
| **f** | **f** | **t** | **f** | **f** | **t** |
| **f** | **t** | **t** | **f** | **t** | **t** |
| **t** | **f** | **f** | **f** | **t** | **f** |
| **t** | **t** | **f** | **t** | **t** | **t** |

▲

To define the meaning of an atom we assign a truth value to it.

**Definition 2.6** A *valuation* of a set of atoms $At$ is any mapping $v: At \rightarrow \{\mathbf{t}, \mathbf{f}\}$ such that $v(\top) = \mathbf{t}$ and $v(\bot) = \mathbf{f}$. ▲

Valuations assign truth values to the atoms of a language. The next result states that each valuation can be extended uniquely to the set of all formulas in such a way that the operations on truth values are preserved. We utilize this observation to assign truth values to arbitrary formulas.

**Theorem 2.7** *Let $v : At \rightarrow \{\mathbf{t}, \mathbf{f}\}$ be a valuation. There exists a unique function $\overline{v} : \mathcal{L}_{At} \rightarrow \{\mathbf{t}, \mathbf{f}\}$ such that:*

1. $\overline{v}(p) = v(p)$, *for every* $p \in At$.
2. $\overline{v}(\neg\varphi) = \neg\overline{v}(\varphi)$.
3. $\overline{v}(\varphi \wedge \psi) = \overline{v}(\varphi) \wedge \overline{v}(\psi)$.
4. $\overline{v}(\varphi \vee \psi) = \overline{v}(\varphi) \vee \overline{v}(\psi)$.
5. $\overline{v}(\varphi \supset \psi) = \overline{v}(\varphi) \supset \overline{v}(\psi)$. ▲

In Theorem 2.7, the symbols $\neg$, $\wedge$, $\vee$ and $\supset$ on the left-hand sides of the equations denote the logical connectives whereas the same symbols on the right-hand sides denote operations on truth values as specified in Definition 2.5.

Since a valuation $v$ of atoms uniquely determines $\overline{v}$ we will usually write $v(\varphi)$ instead of $\overline{v}(\varphi)$. For a set of formulas $T$ we will write $v(T) = \mathbf{t}$ to denote that $v(\varphi) = \mathbf{t}$, for every $\varphi \in T$.

**Definition 2.8** Let $v : At \rightarrow \{\mathbf{t}, \mathbf{f}\}$ be a valuation. We say the a formula $\varphi \in \mathcal{L}$ is *satisfied by* $v$ (or *true in* $v$) if $v(\varphi) = \mathbf{t}$. We will sometimes write $\models_v \varphi$ in such case.

We say that a theory $T$ is *satisfied by* $v$ if $v(T) = \mathbf{t}$. We will sometimes write $\models_v T$ in such case. ▲

Some formulas are true in every valuation.

**Definition 2.9** A formula $\varphi$ is called a *tautology* if for every valuation $v$, $\models_v \varphi$. The set of all tautologies of the language $\mathcal{L}_{At}$ is denoted by $Taut_{At}$. When $At$ is fixed we will omit the subscript $At$. ▲

For every $\varphi, \psi, \vartheta \in \mathcal{L}$ the following formulas are tautologies:

1. $\top$
2. $\neg \bot$
3. $\varphi \vee \neg\varphi$
4. $\varphi \supset \varphi$
5. $\varphi \supset (\psi \supset \varphi)$
6. $(\varphi \supset (\psi \supset \vartheta)) \supset ((\varphi \supset \psi) \supset (\varphi \supset \vartheta))$
7. $\neg\neg\varphi \equiv \varphi$
8. $(\varphi \supset \psi) \supset (\neg\psi \supset \neg\varphi)$
9. $(\varphi \wedge \psi) \equiv (\psi \wedge \varphi)$
10. $((\varphi \wedge \psi) \wedge \vartheta) \equiv (\varphi \wedge (\psi \wedge \vartheta))$
11. $(\varphi \wedge (\psi \vee \vartheta)) \equiv ((\varphi \wedge \psi) \vee (\varphi \wedge \vartheta))$

The notion of satisfaction of a formula by a valuation leads to the concept of semantical entailment or consequence.

**Definition 2.10** Let $T$ be a theory and $\varphi$ a formula. We say that $T$ *semantically entails* $\varphi$ (or that $\varphi$ is a *semantical consequence* of $T$), if for every valuation $v$, $\models_v T$ implies that $\models_v \varphi$. We denote it by writing $T \models \varphi$. We write $\models \varphi$ to denote that $\emptyset \models \varphi$.

The *semantical consequence operator* in the language $\mathcal{L}_{At}$ is defined as:

$$Cn_{At}(T) = \{\varphi \in \mathcal{L}_{At} : T \models \varphi\}.$$

When the set $At$ is fixed, the subscript $At$ is usually omitted. ▲

Clearly, for every valuation $v$ we have that $\models_v \emptyset$. Hence, $\models \varphi$ if and only if $\varphi$ is a tautology. In other words, $Taut = Cn(\emptyset)$.

We will list now several useful properties of the semantical consequence operator $Cn$.

**Theorem 2.11** *For all theories $T_1, T_2 \subseteq \mathcal{L}$:*

1. $T_1 \subseteq Cn(T_1)$.
2. $T_1 \subseteq T_2$ *implies that* $Cn(T_1) \subseteq Cn(T_2)$.
3. $Taut \subseteq Cn(T_1)$.
4. $Cn(Cn(T_1)) = Cn(T_1)$. ▲

**Definition 2.12** We say that a theory $T \subseteq \mathcal{L}$ is *closed under semantical consequence* if $Cn(T) = T$. ▲

**Theorem 2.13** (Closure properties)

1. *For every valuation $v$, the theory $\{\varphi \in \mathcal{L}: v(\varphi) = \mathbf{t}\}$ is closed under semantical consequence.*
2. *Let $T$ be a theory closed under semantical consequence. Then:*
   (a) *If $\varphi \in T$ and $\varphi \supset \psi \in T$ then $\psi \in T$.*
   (b) *If $\varphi \in T$, and $\varphi \supset \psi$ is a tautology then $\psi \in T$.*
   (c) *$\varphi \wedge \psi \in T$ if and only if $\varphi \in T$ and $\psi \in T$/*
3. *For every family $\mathcal{T}$ of theories closed under semantical consequence, the intersection $T$ of the theories in $\mathcal{T}$ is closed under semantical consequence.* ▲

Definition 2.10 yields a simple technique for proving that a formula is not a semantical consequence of a theory.

**Theorem 2.14** *Let $T$ be a theory and $\varphi$ a formula. The formula $\varphi$ does not belong to $Cn(T)$ if and only if there exists a valuation $v$ such that $\models_v T$ and $v(\varphi) = \mathbf{f}$.* ▲

We will now introduce the concept of a complete theory and present some relevant results.

**Definition 2.15** Let $T \subseteq \mathcal{L}$ be closed under semantical consequence. A theory $T \subseteq \mathcal{L}$ is *complete* if for every formula $\varphi \in \mathcal{L}$ at least one of the formulas $\varphi$, $\neg\varphi$ is in $T$. ▲

**Theorem 2.16** *Let $T \subseteq \mathcal{L}$ be closed under semantical consequence. A theory $T \subseteq \mathcal{L}$ is complete if and only if $T$ contains at least one of the literals $p$, $\neg p$, for every atom $p$ in $\mathcal{L}$.* ▲

The concept of a complete theory can be generalized.

**Definition 2.17** Let $\mathcal{C} \subseteq \mathcal{L}$. Let $T \subseteq \mathcal{L}$ be closed under semantical consequence. A theory $T \subseteq \mathcal{L}$ is *complete with respect to $\mathcal{C}$* if for every formula $\varphi \in \mathcal{C}$ at least one of the formulas $\varphi$, $\neg\varphi$ is in $T$. ▲

The following result is a natural extension of Theorem 2.16.

**Theorem 2.18** *Let $P$ be a subset of the set of atoms of $\mathcal{L}$. Let $\mathcal{L}_P$ be the language generated by $P$. A theory $T \subseteq \mathcal{L}$ is complete with respect to $\mathcal{L}_P$ if and only if $T$ contains at least one of the literals $p$, $\neg p$, for every atom $p$ in $P$.* ▲

The whole language $\mathcal{L}$ is, trivially, an example of a complete theory. More interesting examples can be constructed as follows. For a valuation $v$ of $\mathcal{L}$ define

$$Th(v) = \{\varphi : v(\varphi) = \mathbf{t}\}.$$

Clearly, $Th(v)$ is closed under semantical consequence (Theorem 2.13) and, by Theorem 2.16, it is complete. More generally, for a collection $V$ of valuations of $\mathcal{L}$ define

$$Th(V) = \bigcap \{Th(v) : v \in V\}.$$

Let $P$ be the set of atoms which are assigned the same logical value by every valuation in $V$. Then, $Th(V)$ is complete with respect to $P$ or, equivalently, with respect to $\mathcal{L}_P$.

Two formulas $\varphi$ and $\psi$ are *semantically equivalent* if for every valuation $v$, $v(\varphi) = v(\psi)$ or, in other words, if $\varphi \equiv \psi$ is a tautology. It turns out that for an arbitrary formula $\varphi$ there exist formulas of syntactically restricted form that are semantically equivalent to $\varphi$. These formulas are often referred to as *normal form* formulas.

**Definition 2.19** (Normal form formulas)

1. A formula $\varphi$ is in the *conjunctive normal form* (or *CNF* for short) if $\varphi = \vartheta_1 \wedge \ldots \wedge \vartheta_n$ and each $\vartheta_j$, $1 \leq j \leq n$, is a clause.

2. A formula $\varphi$ is in the *disjunctive normal form* (or *DNF* for short) if $\varphi = \psi_1 \vee \ldots \vee \psi_m$ and each $\psi_j$, $1 \leq j \leq m$, is an elementary conjunction.

3. A formula $\varphi$ is in the *double conjunctive form* if $\varphi = \vartheta_1 \wedge \ldots \wedge \vartheta_n$ and each $\vartheta_j$, $1 \leq j \leq n$, is a negation of an elementary conjunction.

4. A formula $\varphi$ is in the *double disjunctive form* if $\varphi = \psi_1 \vee \ldots \vee \psi_m$ and each $\psi_j$, $1 \leq j \leq m$, is a negation of a clause. ▲

**Theorem 2.20** *For every formula $\varphi$ there exist formulas $\psi_1, \psi_2, \psi_3, \psi_4$ in conjunctive, disjunctive, double disjunctive, and double conjunctive normal form, respectively, such that for $i = 1, 2, 3, 4$ the formulas $\varphi$ and $\psi_i$ are semantically equivalent.* ▲

Theorems 2.13 and 2.20 imply that if $T$ is closed under semantical consequence then $T$ contains all the normal forms of its member formulas. Moreover, a formula $\varphi$ in conjunctive normal form belongs to $T$ if and only if all the conjunct clauses of $\varphi$ belong to $T$. Consequently, we obtain the following theorem which is, in fact, the basis for several automated theorem proving techniques.

**Theorem 2.21** *Let $T_1, T_2 \subseteq \mathcal{L}$ be theories closed under semantical consequence. Then $T_1 = T_2$ if and only if $T_1 \cap Clause = T_2 \cap Clause$. In other words, a theory which is closed under semantical consequence is uniquely determined by its set of clauses.* ▲

Finally, we will discuss a situation when new atoms are introduced to a language. Let $At_1$ and $At_2$ be two sets of atoms such that $At_1 \subseteq At_2$. Then the restriction of a valuation of $At_2$ to $At_1$ is a valuation of $At_1$. The truth values of formulas from $\mathcal{L}_{At_1}$ are preserved under such transformation. Two related formalizations of this observation are gathered in the next theorem.

**Theorem 2.22** *Let $At_1, At_2$ be sets of atoms such that $At_1 \subseteq At_2$.*

1. *Let $v$ be a valuation of $At_1$ and $v'$ a valuation of $At_2$ such that for every $p \in At_1$, $v(p) = v'(p)$. Then for every $\varphi \in \mathcal{L}_{At_1}$, $v(\varphi) = v'(\varphi)$.*

2. *Let $v$, $v'$ be two valuations of $At_2$ such that $v|_{At_1} = v'|_{At_1}$. Then for every $\varphi \in \mathcal{L}_{At_1}$, $v(\varphi) = v'(\varphi)$.* ▲

The next theorem states that when a language is extended by new atoms, then the semantical consequences in this language are preserved.

**Theorem 2.23** *Let $At_1, At_2$ be sets of atoms such that $At_1 \subseteq At_2$. Then, for every theory $T \subseteq \mathcal{L}_{At_1}$,*

$$Cn_{At_2}(T) \cap \mathcal{L}_{At_1} = Cn_{At_1}(T).$$

*In particular, $Taut_{At_2} \cap \mathcal{L}_{At_1} = Taut_{At_1}$.* ▲

Finally, our last result in this section deals with the situation when new atoms are added to a theory.

**Theorem 2.24** *Let $At_1 \subseteq At_2$ be two sets of atoms. Let $S \subseteq At_2 \setminus At_1$. Then for every theory $T \subseteq \mathcal{L}_{At_1}$:*

1. *$T$ is consistent if and only if $T \cup S$ is consistent.*
2. *$Cn_{At_2}(T \cup S) \cap \mathcal{L}_{At_1} = Cn_{At_1}(T)$.* ▲

## 2.3 Proof theory for propositional logic

The semantics of propositional logic described in the previous section has a proof-theoretic counterpart. This proof theory is specified by a selection of axiom schemata, inference (or proof) rules, and the notion of proof. We introduce all these concepts below.

**Definition 2.25** (Uniform substitution) Given a formula $\varphi \in \mathcal{L}$, propositional variables $p_1, \ldots, p_k \in At$ and formulas $\psi_1, \ldots, \psi_k \in \mathcal{L}$, the formula $\varphi(p_1/\psi_1, \ldots, p_k/\psi_k)$, called a ***substitution instance*** of $\varphi$, is defined recursively as follows:

1. In the case when $\varphi \in At$, $\varphi(p_1/\psi_1, \ldots, p_k/\psi_k) = \psi_i$ if $\varphi = p_i$, for some $i$, $1 \leq i \leq k$, and $\varphi(p_1/\psi_1, \ldots, p_k/\psi_k) = \varphi$, otherwise.
2. If $\varphi = \neg\beta$, then $\varphi(p_1/\psi_1, \ldots, p_k/\psi_k) = \neg\beta(p_1/\psi_1, \ldots, p_k/\psi_k)$.

3. If $\varphi = \beta_1 \vee \beta_2$, then $\varphi(p_1/\psi_1, \ldots, p_k/\psi_k) = \beta_1(p_1/\psi_1, \ldots, p_k/\psi_k) \vee \beta_2(p_1/\psi_1, \ldots, p_k/\psi_k)$.

4. If $\varphi = \beta_1 \wedge \beta_2$, then $\varphi(p_1/\psi_1, \ldots, p_k/\psi_k) = \beta_1(p_1/\psi_1, \ldots, p_k/\psi_k) \wedge \beta_2(p_1/\psi_1, \ldots, p_k/\psi_k)$.

5. If $\varphi = \beta_1 \supset \beta_2$, then $\varphi(p_1/\psi_1, \ldots, p_k/\psi_k) = \beta_1(p_1/\psi_1, \ldots, p_k/\psi_k) \supset \beta_2(p_1/\psi_1, \ldots, p_k/\psi_k)$.

A theory $T \subseteq \mathcal{L}$ is said to be *closed under uniform substitution* if for every formula $\varphi$ from $T$, every substitution instance of $\varphi$ is in $T$. ▲

The set of tautologies is closed under uniform substitution. That is, if $\varphi$ is a tautology then any substitution instance of $\varphi$ is also a tautology.

**Definition 2.26** (Inference rules)

1. Any $n$-tuple of formulas from $\mathcal{L}$, where $n \geq 1$, is called an *inference (derivation, proof) rule* or, simply, a *rule*.

2. A rule $\langle \varphi_0, \ldots, \varphi_{n-1}, \vartheta \rangle$ is usually written as

$$\frac{\varphi_0, \ldots, \varphi_{n-1}}{\vartheta}. \tag{2.1}$$

   The formulas $\varphi_0, \ldots, \varphi_{n-1}$ are called the *premises* and $\vartheta$ is called the *conclusion* of the rule (2.1).

3. A theory $T$ is *closed under rule* (2.1) if $\varphi_0, \ldots, \varphi_{n-1} \in T$ implies that $\vartheta \in T$.

4. A *substitution instance* of a rule $r$ is any rule of the form

$$\frac{\varphi_0(p_1/\psi_1, \ldots, p_k/\psi_k), \ldots, \varphi_{n-1}(p_1/\psi_1, \ldots, p_k/\psi_k)}{\vartheta(p_1/\psi_1, \ldots, p_k/\psi_k)}.$$

5. A *rule schema* is a set $R$ of rules closed under uniform substitution. That is, for every rule $r \in R$, every substitution instance of the rule $r$ belongs to $R$. ▲

The only rule schema we adopt for the proof system of propositional logic is *modus ponens*, which consists of all inference rules of the form

$$\frac{\varphi, \varphi \supset \psi}{\psi}.$$

To define provability in propositional logic we also need axioms. There are several equivalent choices of axioms possible. One selection is given below.

**Definition 2.27** The following formulas are axioms of propositional logic.

1. $\bot \supset p$
2. $p \supset \top$
3. $p \supset (q \supset p)$
4. $(p \supset (q \supset r)) \supset ((p \supset q) \supset (p \supset r))$
5. $p \wedge q \supset p$
6. $p \wedge q \supset q$
7. $p \supset (q \supset (p \wedge q))$
8. $p \supset p \vee q$
9. $q \supset p \vee q$
10. $(p \supset r) \supset ((q \supset r) \supset ((p \vee q) \supset r))$
11. $(p \supset q) \supset ((p \supset \neg q) \supset \neg p)$
12. $\neg\neg p \supset p$ ▲

We introduce now the notions of a proof in propositional logic, the provability relation and several other related notions.

**Definition 2.28** (Proof system for propositional logic)

1. A *proof* (or *derivation*) of a formula $\varphi \in \mathcal{L}$ from a theory $T \subseteq \mathcal{L}$ (in propositional logic) is a sequence $\psi_1, \ldots, \psi_n$ of formulas of $\mathcal{L}$ such that $\psi_n = \varphi$ and for every $k$, $1 \leq k \leq n$, at least one of the following conditions holds:
   (a) $\psi_k$ is a substitution instance of an axiom of propositional logic from the list given in Definition 2.27.
   (b) $\psi_k \in T$.
   (c) $\psi_k$ is the result of applying the modus ponens rule to formulas $\psi_i, \psi_j$ for some $i, j < k$.

We write $T \vdash \varphi$ to denote that there is a proof of $\varphi$ from $T$. We abbreviate $\emptyset \vdash \varphi$ to $\vdash \varphi$. We call $\vdash$ the *provability* relation.

2. A formula $\varphi$ such that $\vdash \varphi$ is called a *theorem* of propositional logic.
3. A theory $T$ is *consistent* if $\{\varphi : T \vdash \varphi\} \neq \mathcal{L}$. Otherwise, $T$ is *inconsistent*.
4. A theory $T$ is *closed under propositional provability* if $T = \{\varphi : T \vdash \varphi\}$.

▲

The following theorem states a useful property of the provability relation.

**Theorem 2.29** (Deduction theorem) *Let $T$ be a theory and $\varphi, \psi \in \mathcal{L}$. Then the following conditions are equivalent:*

1. $T \cup \{\varphi\} \vdash \psi$.
2. $T \vdash \varphi \supset \psi$. ▲

## 2.4 Completeness and compactness of propositional logic

We have introduced two notions of consequence in propositional logic: semantical entailment and provability. In this section we present a fundamental result of propositional logic, known as the completeness theorem, which states that both notions are equivalent. We also list several consequences of this result.

**Theorem 2.30** (Completeness theorem) *For every theory $T \subseteq \mathcal{L}$, $\{\varphi : T \vdash \varphi\} = Cn(T)$. That is, semantic and syntactic concepts of consequence are equivalent.* ▲

From now on we will use the terms semantical consequence, propositional provability, and propositional consequence interchangeably.

**Corollary 2.31** *A formula is a theorem of propositional logic if and only if it is a tautology.* ▲

Theorem 2.14 and the completeness theorem imply the following result that allows us to show that a formula is not provable from a theory.

**Theorem 2.32** *For every theory $T \subset \mathcal{L}$ and every formula $\varphi \in \mathcal{L}$, $T \nvdash \varphi$ if and only if there exists a valuation $v$ such that $v(T) = \mathbf{t}$ and $v(\varphi) = \mathbf{f}$.* ▲

Theorem 2.32 implies, in particular, that consistent theories are satisfiable.

**Corollary 2.33** *If $\{\varphi : T \vdash \varphi\} \neq \mathcal{L}$ then there exists a valuation $v$ such that $\models_v T$.* ▲

Since every proof involves only finitely many formulas, the completeness theorem implies the following property (called *compactness* or *finitizability*) of the operator $Cn$.

**Theorem 2.34** (Compactness theorem)

$$Cn(T) = \bigcup\{Cn(T') \ : \ T' \subseteq T \ \text{ and } \ T' \ \text{ is finite}\}.$$

▲

## 2.5 Relations, partial orderings

In this section we recall several set-theoretical notions and facts which will be used later in this book. The reader is expected to know elementary facts of set algebra and basic properties of relations and functions. The notation which we use is standard. Symbols $\cup, \cap$ denote the set-theoretical operations of union and intersection. The symbol $\setminus$ denotes the set-theoretical operation of relative complement. The *power set* of a set $X$, that is, the set of all subsets of $X$, will be denoted by $\mathcal{P}(X)$.

**Definition 2.35** (Types of relations) Let $R$ be a binary relation in $X$.

1. The relation $R$ is *reflexive* if $(x, x) \in R$, for every $x \in X$.
2. The relation $R$ is *antisymmetric* if for every $x, y \in X$, $(x, y) \in R$ and $(y, x) \in R$ imply that $x = y$.
3. The relation $R$ is *transitive* if for every $x, y, z \in X$, $(x, y) \in R$ and $(y, z) \in R$ imply that $(x, z) \in R$.
4. The relation $R$ is *connected* if for every $x, y \in X$, $(x, y) \in R$ or $(y, x) \in R$. ▲

**Definition 2.36** (Partially ordered sets)

1. Let $X$ be a set and $\leq$ a binary relation in $X$. A pair $\langle X, \leq \rangle$ is called a *partially ordered set* (or *poset* for short) if $\leq$ is reflexive, antisymmetric and transitive.

2. Let $\langle X, \leq \rangle$ be a poset. A subset $Y$ of $X$ is a *chain* if for all $x, y \in Y$, $x \leq y$ or $y \leq x$.

3. Let $\langle X, \leq \rangle$ be a poset. If $Y \subseteq X$ and $x \in X$ then we say that $x$ is a *lower bound* (*upper bound*) of $Y$ if for all $y \in Y$, $x \leq y$ ($y \leq x$).

4. Let $\langle X, \leq \rangle$ be a poset. An element $x \in X$ is called *minimal* (*maximal*) if for no $y \neq x$, $y \leq x$ ($x \leq y$).

5. Let $\langle X, \leq \rangle$ be a poset. An element $x \in X$ is called *least* (*greatest*) if for every $y \in X$, $x \leq y$ ($y \leq x$). ▲

Not every poset possesses minimal or maximal elements. The following celebrated and useful result provides a sufficient condition for the existence of minimal (maximal) elements.

**Theorem 2.37** (Kuratowski-Zorn lemma)

1. *Let $\langle X, \leq \rangle$ be a poset such that $X \neq \emptyset$ and for every chain $Y \subseteq X$ there is an upper bound. Then $\langle X, \leq \rangle$ possesses a maximal element. In fact, for every $y \in X$ there exists a maximal element $x$ such that $y \leq x$.*

2. *Let $\langle X, \leq \rangle$ be a poset such that $X \neq \emptyset$ and for every chain $Y \subseteq X$ there is a lower bound. Then $\langle X, \leq \rangle$ possesses a minimal element. In fact, for every $y \in X$ there exists a minimal element $x$ such that $x \leq y$.* ▲

Since the inclusion relation $\subseteq$ is a partial ordering of any family $\mathcal{Y} \subseteq \mathcal{P}(X)$, as a corollary to the Kuratowski-Zorn lemma we obtain criteria for the existence of minimal and maximal elements in families of sets.

**Corollary 2.38** (Kuratowski-Zorn lemma in set algebra)

1. *Let $X$ be any set, and let $\mathcal{Y} \subseteq \mathcal{P}(X)$ be a family of subsets of $X$. Assume that for every subfamily $\mathcal{Z} \subseteq \mathcal{Y}$ such that for all $Z_1, Z_2 \in \mathcal{Z}$ $Z_1 \subseteq Z_2$ or $Z_2 \subseteq Z_1$, we have $\bigcup \mathcal{Z} \in \mathcal{Y}$. Then, for every $Z \in \mathcal{Y}$, there is a maximal element in $\mathcal{Y}$ which contains $Z$.*

2. *Let $X$ be any set, and let $\mathcal{Y} \subseteq \mathcal{P}(X)$ be a family of subsets of $X$. Assume that for every subfamily $\mathcal{Z} \subseteq \mathcal{Y}$ such that for all $Z_1, Z_2 \in \mathcal{Z}$ $Z_1 \subseteq Z_2$ or $Z_2 \subseteq Z_1$, we have $\bigcap \mathcal{Z} \in \mathcal{Y}$. Then, for every $Z \in \mathcal{Y}$, there is a minimal element in $\mathcal{Y}$ included in $Z$.* ▲

Finally, we introduce three operations on relations.

**Definition 2.39**

1. Let $R_1 \subseteq A \times B$ and $R_2 \subseteq B \times C$ be two relations. The *composition* of $R_1$ and $R_2$ is a subset $R_1 \circ R_2$ of $A \times C$ such that $(a, c) \in R_1 \circ R_2$ if and only if there exists $b \in B$ such that $(a, b) \in R_1$ and $(b, c) \in R_2$.
2. Let $A, B$ be two disjoint sets and $R_1 \subseteq A \times A$ and $R_2 \subseteq B \times B$. The *concatenation* of relations $R_1$ and $R_2$, denoted $R_1 \odot R_2$, is the relation $R_1 \cup (A \times B) \cup R_2$.
3. Let $R \subseteq A \times A$, and let $B \subseteq A$. The *restriction* of $R$ to $B$, $R|_B$, is the relation $R \cap (B \times B)$. ▲

## 2.6 Elements of predicate logic

In this section we present several results on predicate logic. We will use them in Chapter 6, where we discuss logic programming. We assume familiarity with the concepts defined below; for details the reader is referred to [Fit90]. In our overview we concentrate on the results concerned with Herbrand models, as they play an important role in Chapter 6.

By $\mathcal{L}_{pred}$ we mean a standard language of predicate logic built of:

1. Countable sets $C$, $F$, $P$ and $V$ of constant, function, predicate and variable symbols.
2. Propositional connectives and propositional constants $\bot$, $\top$.
3. Quantifier symbols $\forall$ and $\exists$.
4. Punctuation symbols: ')', '(' and ','.

Each function and predicate symbol has an integer associated with it, called its *arity*. The constant, variable and function symbols together with the punctuation symbols are used to define the *terms* of the language $\mathcal{L}_{pred}$. The terms of the language together with the predicate symbols, the propositional connectives and constants and the punctuation symbols are used to define the set of formulas of the language. This process starts with *atomic formulas* that is formulas of the form $p(t_1, \ldots, p_k)$ (where $p$ is a predicate symbol of our language and $t_j$'s are terms of the language). Subsequently, using propositional connectives and quantifiers we construct all the formulas.

It is clear that the language $\mathcal{L}_{pred}$ depends on the sets $C$, $F$, $P$ and $V$. For instance, predicate symbols of arity 0 can be regarded as propositional variables. Consequently, the predicate language with the empty sets of constant, function and variable symbols and with all predicate symbols of arity 0 is nothing else but a language of propositional logic. Hence, propositional logic is a special case of predicate logic.

We adopt the following conventions. The letters $a$, $b$, $c$ will be used to denote constant symbols. The letters $f$, $g$, $h$, will denote function symbols and the letters $p$ and $q$ will stand for predicate symbols. Variable symbols will be denoted by the letters $x$, $y$ and $z$ and terms by the letters $s$ and $t$. In each case, indices will be used if needed.

We will introduce now the notions of an interpretation for the language $\mathcal{L}_{pred}$ and variable assignment. We follow the approach of [Fit90], where a detailed treatment of semantics for predicate calculus can be found.

**Definition 2.40** An *interpretation* for the language $\mathcal{L}_{pred}$ is a pair $\mathcal{M} = \langle U, I\rangle$, where $U$ is a nonempty set called the *domain* of the interpretation and $I$ is a mapping which assigns:

1. An element $c^I$ from $U$ to every constant symbol $c \in C$.
2. A function $f^I : U^k \rightarrow U$ to every $k$-ary function symbol $f \in F$.
3. A relation $p^I \subseteq U^k$ to every $k$-ary predicate symbol $p \in P$.

A mapping $A$ which assigns an element $x^A$ from $U$ to every variable symbol $x \in V$ is called a *variable assignment*. ▲

Given an interpretation $\mathcal{M} = \langle U, I\rangle$ and a variable assignment $A$ one can associate an element $t^{\mathcal{M},A}$ of $U$ with every term $t$ of the language $\mathcal{L}_{pred}$. The details are given below.

**Definition 2.41** For a term $t$ of the language $\mathcal{L}_{pred}$, we define the element $t^{\mathcal{M},A}$ recursively as follows:

1. For a constant symbol $c \in C$, $c^{\mathcal{M},A} = c^I$.

2. For a variable symbol $x \in V$, $x^{\mathcal{M},A} = x^A$.

3. If $t = f(t_1, \dots, t_k)$, $t^{\mathcal{M},A} = f^I(t_1^{\mathcal{M},A}, \dots, t_k^{\mathcal{M},A})$. ▲

Given an interpretation and a variable assignment we can associate with each formula one of the two truth values **t** and **f**.

**Definition 2.42** Let $\mathcal{M} = \langle U, I\rangle$ be an interpretation for the language $\mathcal{L}_{pred}$, $A$ a variable assignment and $\varphi$ a formula in $\mathcal{L}_{pred}$. To each formula we assign a truth value, denoted, $\varphi^{\mathcal{M},A}$. The definition is recursive:

1. For every atomic formula $p(t_1, \dots, t_k)$, we set $(p(t_1, \dots, t_k))^{\mathcal{M},A} = \mathbf{t}$ if $(t_1^{\mathcal{M},A}, \dots, t_k^{\mathcal{M},A}) \in p^I$, and $(p(t_1, \dots, t_k))^{\mathcal{M},A} = \mathbf{f}$, otherwise.

2. $\top^{\mathcal{M},A} = \mathbf{t}$ and $\bot^{\mathcal{M},A} = \mathbf{f}$.

3. $(\neg\psi)^{\mathcal{M},A} = \neg(\psi)^{\mathcal{M},A}$.

4. $(\psi_1 \vee \psi_2)^{\mathcal{M},A} = (\psi_1)^{\mathcal{M},A} \vee (\psi_2)^{\mathcal{M},A}$.

5. $(\psi_1 \wedge \psi_2)^{\mathcal{M},A} = (\psi_1)^{\mathcal{M},A} \wedge (\psi_2)^{\mathcal{M},A}$.

6. $(\psi_1 \supset \psi_2)^{\mathcal{M},A} = (\psi_1)^{\mathcal{M},A} \supset (\psi_2)^{\mathcal{M},A}$.

7. $((\forall x)\psi)^{\mathcal{M},A} = \mathbf{t}$ if $\psi^{\mathcal{M},A'} = \mathbf{t}$ for every variable assignment $A'$ which coincides with $A$ at every variable except possibly at $x$. Otherwise, $((\forall x)\psi)^{\mathcal{M},A} = \mathbf{f}$.

8. $((\exists x)\psi)^{\mathcal{M},A} = \mathbf{t}$ if $\psi^{\mathcal{M},A'} = \mathbf{t}$ for some variable assignment $A'$ which coincides with $A$ at every variable except possibly at $x$. Otherwise, $((\exists x)\psi)^{\mathcal{M},A} = \mathbf{f}$. ▲

A formula $\varphi$ is *true* in an interpretation $\mathcal{M}$ if $\varphi^{\mathcal{M},A} = \mathbf{t}$ for all variable assignments $A$. We write $\mathcal{M} \models \varphi$ in such case. A formula is *valid* if it is true in every interpretation $\mathcal{M}$ for $\mathcal{L}_{pred}$. We say that an interpretation $\mathcal{M}$ is a *model of a formula* $\varphi$ if $\mathcal{M} \models \varphi$. We say that $\mathcal{M}$ is a model of a theory $I \subseteq \mathcal{L}_{pred}$, which we denote by $\mathcal{M} \models I$, if $\mathcal{M}$ is a model of every formula in $I$.

In Chapter 6, where we make use of the material introduced in this section, we will be interested mainly in the case when formulas do not contain free variables. Such formulas are called *sentences*. It is easy to see that as long as we are interested in truth values of sentences, variable assignment is immaterial. For every sentence $\varphi$, every interpretation $\mathcal{M}$ and every two variable assignments $A$ and $A'$, $\varphi^{\mathcal{M},A} = \varphi^{\mathcal{M},A'}$. In particular, it follows that if $\varphi$ is a sentence, $\mathcal{M} \models \varphi$ or $\mathcal{M} \models \neg\varphi$.

There is a class of interpretations which is especially important in automated theorem proving and in logic programming. It consists of the so-called *Herbrand interpretations.* We will introduce Herbrand interpretations now.

**Definition 2.43** (Substitution)

1. A *substitution* is any mapping $\Theta$ which assigns terms of $\mathcal{L}_{pred}$ to variables in $V$.
2. For a term $t$ we define the term $t\Theta$ recursively as follows:
   (a) If $a$ is a constant, we set $a\Theta = a$.
   (b) If $x$ is a variable, then $x\Theta = \Theta(x)$.
   (c) If $f$ is a function symbol whose arity is $k$, then $f(t_1, \ldots, t_k)\Theta = f(t_1\Theta, \ldots, t_k\Theta)$. ▲

If a substitution $\Theta$ modifies only variables in a finite set $\{x_1, \ldots, x_n\}$, it is commonly written as $(x_1/t_1, \ldots, x_n/t_n)$, where $t_i$ is the term assigned to $x_i$ by $\Theta$.

**Example 2.44** Let $\Theta = (x/f(a), y/f(x), z/g(b, y))$. Then

$$\begin{aligned} g(a, g(x, f(y)))\Theta &= g(a\Theta, g(x, f(y))\Theta) \\ &= g(a\Theta, g(x\Theta, f(y)\Theta)) \\ &= g(a\Theta, g(x\Theta, f(y\Theta))) \\ &= g(a, g(f(a), f(f(x)))). \end{aligned}$$

▲

Next, we define the result of applying a substitution $\Theta$ to a formula of the language. Some care is needed here. The reason is that a variable may have a bounded occurrence in a formula (that is, it may fall under the scope of a quantifier) and a substitution should not have any effect on it.

Consider a substitution $\Theta$. Let $x$ be a variable. By $\Theta_x$ we denote the substitution obtained from $\Theta$ by changing the term it assigns to $x$ to $x$ itself. That is, $\Theta_x(x) = x$ and $\Theta_x(y) = \Theta(y)$, for every variable $y$ such that $y \neq x$.

**Definition 2.45** The action of a substitution $\Theta$ on formulas of the language $\mathcal{L}_{pred}$ is defined as follows:

1. If $p(t_1, \ldots, t_k)$ is an atomic formula (that is, $p$ is a predicate symbol in the language) then $p(t_1, \ldots, t_k)\Theta = p(t_1\Theta, \ldots, t_k\Theta)$.
2. $(\neg\varphi)\Theta = \neg(\varphi\Theta)$.
3. $(\varphi\Box\psi)\Theta = \varphi\Theta\Box\psi\Theta$, for every binary propositional connective $\Box$.
4. $((Qx)\varphi)\Theta = (Qx)(\varphi\Theta_x)$ for $Q = \forall, \exists$. ▲

We say that a substitution $\Theta$ is *ground* if for all $x \in V$, $\Theta(x)$ is a *ground term* of the language, that is, if $\Theta(x)$ has no variables. Let us observe that a ground term $f(t_1, \ldots, t_n)$ is the result of applying a ground substitution $\Theta = (x_1/t_1, \ldots x_n/t_n)$ to a term $f(x_1, \ldots x_n)$.

We define now the concepts of the Herbrand universe and the Herbrand base. They play a crucial role in logic programming and automated theorem proving.

**Definition 2.46** The set of all ground terms of the language $\mathcal{L}_{pred}$ is called the *Herbrand universe* of $\mathcal{L}_{pred}$, and is denoted by $U_H$. ▲

The Herbrand universe of the language $\mathcal{L}_{pred}$ depends on the sets of function and constant symbols of $\mathcal{L}_{pred}$ but not on the set of relational symbols of $\mathcal{L}_{pred}$. If $\mathcal{L}_{pred}$ has at least one constant symbol, then $U_H \neq \emptyset$. If, in addition, $\mathcal{L}_{pred}$ possesses at least one function symbol then $U_H$ is infinite.

**Example 2.47** Let us assume that $\mathcal{L}_{pred}$ contains exactly one constant symbol 0 and exactly two function symbols $f$ and $g$, each of arity 1. The Herbrand universe $U_H$ consists of terms of the form

$$\alpha_1(\alpha_2(\ldots\alpha_n(0)\ldots)),$$

where $n \geq 0$ and each $\alpha_i$ stands for $f$ or $g$. ▲

**Definition 2.48** The set of all ground atomic formulas, that is, formulas of the form

$$p(t_1, \ldots, t_k)$$

where $p$ is a predicate symbol of arity $k$ and $t_1, \ldots, t_k$ are ground terms, is called the *Herbrand base* of $\mathcal{L}_{pred}$, and is denoted by $B_H$. The elements of the Herbrand base are called *ground atoms* of $\mathcal{L}_{pred}$. ▲

**Example 2.49** Consider again the language from Example 2.47 and assume that $p$ is its predicate symbol of arity 2. Then $p(f(g(0)), g(f(0)))$ is a ground atom of the language $\mathcal{L}_{pred}$, that is, an element of the Herbrand base $B_H$. ▲

**Definition 2.50** A *Herbrand interpretation* of the language $\mathcal{L}_{pred}$ is an interpretation $\langle U_H, I\rangle$, where $U_H$ stands for the Herbrand universe and $I$ satisfies the conditions:

1. $c^I = c$ for every constant symbol $c \in C$.
2. $f^I(t_1, \ldots, t_k) = f(t_1, \ldots, t_k)$ for every $k$-ary function symbol $f \in F$ and ground terms $t_1, \ldots, t_k$.

A *Herbrand model* of a set of formulas $I$ is a Herbrand interpretation of $\mathcal{L}_{pred}$ which is a model of $I$. ▲

Herbrand interpretations have several special properties. For example, different constants are interpreted by different elements of the Herbrand universe. Secondly, if the interpretations of two ground terms $f(t_1, \ldots, t_k)$ and $g(s_1, \ldots s_l)$ are the same then $f = g$, $k = l$, and for all $i$, $1 \leq i \leq k$, $t_i = s_i$.

We will now discuss universal theories in $\mathcal{L}_{pred}$ and show the adequacy of Herbrand interpretations for modeling such theories.

**Definition 2.51** A sentence $\varphi$ is called a *universal* sentence if it is of the form

$$(\forall(\forall(\ldots(\forall x_1)\ldots)x_{k-1})x_k)\psi,$$

where $\psi$ is a formula without quantifiers. Such formulas are often written in an abbreviated form as

$$(\forall x_1, \ldots, x_k)\psi.$$

A theory consisting only of universal sentences is called *universal.* ▲

Herbrand interpretations suffice in order to establish consistency of universal theories. Indeed, we have the following well-known theorem.

**Theorem 2.52** *If $T$ is a universal theory then $T$ is consistent if and only if there is a Herbrand model of $T$.* ▲

Next, we will show that consistency checking of universal theories can be reduced to the propositional case.

**Definition 2.53** Let $\varphi$ be a universal sentence of the form

$$(\forall x_1, \ldots, x_k)\psi.$$

where $\psi$ is a formula without quantifiers. By a *propositional grounding* of $\varphi$, we mean a collection of all formulas

$$\psi(x_1/t_1, \ldots, x_k/t_k),$$

where $t_1 \ldots, t_k$ range over all ground terms in the Herbrand universe $U_H$.

For a universal theory $T$, a *propositional grounding* of $T$, *ground*$(T)$, is defined as the union of the propositional groundings of all universal sentences in $T$. ▲

Propositional grounding of a universal theory is a theory in the language of propositional logic generated by the elements of the Herbrand base $B_H$. Valuations or propositional models of theories in this language can be identified with subsets of the Herbrand base. It turns out that there is a *one-to-one* and *onto* correspondence between Herbrand models of a universal theory $T$ and propositional models of its propositional grounding *ground*$(T)$. Each subset $S$ of the Herbrand base determines a Herbrand interpretation $\mathcal{M}_S$. Indeed, by the definition, the universe of a Herbrand interpretation, and its functions and constants, are uniquely determined by the language $\mathcal{L}_{pred}$. It remains to define the relations of the interpretation $\mathcal{M}_S$. Let $p$ be a predicate symbol of $\mathcal{L}_{pred}$ of arity $k$. Define the corresponding relation $p_S$ by

$$p_S = \{(t_1, \ldots, t_k): t_i \in U_H,\ 1 \leq i \leq k, \text{ and } p(t_1, \ldots, t_k) \in S\}.$$

Conversely, each Herbrand interpretation $\mathcal{M}$ determines a subset $S_{\mathcal{M}}$ of the Herbrand base as follows. For every predicate symbol $p$ of arity $k$, the set $S_{\mathcal{M}}$ contains precisely those ground atoms $p(t_1, \ldots, t_k) \in B_H$ for which

$$(t_1, \ldots, t_k) \in p_{\mathcal{M}},$$

where $p_{\mathcal{M}}$ is the relation in the Herbrand interpretation $\mathcal{M}$ corresponding to the predicate symbol $p$.

Mappings $S \mapsto \mathcal{M}_S$ and $\mathcal{M} \mapsto S_{\mathcal{M}}$ establish *one-to-one* and *onto* correspondences between the subsets of the Herbrand base of the language $\mathcal{L}_{pred}$ and the Herbrand interpretations of the language $\mathcal{L}_{pred}$. In addition, we have the following property.

**Theorem 2.54** *For every universal theory $T$, a Herbrand interpretation $\mathcal{M}$ is a model of $T$ if and only if $S_{\mathcal{M}}$ is a propositional model of ground$(T)$.* ▲

From now on we will often think about a subset $S$ of the Herbrand base both as a propositional valuation (model) or as a Herbrand interpretation (model).

A theory $T$ may have many Herbrand models and their structure may be complex.

**Example 2.55** Consider the universal theory

$$T = \{r(a), (\forall x)(p(x) \vee q(x)), (\forall x)(r(x) \supset r(f(x)))\}.$$

This theory has many Herbrand models. In each of these models the extent of (the relation that interprets) $r$ is the same: $\{a, f(a), f(f(a)), \ldots\}$. But the extent of $p$ and $q$ varies. For instance, we can interpret $p$ as

$$\{a, f(f(a)), \ldots \underbrace{f(\ldots(f(a)\ldots)}_{2n}\ldots\}$$

and $q$ as

$$\{f(a), f(f(f(a))), \ldots \underbrace{f(\ldots(f(a)\ldots)}_{2n+1}\ldots\}.$$

In this way we get a Herbrand model of $T$, say $\mathcal{M}_1$.

Similarly, interpreting $q$ as

$$\{a, f(f(a)), \ldots \underbrace{f(\ldots(f(a)\ldots)}_{2n}\ldots\}$$

and $p$ as

$$\{f(a), f(f(f(a))), \ldots \underbrace{f(\ldots(f(a)\ldots)}_{2n+1}\ldots\}$$

gives another model of $T$, say $\mathcal{M}_2$.

The model $\mathcal{M}_1$ is *minimal* in the following sense. No Herbrand model of $T$ exists all of whose relations are contained in the corresponding relations of $\mathcal{M}_1$ and at least one of these containments is proper. We can equivalently define this notion of minimality using the property that Herbrand interpretations correspond to subsets of the Herbrand base. A Herbrand model of a theory is minimal if the corresponding subset of the Herbrand base is inclusion-minimal in the family of the subsets of the Herbrand base corresponding to all Herbrand models of $T$. The model $\mathcal{M}_2$ is minimal in the same sense. ▲

The existence of minimal Herbrand models plays an important role in Chapter 6. We give here two results establishing their existence for universal theories. The first result can be derived by a standard application of the Kuratowski-Zorn lemma. We omit the details. The second result, the one we will actually use later, is a straightforward corollary of the first one.

**Theorem 2.56** *Let $T$ be a universal theory and let $\mathcal{M}$ be a model of $T$. Let $C$ be a class of all models of $T$ with the same universe as $\mathcal{M}$ and with the same interpretation of function and constant symbols. The class $C$ contains a minimal model in the sense of inclusion of all relations. Moreover, there is a minimal model in $C$ all of whose relations are contained in the corresponding relations of $\mathcal{M}$.* ▲

**Corollary 2.57** *Every consistent universal theory possesses a minimal Herbrand model.* ▲

## 2.7 Well-orderings, ordinal numbers

In this section we recall the concepts of a well-ordering and of an ordinal number.

**Definition 2.58** Let $\preceq$ be a binary relation in a set $X$. We say that $\preceq$ *well-orders* $X$ (or, is a *well-ordering* of $X$) if:

1. $\preceq$ is reflexive, antisymmetric and transitive, that is, $\langle X, \preceq \rangle$ is a poset.
2. $\preceq$ is connected.
3. every nonempty subset $Y$ of $X$ has a least element. ▲

If $\preceq$ is a well-ordering of the set $X$ then the pair $\langle X, \preceq \rangle$ is called a *well-ordered set.*

We define now a class of particular well-ordered sets, called ordinal numbers. They serve as "archetypical" well-orderings: every well-ordered set is isomorphic to one of ordinal numbers. The approach we take yields ordinals whose well-ordering coincides with the inclusion relation between their elements. They are sometimes referred to as von Neumann ordinals. Their definition depends on the underlying set theory. In this book we will assume the Zermelo-Fraenkel axiomatization of set theory and the axiom of regularity (Jech [Jec78]).

**Definition 2.59** (Ordinal numbers)

1. A set $x$ is called *transitive* if for every set $y$, if $y \in x$, then $y \subseteq x$.
2. A set $x$ is called a (von Neumann) *ordinal number* (or, simply, ordinal) if $x$ is transitive and the inclusion relation on $x$ is connected. ▲

The set $\emptyset$ is an ordinal. It corresponds to the ordinal number 0. The set $\{\emptyset\}$ is also an ordinal. It corresponds to the ordinal number 1. The set $\{\emptyset, \{\emptyset\}\}$ is an ordinal, corresponding to the ordinal number 2. The set consisting of all finite ordinals, that is, $\{\emptyset, \{\emptyset\}, \{\emptyset, \{\emptyset\}\}, \ldots\}$, is an ordinal. It is the least infinite ordinal and it is denoted by $\omega$.

It is easy to see that if a set $\alpha$ is an ordinal, then so is $\alpha \cup \{\alpha\}$. The ordinal number $\alpha \cup \{\alpha\}$ is called the *successor* of $\alpha$ and is denoted by $\alpha + 1$. There are ordinals which are not successors of any ordinal. For instance, $\omega$ is not a successor of any ordinal. Such ordinals are called *limit* ordinals. The ordinal $\emptyset$ is the only finite limit ordinal.

**Theorem 2.60** (Basic properties of ordinals)

1. *Every ordinal is well-ordered by the inclusion relation.*
2. *For every ordinal $\alpha$, $\alpha = \{\beta : \beta \subset \alpha$ and $\beta$ ordinal$\}$.*
3. *For every well-ordered set $\langle X, \preceq \rangle$ there exists a unique ordinal $\alpha$ such that $\langle X, \preceq \rangle$ is isomorphic to $\langle \alpha, \subseteq \rangle$. Moreover, such an isomorphism is unique.* ▲

(Throughout the book, $\subset$ denotes the relation of proper inclusion of sets.)

We will often write $\alpha \leq \beta$ to denote that $\alpha \subseteq \beta$, and $\alpha < \beta$ to denote that $\alpha \subset \beta$. For any two distinct ordinal numbers $\alpha$ and $\beta$, either $\alpha < \beta$ or $\beta < \alpha$.

**Theorem 2.61** (Transfinite induction principle) *Let $\alpha > 0$ be an ordinal number and let $Z \subseteq \alpha$ be a subset of $\alpha$ possessing these two properties:*

1. $0 \in Z$.
2. *For every $\beta < \alpha$, whenever $\{\gamma : \gamma < \beta\} \subseteq Z$ then $\beta \in Z$.*

*Then $Z = \alpha$.* ▲

Since every ordinal is either a limit or a successor ordinal, transfinite induction is often applied in a slightly different, but equivalent, version.

**Theorem 2.62** *Let $\alpha > 0$ be an ordinal number, and let $Z \subseteq \alpha$ be a subset of $\alpha$ possessing these three properties:*

*1.* $0 \in Z$.

*2.* *For every* $\beta < \alpha$, *such that* $\beta + 1 < \alpha$, *if* $\beta \in Z$ *then* $\beta + 1 \in Z$.

*3.* *Whenever* $\lambda < \alpha$ *is a limit ordinal, and* $\{\xi : \xi < \lambda\} \subseteq Z$ *then* $\lambda \in Z$.

*Then* $Z = \alpha$. ▲

Theorems 2.61 and 2.62 are applied in the following fashion. Assume that $P$ is a property of elements of some ordinal $\alpha$. We will write $P(\beta)$ if an ordinal $\beta < \alpha$ has property $P$. Define $Z = \{\beta : \beta < \alpha \text{ and } P(\beta)\}$. Applying the transfinite induction principle to $Z$ yields a method of proving that all ordinal numbers smaller than $\alpha$ have property $P$.

**Theorem 2.63** *Let* $P$ *be any property, and let* $\alpha$ *be an ordinal number greater than* 0. *Assume that:*

*1.* $P(0)$.

*2.* *Whenever* $\beta < \alpha$, $\beta + 1 < \alpha$ *and* $P(\beta)$, *then* $P(\beta + 1)$.

*3.* *Whenever* $\lambda < \alpha$ *is a limit ordinal and for every ordinal* $\xi < \lambda$ *we have* $P(\xi)$, *then we have* $P(\lambda)$.

*Then, for every* $\beta < \alpha$, $P(\beta)$ *holds.* ▲

As one of the applications of Theorem 2.63 we obtain the following result that allows us to define functions on ordinals.

**Theorem 2.64** *Let* $X$ *be a set and* $a \in X$. *Let* $f : X \to X$ *be a function from* $X$ *to* $X$. *Let* $F : \mathcal{P}(X) \to X$ *be a function from* $\mathcal{P}(X)$ *to* $X$. *Then for every ordinal* $\alpha$ *there exists a unique function* $G : \alpha \to X$ *satisfying the following conditions:*

*1.* $G(0) = a$.

*2.* *For every* $\beta < \alpha$ *such that* $\beta + 1 < \alpha$, $G(\beta + 1) = f(G(\beta))$.

*3.* *For every limit ordinal* $0 < \lambda < \alpha$, $G(\lambda) = F(\{G(\xi) : \xi < \lambda\})$. ▲

We will use Theorem 2.64 to define transfinite sequences of iterations of an operator.

## 2.8 Monotone operators

We will discuss now some elements of the theory of operators in $X$, that is, mappings of $\mathcal{P}(X)$ into itself.

**Definition 2.65** (Types of operators)

1. An *operator* in $X$ is any mapping $F : \mathcal{P}(X) \rightarrow \mathcal{P}(X)$.
2. An operator $F$ is called *monotone* if for all $X_1, X_2 \subseteq X$, $X_1 \subseteq X_2$ implies $F(X_1) \subseteq F(X_2)$.
3. An operator $F$ is called *antimonotone* if for all $X_1, X_2 \subseteq X$, $X_1 \subseteq X_2$ implies $F(X_2) \subseteq F(X_1)$.
4. An operator $F$ is called *finitizable* (or *compact*) if for all $X_1 \subseteq X$,

$$F(X_1) = \bigcup\{F(X_2) : X_2 \subseteq X_1, \text{ and } X_2 \text{ is finite}\}.$$

▲

**Definition 2.66** (Fixpoint) If $F$ is an operator in $X$ then a set $Y \subseteq X$ is called a *fixpoint* of $F$ if $F(Y) = Y$ and a *prefixpoint* of $F$ if $F(Y) \subseteq Y$. ▲

**Example 2.67** Let $X = \omega$ be the set of natural numbers. Define $F(Y) = \{2^y : y \in Y\}$. Clearly $F$ is a monotone and finitizable operator. The set $\emptyset$ is a fixpoint of $F$. In fact it is the only fixpoint of $F$. ▲

Given an operator $F$ in the set $X$ we can associate with $F$ various sequences of subsets of $X$. Correctness of the following definition is guaranteed by Theorem 2.64.

**Definition 2.68** Let $F : \mathcal{P}(X) \rightarrow \mathcal{P}(X)$ be an operator in $X$ and let $Y \subseteq X$.

1. The *upper iteration sequence of $F$ starting at $Y$* is defined by induction as follows:

   $F\uparrow 0(Y) = Y$.
   
   $F\uparrow(\alpha + 1)(Y) = F(F\uparrow\alpha(Y))$.
   
   $F\uparrow\lambda(Y) = \bigcup\{F\uparrow\alpha(Y) : \alpha < \lambda\}$ for a limit ordinal $\lambda$.

2. The *lower iteration sequence of $F$ starting at $Y$* is defined by induction as follows:

   $F{\downarrow}0(Y) = Y$.
   $F{\downarrow}(\alpha+1)(Y) = F(F{\downarrow}\alpha(Y))$.
   $F{\downarrow}\lambda(Y) = \bigcap\{F{\downarrow}\alpha(Y) : \alpha < \lambda\}$ for a limit ordinal $\lambda$. ▲

In our considerations of logic programming and of default logic we will study fixpoints of several operators. Those operators will often be monotone. The following result gives a criterion for the existence of fixpoints of monotone operators.

**Theorem 2.69** (Knaster-Tarski lemma) *Let $F : \mathcal{P}(X) \rightarrow \mathcal{P}(X)$ be a monotone operator.*

1. *The operator $F$ possesses a least fixpoint. This least fixpoint is equal to $F{\uparrow}\alpha(\emptyset)$, for some ordinal $\alpha$.*
2. *If, in addition, $F$ is finitizable then the least fixpoint of $F$ is equal to $F{\uparrow}\omega(\emptyset)$.*
3. *The operator $F$ possesses a largest fixpoint. This largest fixpoint is equal to $F{\downarrow}\alpha(X)$ for some ordinal $\alpha$. Even if $F$ is finitizable, the least such $\alpha$ may be transfinite.* ▲

Example 2.67 shows that a monotone operator may have a unique fixpoint and that this fixpoint can be quite small. Thus, for some monotone operators $F$, there are subsets of $X$ not included in any fixpoint of $F$. The situation is different for the so-called progressive operators. An operator $F$ is *progressive* if $Y \subseteq F(Y)$, for every $Y \subseteq X$. The following result is a variant of the Knaster-Tarski lemma for monotone and progressive operators.

**Theorem 2.70** *If $F$ is a monotone and progressive operator in $X$ then for every $Y \subseteq X$ there exists a least set $Z$ such that $Y \subseteq Z$ and $Z$ is a fixpoint of $F$. This set $Z$ is of the form $F{\uparrow}\alpha(Y)$ for some ordinal $\alpha$. If, in addition, $F$ is finitizable, then this set $Z$ is $F{\uparrow}\omega(Y)$.* ▲

# 3. General default theories

In this chapter we introduce default logic. Default logic is concerned with proof systems which are obtained from propositional logic by adding some nonstandard inference rules called *defaults*. A default differs from an inference rule, as defined in the previous chapter, in that its premises are of two types. Premises of the first type, called *prerequisites*, act in the same way as the premises of an inference rule. In order to apply a default it is necessary that each of its prerequisites be proved. Premises of the second type, called *justifications*, do not need a proof. It is only required that each of them be *possible* in order for a default to be applicable. In this chapter we will formally define defaults and default logic and make precise what is meant by a proof and by being possible. To this end, we will introduce the notion of an *extension* for a default theory. Extensions can be thought of as descriptions of all possible knowledge (belief) sets that are represented by a default theory. They play a similar role to that of the set of all consequences of a theory in propositional logic. In this chapter we study the notion of an extension and provide several characterizations of extensions as well as algorithms to compute them.

Due to the presence of justifications in defaults, which only need to be possible in order for defaults to be applicable, default logic models so-called *defeasible* reasoning, that is, reasoning which may become invalid should more facts become available. For example, one can design a default which states that *if it is possible that an object $x$ has a property $P$ then it should be concluded that $x$ has the property $P$*. Using this default, when no information about $x$ is available, one should conclude that $x$ has the property $P$. But, if one later learns that $x$ does not have a property $P$, the previously valid conclusion has to be withdrawn. Speaking more formally, unlike the consequence operators in standard logics (propositional, predicate, intuitionistic, modal, and so on), the consequence operators associated with default logic and based on the notion of an extension are not monotone.

Logics with nonmonotonic consequence operators, commonly called *nonmonotonic logics*, have been introduced relatively recently by researchers in artificial intelligence for knowledge representation and commonsense reasoning applications. Default logic is one of the most important among them.

We will start our investigations with a review of monotone rules in propositional logic. Next, we introduce default logic informally. A formal definition of the syntax of default logic follows. We introduce the basic structure for default theories, an extension, in the next section. Subsequently we characterize extensions using syntactic and semantic means. We give a number of properties of default extensions, and then show how well-orderings of the set of defaults can be used to construct extensions. This construction as well as results from previous sections are then used to prove the correctness of two algorithms for computation of extensions of finite default theories. Finally, we investigate several other structures that can be associated with default theories: weak extensions, minimal sets closed under default theories and partial extensions.

## 3.1 Monotone rules in propositional logic

Before we introduce default logic we will first study the effect of adding new (monotone) inference rules to the proof system of propositional logic. The concept of a rule of proof has been introduced in Chapter 2. The resulting proof systems play a crucial role in our presentation of default logic in the following sections.

**Definition 3.1** Let $D$ be a set of inference rules. By *a rule proof system* (or, *rule system*) $PC+D$ we mean the proof system obtained by adding the rules in $D$ to the proof system of propositional logic. A *derivation of a formula $\varphi$ from a set of formulas $W$ in the system $PC+D$* is a finite sequence of formulas $\varphi_1, \ldots, \varphi_n$ such that:

1. $\varphi_n = \varphi$.
2. For every $i \leq n$ at least one of the following conditions holds:
   (a) $\varphi_i$ is a substitution instance of an axiom of propositional logic.
   (b) $\varphi_i \in W$.
   (c) For some $j, k < i$, $\varphi_i$ follows from $\varphi_j$ and $\varphi_k$ by modus ponens.
   (d) For some $j_1, \ldots, j_k < i$, the rule $\frac{\varphi_{j_1}, \ldots, \varphi_{j_k}}{\varphi_i}$ belongs to $D$.

By $Cn^D(W)$ we denote the set of all formulas that possess a derivation from $W$ in $PC+D$. ▲

**Remark 3.2** A system $PC + D$ extends the proof system of propositional logic. Hence, a formula $\varphi_1 \wedge \ldots \wedge \varphi_k$ has a derivation from $W$ in $PC + D$ if and only if for every $i$, $1 \leq i \leq k$, $\varphi_i$ has a derivation from $W$ in $PC + D$. Consequently, without loss of generality, we will assume that all the rules in $D$ have just one premise. ▲

Some simple but useful properties of the operator $Cn^D$ are given in the following theorem.

**Theorem 3.3** *Let $D$ be a set of inference rules. For every set $W \subseteq \mathcal{L}$:*

1. $Cn(W) \subseteq Cn^D(W)$.
2. $Cn^D(Cn^D(W)) = Cn^D(W)$. *In other words, $Cn^D(W)$ is closed under propositional provability and all the rules in $D$.*

Proof: Part (1) follows from a simple observation that a proof of a formula from $W$ in propositional logic is also a proof from $W$ in the system $PC + D$.

Since for every $U$, $U \subseteq Cn^D(U)$, it follows that in order to prove (2) it is enough to show that

$$Cn^D(Cn^D(W)) \subseteq Cn^D(W).$$

To this end, we proceed by induction on the length of a derivation of a formula from $Cn^D(W)$ in $PC + D$. Assume that for every formula $\psi$ with a derivation from $Cn^D(W)$ in $PC + D$ of length less than $n$, $\psi \in Cn^D(W)$. Consider a formula $\varphi \in Cn^D(Cn^D(W))$. Let $\varphi_1, \ldots, \varphi_n$ be a derivation of $\varphi$ from $Cn^D(W)$ in $PC + D$. By the induction hypothesis, for every $i$, $1 \leq i \leq n-1$, $\varphi_i \in Cn^D(W)$.

There are several possibilities. First, assume that $\varphi_n (= \varphi)$ is a substitution instance of an axiom of propositional logic or belongs to $W$. Then $\varphi_n \in Cn(W)$. Consequently, $\varphi_n \in Cn^D(W)$, by (1). This argument also establishes the basis for the induction.

Next, assume that $\varphi_n$ is the result of applying modus ponens to $\varphi_i$, $\varphi_j$, for some $i, j < n$. By the induction hypothesis, $\varphi_i, \varphi_j \in Cn^D(W)$. Let $P$ and $Q$ be proofs of $\varphi_i$ and $\varphi_j$, respectively, from $W$ in the system $PC + D$. Then the sequence of formulas $P, Q, \varphi_n$ is a proof of $\varphi_n$ from $W$ in $PC + D$. Hence, $\varphi_n \in Cn^D(W)$.

Finally, assume that $\varphi_n$ is the result of applying one of the rules from $D$ to a formula $\varphi_i$, for some $i < n$ (we assume that each rule in $D$ has a single premise). Since $\varphi_i \in Cn^D(W)$, there is a proof $P$ of $\varphi_i$ from $W$ in $PC + D$. Then $P, \varphi_n$ is a proof of $\varphi_n$ from $W$ in $PC + D$. Hence, $\varphi_n \in Cn^D(W)$. ■

An inference rule

$$\frac{\varphi}{\psi} \tag{3.1}$$

where $\varphi, \psi \in \mathcal{L}$, is somewhat similar in its effects to the implication

$$\varphi \supset \psi. \tag{3.2}$$

In the presence of propositional logic, given $\varphi$, both the rule (3.1) and the implication (3.2) allow us to derive $\psi$. But there are important differences. Several "side-effects" of having the implication (3.2) in a theory are avoided if, instead, the rule (3.1) is used. Specifically, the implication (3.2) allows us to derive $\neg\varphi$, given $\neg\psi$, whereas the rule (3.1) does not. Similarly, the formula $\alpha \vee \varphi$ and the implication (3.2) imply $\alpha \vee \psi$. The same conclusion cannot be reached if the implication (3.2) is replaced with the rule (3.1).

**Definition 3.4** A rule $\frac{\varphi_1, \ldots, \varphi_n}{\psi}$ is *valid* if the formula $\varphi_1 \wedge \ldots \wedge \varphi_n \supset \psi$ is a tautology. In particular, a rule $\frac{\varphi}{\psi}$ is valid if the formula $\varphi \supset \psi$ is a tautology. ▲

**Theorem 3.5** *Let $D$ be a set of valid inference rules. Then:*

1. *For every theory $W$, $Cn(W) = Cn^D(W)$.*
2. *For every consistent theory $W$, $Cn^D(W)$ is consistent.* ▲

Theorem 3.5 states that adding valid rules to propositional logic has no effect on the consequence operator. Although the consequence operator does not change, the class of proofs does. In particular, adding valid rules may result in shorter derivations of some formulas.

The assertion of Theorem 3.5 fails for rules which are not valid.

**Theorem 3.6** *Let $r$ be an inference rule which is not valid. Then there exists a consistent theory $W$ such that $Cn^{\{r\}}(W)$ is inconsistent.*

Proof: Let $r = \frac{\varphi}{\psi}$. Define $W = \{\varphi, \neg\psi\}$. Since $r$ is not valid, $W$ is consistent. Since $\psi \in Cn^{\{r\}}(W)$, and $W \subseteq Cn^{\{r\}}(W)$, $Cn^{\{r\}}(W)$ is inconsistent. ■

Theorem 3.6 states that if a proof system extending propositional logic is to be used to derive consequences of *arbitrary* theories then all rules of the system must be valid. Otherwise consistency is not preserved. On the other hand, we often design proof systems in order to reason about specific domains of knowledge. Consequently, such systems are only used to reason from a restricted class of theories and Theorem 3.6 is not applicable. That is, we are no longer required to use only valid rules of proof. In fact, Theorem 3.5 implies that, in order to obtain a reasoning system stronger than propositional logic, it is necessary that some rules in the system express domain-specific knowledge and, hence, are not universally valid.

Since $Cn^D$ is nothing else but the consequence operator of a certain proof system, it can be described in terms of the least theories closed under its inference rules.

**Theorem 3.7** *Let $D$ be a set of inference rules. For every set $W$, $Cn^D(W)$ is the least set $U$ containing $W$ and closed under propositional provability and the rules from $D$.*

Proof: By Theorem 3.3, $Cn^D(W)$ contains $W$ and is closed under propositional provability and the rules in $D$. Let $U$ be any set such that $W \subseteq U$ and $U$ is closed under propositional provability and the rules in $D$. We will prove that $Cn^D(W) \subseteq U$. We will proceed by induction on the length of a derivation. Let us assume that the claim is true for every formula $\psi \in Cn^D(W)$ with a derivation from $W$ in $PC + D$ of length less than $n$. Let $\varphi \in Cn^D(W)$ be a formula with a derivation $\varphi_1, \ldots, \varphi_n$ from $W$ in $PC + D$. By the induction hypothesis, for every $i$, $1 \leq i \leq n-1$, $\varphi_i \in U$.

There are several possibilities. If $\varphi_n (= \varphi)$ is a substitution instance of an axiom of propositional logic or belongs to $W$, then $\varphi_n \in Cn(W)$. Since $W \subseteq U$ and $U$ is closed under propositional provability, $\varphi_n \in U$. This argument also establishes the basis for the induction.

If $\varphi_n$ is the result of applying modus ponens to $\varphi_i$, $\varphi_j$, for some $i, j < n$ then, since $\varphi_i, \varphi_j \in U$ and $U$ is closed under propositional provability, $\varphi_n \in U$.

Finally, if $\varphi_n$ is the result of applying one of the rules from $D$ to some formula $\varphi_i$, $i < n$, then, since $\varphi_i \in U$ and $U$ is closed under the rules from $D$, $\varphi_n \in U$. ■

We will now characterize $Cn^D$ in terms of a fixpoint of a monotone operator. Define:

$$R^D(U) = Cn(U \cup \{\gamma : \frac{\alpha}{\gamma} \in D \text{ and } \alpha \in U\}) \tag{3.3}$$

The next theorem gathers some important properties of the operator $R^D$. The proof is straightforward and is omitted.

**Theorem 3.8** *The operator $R^D$ is monotone, finitizable and progressive.* ▲

Theorem 3.8 allows us to use Theorem 2.70 and to describe $Cn^D(W)$ as the least fixpoint of the operator $R^D$ containing $W$.

**Theorem 3.9** *For every theory $W \subseteq \mathcal{L}$, $Cn^D(W) = R^D{\uparrow}\omega(W)$. That is, the theory $Cn^D(W)$ is the least fixpoint of the operator $R^D$ which contains $W$.*

Proof: It follows from Theorems 3.8 and 2.70 that the least fixpoint of the operator $R^D$ which contains $W$ is given by $R^D{\uparrow}\omega(W)$. Hence, to prove the assertion we need to show that

$$Cn^D(W) = R^D{\uparrow}\omega(W).$$

We will first show that $Cn^D(W) \subseteq R^D{\uparrow}\omega(W)$. Consider a formula $\varphi \in Cn^D(W)$. Then $\varphi$ has a derivation from $W$ in $PC + D$. We will prove, by induction on the length of such a derivation, that $\varphi \in R^D{\uparrow}\omega(W)$.

Let $\varphi_1, \ldots, \varphi_n$ be a derivation of $\varphi$ from $W$ in $PC+D$. By the induction hypothesis, for every $i$, $1 \leq i \leq n-1$, $\varphi_i \in R^D{\uparrow}\omega(W)$. Then there exists $m \geq 1$ such that for every $i$, $1 \leq i \leq n-1$, $\varphi_i \in R^D{\uparrow}m(W)$.

There are several possibilities. First, assume that $\varphi_n$ is a substitution instance of an axiom of propositional logic or that $\varphi_n \in W$. Then we have $\varphi_n \in R^D{\uparrow}1(W)$ and, consequently, $\varphi_n \in R^D{\uparrow}\omega(W)$ (this argument also establishes the basis for the induction).

Next, assume that $\varphi_n$ results from an application of modus ponens to formulas $\varphi_i$, $\varphi_j$, for some $i, j \leq n-1$. Since $m \geq 1$, $R^D{\uparrow}m(W)$ is closed under propositional provability. Hence, $\varphi_n \in R^D{\uparrow}m(W)$ and, consequently, $\varphi_n \in R^D{\uparrow}\omega(W)$.

Finally, assume that $\varphi_n$ is the result of an application of some rule in $D$, say $\frac{\varphi_i}{\varphi_n}$, to a formula $\varphi_i$, for some $i \leq n-1$. Since $\varphi_i \in R^D{\uparrow}m(W)$, it follows from the definition of $R^D{\uparrow}(m+1)(W)$ that $\varphi_n \in R^D{\uparrow}(m+1)(W)$.

Hence, in each possible case we have $\varphi_n \in R^D{\uparrow}\omega(W)$.

To prove the inclusion $R^D{\uparrow}\omega(W) \subseteq Cn^D(W)$ we show, by induction on $n$, that for every non-negative integer $n$, $R^D{\uparrow}n(W) \subseteq Cn^D(W)$ holds. Since $R^D{\uparrow}0(W) = W$, the basis of the induction follows by Theorem 3.3.

Assume next that $R^D{\uparrow}n(W) \subseteq Cn^D(W)$. Consider a formula $\gamma$ such that $\frac{\alpha}{\gamma} \in D$ and $\alpha \in R^D{\uparrow}n(W)$. Then, by the inductive hypothesis, $\alpha \in Cn^D(W)$. Since $Cn^D(W)$ is closed under the rules in $D$, $\gamma \in Cn^D(W)$. Hence,

$$R^D{\uparrow}n(W) \cup \left\{\gamma\colon \frac{\alpha}{\gamma} \in D \text{ and } \alpha \in R^D{\uparrow}n(W)\right\} \subseteq Cn^D(W).$$

Since $Cn^D(W)$ is closed under propositional provability and

$$R^D{\uparrow}(n+1)(W) = Cn(R^D{\uparrow}n(W) \cup \{\gamma\colon \frac{\alpha}{\gamma} \in D \text{ and } \alpha \in R^D{\uparrow}n(W)\}),$$

the inclusion $R^D{\uparrow}(n+1)(W) \subseteq Cn^D(W)$ follows. ■

Finally, we will provide one more characterization of the operator $Cn^D$. In the case when $D$ and $W$ are finite, it provides a finite representation for $Cn^D(W)$ and is useful in determining whether a formula $\varphi$ belongs to $Cn^D(W)$ or not. To this end we need yet another operator.

**Definition 3.10** For a set $D$ of rules and a theory $W$ put

$$B^D(W) = W \cup \left\{\gamma\colon \frac{\alpha}{\gamma} \in D \text{ and } \alpha \in Cn(W)\right\}.$$

The operator $B^D$ is called the *base* operator. ▲

There is a subtle aspect to the definition of the operator $B^D$. Notice that in the definition we stipulate that $\alpha \in Cn(W)$ and not to $W$ alone. This is a critical requirement.

The following property relates the operators $B^D$, $R^D$ and $Cn^D$.

**Theorem 3.11** *For every set of rules $D$ and for every theory $W \subseteq \mathcal{L}$,*

$$Cn^D(W) = R^D{\uparrow}\omega(W) = Cn(B^D{\uparrow}\omega(W)).$$

Proof: By Theorem 3.9, it suffices to prove the second equality. First, observe that $Cn(B^D{\uparrow}\omega(W))$ is closed under propositional provability and the rules from $D$. In addition, $W \subseteq Cn(B^D{\uparrow}\omega(W))$. Since $R^D{\uparrow}\omega(W)$ is the least theory closed under the rules from $D$ and containing $W$, therefore we have $R^D{\uparrow}\omega(W) \subseteq Cn(B^D{\uparrow}\omega(W))$.

Since $R^D{\uparrow}\omega(W)$ is closed under propositional provability, to prove the converse inclusion it is enough to show that

$$B^D{\uparrow}\omega(W) \subseteq R^D{\uparrow}\omega(W).$$

To this end, we prove by induction on $n$ that for every non-negative integer $n$, $B^D{\uparrow}n(W) \subseteq R^D{\uparrow}\omega(W)$.

---

$B{\uparrow}\omega(D, W)$
*Input:* A finite set $D$ of rules and a finite theory $W$;
*Output:* The set $B^D{\uparrow}\omega(W)$.

$B' := W$;
**repeat**
  $B := B'$;
  $AR := \{\frac{\alpha}{\gamma} \in D: \ \alpha \in Cn(B)\}$;
  $B' := B \cup \{\gamma: \frac{\alpha}{\gamma} \in AR\}$;
  $D := D \setminus AR$;
**until** $AR = \emptyset$;
**return**$(B)$.

---

**Fig. 3.1.** Algorithm for computing $B^D{\uparrow}\omega(W)$

First, observe that $B^D{\uparrow}0(W) = W$. Hence, $B^D{\uparrow}0(W) \subseteq R^D{\uparrow}\omega(W)$. Assume that $B^D{\uparrow}n(W) \subseteq R^D{\uparrow}\omega(W)$ and consider $\gamma \in B^D{\uparrow}(n+1)(W)$. There are two possibilities. If $\gamma \in B^D{\uparrow}n(W)$ then, by the induction hypothesis, $\gamma \in R^D{\uparrow}\omega(W)$. If not, then there is a rule $\frac{\alpha}{\gamma} \in D$ such that $\alpha \in Cn(B^D{\uparrow}n(W))$. Since $R^D{\uparrow}\omega(W)$ is closed under propositional provability, the induction hypothesis implies that $\alpha \in R^D{\uparrow}\omega(W)$. Since $R^D{\uparrow}\omega(W)$ is closed under the rules from $D$, $\gamma \in R^D{\uparrow}\omega(W)$. ■

**Corollary 3.12** *For every set of rules D and for every theory $W \subseteq \mathcal{L}$, there is a set $U \subseteq \{\gamma: \frac{\alpha}{\gamma} \in D\}$ such that $Cn^D(W) = Cn(W \cup U)$.* ▲

The operator $B^D$ is similar to the *one-step provability* operator of van Emden-Kowalski [vEK76] and plays an analogous role.

Next, we study a straightforward algorithm that, given a finite set rules $D$ and a finite theory $W$, computes $B^D{\uparrow}\omega(W)$. The algorithm is shown in Figure 3.1. It follows the definition of the operator $B^D{\uparrow}\omega$ and is based on the observation that if for some integer $n$, $B^D{\uparrow}n(W) = B^D{\uparrow}(n+1)(W)$, then $B^D{\uparrow}\omega(W) = B^D{\uparrow}n(W)$. In the iteration $n$ of the **repeat** loop, $AR$ contains rules that are used to derive new formulas out of $B$. Once these rules are used, they are not needed any more and can be removed. After the iteration $n$, $B = B^D{\uparrow}n(W)$ and $B' = B^D(B) = B^D{\uparrow}(n+1)(W)$.

The algorithm is computationally complex. In the worst case, it requires $O(|D|^2)$ calls to a procedure testing whether a formula $\alpha$ is provable from a collection $B$ of formulas.

The last topic we discuss in this section is the use of well-orderings of the set $D$ of rules to compute the set $Cn^D(W)$. In this approach, the set $Cn^D(W)$ is obtained not as a fixpoint of an operator but rather as a "limit" of a construction. A well-ordering $\preceq$ of $D$ is used to determine the next rule to be applied. When no rule can be used any more, we end the construction. The result of this construction is the set $Cn^D(W)$. A similar technique will be later applied to compute extensions of default theories.

We assume that the set of atoms of $\mathcal{L}$ is denumerable. Therefore the set of all rules is denumerable.

**Definition 3.13** Let $D$ be a set of rules and let $W \subseteq \mathcal{L}$. Let $\preceq$ be a well-ordering of $D$.

We define an ordinal $\eta_\preceq$. For every $\xi < \eta_\preceq$ we define a set of rules $GR_\xi$ and a rule $d_\xi$. We also define a set of rules $GR_\preceq \subseteq D$ . We proceed as follows: If for all $\xi < \alpha$, $GR_\xi$ have been defined, but $\eta_\preceq$ has not been defined, then:

1. If there is no rule $d \in D \setminus (\bigcup_{\xi<\alpha} GR_\xi)$ such that

   $p(d) \in Cn(W \cup c(\bigcup_{\xi<\alpha} GR_\xi))$

   then $\eta_\preceq = \alpha$.

2. Otherwise, define $d_\alpha$ to be the $\preceq$-least rule in the set $D \setminus (\bigcup_{\xi<\alpha} GR_\xi)$ such that

   $p(d) \in Cn(W \cup c(\bigcup_{\xi<\alpha} GR_\xi))$.

   Then set $GR_\alpha = (\bigcup_{\xi<\alpha} GR_\xi) \cup \{d_\alpha\}$.

3. Put $GR_\preceq = \bigcup_{\xi<\alpha} GR_\xi$. ▲

One can easily show that the ordinal $\eta_\preceq$ is well-defined.

**Theorem 3.14** *Let $W \subseteq \mathcal{L}$. Let $D$ be a set of rules. For every well-ordering $\preceq$ of $D$ we have:*

1. *The set $GR_\preceq$ consists precisely of rules $d \in D$ such that $p(d) \in Cn^D(W)$.*
2. *$Cn(W \cup c(GR_\preceq)) = Cn^D(W)$.* ▲

It follows that $GR_\preceq$ does not depend on the particular ordering $\preceq$, and that the result of our construction is always $Cn^D(W)$. Even though every well-ordering $\preceq$ generates the same set $GR_{\preceq}$, the ordinal number $\eta_{\preceq}$ does not need to be the same. Hence one needs to ask what are the minimal values attained by ordinal $\eta_{\preceq}$.

**Theorem 3.15** *If $\preceq$ is a well-ordering of $D$ in type less than or equal to $\omega$, then $\eta_{\preceq}$ is less than or equal to $\omega$.* ▲

We did not prove Theorem 3.15 since its assertion will not be directly used in the sequel. On the other hand, an analogous result for default rules will be proved later in this chapter, and its proof indicates how one can prove Theorem 3.15.

## 3.2 Default logic — motivation

In this section we will discuss examples of reasoning with incomplete information. We will argue that the formalism of rule proof systems does not properly capture those cases. We will then, informally, introduce a new method of reasoning, called default logic, which properly handles our examples. Formal investigations of default logic are among the main topics of this book.

**Example 3.16** An agent reasons about page layouts. There is a rule which says that a page has a capacity of 24 lines. Another rule says that the word capacity of a page is at most 12 times bigger than its line capacity. Using the terminology of the previous section we formulate these rules as:

$$r_1 = \frac{page(x)}{line\_capacity(x, 24)}$$

and

$$r_2 = \frac{line\_capacity(x, y)}{word\_capacity(x, y * 12)}.$$

We want to reason about page number 7. So, we provide the agent with the data $\{page(7)\}$. Using the theory described in Section 3.1 the agent will be able to deduce these statements:

$$line\_capacity(7, 24) \text{ and } word\_capacity(7, 288).$$

Since the agent can use the proving mechanism of classical logic, the belief set of the agent is

$$S = Cn(\{page(7), line\_capacity(7, 24), word\_capacity(7, 288)\}).$$

▲

Assume now that we want the agent to be able to handle exceptions, that is, occasional cases when a page will not conform to the standards. In such a situation, unless the agent is able to establish that page 7 is exceptional, she should derive the same properties of that page as in Example 3.16. But in exceptional cases these deductions should not be allowed. More formally, as long as we provide the agent with data insufficient for establishing $exception(7)$, that is, as long as $\neg exception(7)$ is consistent, then the deductions of Example 3.16 should remain valid. Otherwise, they should be blocked.

Hence, for the initial data $W_1 = \{page(7)\}$ the agent should obtain the same consequences as above, that is:

$$S_1 = Cn(\{page(7), line\_capacity(7, 24), word\_capacity(7, 288)\}).$$

On the other hand, for a larger set of initial assumptions

$$W_2 = \{page(7), exception(7)\}$$

the agent should be able to derive only the logical consequences of $W_2$, that is:

$$S_2 = Cn(\{page(7), exception(7)\}).$$

It is quite clear that such a reasoning system must use rules different from those introduced in the previous section. It follows directly from Definition 3.1 that the operator $Cn^D$ is *monotone*, that is, for every set $D$ of inference rules and for all theories $W_1, W_2 \subseteq \mathcal{L}$, if $W_1 \subseteq W_2$ then $Cn^D(W_1) \subseteq Cn^D(W_2)$. As our discussion shows, reasoning with exceptions is not monotone, that is, a larger set of initial assumptions does not necessarily produce a larger set of consequences.

In order to handle exceptions a reasoning mechanism must have a feature that allows it to invalidate deductions from the initial assumptions and from conclusions of derivations. Let us return to page-layouts with exceptions. The solution offered by default logic is to replace the rule $r_1$ by an expression of the form:

$$r_1' = \frac{page(x) : M\neg exception(x)}{line\_capacity(x, 24)}$$

with the following informal understanding. If $x$ is a page then, as long as it is consistent to assume $\neg exception(x)$, deduce that the line capacity of $x$ is

24. If $x$ is a page such that $exception(x)$ holds, then such derivation is not allowed. Rules of this type will be called *defaults* or *default rules.*

**Example 3.17** Let us provide an agent with the rules $r_1'$ and $r_2$. In the case of the initial assumptions $W_1 = \{page(7)\}$ the agent can use the rule $r_1'$ in the very same way as the rule $r_1$ is used in Example 3.16. Subsequently, the rule $r_2$ is applicable and the agent computes

$$Cn(\{page(7), line\_capacity(7, 24), word\_capacity(7, 288)\}$$

as her belief set.

In the case of initial assumptions $W_2 = \{page(7), exception(7)\}$ the agent must not use the rule $r_1'$ since the sentence $\neg exception(7)$ is no longer consistent. Thus the conclusion $line\_capacity(7, 24)$ cannot be reached and $word\_capacity(7, 288)$ cannot be deduced either. Hence, the agent computes the belief set

$$Cn(\{page(7), exception(7)\}).$$

▲

Example 3.17 shows that a default

$$r = \frac{\alpha : M\beta}{\omega}$$

or, more generally,

$$r = \frac{\alpha : M\beta_1, \ldots, M\beta_n}{\omega},$$

can be understood as: "Normally, from $\alpha$ deduce $\omega$". That is, it acts as a quantifier. The role of $\beta_1, \ldots, \beta_n$ is that they (or to be precise the formulas $\neg\beta_1, \ldots, \neg\beta_n$) describe *all* abnormalities blocking applicability of $r$. That is as long as none of $\neg\beta_1, \ldots, \neg\beta_n$ is detected in a context $S$ the situation described by $S$ is considered normal and $r$ is applicable in proofs.

In our next example we use rules of the same type as $r_1'$ to model a situation in which, in the presence of incomplete information, the agent is forced to establish more than one belief set.

**Example 3.18** Consider an agent reasoning about English first names. The initial data of the agent is given by a theory

$$W = \{first\_name(jones, leslie), male(jones) \lor female(jones),$$
$$\neg(male(jones) \land female(jones))\}$$

which, roughly speaking, says that the first name of Jones is Leslie and that Jones is a male or a female, but not both. Next, the agent is given the rules:

$$r_1 = \frac{first\_name(jones, leslie) : M(male(jones))}{male(jones)},$$

$$r_2 = \frac{first\_name(jones, leslie) : M(female(jones))}{female(jones)}.$$

Assume that the agent reasons as in Example 3.17. In the first step she notices that both $male(jones)$ and $female(jones)$ are consistent with $W$ (each separately) so she derives them both and obtains an inconsistent theory. It is quite clear, however, that the effect of her reasoning should be two different belief sets:

$$S_1 = Cn(W \cup \{male(jones)\})$$

and

$$S_2 = Cn(W \cup \{female(jones)\}).$$ ▲

It follows that the reasoning mechanism that has been described in Example 3.17 is not appropriate for the situation of Example 3.18. As a result of deriving the statements $male(jones)$ and $female(jones)$ the set of beliefs of the agent grows. The larger belief set makes the rules $r_1$ and $r_2$ inapplicable and invalidates deductions requiring them. This phenomenon did not appear in Example 3.17 since the applicability of the rule $r_1'$ was retained after new beliefs were generated.

The problems we encounter in Example 3.18 are caused by our misuse of the notion of consistency. In particular, when we talk about the consistency of a formula, we do not specify with what theory it is to be consistent. Can this theory change within a single proof? Can it change from one proof to another? Example 3.18 shows that varying the theory used to evaluate consistency leads to an ill-defined notion of derivation.

A possible solution, and the one adopted in this book, is to use the same theory, chosen *beforehand*, in all derivations. That is, in order to precisely define the notion of derivability we need an additional parameter, a theory called *context*, which tells us whether a default can or cannot be applied. Consequently, not one but a whole family of notions of derivability, parameterized with context theories, is obtained. The issue we will discuss next is which contexts are appropriate for constructing belief sets.

Consider a set of defaults $D$ and a collection of initial assumptions $W$. Assume that a theory $S$ describes a belief set grounded in $D$ and $W$. Let us use $S$ as the context. The context $S$ determines a subset of $D$ consisting of those defaults which are available for application and determines the notion of derivability. We will use the existence or nonexistence of a proof as a criterion for deciding whether a formula is a belief or not. All formulas that possess a proof from $W$ by means of propositional logic and applicable (with respect to $S$) default rules must be believed. Consequently, they must belong to the belief set $S$. Conversely, if a formula has no proof, how can it be justified? Such formulas should not belong to $S$. Hence, a belief set must consist of precisely those formulas that possess a proof using the belief set itself as a context.

Conversely, consider a theory $S$ which consists of those formulas which possess a proof from $W$ by means of applicable (with respect to $S$) default rules from $D$ and propositional logic. Then the criterion of existence or nonexistence of a proof implies that $S$ can be adopted as a belief set grounded in $D$ and $W$.

Summarizing, belief sets *grounded* in $D$ and $W$ are exactly those contexts $S$ which have the property that they consist of precisely those formulas that possess a proof using $S$ itself as a context.

Let us apply this idea to Example 3.17. It is easy to see that the theory $S_1$, found in Example 3.17, is indeed a proper context for $\{r_1', r_2\}$ and $W_1$ and should be adopted as a belief set. Similarly, $S_2$ is a belief set for $\{r_1', r_2\}$ and $W_2$.

Let us turn our attention to Example 3.18. There are two belief sets grounded in $\{r_1, r_2\}$ and $W$: $S_1 = Cn(W \cup \{male(jones)\})$ and $S_2 = Cn(W \cup \{female(jones)\})$.

Let us check the case of $S_1$. Indeed, with the context $S_1$ we can prove $\neg female(jones)$ so the default $r_2$ becomes inapplicable. But the default rule $r_1$ is applicable since $S_1 \not\vdash \neg male(jones)$. Hence the default rule $r_1$ can be used in a deduction and it is clear that all elements of $S_1$ (and only elements of $S_1$) can be proved from $W$, by means of propositional logic and the default $r_1$.

Example 3.18 shows that defaults can be used to form belief sets when more than one is appropriate.

Below we formalize the concept of context-dependent reasoning with defaults as an $S$-proof, and the notion of a belief set as an *extension* for a default theory.

## 3.3 Default logic — syntax

We will now introduce the syntax of default logic. The language of default logic consists of elements of two types: well-formed formulas of the propositional language $\mathcal{L}$ and defaults. Formally, a *default* is a triple

$$\langle \alpha, J, \gamma \rangle, \tag{3.4}$$

where $\alpha$ and $\gamma$ are formulas from $\mathcal{L}$ and $J$ is a finite subset of $\mathcal{L}$. The formula $\alpha$ is called the *prerequisite*, the formulas in $J$ are called the *justifications* and $\gamma$ is called the *consequent* of the default (3.4). The prerequisite of a default $d$ will be denoted by $p(d)$, the set of justifications by $J(d)$ and the consequent by $c(d)$. For a set of defaults $D$, by $p(D)$ we denote the set of the prerequisites of all defaults in $D$. The terms $J(D)$ and $c(D)$ are defined analogously.

When $J = \{\beta_1, \ldots, \beta_k\}$, a default $\langle \alpha, J, \gamma \rangle$ is usually written as

$$\frac{\alpha : M\beta_1, \ldots, M\beta_k}{\gamma}. \tag{3.5}$$

If $\alpha = \top$, a default (3.4) is called *prerequisite-free* and it is usually written as

$$\frac{: M\beta_1, \ldots, M\beta_k}{\gamma}.$$

If $J = \emptyset$, a default (3.4) is called *justification-free* and is written as

$$\frac{\alpha :}{\gamma}.$$

A *default theory* is a pair $(D, W)$, where $D$ is a set of defaults and $W$ is a theory in the language $\mathcal{L}$. If $p(D) = \emptyset$, that is, all defaults in $D$ are prerequisite-free, then a default theory $(D, W)$ is called *prerequisite-free*. The notion of a *justification-free* default theory is defined similarly.

## 3.4 Extensions for default theories

We will now formalize our considerations from Section 3.2. We will define the notion of derivability used in default logic. Then we will introduce the notion of an extension, as a formal representation of a *belief set* an agent can create on the basis of a default theory.

**Definition 3.19** Let $D$ be a set of defaults. By a *default proof system* (or, *default system*) $PC + D$ we mean the proof system of propositional logic extended by the set of defaults $D$. The concept of a proof in a default proof system $PC + D$ is defined relative to a theory $S \subseteq \mathcal{L}$. An *S-derivation* (*S-proof*) of a formula $\varphi$ from $W$ in $PC + D$ is a finite sequence $\varphi_1, \ldots, \varphi_n$ such that $\varphi_n = \varphi$ and, for every $i$, $1 \leq i \leq n$, at least one of the following conditions holds:

1. $\varphi_i \in W$.

2. $\varphi_i$ is a substitution instance of an axiom of propositional calculus.

3. $\varphi_i$ is the result of applying modus ponens to formulas $\varphi_j$, $\varphi_k$ for some $j, k < i$.

4. There is a default $d = \frac{\alpha:M\beta_1,\ldots,M\beta_k}{\varphi_i}$ such that $d \in D$, $\alpha = \varphi_j$ for some $j < i$, and $\neg\beta_1 \notin S, \ldots, \neg\beta_k \notin S$.

The theory $S$ is called a *context*. By $Cn^{D,S}(W)$ we denote the set of all formulas having an $S$-proof from $W$ in $PC + D$. ▲

The notion of a default proof system is a substantial departure from the standard notion of a proof system. Default systems are endowed with a whole family of consequence operators parameterized with a context $S$, while standard proof systems, like rule proof systems discussed in Section 3.1, have only one consequence operator associated with them. This multitude of consequence operators is a formal reflection of the fact that default rules allow an agent to make assumptions when full information is not available. Consequently, depending on the assumptions the agent makes, different belief (consequence) sets can be produced for a theory $W$ of initial assumptions.

We are using the same symbol, $PC + D$, to denote a rule proof system, if $D$ is a set of inference rules, and a default proof system, if $D$ is a set of defaults. It is not a coincidence. We will show now that inference rules can be treated as justification-free defaults and, consequently, a default proof system can be regarded as a natural generalization of a rule proof system. First, for a set $D$ of defaults define

$$D_m = \left\{ \frac{\alpha}{\gamma} : \frac{\alpha:}{\gamma} \in D \right\}. \tag{3.6}$$

We now have the following result

**Theorem 3.20** *Let $D$ be a set of justification-free defaults. For every context $S \subseteq \mathcal{L}$ and for every theory $W \subseteq \mathcal{L}$*

$$Cn^{D_m}(W) = Cn^{D,S}(W).$$

*In other words, all consequence operators associated with the default system $PC + D$ coincide and are equal to the consequence operator of the rule system $PC + D_m$.*

Proof: Since all defaults in $D$ are justification-free, it follows directly from Definitions 3.1 and 3.19 that the notion of an $S$-derivation from a theory $W$ in the default proof system $PC + D$ does not depend on $S$ and coincides with the notion of a derivation from $W$ in the rule proof system $PC + D_m$. ■

The next theorem collects monotonicity properties of the operator $Cn^{D,S}(W)$ with respect to the argument $W$ and the parameters $D$ and $S$.

**Theorem 3.21** *The operator $Cn^{D,S}(W)$ is monotone in $D$ and $W$ and antimonotone in $S$. That is:*

1. *For every two theories $W_1 \subseteq W_2$, $Cn^{D,S}(W_1) \subseteq Cn^{D,S}(W_2)$.*
2. *For every two sets of defaults $D_1 \subseteq D_2$, $Cn^{D_1,S}(W) \subseteq Cn^{D_2,S}(W)$.*
3. *For every two contexts $S_1 \subseteq S_2$, $Cn^{D,S_2}(W) \subseteq Cn^{D,S_1}(W)$.*

Proof: Definition 3.19 implies that if $W_1 \subseteq W_2$, then every $S$-derivation from $W_1$ in $PC+D$ is an $S$-derivation from $W_2$ in $PC+D$. Hence, part (1) follows. Similarly, it is easy to see that if $D_1 \subseteq D_2$, then every $S$-derivation from $W$ in $PC+D_1$ is an $S$-derivation from $W$ in $PC+D_2$, which proves part (2). Part (3) follows from the observation that if $S_1 \subseteq S_2$ and $\beta \notin S_2$ then $\beta \notin S_1$. Consequently, every $S_2$-derivation from $W$ in $PC + D$ is an $S_1$-derivation from $W$ in $PC + D$. ■

Given a default theory $(D, W)$, an agent is given a reasoning mechanism, that is, a default proof system $PC + D$, and a collection $W$ of initial assumptions. The goal of an agent is to use these components to produce a number of possible belief sets. To this end, the agent will use consequence operators $Cn^{D,S}$ associated with the default proof system. But, are all these consequence operators equally appropriate for computing belief sets grounded in $W$? In Section 3.2 we have informally argued that the answer to this question is negative and we have described a method to select contexts $S$ which can be used for reasoning from a default theory $(D, W)$. To recount the result of our considerations, a context $S$ can be selected if $S$ is precisely the set of formulas that have a support in the theory $(D, W)$, that is, can be derived from $W$ by means of default derivations with respect to $S$. We will now express this criterion in terms of $S$-derivability and the consequence operator $Cn^{D,S}$ and will formally introduce the key concept of default logic, that of an extension for a default theory.

**Definition 3.22** Let $(D, W)$ be a default theory. We say that a theory $S$ is a *default extension* for $(D, W)$ if

$$S = Cn^{D,S}(W). \tag{3.7}$$

▲

Extensions can be looked at in two ways. First, they are contexts that determine which consequence operators in the family $\{Cn^{D,S} : S \subseteq \mathcal{L}\}$ can be used by an agent to construct belief sets. At the same time, extensions are exactly the belief sets that the agent will compute in this fashion.

It follows from Definition 3.22 that the process of finding an extension consists of two phases. In the first phase we guess the context (belief set) $S$. In the second we check if $S$ satisfies the equation (3.7). The whole process simplifies significantly if all defaults in a default theory are justification-free. Each such default theory possesses exactly one extension and it can be found in a constructive manner, without the need to guess a context first.

**Theorem 3.23** *Let $D$ be a set of justification-free defaults. For every theory $W \subseteq \mathcal{L}$, the theory $Cn^{D_m}(W)$ is the only extension for $(D, W)$, where $D_m$ is defined by (3.6).*

Proof: Let $S = Cn^{D_m}(W)$. It follows from Theorem 3.20 that

$$S = Cn^{D_m}(W) = Cn^{D,S}(W).$$

Hence, $S$ is an extension for $(D, W)$. To prove that $S$ is the unique extension for $(D, W)$ let us consider an extension $S'$ for $(D, W)$. Then, by Theorem 3.20,

$$S' = Cn^{D,S'}(W) = Cn^{D_m}(W) = S.$$

■

Theorem 3.23 says that from the perspective of extensions there is no difference between justification-free defaults and corresponding monotone inference rules. That is, if $D$ is a set of justification-free defaults and $D_m$ is the corresponding set of inference rules as defined in Theorem 3.23, then the set of consequences of $W$ in the system $PC + D_m$, $Cn^{D_m}(W)$, is precisely the unique extension of the default theory $(D, W)$.

Theorems 3.20 and 3.23 are based on the observation that if $D$ consists of justification-free defaults, then the consequence operator $Cn^{D,S}$ can be expressed in terms of the consequence operator $Cn^{D_m}$, for an appropriately defined set $D_m$ of rules. Now we will extend this result to the case of arbitrary defaults and use it to provide several characterizations of extensions for default theories.

**Definition 3.24** Given a set of default rules $D$ and a subset $S \subseteq \mathcal{L}$, the *reduct of $D$ with respect to $S$*, denoted by $D_S$, is the set of inference rules defined as follows:

$$D_S = \left\{ \frac{\alpha}{\gamma} \colon \frac{\alpha \colon M\beta_1, \ldots, M\beta_k}{\gamma} \in D \text{ and } \neg\beta_1 \notin S, \ldots, \neg\beta_k \notin S \right\}. \quad (3.8)$$

▲

Speaking informally, the reduct $D_S$ contains exactly the information needed for constructing $S$-derivations in the default system $PC + D$. If a justification of a default $d$ is inconsistent with $S$, then $d$ is never used in $S$-derivations. Such defaults do not contribute to the reduct. On the other hand, if all justifications of a default $d$ are consistent with $S$ then its applicability hinges entirely on its prerequisite. Hence, in the presence of $S$, such a default is equivalent to the rule $\frac{p(d)}{c(d)}$, which is exactly what $d$ contributes to the reduct.

The concept of a reduct allows us to characterize the operator $Cn^{D,S}$ of a default proof system in terms of the consequence operator of the rule proof system.

**Theorem 3.25** *For every context $S$, for every theory $W$ and for every set of defaults $D$, $Cn^{D,S}(W) = Cn^{D_S}(W)$.*

Proof: It follows easily from Definitions 3.1 and 3.19 that every derivation from $W$ in the rule system $PC + D_S$ is an $S$-derivation from $W$ in the default system $PC + D$, and conversely. ■

We will describe now the operators $Cn^{D,S}$ and $Cn^{D_S}$ in terms of the corresponding "one-step provability" operators. By Theorem 3.9,

$$Cn^{D_S}(W) = R^{D_S}\uparrow\omega(W).$$

That is, $Cn^{D_S}(W)$ is the least fixpoint of the operator $R^{D_S}$ containing $W$. In a similar way the operator $Cn^{D,S}$ can be described in terms of a one-step derivability operator $R^{D,S}$. For a theory $W \subseteq \mathcal{L}$, define

$$R^{D,S}(W) = Cn(W \cup \{c(d) : d \in D,\ p(d) \in W \text{ and for every } \beta \in J(d),\ \neg\beta \notin S\}).$$

Clearly, $R^{D,S}$ is monotone and progressive. Hence, for any set $W \subseteq \mathcal{L}$, $R^{D,S}$ possesses a fixpoint containing the set $W$. In fact, it is easy to see that

$$R^{D_S}(W) = R^{D,S}(W).$$

Consequently, $Cn^{D,S}(W)$ is the least fixpoint of $R^{D,S}$ containing $W$ or, in other terms,

$$Cn^{D,S}(W) = R^{D,S}\uparrow\omega(W). \tag{3.9}$$

There is a subtle difference between default reasonings by means of $Cn^{D,S}$ and $Cn^{D_S}$ even though, for every $W \subseteq \mathcal{L}$, $Cn^{D,S}(W) = Cn^{D_S}(W)$. In the operator $Cn^{D_S}$ we first reduce defaults in $D$, eliminating all inapplicable defaults in one step and prior to the phase when derivations are computed.

In the operator $Cn^{D,S}$ we test the applicability of defaults as we derive. One needs to realize a deeper difference between both approaches. Although the formulas proved in $PC + D_S$ are precisely the formulas which possess $S$-proofs in $PC + D$, the concept of $S$-provability carries more information and tells us more about the nature of default reasoning. When we look at an $S$-proof $\varphi_1, \ldots, \varphi_n$ then, by considering the defaults which actually have been applied, we can isolate a *finite* fragment of the complement of $S$, which guarantees the applicability of all the default rules used in the derivation $\varphi_1, \ldots, \varphi_n$. More specifically, we can find a *finite* subset $Z \subseteq \mathcal{L} \setminus S$ such that for every $S'$, if $S' \cap Z = \emptyset$ then $\varphi_1, \ldots, \varphi_n$ is an $S'$-proof. That is, whether a sequence of formulas is an $S$-proof depends on a finite set of formulas. This property of default reasonings is less evident when they are modeled as monotone proofs in $PC + D_S$. In the investigation of the expressive power of default reasoning, this property of "localization" of negative information carried by an $S$-proof can be used to construct default theories with complex extensions (see [MNR90]). On the other hand, preprocessing of rules from $D$ to $D_S$ permits us to simplify the technical discussion of default logic — we do not need to test the applicability of a default rule every time we want to apply it. In addition, the operator $Cn^{D_S}$ is monotone, finitizable and progressive. This allows us to use the results described in Chapter 2.

As corollaries to Theorem 3.25 we now obtain several properties and characterizations of extensions.

**Corollary 3.26** *A theory $S$ is an extension for $(D, W)$ if and only if $S = Cn^{D_S}(W)$.* ▲

To formulate the next corollary we need more notation. For a set of monotone inference rules $D$ we define the set of corresponding defaults *default*($D$) as follows:

$$\mathit{default}(D) = \left\{ \frac{\alpha :}{\gamma} : \frac{\alpha}{\gamma} \in D \right\}.$$

We now have the following property of extensions.

**Corollary 3.27** *Let $S \subseteq \mathcal{L}$ be an extension for a default theory $(D, W)$. Then $S$ is the unique extension for the default theory $(\mathit{default}(D_S), W)$. Conversely, if $S$ is an extension for $(\mathit{default}(D_S), W)$ then $S$ is an extension for $(D, W)$.*

Proof: Clearly, $(\mathit{default}(D_S))_S = D_S$. Consequently, it follows that

$$S = Cn^{D_S}(W) \text{ if and only if } S = Cn^{(\mathit{default}(D_S))_S}(W).$$

Hence, by Corollary 3.26, $S$ is an extension for $(D, W)$ if and only if $S$ is an extension for $(\mathit{default}(D_S), W)$. Since *default*($D_S$) consists of justification-free defaults, it follows from Theorem 3.23 that $S$ is the unique extension of $(default(D_S), W)$. Hence, the assertion follows. ■

**Corollary 3.28** *A theory $S$ is an extension for $(D, W)$ if and only if $S$ is the least fixpoint of the operator $R^{D_S}$ containing $W$, that is, if and only if $S = R^{D_S}{\uparrow}\omega(W)$.*

Proof: The assertion follows from Corollary 3.26 and Theorem 3.9. ■

The identity (3.9) entails the following related characterization of extensions.

**Corollary 3.29** *A theory $S$ is an extension for $(D, W)$ if and only if $S$ is the least fixpoint of the operator $R^{D,S}$ containing $W$, that is, if and only if $S = R^{D,S}{\uparrow}\omega(W)$.* ▲

The connection between the fixpoints of the operator $Cn^{D_S}$ and the base operator $B^{D_S}$, given in Theorem 3.11, allows us to derive a characterization of extensions that is especially useful for algorithm design.

**Corollary 3.30** *A theory $S$ is an extension for a default theory $(D, W)$ if and only if $S = Cn(B^{D_S}{\uparrow}\omega(W))$.*

Proof: The assertion follows from Corollary 3.28 and Theorem 3.11. ■

Let $D$ be a set of defaults. Recall that $c(D) = \{c(d) : d \in D\}$, that is, $c(D)$ consists of the consequents of all defaults in $D$.

**Corollary 3.31** *A theory $S$ is an extension for a default theory $(D, W)$ if and only if there is a set $U \subseteq c(D)$ such that $S = Cn(W \cup U)$ and $W \cup U = B^{D_{Cn(U \cup W)}}{\uparrow}\omega(W)$.*

Proof: The assertion follows easily from Corollaries 3.30 and 3.12. ■

Corollary 3.31 is the basis for algorithms that compute extensions for finite default theories. Two such algorithms will be described in Section 3.9. To give an idea how these algorithms work and to illustrate some basic properties of extensions, we will now consider several examples.

**Example 3.32** Consider the default theory $(D, W)$, where

$$D = \left\{ \frac{: M\neg p}{p} \right\} \quad \text{and} \quad W = \emptyset.$$

According to Corollary 3.31, there are only two candidate theories for an extension: $S_1 = Cn(\emptyset)$ and $S_2 = Cn(\{p\})$.

Let us consider $S_1$ first. Clearly, $\neg\neg p \notin S_1$ ($\neg p$ is consistent with $S_1$). Consequently,

$$D_{S_1} = \left\{ \frac{\top}{p} \right\}$$

(recall that $\frac{:M\neg p}{p}$ denotes the default $\frac{\top : M\neg p}{p}$). Now, using the definition of the operator $B^{D_{S_1}}$ one easily computes:

$$B^{D_{S_1}}\uparrow 1(W) = \{p\}$$

and

$$B^{D_{S_1}}\uparrow 2(W) = \{p\}.$$

Consequently,

$$B^{D_{S_1}}\uparrow \omega(W) = \{p\}.$$

Since $S_1 \neq Cn(B^{D_{S_1}}\uparrow \omega(W))$, $S_1$ is not an extension for $(D, W)$. Speaking informally, $S_1$ is not an extension (belief set) for $(D, W)$ because under the assumption that $S_1$ is a belief set, formulas not assumed as beliefs (namely, $p$) can be derived.

Let us consider the second candidate, $S_2$. Since $\neg\neg p \in S_2$ ($\neg p$ is not consistent with $S_2$)

$$D_{S_2} = \emptyset.$$

Consequently,

$$B^{D_{S_2}}\uparrow \omega(W) = \emptyset.$$

Since $S_2 \neq Cn(B^{D_{S_2}}\uparrow \omega(W))$, $S_2$ is not an extension for $(D, W)$ either. Informally, $S_2$ is not an extension for $(D, W)$ because not all formulas assumed as beliefs (namely, $p$) can be derived. Hence, the theory $(D, W)$ has no extension. ▲

Example 3.32 shows that a default theory may have no extensions. In other words, given a set of defaults $D$, which specify how to reason with incomplete information, there may be no way of extending a consistent theory $W$ of initial assumptions to a total set of beliefs. Such default theories are, in a sense, inconsistent (they determine no consistent belief set). But a word of caution is necessary. There is yet another natural notion of inconsistency that is associated with default theories. Let us consider another example.

**Example 3.33** Consider the default theory $(D, W)$, where

$$D = \left\{\frac{p:}{\neg p}\right\} \quad \text{and} \quad W = \{p\}.$$

By Corollary 3.31, there are two theories that need to be considered in order to find all extensions for $(D, W)$: $S_1 = Cn(\{p\})$ and $S_2 = Cn(\{p, \neg p\}) = \mathcal{L}$. Reasoning as in Example 3.32, we see that $S_1$ is not an extension. Let us consider now the case of $S_2$. Then,

$$D_{S_2} = \left\{\frac{p}{\neg p}\right\}$$

and

$$B^{D_{S_2}}{\uparrow}\omega(W) = \{p, \neg p\}.$$

Consequently, $S_2$ is the only extension for $(D, W)$. ▲

Example 3.33 shows that a default theory may have an inconsistent extension. Also such theories can be regarded as inconsistent. To summarize, there are two different cases when a default theory does not determine a consistent belief set. First, when it does not determine any belief set at all and, secondly, when it determines an inconsistent belief set.

**Example 3.34** Let $D = \{\frac{:M\neg p}{q}\}$ and $W = \emptyset$, where $p$ and $q$ are two different atoms. Then there are two candidates for extensions for $(D, W)$: $S_1 = Cn(\emptyset)$ and $S_2 = Cn(\{q\})$. Clearly, $D_{S_1} = D_{S_2} = \{\frac{\top}{q}\}$ and

$$B^{D_{S_1}}{\uparrow}\omega(W) = B^{D_{S_2}}{\uparrow}\omega(W) = \{q\}.$$

Hence, $S_2$ is the only extension for $(D, W)$. ▲

Example 3.34 shows that a default theory may have a unique consistent extension. The next example illustrates a situation where a default theory has several different extensions.

**Example 3.35** Let $D = \{\frac{p:M\neg q}{r}, \frac{p:M\neg r}{q}\}$ and $W = \{p\}$. In this case, we have four candidates for extensions: $S_1 = Cn(\{p\})$, $S_2 = Cn(\{p, q\})$, $S_3 = Cn(\{p, r\})$ and $S_4 = Cn(\{p, q, r\})$. First, for $i = 1, 2, 3, 4$, we compute:

1. $D_{S_1} = \{\frac{p}{r}, \frac{p}{q}\}$
2. $D_{S_2} = \{\frac{p}{q}\}$
3. $D_{S_3} = \{\frac{p}{r}\}$

4. $D_{S_4} = \emptyset$

Next, we compute:

1. $B^{D_{S_1}}\uparrow\omega(W) = \{p, q, r\}$
2. $B^{D_{S_2}}\uparrow\omega(W) = \{p, q\}$
3. $B^{D_{S_3}}\uparrow\omega(W) = \{p, r\}$
4. $B^{D_{S_4}}\uparrow\omega(W) = \{p\}$

Hence, $S_2$ and $S_3$ are extensions for $(D, W)$ while $S_1$ and $S_4$ are not. ▲

The next group of examples will illustrate several aspects of nonmonotonicity of extensions.

**Example 3.36** Consider the default theory $(D, W)$ from Example 3.32. This theory has no extension at all. Let us add to $W$ a formula $p$ and denote the result by $W'$. Let $S = Cn(\{p\})$. It is easy to see that $D_S = \emptyset$ and $B^{D_S}\uparrow\omega(W') = \{p\}$. Hence, $S$ is an extension for $(D, W')$. This example shows that by adding new formulas to $W$ we may convert a default theory without extensions (that is, an inconsistent theory, in a sense) into a theory which possesses consistent extensions (that is, into a consistent default theory). ▲

**Example 3.37** Consider the default theory $(D, W)$ from Example 3.34. It has exactly one extension $Cn(\{q\})$. Let us add to $W$ the formula $p$ and denote the result by $W'$. Clearly, neither the extension $Cn(\{q\})$ for the "old" theory nor the theory $Cn(\{p, q\})$ are extensions for $(D, W')$. Instead, it is easy to compute that $Cn(\{p\})$ is the only extension for $(D, W')$. Thus, it may happen that an extension of an enlarged theory does not contain any extension of the original theory. ▲

**Example 3.38** Consider now the default theory $(D, W)$ from Example 3.35. The theory $(D, W)$ has two extensions: $Cn(\{p, q\})$ and $Cn(\{p, r\})$. Now replace $W$ by a larger theory $W' = \{p, q\}$. There are now only two candidates for extensions: $S_1 = Cn(\{p, q\})$ and $S_2 = Cn(\{p, q, r\})$. Clearly, $D_{S_1} = \{\frac{p}{q}\}$ and $D_{S_2} = \emptyset$. Hence, $B^{D_{S_1}}\uparrow\omega(W') = \{p, q\}$ and $B^{D_{S_2}}\uparrow\omega(W') = \{p, q\}$, too. Consequently, only $Cn(\{p, q\})$ is an extension of the enlarged theory $(D, W')$. ▲

Similar effects can be observed when new defaults are added.

**Example 3.39** Consider again the theory $(D, W)$ from Example 3.34. Replace $D$ by $D' = D \cup \{\frac{:M\neg r}{p}\}$. The theory $(D', W)$ has only one extension, $Cn(\{p\})$, and it does not contain the only extension for $(D, W)$, $Cn(\{q\})$.

Finally, let $(D, W)$ be the default theory from Example 3.35. Replace $D$ by $D' = D \cup \{\frac{q:}{r}\}$. The same four theories $S_i$, $i = 1, 2, 3, 4$, are the only candidates for extensions for $(D', W)$. We have:

1. $D'_{S_1} = \{\frac{p}{r}, \frac{p}{q}, \frac{q}{r}\}$
2. $D'_{S_2} = \{\frac{p}{q}, \frac{q}{r}\}$
3. $D'_{S_3} = \{\frac{p}{r}, \frac{q}{r}\}$
4. $D'_{S_4} = \{\frac{q}{r}\}$

Next, we compute:

1. $B^{D'_{S_1}}{\uparrow}\omega(W) = \{p, q, r\}$
2. $B^{D'_{S_2}}{\uparrow}\omega(W) = \{p, q, r\}$
3. $B^{D'_{S_3}}{\uparrow}\omega(W) = \{p, r\}$
4. $B^{D'_{S_4}}{\uparrow}\omega(W) = \{p\}$

Hence, $(D', W)$ has only one extension, $Cn(\{p, r\})$. ▲

## 3.5 Two more characterizations of extensions

In this section we will show that the definition of extensions introduced in the previous section is equivalent to the original definition of Reiter. We will also provide a characterization of extensions in terms of sets of valuations discovered by Guerreiro and Casanova.

We start with the approach of Reiter. It is based on the following simple property of default theories.

**Theorem 3.40** *Let $(D, W)$ be a default theory. For every theory $S \subseteq \mathcal{L}$, there exists a least theory $U \subseteq \mathcal{L}$ such that:*

1. *$W \subseteq U$.*
2. *$Cn(U) = U$ (that is. $U$ is closed under propositional provability).*

*3.* *For every default* $d \in D$, *if* $p(d) \in U$ *and, for every* $\beta \in J(d)$, $\neg\beta \notin S$, *then* $c(d) \in U$.

Proof: Define

$$\mathcal{V} = \{U \subseteq \mathcal{L}: U \text{ satisfies conditions (1), (2) and (3)}\}.$$

Clearly, $\mathcal{V} \neq \emptyset$ (because $\mathcal{L} \in \mathcal{V}$). Define $S' = \bigcap \mathcal{V}$. To complete the proof it suffices to show that $S'$ satisfies conditions (1), (2) and (3).

Since for every $U \in \mathcal{V}$, $W \subseteq U$, it follows that $W \subseteq S'$. Hence, $S'$ satisfies (1).

Since the intersection of theories closed under propositional provability is closed under propositional provability (Theorem 2.13), $Cn(S') = S'$. Hence, $S'$ satisfies (2).

Finally, consider a default $d \in D$. Assume that $p(d) \in S'$ and that, for every $\beta \in J(d)$, $\neg\beta \notin S$. Let $U \in \mathcal{V}$. Then, $S' \subseteq U$ and, therefore, $p(d) \in U$. Since $U$ satisfies (3), $c(d) \in U$. Since $U$ is an arbitrary element of $\mathcal{V}$, $c(d) \in S'$. Consequently, $S'$ satisfies (3). ■

Theorem 3.40 guarantees the correctness of the following definition of Reiter's operator.

**Definition 3.41** Let $(D, W)$ be a default theory. *Reiter's operator* $\Gamma^{D,W}$: $\mathcal{P}(\mathcal{L}) \rightarrow \mathcal{P}(\mathcal{L})$ is defined by:

$$\Gamma^{D,W}(S) = \text{the least theory } U \text{ satisfying conditions (1), (2) and (3) of Theorem 3.40,}$$

▲

The operator $\Gamma^{D,W}$ assigns a belief set to a context. The operator $Cn^{D,S}$ assigns a belief set to the "hard facts" $W$ of a default theory $(D, W)$. The next result shows that the belief sets produced by both operators coincide.

**Theorem 3.42** *Let* $(D, W)$ *be a default theory. For every* $S \subseteq \mathcal{L}$,

$$\Gamma^{D,W}(S) = Cn^{D_S}(W) = Cn^{D,S}(W).$$

Proof: It is clear that condition (3) of Theorem 3.40 is equivalent to the condition that $U$ be closed under the rules in $D_S$. Thus, $\Gamma^{D,W}(S)$ is the least set containing $W$ that is closed under propositional provability and closed under the rules in $D_S$. By Theorem 3.7, $\Gamma^{D,W}(S) = Cn^{D_S}(W)$. Hence, the result follows by Theorem 3.25. ■

Reiter defined extensions for a default theory $(D, W)$ as those theories $S$ which satisfy the equation

$$S = \Gamma^{D,W}(S).$$

It follows immediately from Theorem 3.42 that Reiter's definition of extension is equivalent to Definition 3.22.

**Theorem 3.43** *Let $(D, W)$ be a default theory. Then a theory $S \subseteq \mathcal{L}$ is an extension of $(D, W)$ if and only if $S = \Gamma^{D,W}(S)$.* ▲

We will establish now yet another characterization of extensions, this time in terms of sets of valuations rather than provability operators. In the propositional case it is not particularly deep and can be obtained from proof-theoretic characterizations of extensions by a simple application of the completeness theorem. We describe it here because it leads to the interesting extension of default logic to the predicate case discovered recently by Lifschitz.

Let us recall that for a valuation $v$ of $\mathcal{L}$ we denote

$$Th(v) = \{\varphi : \models_v \varphi\}$$

and that for a set $V$ of valuations we denote

$$Th(V) = \{\varphi : \models_v \varphi, \text{ for every } v \in V\}.$$

Clearly,

$$Th(V) = \bigcap \{Th(v) : v \in V\}.$$

Finally, for a theory $S \subseteq \mathcal{L}$, define

$$Mod(S) = \{v : \models_v S\}.$$

It follows directly from the definitions of the operators $Th$ and $Mod$, and from the definition of the operator $Cn$ (Definition 2.10) that for every $W \subseteq \mathcal{L}$,

$$Th(Mod(W)) = Cn(W).$$

Hence, if $W$ is closed under propositional provability,

$$Th(Mod(W)) = W. \tag{3.10}$$

In order to characterize extensions of default theories in terms of valuations, for a default theory $(D, W)$ we introduce an *operator* $\Sigma^{D,W}$ which assigns sets of valuations to sets of valuations. Our definition relies on the following corollary to Theorem 3.40.

**Theorem 3.44** *Let $(D, W)$ be a default theory. For every set of valuations $V$ the set $Mod(\Gamma^{D,W}(Th(V)))$ is the largest set $V'$ of valuations satisfying the following conditions:*

1. *$V' \subseteq Mod(W)$.*
2. *For every $d \in D$, if $p(d) \in Th(V')$ and, for every $\beta \in J(d)$, $\neg\beta \notin Th(V)$, then $c(d) \in Th(V')$.*

Proof: First, we will show that $Mod(\Gamma^{D,W}(Th(V)))$ satisfies conditions (1) and (2). Since $W \subseteq \Gamma^{D,W}(Th(V))$, $Mod(\Gamma^{D,W}(Th(V))) \subseteq Mod(W)$. Hence, $Mod(\Gamma^{D,W}(Th(V)))$ satisfies condition (1). Consider now a default $d \in D$. Assume that $p(d) \in Th(Mod(\Gamma^{D,W}(Th(V))))$ and, for every $\beta \in J(d)$, $\neg\beta \notin Th(V)$. Since $\Gamma^{D,W}(Th(V))$ is closed under propositional provability, it follows by (3.10) that $p(d) \in \Gamma^{D,W}(Th(V))$. By the definition of the operator $\Gamma^{D,W}$, $c(d) \in \Gamma^{D,W}(Th(V))$. Consequently, again by (3.10), $c(d) \in Th(Mod(\Gamma^{D,W}(Th(V))))$. Hence, $Mod(\Gamma^{D,W}(Th(V)))$ satisfies condition (2).

To complete the proof we will show that for every set $V'$ of valuations which satisfies (1) and (2), we have $V' \subseteq Mod(\Gamma^{D,W}(Th(V)))$. To this end, consider a set of valuations satisfying conditions (1) and (2). Observe that the theory $Th(V')$ satisfies conditions (1), (2) and (3) of Theorem 3.40 with $S = Th(V)$ and $U = Th(V')$. Hence, by the definition of $\Gamma^{D,W}(Th(V))$,

$$Th(V') \supseteq \Gamma^{D,W}(Th(V)).$$

Consequently,

$$V' \subseteq Mod(Th(V')) \subseteq Mod(\Gamma^{D,W}(Th(V))),$$

as required. ■

Let $(D, W)$ be a default theory and let $V$ be a set of valuations. We define $\Sigma^{D,W}(V)$ to be the largest set $V'$ of valuations satisfying the conditions (1) and (2) of Theorem 3.44. That is,

$$\Sigma^{D,W}(V) = Mod(\Gamma^{D,W}(Th(V))).$$

We have the following characterization of extensions in terms of fixpoints of the operator $\Sigma^{D,W}$.

**Theorem 3.45** *Let $(D, W)$ be a default theory. Then a theory $S$ is an extension for $(D, W)$ if and only if $S = Th(V)$, for some set of valuations $V$ such that $V = \Sigma^{D,W}(V)$.*

Proof: Assume first that $S$ is an extension for $(D, W)$. Define $V = Mod(S)$. By Theorem 3.43, $S = \Gamma^{D,W}(S)$. Hence,

$$V = Mod(S) = Mod(\Gamma^{D,W}(S)). \tag{3.11}$$

Next, since $S$ is closed under propositional provability,

$$S = Th(Mod(S)) = Th(V).$$

Hence, by (3.11) and the definition of the operator $\Sigma^{D,W}$,

$$V = Mod(\Gamma^{D,W}(Th(V))) = \Sigma^{D,W}(V).$$

Conversely, assume that $S = Th(V)$, where $V = \Sigma^{D,W}(V)$. Then

$$S = Th(\Sigma^{D,W}(V)) = Th(Mod(\Gamma^{D,W}(Th(V)))).$$

Hence, by (3.10),

$$S = \Gamma^{D,W}(Th(V)) = \Gamma^{D,W}(S).$$

By Theorem 3.43, $S$ is an extension for $(D, W)$. ■

We conclude this section by gathering all characterizations of extensions in one result.

**Corollary 3.46** *Let $(D, W)$ be a default theory. For every theory $S \subseteq \mathcal{L}$ the following properties are equivalent:*

1. *$S$ is an extension of $(D, W)$, that is, $S = Cn^{D,S}(W)$.*
2. *$S$ is the least fixpoint of the operator $R^{D,S}$ containing $W$.*
3. *$S = Cn^{D_S}(W)$.*
4. *$S$ is the least fixpoint of the operator $R^{D_S}$ containing $W$.*
5. *$S = Cn(B^{D_S}\uparrow\omega(W))$.*
6. *$S$ is a fixpoint of the operator $\Gamma^{D,W}$, that is, $\Gamma^{D,W}(S) = S$.*
7. *$S = Th(V)$, where $V$ is a set of valuations such that $V = \Sigma^{D,W}(V)$.*

▲

## 3.6 Properties of default theories

This section gathers several basic properties of default theories. First, let us observe that extensions are closed under propositional provability.

**Theorem 3.47** *Let a theory $S$ be an extension for a default theory $(D, W)$. Then $S$ is closed under propositional provability.*

Proof: The result follows immediately from Corollary 3.46. ■

Next, we will generalize the concept of closure under inference rules to the case of defaults and show that extensions are closed under defaults.

**Definition 3.48** A theory $S$ is *closed under a default* $d$, if whenever $p(d) \in S$ and for every $\beta \in J(d)$, $\neg\beta \notin S$ then $c(d) \in S$. A theory $S$ is *closed under a set of defaults* $D$ if $S$ is closed under every default in $D$. ▲

**Theorem 3.49** *Let a theory $S$ be an extension for a default theory $(D, W)$. Then $S$ is closed under the set of defaults $D$.*

Proof: Clearly, a theory $S$ is closed under the defaults in $D$ if and only if $S$ is closed under the inference rules in $D_S$. By Corollary 3.26, $S = Cn^{D_S}(W)$. Hence $S$ is closed under the rules in $D_S$. Consequently, $S$ is closed under the defaults in $D$. ■

We will now study the so-called *confirmation of evidence* property. The consequence operator of propositional logic has the property that, for every theory $W$, $Cn(Cn(W)) = Cn(W)$. It follows that if $Z \subseteq Cn(W)$, then

$$Cn(W \cup Z) = Cn(W).$$

A similar property holds for extensions of default theories.

**Theorem 3.50** *Let $S$ be an extension of a default theory $(D, W)$. For every theory $Z$ such that $Z \subseteq S$, $S$ is an extension for $(D, W \cup Z)$.*

Proof: By Theorem 3.21, we have:

$$S = Cn^{D,S}(W) \subseteq Cn^{D,S}(W \cup Z).$$

Next, again by Theorem 3.21,

$$Cn^{D,S}(W \cup Z) \subseteq Cn^{D,S}(S) = Cn^{D,S}(Cn^{D,S}(W)).$$

Now, by Theorems 3.25 and 3.3,

$$Cn^{D,S}(Cn^{D,S}(W)) = Cn^{D_S}(Cn^{D_S}(W)) = Cn^{D_S}(W) = S.$$

Hence, $S = Cn^{D,S}(W \cup Z)$, that is, $S$ is an extension for $(D, W \cup Z)$. ■

Theorem 3.50 says that if on the basis of $(D, W)$ we accept $S$ as a belief set and then obtain a confirming evidence $Z \subseteq S$ then $S$ remains a belief set for the new default theory $(D, W \cup Z)$.

A word of caution is necessary. It is possible that a default theory $(D, W)$ has an extension $S$ but, after adding to $W$ a theory $Z \subseteq S$, new extensions arise and some old ones disappear. In other words, the confirmation of evidence property guarantees only that extensions containing the new facts added to $W$ will be preserved and says nothing about other extensions.

**Example 3.51** Let $W = \emptyset$ and

$$D = \left\{ \frac{:M\neg b, M\neg d}{a}, \frac{:M\neg b, M\neg d}{c}, \frac{:M\neg a, M\neg c}{d}, \frac{a : M\neg c}{b} \right\}.$$

There are 16 candidates for extensions for $(D, W)$. By checking all of them, one determines that $(D, W)$ has exactly two extensions: $Cn(\{d\})$ and $Cn(\{a, c\})$. Suppose that we are given $a$ as a confirmation of evidence for the extension $Cn(\{a, c\})$. It is easy to see that the default theory $(D, W \cup \{a\})$ has also two extensions. One of them, in agreement with Theorem 3.50, is the "old" extension $Cn(\{a, c\})$. The other one is $Cn(\{a, b\})$, in no way related to the other extension for $(D, W)$, $Cn(\{d\})$. ▲

Next, we will prove that extensions of a default theory form an antichain, that is, no extension is properly contained in another.

**Theorem 3.52** *Let $S$ and $S'$ be two extensions for a default theory $(D, W)$. If $S \subseteq S'$, then $S = S'$.*

Proof: By Theorem 3.21 we have

$$S \subseteq S' = Cn^{D,S'}(W) \subseteq Cn^{D,S}(W) = S.$$

Consequently, $S = S'$. ■

The fact that extensions of a default theory form an antichain does not imply that they are incompatible, that is, that their union is inconsistent. The theory $(D, W)$ from Example 3.35 has two extensions: $Cn(\{p, q\})$, and $Cn(\{p, r\})$. Their union is contained in a consistent theory $Cn(\{p, q, r\})$ and is therefore consistent. In the next chapter we will consider an important class of default theories, the so-called normal default theories, and we will show that different extensions of normal default theories are incompatible.

**Definition 3.53** A default $d$ is *generating for a context* $S \subseteq \mathcal{L}$ if $p(d) \in S$ and, for every $\beta \in J(d)$, $\neg\beta \notin S$. Given a set of defaults $D$ and a context $S$, we denote by $GD(D,S)$ the collection of all generating defaults for $S$. ▲

The next theorem shows that, given an extension $S$, eliminating non-generating defaults leaves $S$ as a unique extension of the resulting theory.

**Theorem 3.54** *Let $S \subseteq \mathcal{L}$ be an extension for a default theory $(D,W)$. Then $S$ is the unique extension for the default theory $(GD(D,S),W)$.*

Proof: Let us define $D' = GD(D,S)$. Clearly, $D'_S \subseteq D_S$. Hence,

$$Cn^{D'_S}(W) \subseteq Cn^{D_S}(W).$$

Since $S$ is an extension for $(D,W)$, $S = Cn^{D_S}(W)$. Consequently, it follows that $Cn^{D'_S}(W) \subseteq S$. Consider now a rule $\frac{\alpha}{\gamma} \in D_S$. We will prove that $Cn^{D'_S}(W)$ is closed under $\frac{\alpha}{\gamma}$. To this end, assume that $\alpha \in Cn^{D'_S}(W)$. Then $\alpha \in S$. Hence, $\frac{\alpha}{\gamma} \in D'_S$. Consequently, $Cn^{D'_S}(W)$ is closed under $\frac{\alpha}{\gamma}$, as claimed. Since $W \subseteq Cn^{D'_S}(W)$, it follows that

$$Cn^{D_S}(W) \subseteq Cn^{D'_S}(W).$$

Hence, $S = Cn^{D'_S}(W)$ and $S$ is an extension for $(GD(D,S),W)$.

To prove that $S$ is a unique extension observe that if $S'$ is an extension for $(GD(D,S),W)$ then $S' = Cn(W \cup U)$, where $U \subseteq c(GD(D,S))$. Hence, $S' \subseteq S$ and, consequently, $S = S'$ (Theorem 3.52). ■

Once an extension is guessed, only the generating defaults for this extension are needed for its reconstruction.

**Theorem 3.55** *A theory $S \subseteq \mathcal{L}$ is an extension for a default theory $(D,W)$ if and only if $S = Cn^{GD(D,S)_S}(W)$.*

Proof. If $S$ is an extension for $(D,W)$ then, by Theorem 3.54, $S$ is an extension for $(GD(D,S),W)$. Hence, $S = Cn^{GD(D,S)_S}(W)$.

To prove the converse implication, consider a rule $\frac{\alpha}{\gamma} \in D_S$. If $\alpha \in S$ then $\frac{\alpha}{\gamma} \in GD(D,S)_S$. Consequently, $\gamma \in S$. Hence, $S$ is closed under the rules in $D_S$.

Since $W \subseteq S$, $Cn^{D_S}(W) \subseteq S$. On the other hand, $GD(D,S)_S \subseteq D_S$. Hence, $S = Cn^{GD(D,S)_S}(W) \subseteq Cn^{D_S}(W)$. Thus, $S = Cn^{D_S}(W)$. ■

Not all generating defaults are necessary for the reconstruction of an extension.

**Example 3.56** Let $W = \{p_{n+1} \supset p_n : n = 1, 2\ldots.\}$ and

$$D = \left\{ \frac{: Mq_n}{p_n} : n = 1, 2\ldots. \right\}.$$

This default theory has a unique extension $S = Cn(\{p_n : n = 1, 2\ldots.\})$. All defaults in $D$ are generating defaults but not all of them are necessary to generate $S$. It is easy to see that $S$ can be reconstructed from $W$ and every set of defaults of the form

$$\left\{ \frac{: Mq_n}{p_n} : n = k, k+1\ldots. \right\},$$

where $k$ is a positive integer. ▲

A natural question to ask is whether for every default theory $(D, W)$ and for every extension $S$ of $(D, W)$ there exists a minimal subset $D'$ of $D$ such that $S$ is an extension of $(D', W)$. Example 3.56 shows that the answer to this question is negative. However, for theories with a finite set of defaults such minimal sets always exist.

Next, we will relate generating defaults for an extension $S$ of a default theory $(D, W)$ and the operator $B^{D_S}$. In the proof we will use the following simple consequence of Definition 3.53:

$$c(GD(D,S)) = \left\{ \gamma : \frac{\alpha}{\gamma} \in D_S \text{ and } \alpha \in S \right\}. \tag{3.12}$$

**Theorem 3.57** *Let $(D, W)$ be a default theory. If a theory $S$ is an extension for $(D, W)$, then*

$$B^{D_S}\uparrow\omega(W) = W \cup c(GD(D,S)) \tag{3.13}$$

*and*

$$S = Cn(W \cup c(GD(D,S))). \tag{3.14}$$

Proof: By Corollary 3.30, it suffices to prove (3.13). Clearly,

$$W \subseteq B^{D_S}\!\uparrow\!\omega(W).$$

Let $\gamma \in c(GD(D,S))$. By (3.12), there is a rule $\frac{\alpha}{\gamma} \in D_S$ such that $\alpha \in S$. Since $S = Cn(B^{D_S}\!\uparrow\!\omega(W))$, by the compactness theorem (Theorem 2.34) it follows that there is an integer $m$ such that $\alpha \in Cn(B^{D_S}\!\uparrow\! m(W))$. Hence, $\gamma \in B^{D_S}\!\uparrow\!(m+1)(W)$ and, consequently, $\gamma \in B^{D_S}\!\uparrow\!\omega(W)$. Thus,

$$W \cup c(GD(D,S)) \subseteq B^{D_S}\!\uparrow\!\omega(W).$$

To prove the converse implication we will show, by induction, that for every non-negative integer $n$,

$$B^{D_S}\!\uparrow\! n(W) \subseteq W \cup c(GD(D,S)).$$

Since $B^{D_S}\uparrow 0(W) = W$, the basis of the induction follows. Assume that $B^{D_S}\!\uparrow\! n(W) \subseteq W \cup c(GD(D,S))$ and consider a formula

$$\gamma \in B^{D_S}\!\uparrow\!(n+1)(W).$$

There are two possibilities. If $\gamma \in B^{D_S}\!\uparrow\! n(W)$ then $\gamma \in W \cup c(GD(D,S))$. If not, then there is a rule $\frac{\alpha}{\gamma} \in D_S$ such that $\alpha \in Cn(B^{D_S}\!\uparrow\! n(W))$. By Corollary 3.30, $\alpha \in S$. Hence, by (3.12), $\gamma \in c(GD(D,S))$. ■

In Section 3.10 we will study theories $S$ that are solutions to the equation (3.14). We will call them weak extensions. Theorem 3.57 implies that extensions are weak extensions. In general, this implication cannot be reversed, that is, there are theories $S$ which satisfy the equation (3.14) but which are not extensions for a default theory $(D,W)$.

Finally, we will study default theories with an inconsistent extension. As a consequence of Theorem 3.52 we obtain that if a default theory $(D,W)$ has an inconsistent extension, then $(D,W)$ has no other extensions.

**Corollary 3.58** *If $(D,W)$ has an inconsistent extension $S$ then $S = \mathcal{L}$ and $(D,W)$ has no other extensions.* ▲

Next, we provide a complete description of default theories with an inconsistent extension. Recall that

$$D_m = \left\{\frac{\alpha}{\gamma} : \frac{\alpha :}{\gamma} \in D\right\}.$$

That is, $D_m$ consists of the inference rules that naturally correspond to justification-free defaults in $D$ (the subscript $m$ stands for monotone). It is easy to see that, in the terminology of reducts, $D_m = D_{\mathcal{L}}$.

**Theorem 3.59** *A default theory $(D,W)$ possesses an inconsistent extension if and only if $Cn^{D_m}(W)$ is inconsistent.*

Proof: We have that $\mathcal{L}$ is an extension of $(D, W)$ if and only if

$$Cn^{D_{\mathcal{L}}}(W) = \mathcal{L}.$$

Since $D_m = D_{\mathcal{L}}$ and since $Cn^{D_{\mathcal{L}}}(W)$ is closed under propositional provability, $\mathcal{L}$ is an extension of $(D, W)$ if and only if $Cn^{D_m}(W)$ is inconsistent. ■

In this book we allow for defaults without justifications. Reiter's original definition of defaults requires that a default always has at least one justification. For default theories with no justification-free defaults Theorem 3.59 simplifies.

**Corollary 3.60** *Let $(D, W)$ be a default theory such that no default in $D$ is justification-free. Then, $(D, W)$ has an inconsistent extension if and only if $W$ is inconsistent.*

Proof: The assertion follows easily from Theorem 3.59 once we observe that $D_m = \emptyset$. ■

## 3.7 Extensions and well-orderings

In this section we will study a "naive" method of reasoning from a default theory $(D, W)$. We will be constructing an extension in steps. In each step we will apply one default. The applicability of the default will be determined with respect to the theory constructed so far — a current context $T$. Once a default $d$ is found such that $p(d) \in T$ and $\neg\beta \notin T$ for every $\beta \in J(d)$, the new context $Cn(\{c(d)\} \cup T)$ will be formed. The construction goes on until a default applicable with respect to the current context can no longer be found.

Clearly, the result of such a construction depends on the ordering of defaults. Applying a default $d_1$ and adding its consequent to the current context may render another default, $d_2$, inapplicable. On the other hand, applying $d_2$ first and adding the consequent of $d_2$ to the current context may make $d_1$ inapplicable. Thus, an ordering, in fact a well-ordering, of the defaults in $D$ must be given as an input parameter for our construction.

Two main questions that need to be investigated are:

1. Is this method sound? That is, does it generate an extension for every ordering of the defaults in $D$?
2. Is this method complete? That is, does it generate all extensions if all orderings of the defaults are tried? If so, do we have to consider all orderings to be certain that all extensions are constructed?

We might suspect that the method is not sound. After all, the current context grows from step to step and defaults applicable with respect to small contexts need no longer be applicable after the context grows. Whether the method is complete or not is much less obvious.

In this section these questions will be studied in detail. We will make precise the construction outlined above. We will show that the method is not sound, but is complete. For each extension $S$ of $(D, W)$ we will show several orderings of the set $D$ of defaults which generate $S$. We will also establish a condition which allows us to eliminate those theories produced by the method which are not extensions. These results will be utilized in Section 3.9 to design an algorithm for generating all extensions of a *finite* default theory and in Section 3.8 where we discuss prioritized default theories. In Chapter 4, they will serve as the main tool in our investigations of normal default theories.

As usual, we will assume that the set $At$ of the atoms of our language $\mathcal{L}$ is denumerable. That is, $|At| \leq \omega$. Consequently, the set of all defaults over the language $\mathcal{L}$ is denumerable.

**Definition 3.61** Let $(D, W)$ be a default theory and $\preceq$ a well-ordering of $D$. We define an ordinal $\eta_{\preceq}$. For every $\xi < \eta_{\preceq}$ we define a set of defaults $AD_\xi$ and a default $d_\xi$. We also define a set of defaults $AD_{\preceq}$. We proceed as follows: If the sets $AD_\xi$, $\xi < \alpha$, have been defined but $\eta_{\preceq}$ has not been defined then

1. If there is no default $d \in D \setminus \bigcup_{\xi<\alpha} AD_\xi$ such that:
   (a) $W \cup c(\bigcup_{\xi<\alpha} AD_\xi) \vdash p(d)$
   (b) $W \cup c(\bigcup_{\xi<\alpha} AD_\xi) \not\vdash \neg\beta$, for every $\beta \in J(d)$

   then $\eta_{\preceq} = \alpha$.
2. Otherwise, define $d_\alpha$ to be the $\preceq$-least default $d \in D \setminus \bigcup_{\xi<\alpha} AD_\xi$ such that:
   (a) $W \cup c(\bigcup_{\xi<\alpha} AD_\xi) \vdash p(d)$.
   (b) $W \cup c(\bigcup_{\xi<\alpha} AD_\xi) \not\vdash \neg\beta$, for every $\beta \in J(d)$.

   Then set $AD_\alpha = \bigcup_{\xi<\alpha} AD_\xi \cup \{d_\alpha\}$.
3. Put $AD_{\preceq} = \bigcup_{\xi<\eta_{\preceq}} AD_\xi$.

The theory $Cn(W \cup c(AD_{\preceq}))$ will be called *generated by* the well-ordering $\preceq$. ▲

Usually we will write $\eta$ for $\eta_{\preceq}$. Our first result shows that the ordinal $\eta$ is well-defined, that is, that the construction stops.

**Theorem 3.62** *The ordinal $\eta$ is well-defined. In fact, $\eta < \omega_1$, where $\omega_1$ is the least uncountable ordinal number.*

Proof: Assume that $d_\xi$ is defined for every $\xi < \omega_1$. Clearly, $d_\xi \in D$, for every $\xi < \omega_1$, and all $d_\xi$ are pairwise distinct. Hence, $|D| \geq \omega_1$, a contradiction. ∎

We will now rephrase the construction of Definition 3.61 in slightly different terms. For every $\xi < \eta$, let us define

$$T_\xi = Cn(W \cup c(AD_\xi))$$

and, for every $\xi \leq \eta$, define

$$AD_{<\xi} = \bigcup_{\alpha<\xi} AD_\alpha.$$

This implies that $AD_{<0} = \emptyset$ and

$$T_{<\xi} = Cn(W \cup c(AD_{<\xi})).$$

Finally, define

$$T_\preceq = Cn(W \cup c(AD_\preceq)).$$

It follows directly from the definitions that

$$AD_\preceq = AD_{<\eta} \text{ and } T_\preceq = T_{<\eta}.$$

This notation will be used throughout the remainder of this section.

The defaults in $AD_{<\xi}$ are exactly those defaults that have been applied by the construction of Definition 3.61 prior to step $\xi$. The defaults in $AD_\xi$ are those used in the first $\xi$ steps. The theory $T_{<\xi}$ is the context used in step $\xi$ and the theory $T_\xi$ is the next context (after $\xi$ steps of the construction). In each step of the construction, the defaults that have not been applied yet and are generating (applicable) with respect to the current context are considered. These are the defaults in $GD(D \setminus AD_{<\xi}, T_{<\xi})$. The $\preceq$-least default in this set is selected and used in the derivation. That is, its consequent is appended to the context. The construction terminates when no generating defaults with respect to the current context can be found among those that have not been applied yet.

We will now study the properties of the construction described in Definition 3.61. First, we present several simple facts. For every $\xi < \eta$ we have

$$GD(D \setminus AD_{<\xi}, T_{<\xi}) \neq \emptyset, \tag{3.15}$$

$$p(d_\xi) \in T_{<\xi}, \tag{3.16}$$

$$AD_{<\xi} = \{d_\alpha : \alpha < \xi\}, \tag{3.17}$$

$$AD_\xi = \{d_\alpha : \alpha \leq \xi\}, \tag{3.18}$$

and,

$$\neg\beta \notin T_{<\xi}, \text{ for every } \beta \in J(d_\xi). \tag{3.19}$$

We also have that

$$d_\xi \in GD(D \setminus AD_{<\xi}, T_{<\xi}) \tag{3.20}$$

and

$$GD(D \setminus AD_{\preceq}, T_{\preceq}) = \emptyset. \tag{3.21}$$

This last identity implies that

$$GD(D, T_{\preceq}) \subseteq AD_{\preceq}. \tag{3.22}$$

Next, we will investigate more complex properties of the construction described in Definition 3.61. First, we will show that the method is not sound. That is, the theories generated by some well-orderings are not extensions.

**Example 3.63** Let $D = \{\frac{:M\neg p}{p}\}$ and $W = \emptyset$. Then there is exactly one well-ordering $\preceq$ of $D$. It is the ordering in which the only default in $D$ is related to itself. Clearly $AD_{\preceq}$ coincides with $D$, since $\frac{:M\neg p}{p}$ is applicable with respect to the original context $T_{<0} = Cn(W)$. Hence, the theory generated by $\preceq$ is $Cn(\{p\})$. On the other hand, $Cn(\{p\})$ is not an extension of $(D, W)$. ▲

We will now establish a criterion for $AD_{\preceq}$ to generate an extension of $(D, W)$. To this end we will need the following lemma. Before we formulate it, we introduce one more piece of notation. For a set of defaults $D$ define

$$Mon(D) = \left\{\frac{p(d)}{c(d)} : d \in D\right\}.$$

Informally, $Mon(D)$ consists of monotone inference rules obtained as the result of eliminating justifications from every default in $D$.

**Lemma 3.64** *Under the notation introduced above,*

$$Cn^{Mon(AD_{\preceq})}(W) = T_{\preceq}.$$

Proof: It follows from Corollary 3.12 that

$$Cn^{Mon(AD_{\preceq})}(W) \subseteq Cn(W \cup c(AD_{\preceq})) = T_{\preceq}.$$

To prove the converse inclusion, consider a default $d_\xi$, where $\xi < \eta$. We use induction on $\xi$. Assume that for every $\alpha < \xi$, $c(d_\alpha) \in Cn^{Mon(AD_{\preceq})}(W)$. It follows that $W \cup c(AD_{<\xi}) \subseteq Cn^{Mon(AD_{\preceq})}(W)$. Hence,

$$T_{<\xi} \subseteq Cn^{Mon(AD_{\preceq})}(W).$$

By (3.16), $p(d_\xi) \in T_{<\xi}$. Consequently, $p(d_\xi) \in Cn^{Mon(AD_{\preceq})}(W)$. Since $\frac{p(d_\xi)}{c(d_\xi)} \in Mon(AD_{\preceq})$, $c(d_\xi) \in Cn^{Mon(AD_{\preceq})}(W)$.

Hence, for every $\xi < \eta$, $c(d_\xi) \in Cn^{Mon(AD_{\preceq})}(W)$. Thus, it follows that $W \cup c(AD_{\preceq}) \subseteq Cn^{Mon(AD_{\preceq})}(W)$. Since $Cn^{Mon(AD_{\preceq})}(W)$ is closed under propositional provability, the assertion follows. ∎

We are ready to present a sufficient condition for $T_{\preceq}(= Cn(W \cup c(AD_{\preceq})))$ to be an extension for $(D, W)$.

**Theorem 3.65** *Let $(D, W)$ be a default theory and let $\preceq$ be a well-ordering of $D$. If*

$$\text{for every } \beta \in J(AD_{\preceq}),\ W \cup c(AD_{\preceq}) \not\vdash \neg\beta, \tag{3.23}$$

*then $T_{\preceq}$ is an extension for $(D, W)$.*

Proof: By (3.16), for every $\xi < \eta$, $p(d_\xi) \in T_{<\xi}$. Consequently, for every $\xi < \eta$, $p(d_\xi) \in T_{\preceq}$. In addition, by (3.23), for every $\beta \in J(AD_{\preceq})$, $\neg\beta \notin T_{\preceq}$. Hence,

$$AD_{\preceq} \subseteq GD(D, T_{\preceq}).$$

By (3.22), it follows that

$$AD_{\preceq} = GD(D, T_{\preceq}).$$

It is easy to see that for every theory $T$,

$$GD(D,T)_T = Mon(GD(D,T)).$$

In particular, $Mon(AD_{\preceq}) = Mon(GD(D, T_{\preceq})) = GD(D, T_{\preceq})_{T_{\preceq}}$. Hence, by Lemma 3.64,

$$Cn^{GD(D,T_{\preceq})_{T_{\preceq}}}(W) = Cn^{Mon(AD_{\preceq})}(W) = T_{\preceq}.$$

By Theorem 3.55, $T_{\preceq}$ is an extension for $(D, W)$. ∎

The condition (3.23) in Theorem 3.65 is sufficient but not necessary. That is, there exist theories $(D, W)$ and orderings $\preceq$ of $D$ such that for some $\beta \in J(AD_{\preceq})$, $W \cup c(AD_{\preceq}) \vdash \neg\beta$ and, yet, $Cn(W \cup c(AD_{\preceq}))$ is an extension for $(D, W)$.

**Example 3.66** Let $W = \emptyset$ and $D = \{d_0, d_1\}$, where $d_0 = \frac{:M\neg p}{p}$ and $d_1 = \frac{:M\neg q}{p}$. Consider the ordering $\preceq$ of $D$ such that $d_0 \preceq d_1$. It is easy to see that $AD_{\preceq} = \{d_0, d_1\}$. Consequently, $J(AD_{\preceq}) = \{\neg p, \neg q\}$ and $W \cup c(AD_{\preceq}) = \{p\}$.

Hence, $W \cup c(AD_{\preceq}) \vdash \neg\neg p$. At the same time, the theory $Cn(\{p\})$ is an extension of $(D, W)$.

Next, let us consider the ordering $\preceq'$ such that $d_1 \preceq' d_0$. Clearly, $d_1$ is applicable with respect to $T_{<0} = Cn(W)$. Hence, $AD_0 = \{d_1\}$ and the context $T_0$ after the $0^{\text{th}}$ step of the construction equals $Cn(\{p\})$. The default $d_0$ is not applicable with respect to the context $T_0$. Hence, the construction stops and $AD_{\preceq'} = \{d_1\}$.The set $AD_{\preceq'}$ satisfies the condition (3.23). Consequently, $Cn(W \cup c(AD_{\preceq'})) = Cn(\{p\})$ is an extension for $(D, W)$.

The moral of this example is that the same theory may be generated by several orderings. For some of them, the corresponding set of defaults $AD_{\preceq}$ may satisfy the condition of Theorem 3.65 and for some others it may fail. ▲

Our example shows that the criterion provided in Theorem 3.65 may sometimes fail to detect an extension. That is, an extension will be constructed but the condition (3.23) will not be met. It turns out, however, that if $S$ is an extension, then there is an ordering $\preceq$ such that $S$ is generated by $\preceq$ and the corresponding set of defaults $AD_{\preceq}$ satisfies the condition (3.23). We will present several results to this effect.

Our first result is closely related to the characterization of extensions given in Corollary 3.28. That corollary describes extensions for a default theory $(D, W)$ as exactly those theories $S$ that satisfy

$$S = \bigcup_{n=0}^{\infty} R^{D,S}\!\uparrow\! n(W).$$

The set of all defaults over the language $\mathcal{L}$ is denumerable. Hence, it can be ordered in type $\omega$. Let $\preceq$ be any such ordering. Let $S$ be an extension for a default theory $(D, W)$. We now define sets of defaults $D_n$, $n = 1, 2, \ldots$, as follows:

$$D_1 = \{d \in D : p(d) \in Cn(W) \ \text{ and } \ \neg\beta \notin S, \ \text{ for } \ \beta \in J(d)\},$$

and, for $n \geq 1$,

$$D_{n+1} = \left\{ d \in D \setminus \bigcup_{i=1}^{n} D_n : p(d) \in R^{D,S}\!\uparrow\! n(W) \text{ and } \neg\beta \notin S, \text{ for } \beta \in J(d) \right\},$$

Less formally, the set $D_{n+1}$ consists of the defaults that have been used in the construction of $R^{D,S}\!\uparrow\!(n+1)(W)$ out of $R^{D,S}\!\uparrow\! n(W)$. The sets $D_n$ are pairwise disjoint and their union coincides with the set of generating defaults for $S$.

For every positive integer $n$, the ordering $\preceq$ induces in the set $D_n$ an ordering $\preceq_n$ of order type at most $\omega$. It also induces an ordering $\preceq_\infty$, of type at most $\omega$, in the set $D \setminus \bigcup_{n=1}^{\infty} D_n$.

Define now the ordering $\preceq_S$ of $D$, as follows. For every two defaults $d, d' \in D$, $d \preceq_S d'$ if at least one of the following conditions holds:

1. For some positive integer $i$, $d, d' \in D_i$ and $d \preceq_i d'$.
2. $d, d' \in D \setminus \bigcup_{n=1}^{\infty} D_n$ and $d \preceq_\infty d'$.
3. For some positive integers $i, j$ such that $i < j$, $d \in D_i$ and $d' \in D_j$.
4. For some positive integer $i$, $d \in D_i$ and $d' \in D \setminus \bigcup_{n=1}^{\infty} D_n$.

The ordering $\preceq_S$ is of type at most $\omega \times \omega + \omega$. The ordering $\preceq_S$ orders the generating defaults for $S$ according to the order in which they are used during the construction of $R^{D,S} \uparrow \omega(W)$. The defaults in $GD(D, S)$ form an initial segment in the ordering $\preceq_S$. All other defaults form a final segment of $\preceq_S$.

**Theorem 3.67** *If $S$ is an extension of $(D, W)$ and $\preceq_S$ is the ordering of $D$ defined above, then $S = Cn(W \cup c(AD_{\preceq_S}))$, that is, $S$ is generated by $\preceq_S$. Moreover, $AD_{\preceq_S} = GD(D, S)$ and the condition (3.23) is satisfied.* ▲

**Corollary 3.68** *Let $S$ be an extension for a default theory $(D, W)$. There exists a well-ordering $\preceq$ of $D$ such that $S$ is generated by $\preceq$ and $AD_{\preceq}$ satisfies the condition (3.23).* ▲

We will not prove this result. Soon two stronger results will be stated and proved. We will only discuss some consequences of Theorem 3.67. First of all, it implies that the construction described in Definition 3.61 is complete. That is, for every extension $S$ of a default theory $(D, W)$ there exists a well-ordering of $D$ which generates $S$. Secondly, it implies that in the search for the orderings that generate extensions we can restrict our attention to orderings of type at most $\omega \times \omega + \omega$. Finally, Theorem 3.67 implies that if the condition (3.23) of Theorem 3.65 is used as the sieve to eliminate the theories generated by some orderings then every extension will be accepted and all the theories which are not extensions will be rejected.

The remaining results of this section are similar to Theorem 3.67 in that for an extension $S$ they exhibit an ordering generating $S$ for which the condition (3.23) of Theorem 3.65 is satisfied. Two of them, Corollary 3.71 and Theorem 3.73, exhibit orderings generating an extension $S$ of types at most $\omega + \omega$ and $\omega$, respectively.

First, we will establish a simple lemma.

**Lemma 3.69** *Let $S$ be an extension of $(D,W)$, and let $S_1$ be a theory closed under propositional provability and such that $W \subseteq S_1 \subseteq S$. If $S_1 \neq S$, then there exists a default $d \in GD(D,S)$ such that $p(d) \in S_1$, but $c(d) \notin S_1$.*

Proof: Assume that for every $d \in GD(D,S)$, if $p(d) \in S_1$ then $c(d) \in S_1$. Then it follows that $S_1$ is closed under the rules in the reduct $GD(D,S)_S$. Since $S_1$ is closed under propositional provability and contains $W$, we have $Cn^{GD(D,S)_S}(W) \subseteq S_1$. By Theorem 3.55, $S = Cn^{GD(D,S)_S}(W)$. Hence, $S \subseteq S_1$, a contradiction. ∎

**Theorem 3.70** *Let $S$ be an extension for a default theory $(D,W)$. Let $\preceq$ be any well-ordering of $D$ such that the defaults in $GD(D,S)$ precede the defaults in $D \setminus GD(D,S)$. Then $S$ is generated by $\preceq$. Moreover, $AD_{\preceq} = GD(D,S)$ and the condition (3.23) is satisfied.*

Proof: Let $\eta$ be the ordinal number specified by Definition 3.61 for the ordering $\preceq$. First, we prove by induction that for every ordinal number $\xi < \eta$, $AD_\xi \subseteq GD(D,S)$. To this end, assume that for all $\alpha < \xi$ $AD_\alpha \subseteq GD(D,S)$. Recall that we have defined $AD_{<\xi} = \bigcup_{\alpha<\xi} AD_\alpha$ and $T_{<\xi} = Cn(W \cup c(AD_{<\xi}))$. The inductive hypothesis implies that $T_{<\xi} \subseteq S$. By (3.16) and (3.19), $d_\xi \in GD(D, T_{<\xi})$.

Assume that $T_{<\xi} = S$. Then $d_\xi \in GD(D,S)$. Consequently, $AD_\xi \subseteq GD(D,S)$.

Assume next that $T_{<\xi} \neq S$. By Lemma 3.69, there is a default $d$ in $GD(D,S)$ such that $p(d) \in T_{<\xi}$ and $c(d) \notin T_{<\xi}$. Since $T_{<\xi} \subseteq S$, $d \in GD(D,T_{<\xi})$. Since $c(d) \notin T_{<\xi}$, $d \notin AD_{<\xi}$. Hence, $d \in GD(D \setminus AD_{<\xi}, T_{<\xi})$. By the definition of $d_\xi$, $d_\xi \preceq d$. In addition, by the definition of $\preceq$, $d_\xi \in GD(D,S)$. Consequently, $AD_\xi \subseteq GD(D,S)$. That completes our induction.

Hence, it follows that $AD_{\preceq} \subseteq GD(D,S)$. Consequently, $T_{\preceq} = Cn(W \cup c(AD_{\preceq})) \subseteq Cn(W \cup c(GD(D,S)))$. By Theorem 3.57, $T_{\preceq} \subseteq S$. Assume that $T_{\preceq} \neq S$. By Lemma 3.69 there exists a default $d \in GD(D,S)$ such that $p(d) \in T_{\preceq}$ and $c(d) \notin T_{\preceq}$. Since $c(d) \notin T_{\preceq}$, $d \notin AD_{\preceq}$. On the other hand, since for every $\beta \in J(d), \neg\beta \notin S$, therefore for every $\beta \in J(d)$, $\neg\beta \notin T_{\preceq}$. Existence of the default $d$ contradicts the definition of $\eta_{\preceq}$.

Therefore, $T_{\preceq} = S$ and, by (3.22), $GD(D,S) = AD_{\preceq}$. In other words, $S$ is generated by the well-ordering $\preceq$. Since $AD_{\preceq} = GD(D,S)$, the condition (3.23) follows. ∎

Since $GD(D,S)$ and $D \setminus GD(D,S)$ are denumerable and can be ordered in type at most $\omega$, it follows from Theorem 3.70 that an extension $S$ can be generated by an ordering of type at most $\omega + \omega$. Hence, we obtain the following result.

**Corollary 3.71** *Let $S$ be an extension for a default theory $(D,W)$. There exists a well-ordering $\preceq$ of $D$ of type at most $\omega + \omega$ such that $S$ is generated by $\preceq$ and $AD_{\preceq}$ satisfies the condition (3.23).* ▲

Hence we have proved that every extension is generated by an ordering $\preceq$ of type not exceeding $\omega + \omega$. We will show that this upper bound can be lowered to orders of type not exceeding $\omega$. The importance of Theorem 3.67 and Corollary 3.71 lies in their numerous applications, especially in our presentation of normal default logic.

Our last result in this section shows that, in fact, in order to generate all extensions it is enough to restrict to orderings in type at most $\omega$.

**Lemma 3.72** *Let $(D,W)$ be a default theory, and let $S$ be an extension of $(D,W)$. Moreover assume that $D = GD(D,S)$. Then there exists a well-ordering $\preceq$ of $D$ such that $\eta_{\preceq} \leq \omega$ and $d_k$ is the $k^{th}$ element of $D$ in the ordering $\preceq$.*

Proof: Assume that $|D| = \omega$ (if $|D| < \omega$ the lemma is obvious). Let $D_0 = \{d \in D : p(d) \in W\}$, $T_0 = Cn(W \cup c(D_0))$. Next, $D_{n+1} = \{d \in D : p(d) \in T_n\} \setminus \bigcup_{m \leq n} D_m$, $T_{n+1} = Cn(W \cup c(\bigcup_{m \leq n+1} D_m))$.

The set $D_n$ consists of all the rules used to construct $R^{D,S}{\uparrow}n(W)$ from $R^{D,S}{\uparrow}(n-1)(W)$. Clearly, the sets $D_n$ are pairwise disjoint.

We assign now to a default rule $d \in D$ a finite set $X_d$ of defaults in $D$ with the following properties:

1. If $d \in D_0$ then $X_d$ is empty, and if $d \in D_{n+1}$ then $X_d \subseteq \bigcup_{m \leq n} D_m$.

2. $p(d) \in Cn(W \cup c(X_d))$.

Clearly, such a set $X_d$ can be found. It is enough to consider the set of all rules from $D$ appearing in an $S$-proof of $p(d)$ from $W$ in $PC + D$ with the least number of rules from $D$.

We will construct a well-ordering $\preceq$ by putting all default rules in "slots" enumerated with natural numbers. Every default will be assigned a unique slot, but not every slot will be filled. If the default $d$ is put in a slot $k$ then we write $s(d) = k$.

We will reserve slots with indices of the form $2^n \cdot (2k+1)$ for the defaults from $D_n$.

Label the defaults of $D_n$ as $g_0^n, g_1^n, g_2^n \ldots$. The set $D_n$ may be infinite or finite; it will make no difference. The default rules from $D_0$ fill the slots $1, 3, 5, \ldots$ in natural order. That is, the first default of $D_0$ is put in the slot with index 1, the second in the slot with index 3 etc.

Assume that we have already assigned slots to elements of $D_j$, $j \leq n$. We will now show how to assign a slot to every element of $D_{n+1}$. We do this inductively. Consider the default rule $g_k^{n+1}$. Assume that for every default $g_i^{n+1}$, $i < k$, a slot has been assigned. Let $t$ be the least integer of the form $2^{n+1} \cdot (2l+1)$ bigger than all the elements of the set

$$\{s(d) : d \in X_{g_k^{n+1}}\} \cup \{s(g_i^{n+1}) : i < k\}.$$

We assign to $g_k^{n+1}$ the slot $t$. That is, $s(g_k^{n+1}) = t$. It follows that our construction assigns to every default rule in $D$ a slot.

We now define $d \preceq d'$ as $s(d) < s(d')$. Since all the slots are natural numbers, it follows that the order type of $\preceq$ is $\omega$.

We now prove, by induction on $k$, that if $d$ is the $k^{th}$ default rule of order $\preceq$ then $d = d_k$, that is $d$ is selected precisely at the $k^{th}$ step of the construction described in Definition 3.61. Since all the defaults in $D$ belong to $GD(D, S)$, the theories we construct are all included in $S$. Moreover justifications of defaults $d$ are consistent with $S$.

The base step is obvious, as the first slot is filled with an element of $D_0$. In the inductive step, if we consider the rule $d$ which is $k^{th}$ in the ordering $\preceq$ then by the inductive hypothesis $d$ is the $\preceq$-least rule in the set $D \setminus \{d_0, \ldots, d_{k-1}\}$. By the definition of $\preceq$, $X_d \subseteq \{d_0, \ldots, d_{k-1}\}$. Therefore $p(d) \in Cn(W \cup c(\{d_0, \ldots, d_{k-1}\}))$. Hence $d_k = d$. Since the rules are ordered in type $\omega$, at step $\omega$ there are no rules left. Therefore the ordinal $\eta_{\preceq}$ is $\omega$. ■

**Theorem 3.73** *Let $(D, W)$ be a default theory and let $S$ be an extension of $(D, W)$. Then there exists a well-ordering $\preceq$ of $D$ of type at most $\omega$ such that $S$ is generated by $\preceq$. Moreover, $AD_{\preceq} = GD(D, S)$ and the condition (3.23) is satisfied.*

Proof: If $GD(D, S)$ is finite then we use Corollary 3.71. We order $GD(D, S)$ in a finite type, and $D \setminus GD(D, S)$ in type at most $\omega$, and concatenate these orderings. The resulting ordering is of type at most $\omega$ and by Corollary 3.71 satisfies our thesis.

Hence, assume that $GD(D,S)$ is infinite. Denote $D' = GD(D,S)$. By Theorem 3.54, $S$ is an extension of $(D',W)$. Since $GD(GD(D,S),S) = GD(D,S)$ Lemma 3.72 is applicable. Let $\preceq'$ be the ordering of $D'$ in type $\omega$ constructed in the proof of Lemma 3.72. Then the $k^{th}$ element of $D'$ in the ordering $\preceq'$ is precisely $d'_k$ (where $d'_k$ are defaults selected using the ordering $\preceq'$ according to Definition 3.61). In particular, the ordinal $\eta_{\preceq'}$ is equal to $\omega$.

Let $\{g_0, g_1, \ldots\}$ be the listing of all defaults in $GD(D,S)$ in the order of $\preceq'$. Let $\{h_0, h_1, \ldots\}$ be any listing of $D \setminus GD(D,S)$. We will construct a desired ordering $\preceq$ by assigning to every default $d \in D$ a slot. The fact that $d$ is assigned a slot $t$ is denoted by $s(d) = t$. As in the proof of Lemma 3.72 not all slots will be filled. Default $g_i$ will be assigned the slot $2i$. Assigning slots to the remaining defaults will be more complicated.

Assume that defaults $h_0, \ldots, h_{k-1}$ have already been assigned slots. If $p(h_k) \notin S$ then we assign to $h_k$ the first available slot with odd index. But if $p(h_k) \in S$ then, since $h_k \notin GD(D,S)$, there must be a formula $\beta \in J(h_k)$ such that $S \vdash \neg\beta$. By the compactness theorem there is $l$ such that $Cn(W \cup c(\{g_0, \ldots, g_l\})) \vdash \neg\beta$. Let $t$ be the least odd number bigger than $2l$ and all the slots assigned to $h_0, \ldots, h_{k-1}$. We assign $t$ to $h_k$. That is, $s(h_k) = t$.

It should be clear that every default in $D$ has been assigned a slot. Define

$$d \preceq d' \text{ if and only if } s(d) \leq s(d')$$

The order type of $\preceq$ is $\omega$. The ordering $\preceq$ restricted to $GD(D,S)$ is isomorphic to $\preceq'$.

We show that when we execute the construction of $AD_n$ (as in Definition 3.61) with respect to the ordering $\preceq$ then in step $k$ the default $d_k$ is precisely $g_k$. This is proved by induction on $k$. The base step is obvious since the $0^{th}$ element of $\preceq$ is $g_0$ and by construction of $\preceq'$, $p(g_0) \in Cn(W)$.

In the inductive step assume that for $i < k$, $d_i = g_i$. We show that $d_k = g_k$. Let $T_{k-1} = Cn(W \cup c(\{g_0, \ldots, g_{k-1}\}))$. Let $d$ be the $\preceq$-least default in $D \setminus \{g_0, \ldots, g_{k-1}\}$ such that $p(d) \in T_{k-1}$ and for all $\beta \in J(d)$, $T_{k-1} \nvdash \neg\beta$. In particular, $p(d) \in S$. Since $g_k$ satisfies both conditions (by the choice of ordering $\preceq'$), $s(d) \leq 2k$. Since $d_i = g_i$ for $i < k$, either $d = g_k$, or $d \notin GD(D,S)$.

In the first case we have $d_k = g_k$. We show that the second case is impossible. Hence, let us assume that $d \notin GD(D,S)$. Since $p(d) \in S$, it follows from the definition of $s(d)$ that there is an integer $r$ and a formula $\beta \in J(d)$ such that:

1. $s(d) \geq s(g_r)$.
2. $Cn(W \cup c(\{g_0, \ldots, g_r\})) \vdash \neg\beta$.

Since $s(d)$ is odd, $s(d) \geq 2r+1$. Thus $2k \geq s(d) \geq 2r+1$. Therefore $r \leq k-1$. This implies that for the formula $\beta$, $T_{k-1} \vdash \neg\beta$. A contradiction.

Hence, we have proved that for every $k \in \omega$, $d_k = g_k$. This implies that $\bigcup_{n=0}^{\infty} AD_n = GD(D,S)$. Since $S$ is an extension of $(D,W)$, $S = Cn(W \cup c(GD(D,S)))$. Consequently, at stage $\omega$ we test the applicability of defaults with respect to $S$. Since none of the defaults that are left at this stage belong to $GD(D,S)$, none is applicable, and so $\eta_{\preceq} = \omega$. Hence $\bigcup_{n<\omega} A_n = AD_{\preceq} = GD(D,S)$ and so the condition (3.23) is satisfied. ■

Theorem 3.73 generalizes Theorem 3.15. In particular, the proofs of Lemma 3.72 and of Theorem 3.73 can be used to derive Theorem 3.15.

## 3.8 Prioritized default theories

Often the defaults of a default theory are assigned priorities or, more generally, are ordered by some partial ordering relation. In this way we attempt to represent the fact that some defaults may be more preferable to others and, assuming they are applicable, should be used first. A good example of such a situation is when a set of defaults is created by several experts who themselves are arranged into a hierarchy. Then, it is often natural to treat defaults created by experts who are higher up in the hierarchy as preferable to those created by their subordinates. While it is hard to see how the original definition of an extension could be generalized to such a setting, the relationship between extensions and well-orderings of defaults, which we studied in Section 3.7, suggests a possible approach.

**Definition 3.74** A *prioritized default theory* is a triple $(D, W, \leq)$, where $D$ is a set of defaults, $W$ is a set of formulas and $\leq$ is a well-founded partial ordering of the set $D$. ▲

Intuitively, the relation $\leq$ should be understood as follows: $d \leq d'$ if and only if $d'$ has higher priority than $d$ (is preferable). Let us observe that we do not require that the set of defaults be finite. Instead, in order to guarantee that every nonempty subset of $D$ has always a minimal element with respect to $\leq$, we require that $\leq$ be well-founded, that is, that there are no infinite descending chains $\ldots < d_n < d_{n-1} \ldots < d_1$ of defaults.

Results of Section 3.7 establish a correspondence between extensions of a default theory and well-orderings of its set of defaults. Given a well-ordering of defaults we proceed as follows. We start with the empty theory $S$ and then, in each step we apply a default (as long as it is possible) and update $S$ by adding to it the consequent of the selected default. The key idea of this method is to use the well-ordering to break ties between defaults if more than one are applicable at a certain step. When the construction terminates, an additional check is used to decide whether a theory that has been produced is indeed an extension. If all well-orderings of the set of defaults are used, all extensions will be found.

Consider now a prioritized default theory $(D, W, \leq)$ and let $\preceq$ be any well-ordering such that for every defaults $d, d' \in D$, if $d \leq d'$ then $d \preceq d'$. Clearly, at any step of the construction described above whenever there are several defaults that are applicable, the one that is selected is minimal with respect to the well-ordering $\preceq$ (has maximal priority). Suppose that the theory produced, say $S$, happens to be an extension of the default theory $(D, W)$. Since $S$ can be produced by processing defaults according to their priorities and since it is an extension of $(D, W)$, $S$ is a natural candidate for an extension of the prioritized default theory $(D, W, \preceq)$.

**Definition 3.75** Let $(D, W, \leq)$ be a prioritized default theory. A theory $S$ is an extension for $(D, W, \leq)$ if $S$ is an *extension* for $(D, W)$ generated by a well-ordering $\preceq$ *compatible* with $\leq$, that is, such that for every two defaults $d, d' \in D$, if $d \leq d'$ then $d \preceq d'$. ▲

**Example 3.76** Let $W = \emptyset$ and $D = \{d_1, \ldots, d_4\}$, where

$$d_1 = \frac{a : M\neg b}{d},\ d_2 = \frac{: M\neg b}{a},\ d_3 = \frac{: M\neg d}{c},\ d_4 = \frac{: M\neg a}{b}.$$

In addition, let $\leq$ be defined as follows: $d_1 \leq d_i$, $i = 2, 3, 4$, $d_i \leq d_4$, $i = 1, 2, 3$, and $d_2$ and $d_3$ are not related by $\leq$.

It is easy to see that there are two well-orderings that are compatible with $\leq$: $d_1, d_2, d_3, d_4$ and $d_1, d_3, d_2, d_4$. If the method of Section 3.7 is used for the first of them, the theory $S_1 = Cn(\{a, d\})$ is generated. Since the justifications of all the defaults used in the process ($d_2$ and $d_1$) are consistent with $S_1$, $S_1$ is an extension for the default theory $(D, W)$, and also an extension for the prioritized default theory $(D, W, \leq)$. When the second well-ordering is used, the theory $S_2 = Cn(\{a, c, d\})$ is generated. Since the justification of one of the defaults used to generate $S_2$ ($d_3$) is inconsistent with $S_2$ we are not guaranteed that $S_2$ is an extension for $(D, W)$ and, in fact, it is not. Hence, $(D, W, \leq)$ has exactly one extension: $S_1$. On the other hand, the default theory $(D, W)$ has an additional extension, $Cn(\{b, c\})$. This extension is eliminated when priorities on the defaults are imposed by $\leq$. ▲

## 3.9 Computing extensions

In this section we will present two methods to compute extensions for a given finite default theory $(D, W)$. The first algorithm is based on the "guess-and-check" approach to extensions. The second algorithm generates all extensions from different orderings of defaults.

Both algorithms require exponentially many (in the size of $(D, W)$) calls to a procedure testing whether a formula is a propositional consequence of a collection of formulas. Since the satisfiability problem in propositional logic is *NP*-complete, any procedure that determines whether a formula is a propositional consequence of a set of formulas is likely to be computationally complex and no efficient procedures for this task are known so far. Hence, also our algorithms for generating extensions, which make heavy use of the procedure for testing propositional provability, are computationally complex.

The high computational complexity of our algorithms is not surprising. In Chapter 13 we will show that the problem of determining the existence of an extension for a default theory is $\Sigma_2^P$-complete. We will also show that even when strong syntactic restrictions are imposed on default theories the problem remains *NP*-complete.

Our first algorithm, called **All_Extensions_1**, simulates the "guess-and-check" approach to extensions as presented in Section 3.4. Let us recall that under this approach a context (a possible extension) $S$ is "guessed" and then it is checked whether $S$ is indeed an extension by means of the notion of $S$-derivability. Corollary 3.31 allows us to restrict the class of possible "guesses" to the theories of the form $Cn(W \cup U)$, where $U \subseteq c(D)$ (recall that $c(D)$ is the collection of the consequents of all defaults in $D$). The algorithm **All_Extensions_1** generates all possible subsets $U \subseteq c(D)$ and for each of them it verifies whether the theory $Cn(W \cup U)$ is an extension or not. If so, the set $W \cup U$ of *generators* of the extension is output. The algorithm is described formally in Figure 3.2.

The correctness of the algorithm **All_Extensions_1** follows directly from Corollary 3.31. In order to determine whether or not $Cn(W \cup U)$, where $U \subseteq c(D)$ is an extension, the reduct $D' = D_{Cn(W \cup U)}$ is computed. According to Corollary 3.31 the condition $W \cup U = B^{D'}\uparrow\omega(W)$ is necessary and sufficient for $Cn(W \cup U)$ to be an extension for $(D, W)$. To verify whether $W \cup U = B^{D'}\uparrow\omega(W)$ holds, $B^{D'}\uparrow\omega(W)$ is computed by calling the algorithm $B\uparrow\omega$, described in Section 3.1, for $D'$ and $W$.

The algorithm **All_Extensions_1** can easily be modified to a decision procedure that determines whether a given theory has at least one extension or not. To this end, one must make sure that the main loop terminates as soon as a set $U \subseteq c(D)$ is found such that $W \cup U = B^{D'}\uparrow\omega(W)$. In this case,

---

**All_Extensions_1**$(D, W)$
*Input:* A finite default theory $(D, W)$;
*Output:* The list of all extensions of $(D, W)$;

**for** $U \subseteq c(D)$ **do**
    $D' := D_{Cn(W \cup U)}$;
    **if** $W \cup U = B{\uparrow}\omega(D', W)$ **then output**$(W \cup U)$
**rof**

---

**Fig. 3.2.** Algorithm for computing extensions of default heories

the answer **yes** should be returned. Otherwise, the answer **no** is returned. The algorithm **All_Extensions_1** can also be modified to determine whether a default theory has a unique extension.

The algorithm **All_Extensions_1** is computationally complex. Both to produce the reduct $D'$ and in the algorithm $B{\uparrow}\omega$, a procedure for propositional provability testing is needed. Let us denote by $n$ the number of defaults in $D$, and by $m$ the total number of justifications in all defaults in $D$. There are $2^n$ iterations of the **for** loop. Within each iteration, in order to construct the reduct $D'$, there are $m$ calls to the propositional provability testing procedure. An additional $O(|D'|^2)$ (which can be as large as $n^2$) calls to the propositional provability testing procedure are then executed by the algorithm $B{\uparrow}\omega$. In summary, the number of calls to the propositional provability testing procedure that are made by the algorithm **All_Extensions_1** is bounded from below by $m2^n$ and from above by $(m + n^2)2^n$.

Our second algorithm for computing extensions is based on the idea of deriving extensions by applying defaults in a prespecified order and with respect to the changing context. The theoretical foundation for this approach was laid in Section 3.4. The idea of the algorithm is as follows. An ordering of defaults is chosen. Then, according to this ordering, a candidate theory for an extension is generated. When defaults are applied their consequents are added to the context. Since the context grows, a default rule that has been applicable in an earlier stage of the construction may become inapplicable at a later stage. This means that some conclusions may become invalid as we construct a putative extension. Therefore there is no guarantee that the theory we obtain in the process will actually be an extension. To weed out those theories that are not extensions an additional check is performed. Theorem 3.70 guarantees that if all orderings of $D$ are tried, then all extensions will be generated, and Theorem 3.65 gives a condition that allows us to

---

**Derive_Candidate**$(D, W, \preceq, AD_{\preceq})$
*Input:* A finite default theory $(D, W)$ and a total ordering $\preceq$ of the defaults in $D$;
*Output:* A set $AD_{\preceq}$ of defaults;

$U := W$;
$AD_{\preceq} := \emptyset$;
**while** $GD(D \setminus AD_{\preceq}, Cn(U)) \neq \emptyset$ **do**
  $d :=\preceq$-least default in $GD(D \setminus AD_{\preceq}, Cn(U))$;
  $U := U \cup \{c(d)\}$;
  $AD_{\preceq} := AD_{\preceq} \cup \{d\}$;
**od**

---

**Fig. 3.3.** Algorithm for computing a candidate theory for an extension, given an ordering of defaults

eliminate from among the generated theories all those which are not extensions.

The description of the algorithm is given in Figures 3.3 and 3.4. Figure 3.3 provides a description of an auxiliary algorithm, **Derive_Candidate**. This algorithm produces, given a well-ordering $\preceq$ of the defaults in $D$, the collection of defaults $AD_{\preceq}$, according to the construction specified in Definition 3.61.

In each iteration of the **while** loop one default is added to the set $AD_{\preceq}$. Hence, it follows that the algorithm **Derive_Candidate** terminates after at most $n$ iterations of the **while** loop, where $n = |D|$. In each iteration, $O(n + m)$ calls to the procedure for propositional provability testing are necessary to find the $\preceq$-first default in $GD(D \setminus AD_{\preceq}, Cn(U))$ or to determine that $GD(D \setminus AD_{\preceq}, Cn(U))$ is empty. Hence, the algorithm **Derive_Candidate** performs $O(n(n + m))$ calls to the procedure for propositional provability testing.

In the previous section we proved that all extensions for a default theory $(D, W)$ can be found among the theories $Cn(W \cup c(AD_{\preceq}))$, where $\preceq$ ranges over all well-orderings of $D$. In addition, we found a condition which allows us to eliminate all those theories $Cn(W \cup c(AD_{\preceq}))$ which are not extensions. As a result, we get another algorithm to generate all extensions for a finite default theory, called **All_Extensions_2** and described in Figure 3.4.

The algorithm, as it is described, requires $n!$ iterations of the main loop, where $n$ is the number of defaults in $D$. To check whether each $\beta \in J$ is

**All_Extensions_2**$(D, W)$
*Input:* A finite default theory $(D, W)$;
*Output:* The list of all extensions of $(D, W)$;

**for** all well-orderings $\preceq$ of $D$ **do**
  **Derive_Candidate**$(\mathbf{D}, \mathbf{W}, \preceq, \mathbf{AD}_{\preceq})$;
  $J := J(AD_{\preceq})$;
  (*$J$ consists of all justifications of defaults in $AD_{\preceq}$*)
  $U = c(AD_{\preceq})$;
  **if** for every $\beta \in J$, $W \cup U \not\vdash \neg\beta$ **then write**$(W \cup U)$
**rof**

**Fig. 3.4.** Algorithm for computing extensions of default theories

consistent with $W \cup U$, $O(m)$ calls to the procedure for testing propositional provability are needed (recall that $m$ stands for the total number of justifications of all defaults in $D$). Summarizing, the algorithm **All_Extensions_2** requires $O(n!n(n+m))$ calls to the procedure for testing propositional provability.

The algorithm can easily be modified to improve its efficiency. To see this, let us look again at the Theorem 3.70. It follows that if a theory $S$ is an extension for $(D, W)$ then it can be generated from the well-ordering $\preceq_S$ of $D$, in which it is only important that the defaults in $GD(D, S)$ precede the defaults in $D \setminus GD(D, S)$. The ordering of defaults in each of the sets $GD(D, S)$ and $D \setminus GD(D, S)$ is immaterial. So, to generate all extensions it is enough to consider $2^n$ orderings, one for each subset $D'$ of $D$, in which the defaults from $D'$, arranged in some arbitrary order, precede the defaults in $D \setminus D'$, also arranged arbitrarily. Thus, the modified variant of the algorithm **All_Extensions_2** needs only $O(2^n n(n+m))$ calls to the procedure for testing propositional provability.

## 3.10 Other structures associated with default theories

There are structures other than extensions that can naturally be associated with default logic. In this section, we will study two such structures, *weak extensions* and minimal sets closed under a default theory. Another concept, *partial extensions*, will be discussed in Section 3.11.

The notion of an extension is a natural consequence of the following intuitive interpretation of a default $r$,

$$r = \frac{\alpha : M\beta_1, \ldots, M\beta_n}{\gamma}. \tag{3.24}$$

*If $\alpha$ has been derived and every $\beta_i$, $1 \leq i \leq n$, is consistent, then conclude* $\gamma$. Consistency is checked with respect to a theory $S$, called context, which consists of formulas an agent thinks might be entailed by a default theory $(D, W)$. Once a hypothetical consequence set $S$ of $(D, W)$ (a context) is "guessed", it is verified whether all formulas in $S$, and only those formulas, can be justified by means of an $S$-derivation. If so, $S$ is an extension.

Let us slightly modify this approach. We will still "guess" a context $S$ but we will use a much weaker notion of derivability to justify our guess. It will be helpful to think that we are constructing *belief* sets by interpreting a default (3.24) as a rule saying: *if $\alpha$ is a belief and every $\beta_i$, $1 \leq i \leq n$, is consistent, then derive $\gamma$ as a belief.* Now, assume that we "guessed" $S$ as a possible belief set. Under what conditions should one be able to use a default (3.24) in a derivation? According to our modified interpretation of defaults, exactly when $\alpha$ is a belief, that is, $\alpha \in S$, and $\beta_i$ is consistent with $S$, for every $i$, $1 \leq i \leq n$. The first requirement, $\alpha \in S$, is weaker than the requirement that $\alpha$ be $S$-derivable used in the definition of extensions. Let us now formalize these ideas.

**Definition 3.77** Let $(D, W)$ be a default theory and let $S \subseteq \mathcal{L}$. A *weak $S$-derivation* (*weak $S$-proof*) of a formula $\varphi$ from $(D, W)$ is a finite sequence $\psi_1, \ldots, \psi_n$ such that $\psi_n = \varphi$ and, for every $i$, $1 \leq i \leq n$, at least one of the following conditions holds:

1. $\psi_i \in W$.
2. $\psi_i$ is a substitution instance of an axiom of propositional calculus.
3. $\psi_i$ is the result of applying modus ponens to formulas $\psi_j$, $\psi_k$ for some $j, k < i$.
4. There is a rule $d = \frac{\alpha : M\beta_1, \ldots, M\beta_k}{\psi_i}$ such that $d \in D$, $\alpha \in S$ and $\neg\beta_1 \notin S, \ldots, \neg\beta_k \notin S$.

By $Cn_w^{D,S}(W)$ we denote the set of all formulas that possess a weak $S$-derivation from $(D, W)$. ▲

Let us note here that the formulas of $S$ do not necessarily possess a weak derivation.

The crucial difference between weak $S$-derivations and $S$-derivations is in the way defaults are used. In the case of weak $S$-derivations, instead of having to "prove" the prerequisite $p(d)$ of a default $d$, it is enough that we have assumed $p(d)$ to be in the context $S$ (in other words, we have chosen to believe $p(d)$). Consequently, it can be proved that the class of weak $S$-derivations contains, for all reasonable contexts $S$, the class of $S$-derivations. That is, the concept of a weak $S$-derivation is indeed a weakening of the concept of an $S$-derivation. A precise formulation of the relationship between the two notions is given in the next theorem. The proof (by induction on the length of $S$-proof) is straightforward and is left to the reader.

**Theorem 3.78** *Let $(D, W)$ be a default theory and let $S \subseteq \mathcal{L}$ be such that $Cn^{D,S}(W) \subseteq S$. Then every $S$-derivation from $W$ is a weak $S$-derivation from $W$ and*

$$Cn^{D,S}(W) \subseteq Cn_w^{D,S}(W).$$

▲

**Definition 3.79** Let $(D, W)$ be a default theory, and let $S \subseteq \mathcal{L}$. The theory $S$ is a *weak extension* for $(D, W)$ if $S = Cn_w^{D,S}(W)$. ▲

In Chapter 12 we will discuss the most important property of weak extensions — their exact correspondence with the notion of stable (autoepistemic) expansions. More precisely, we will show that the logic of weak extensions is, under a very simple mapping, isomorphic to autoepistemic logic (or equivalently, the modal nonmonotonic logic **kd45**). This result is not surprising if one takes into account that autoepistemic logic was designed specifically to deal with *belief* sets of an agent reasoning with incomplete information. In this section we will present some simple properties of weak extensions. We start with a characterization of weak extensions in terms of the propositional provability operator and generating defaults. It turns out that weak extensions are exactly those theories which satisfy the equation (3.14) discussed in Section 3.6.

**Theorem 3.80** *Let $(D, W)$ be a default theory. A theory $S$ is a weak extension for $(D, W)$ if and only if $S = Cn(W \cup c(GD(D, S)))$.*

Proof: The assertion follows immediately from the simple observation that $Cn_w^{D,S}(W) = Cn(W \cup c(GD(D, S)))$. ■

**Corollary 3.81** *If $S$ is a weak extension of $(D, W)$ then there exists a subset $U \subseteq c(D)$ such that $S = Cn(W \cup U)$.* ▲

**Theorem 3.82** *If a theory $S \subseteq \mathcal{L}$ is an extension for a default theory $(D, W)$ then $S$ is a weak extension for $(D, W)$.*

Proof: The assertion follows from Theorems 3.57 and 3.80. ■

The converse implication does not hold. Let us consider the following example.

**Example 3.83** Let $D = \{\frac{p:\neg q}{p}\}$, $W = \emptyset$. Then $S_1 = Cn(\emptyset)$ is the only extension of $(D, W)$. On the other hand, the theory $(D, W)$ possesses two weak extensions. Indeed, by Corollary 3.81, if $S$ is a weak extension for a default theory $(D, W)$ then $S = Cn(W \cup U)$, for some $U \subseteq c(D)$. Hence, there are two candidates for a weak extension: $S_1 = Cn(\emptyset)$ and $S_2 = Cn(\{p\})$. Since $S_1$ is an extension, it is a weak extension as well. As for $S_2$, it is easy to see that $GD(D, S_2) = D$. Consequently, $Cn(W \cup c(GD(D, S_2))) = Cn(\{p\}) = S_2$. Hence, $S_2$ is a weak extension for $(D, W)$. ▲

We have seen examples of default theories which have no extensions. Since the class of weak extensions is bigger, one might hope that a default theory always possess a weak extension. This is not the case, as shown in the next example.

**Example 3.84** Consider the default theory $(D, W)$, where

$$D = \left\{ \frac{: M\neg p}{p} \right\} \text{ and } W = \emptyset.$$

This theory was considered in Example 3.32, where we argued that it has no extensions. It has no weak extensions either. There are two candidates for a weak extension: $S_1 = Cn(\emptyset)$ and $S_2 = Cn(\{p\})$. It is easy to see that $GD(D, S_1) = D$ and $GD(D, S_2) = \emptyset$. Hence $Cn(W \cup c(GD(D, S_1))) = Cn(\{p\}) \neq S_1$ and $Cn(W \cup c(GD(D, S_2))) = Cn(\emptyset) \neq S_2$. Hence, neither $S_1$ nor $S_2$ is an extension for $(D, W)$. ▲

Example 3.83 also shows that the property that extensions of a default theory form an antichain does not hold for weak extensions. But weak extensions share another property of extensions: they are closed under defaults.

**Theorem 3.85** *If a theory $S \subseteq \mathcal{L}$ is a weak extension for $(D, W)$, then $S$ is closed under defaults in $D$.*

Proof: Let $S$ be a weak extension of $(D, W)$. Consider a default $d \in D$. If $p(d) \in S$ and $\neg\beta \notin S$, for every $\beta \in J(d)$, then $d \in GD(D, S)$. Hence, $c(d) \in c(GD(D, S))$ and, consequently, $c(d) \in S$. ■

We have seen that the notions of an extension and weak extension are different. There is, however, a wide class of default theories for which these two notions coincide. Let us recall that a default $d$ is called prerequisite-free if $p(d) = \top$. Our next result shows that extensions and weak extensions for default theories with all defaults prerequisite-free coincide.

**Theorem 3.86** *Let $(D, W)$ be a default theory such that every default in $D$ is prerequisite-free. A theory $S$ is an extension for $(D, W)$ if and only if $S$ is a weak extension for $(D, W)$.*

Proof: Since extensions are always weak extensions, it suffices to show that if $S$ is a weak extension for $(D, W)$, where all defaults in $D$ are prerequisite-free, then $S$ is an extension for $(D, W)$.

Since all defaults in $D$ are prerequisite-free, $Cn^{D,S}(W) = R^{D_S}\uparrow 1(W)$. Moreover, it is easy to see that $R^{D_S}\uparrow 1(W) = Cn(W \cup c(GD(D, S)))$. Since $S$ is a weak extension, $Cn(W \cup c(GD(D, S))) = S$. Therefore, we have $S = Cn^{D,S}(W)$ and, consequently, $S$ is an extension for $(D, W)$. ■

We will now discuss another structure associated with default theories. In a classical proof system the set of consequences $C(W)$ of a theory $W$ can be defined syntactically in two alternative but equivalent ways:

1. By means of the notion of proof: given a set of formulas $W$, $C(W)$ is defined as the collection of all formulas having a proof from $W$ in the system.

2. By means of sets closed under the inference rules of the system: $C(W)$ is defined as the least set of formulas closed under the rules and containing $W$. The correctness of this definition relies on the property of classical proof systems that such a least set always exists.

We have presented this dual approach to the consequence operator in Section 3.1 in the case of the system $PC+D$, where $D$ is a set of (monotone) inference rules (see Definition 3.1 and Theorem 3.7).

In our development of default logic we have considered so far only the first approach. We have defined two different notions of proof and for each of them a class of theories for describing the consequences of a default theory: extensions and weak extensions, respectively. We now investigate the second approach. We have already introduced the notion of a theory closed under defaults and propositional provability. It naturally generalizes the notion of a theory closed under monotone inference rules. For a default theory $(D, W)$ we now consider theories containing $W$ and closed under defaults in $D$ and relate them to extensions and weak extensions. We will see that in default logic, unlike in the classical case, the least set closed under defaults may not exist. On the other hand, minimal sets containing $W$ and closed under propositional provability and the defaults in $D$ exist but need not coincide with extensions or weak extensions. In particular, in default logic the two syntactic approaches to the notion of logical consequence are not equivalent.

**Definition 3.87** Let $(D, W)$ be a default theory, and let $S \subseteq \mathcal{L}$. Theory $S$ is called a *minimal set for* $(D, W)$ if $S$ is minimal with respect to inclusion among all sets $S'$ such that:

1. $W \subseteq S'$.
2. $S'$ is closed under propositional provability.
3. $S'$ is closed under the defaults in $D$. ▲

The Kuratowski-Zorn lemma implies that every default theory possesses at least one minimal set.

**Theorem 3.88** *Let $(D, W)$ be a default theory. For every $S \subseteq \mathcal{L}$ such that $W \subseteq S$ and $S$ is closed under propositional consequence and under the defaults in $D$, there exists a minimal set $S'$ for $(D, W)$ such that $S' \subseteq S$. In particular, there exists at least one minimal set for $(D, W)$.*

Proof: We use the Kuratowski-Zorn lemma in set algebra (Corollary 2.38). Let $\mathcal{Y}$ be the family of subsets of $\mathcal{L}$ which contain $W$ and are closed under propositional provability and under the defaults in $D$. Let $\mathcal{Z}$ be any subfamily of $\mathcal{Y}$ such that for every $Z_1, Z_2 \in \mathcal{Z}$, $Z_1 \subseteq Z_2$ or $Z_2 \subseteq Z_1$. To apply the Kuratowski-Zorn lemma we need to show that $T = \bigcap \mathcal{Z}$ is in $\mathcal{Y}$.

Since each set in $\mathcal{Z}$ contains $W$ and is closed under propositional provability, it follows that $T$ has these two properties, too.

Let us consider now an arbitrary default $d \in D$, say

$$d = \frac{\alpha : M\beta_1, \ldots, M\beta_n}{\gamma}.$$

Assume that $\alpha \in T$ and that for every $i$, $1 \leq i \leq n$, $\neg\beta_i \notin T$. Clearly, $\alpha \in Z$, for every set $Z \in \mathcal{Z}$. Also, for every $i$, $1 \leq i \leq n$, there is a set $Z_i \in \mathcal{Z}$ such that $\neg\beta_i \notin Z_i$. Since $\mathcal{Z}$ is totally ordered by inclusion, there is $Z \in \mathcal{Z}$ such that $\neg\beta_i \notin Z$ for every $i$, $1 \leq i \leq n$. Since $Z$ is closed under the defaults in $D$, $\gamma \in Z$.

Consider now any set $Z' \in \mathcal{Z}$. Then, $Z \subseteq Z'$ or $Z' \subseteq Z$. In the former case, $\gamma \in Z'$ because $\gamma \in Z$. In the latter, $\neg\beta_i \notin Z'$ for every $i$, $1 \leq i \leq n$. Since $Z'$ is closed under the defaults in $D$, $\gamma \in Z'$. Hence, $\gamma \in T$.

The second part of the assertion follows from the observation that $\mathcal{L}$ contains $W$ and is closed under propositional provability and under the defaults in $D$. ■

We will now derive a simple representation result for minimal sets which allows us to find minimal sets for a given default theory.

**Lemma 3.89** *Let $(D, W)$ be a default theory. Let $S \subseteq \mathcal{L}$ contain $W$ and be closed under propositional provability and the defaults in D. Then there exists $U \subseteq c(D) \cup \{\neg\beta : \beta \in J(D)\}$ such that $Cn(W \cup U)$ is closed under the defaults in $D$ and $Cn(W \cup U) \subseteq S$.*

Proof: Since $S$ is closed under the defaults in $D$, for every default $d$ such that $p(d) \in S$, at least one of the formulas in $\{\neg\beta : \beta \in J(d)\} \cup \{c(d)\}$ is in $S$. Let $\varphi_d$ denote any such formula. Define $U = \{\varphi_d : d \in D,\ p(d) \in S\}$. Clearly, $U \subseteq S$. Consequently, $Cn(W \cup U) \subseteq S$. In addition, $Cn(W \cup U)$ is closed under the defaults in $D$.

Indeed, consider a default $d \in D$. Assume that $p(d) \in Cn(W \cup U)$ and that $\neg\beta \notin Cn(W \cup U)$, for every $\beta \in J(d)$. Then $p(d) \in S$. Consequently, $\varphi_d \in U$. Since $\neg\beta \notin Cn(W \cup U)$, for every $\beta \in J(d)$, it follows that $\varphi_d = c(d)$. Thus $c(d) \in Cn(W \cup U)$. ■

**Corollary 3.90** *If $S$ is a minimal set for a default theory $(D, W)$ then $S = Cn(W \cup U)$, for some $U \subseteq c(D) \cup \{\neg\beta : \beta \in J(D)\}$.* ▲

**Example 3.91** Let $D = \{\frac{:M\neg p}{q}\}$ and $W = \emptyset$. According to Corollary 3.90 there are four candidates for minimal sets for $(D, W)$: $S_1 = Cn(\emptyset)$, $S_2 = Cn(\{p\})$, $S_3 = Cn(\{q\})$ and $S_4 = Cn(\{p, q\})$. $S_1$ is not closed under the defaults in $D$. The remaining three theories are closed under $D$. Consequently, $S_2$ and $S_3$ are minimal sets for $(D, W)$. The theory $S_3$ is an extension of $(D, W)$, hence, a weak extension for $(D, W)$, as well. On the other hand, the minimal set $S_2$ is neither an extension nor a weak extension for $(D, W)$. ▲

As Example 3.91 shows, there are default theories $(D, W)$ such that there is no least theory containing $W$, closed under propositional provability and under defaults in $D$. At the same time, Example 3.91 demonstrates that the notion of a minimal set coincides neither with the notion of an extension nor with the notion of a weak extension. However it is true that every extension is a minimal set.

**Theorem 3.92** *Let $(D, W)$ be a default theory, and let $S$ be an extension of $(D, W)$. Then $S$ is a minimal set for $(D, W)$.*

Proof: Clearly, $S$ contains $W$ and is closed under propositional consequence. By Theorem 3.49, $S$ is also closed under $D$. By Theorem 3.88, there is a minimal set $S'$ for $(D, W)$ such that $S' \subseteq S$. In particular, it follows that $D_S \subseteq D_{S'}$. Since $S$ is an extension of $(D, W)$, $S = Cn^{D_S}(W)$. Therefore,

$$S = Cn^{D_S}(W) \subseteq Cn^{D_{S'}}(W) \subseteq S'.$$

Hence $S \subseteq S'$ and, consequently, $S = S'$. ■

We proved that every extension of $(D, W)$ is both a weak extension of $(D, W)$ and a minimal set for $(D, W)$. However, the converse does not hold in general.

**Example 3.93** Let $D = \{\frac{p:}{q}, \frac{:M\neg p}{p}\}$ and $W = \{q \supset p\}$. The theory $(D, W)$ has no extensions but $Cn(\{p, q\})$ is both a weak extension for $(D, W)$ and a minimal set for $(D, W)$. ▲

Finally, notice that a weak extension of a default theory may be contained in another weak extensions of the same default theory (Example 3.83). whereas this is not possible for minimal sets. Therefore the notions of weak extension and minimal set do not coincide.

Minimal sets for a default theory $(D, W)$ will resurface again in Chapter 12. We will show there that they correspond exactly to minimal stable theories containing a modal interpretation of $(D, W)$.

## 3.11 Partial extensions

In this section we discuss yet another structure associated with default theories. This structure, called partial extension, is closely related to the construction of Section 3.7. Given a well-ordering $\preceq$ of $D$, we constructed there a theory $T_{\preceq}$. It could happen, however, that the theory $T_{\preceq}$ was too big. That is, in the process of constructing $T_{\preceq}$ we could add formulas which make invalid application of defaults previously used in the construction. We will modify now the construction and eliminate this possibility. The resulting structures, partial extensions, will always exist. Every well-ordering of the set $D$ of default rules produces a partial extension. Moreover it turns out that every extension of $(D, W)$ is a partial extension of $(D.W)$. The construction of a partial extension differs from that described in Definition 3.61. The difference is that in addition to the set of formulas *accepted* by the stage $\alpha$ we will also construct a set of formulas *rejected* at that stage. Also, the selection of the default $d_\alpha$ will be more stringent. Denoting the sets of accepted formulas and rejected formulas by $A_\alpha$ and $R_\alpha$, respectively, we will make sure that

$$A_\alpha \cap R_\alpha = \emptyset.$$

We denote by $A_{\preceq}$ the union of the sets $A_\alpha$ constructed in this fashion. The resulting set is called a partial extension of $(D, W)$.

At the end of the construction we may face a situation that some rules are inconsistent. This means that they either invalidate their own applicability or of some rule applied during the construction. When we eliminate the set $I_{\preceq}$ composed of inconsistent rules we find that $A_{\preceq}$ is an extension of the theory $(D \setminus I_{\preceq}, W)$.

Thus a partial extension is a not necessarily an extension of the original $(D, W)$ but rather of a theory with a smaller set of defaults. On the other hand, every extension is a partial extension. That is, for every extension $S$ of $(D, W)$ there is a well-ordering $\preceq$ such that $S = A_{\preceq}$.

In our presentation of the modified construction we will follow very closely the presentation of Section 3.7.

**Definition 3.94** Let $(D, W)$ be a default theory and $\preceq$ a well-ordering of $D$. We define an ordinal $\xi_{\preceq}$. For every $\rho < \xi_{\preceq}$ we define two sets of formulas $A_\rho$ and $R_\rho$ and a default $d_\rho$. We also define two sets of formulas, $A_{\preceq}$ and $R_{\preceq}$. We proceed as follows: First we set $A_0 = Cn(W)$ and $R_0 = \emptyset$. If the sets $A_\rho$, and $R_\rho$, $\rho < \alpha$, have been defined but $\xi_{\preceq}$ has not been defined then:

1. If there is no default $d \in D$ such that:

   (a) $c(d) \notin \bigcup_{\rho<\alpha} A_\rho$

   (b) $p(d) \in \bigcup_{\rho<\alpha} A_\rho$

   (c) $Cn(\bigcup_{\rho<\alpha} A_\rho \cup \{c(d)\}) \cap (\bigcup_{\rho<\alpha} R_\rho \cup \{\neg\beta : \beta \in J(d)\}) = \emptyset$

   then $\xi_{\preceq} = \alpha$.

2. Otherwise, define $d_\alpha$ to be the $\preceq$-least default in $D$ such that:

   (a) $c(d) \notin \bigcup_{\rho<\alpha} A_\rho$.

   (b) $p(d) \in \bigcup_{\rho<\alpha} A_\rho$.

   (c) $Cn(\bigcup_{\rho<\alpha} A_\rho \cup \{c(d)\}) \cap (\bigcup_{\rho<\alpha} R_\rho \cup \{\neg\beta : \beta \in J(d)\}) = \emptyset$.

   Then set

$$A_\alpha = Cn(\bigcup_{\rho<\alpha} A_\rho \cup \{c(d_\alpha)\}), \qquad R_\alpha = \bigcup_{\rho<\alpha} R_\rho \cup \{\neg\beta : \beta \in J(d_\alpha)\}.$$

The theory $\bigcup_{\rho<\xi_\preceq} A_\rho$ is called a *partial extension generated by the well-ordering* $\preceq$. A *partial extension* of $(D, W)$ is a set of formulas which is a partial extension generated by some well-ordering of $D$. ▲

First, we have the following property of hierarchies defined above. The symbol $\subset$ denotes here the proper inclusion.

**Theorem 3.95** *For every* $\rho_1, \rho_2 < \xi_\preceq$*:*

1. *If* $\rho_1 < \rho_2$ *then* $A_{\rho_1} \subset A_{\rho_2}$ *and* $R_{\rho_1} \subseteq R_{\rho_2}$.

2. $A_{\rho_1} \cap R_{\rho_1} = \emptyset$.

3. $A_\preceq \cap R_\preceq = \emptyset$.

Proof: The definition of theories $A_\alpha$ and of the ordinal $\xi_\preceq$ imply that for $\alpha < \xi_\preceq$, $\bigcup_{\rho<\alpha} A_\rho \subset A_\alpha$, and $\bigcup_{\rho<\alpha} R_\rho \subseteq R_\alpha$. This implies (1). Next, by induction on $\rho_1$ we check that (2) holds. The property (3) follows from (1) and (2) immediately. ■

We now introduce the set of defaults *inconsistent* with the well-ordering $\preceq$.

**Definition 3.96**

1. A default $d$ is *inconsistent* with $\preceq$ if $p(d) \in A_\preceq$, $c(r) \notin A_\preceq$, and

$$Cn(A_\preceq \cup \{c(d)\}) \cap (R_\preceq \cup \{\neg\beta : \beta \in J(d)\}) \neq \emptyset.$$

2. $I_\preceq$ is the set of all defaults inconsistent with $\preceq$. ▲

Let us see what it means that the default $r$ is inconsistent with $\preceq$. The default $d$ must be ready for application ($p(d) \in A_{\preceq}$) but if we attempt to apply it, then either we disprove one of its justifications (this happens when $Cn(A_{\preceq} \cup \{c(d)\}) \cap \{\neg\beta : \beta \in J(d)\} \neq \emptyset$) or we invalidate one of derivations made in the process of computing $A_{\preceq}$ (this happens when $Cn(A_{\preceq} \cup \{c(r)\}) \cap R_{\preceq} \neq \emptyset$).

**Definition 3.97** $D_{\preceq} = D \setminus I_{\preceq}$. ▲

The defaults in $D_{\preceq}$ are those that are *not* inconsistent with $\preceq$. We will call such defaults *consistent* with $\preceq$.

**Lemma 3.98** $\{d_\alpha : \alpha < \xi_{\preceq}\} \subseteq D_{\preceq}$.

Proof: First, we notice that $p(d_\alpha) \in \bigcup_{\rho<\alpha} A_\rho \subseteq A_{\preceq}$. Thus $p(d_\alpha) \in A_{\preceq}$. Similarly, $c(d_\alpha) \in A_{\preceq}$. $A_{\preceq}$ is closed under propositional consequence since it is the union of the monotone family $\langle A_\rho : \rho < \xi_{\preceq}\rangle$ and each $A_\rho$ is closed under propositional consequence). Therefore $Cn(A_{\preceq} \cup \{c(d_\alpha)\}) = A_{\preceq}$. Also, $J(d_\alpha) \subseteq R_\alpha \subseteq R_{\preceq}$. Since $A_{\preceq} \cap R_{\preceq} = \emptyset$, $d_\alpha$ is not inconsistent. ■

Our first result states that a partial extension is an extension of a subtheory of $(D, W)$.

**Theorem 3.99** *$A_{\preceq}$ is an extension of $(D_{\preceq}, W)$.*

Proof: First notice that by Lemma 3.98 all defaults of the form $d_\alpha$ for $\alpha < \xi_{\preceq}$ belong to $D_{\preceq}$. Those rules have justifications consistent with $A_{\preceq}$. Now, by transfinite induction we prove that $c(d_\alpha)$ belongs to $Cn^{D_{\preceq}, A_{\preceq}}(W)$. It should be obvious that $A_{\preceq}$ is generated by $W$ and the conclusions of all the defaults $d_\alpha$ for $\alpha < \xi_{\preceq}$. Therefore $A_{\preceq} \subseteq Cn^{D_{\preceq}, A_{\preceq}}(W)$.

Conversely, if $\varphi \in Cn^{D_{\preceq}, A_{\preceq}}(W)$ then there exists an $A_{\preceq}$-derivation of $\varphi$. By induction on the length of such derivation we show that the conclusion of this derivation, belongs to $A_{\preceq}$. Since $A_{\preceq}$ is closed under the propositional consequence we only need to take care of the case when the last rule used in the derivation is a default from $D_{\preceq}$ (that is we do not need to consider the case when that rule is modus ponens). Let that default be

$$d = \frac{\psi : M\beta_1, \ldots, M\beta_r}{\varphi}.$$

The formula $\psi$ has a shorter derivation and so $\psi$ belongs to $A_{\preceq}$ by the inductive assumption. Also, $\neg\beta_1, \ldots, \neg\beta_r \notin A_{\preceq}$. Moreover, the rule $d$ is not

inconsistent. There are two cases that need to be considered.
(1) $d$ is $d_\alpha$ for some $\alpha < \xi_{\preceq}$. Then $\varphi \in A_\alpha \subseteq A_{\preceq}$.
(2) $d$ does belong to $\{d_\alpha : \alpha < \xi_{\preceq}\}$. Since $d$ is not inconsistent with respect to $\preceq$ and $\psi \in A_{\preceq}$, it must be the case that

$$Cn(A_{\preceq} \cup \{\varphi\}) \cap (R_{\preceq} \cup \{\neg\beta_1, \ldots, \neg\beta_r\}) = \emptyset.$$

This implies that we have $\varphi \in A_{\preceq}$ (otherwise we would be able to define $r_{\xi_{\preceq}}$, which is impossible). ■

**Corollary 3.100** *If $I_{\preceq} = \emptyset$ then $A_{\preceq}$ is an extension of $(D, W)$.* ▲

Next, we prove that all extensions of $(D, W)$ are partial extensions of $(D, W)$.

**Theorem 3.101** *Let $(D, W)$ be a default theory, let $S$ be an extension of $(D, W)$. Then there exists a well-ordering $\preceq$ such that $A_{\preceq} = S$ and $I_{\preceq} = \emptyset$.*

Proof: As before, we take as $\preceq$ any well-ordering of $D$ in which $GD(D, S)$ forms an initial segment. This guarantees that all the defaults ever applied in the construction belong to $GD(D, S)$. This, in turn, implies that $A_{\preceq} \subseteq S$.

Conversely, we prove that $Cn^{D,S}(W) \subseteq A_{\preceq}$. We prove this by induction on the length of $S$-derivation. This argument directly follows the case (2) of Theorem 3.99. Since $S = Cn^{D,S}(W)$, the inclusion $S \subseteq A_{\preceq}$ holds. Thus $S = A_{\preceq}$.

Finally, since every extension is closed under the rules of $D$, it must be the case that $I_{\preceq} = \emptyset$. ■

Let us now look at some examples.

**Example 3.102** Let $D = \{\frac{q:M\neg p}{p}\}$, $W = \{q\}$. Then $A_0 = Cn(\{q\})$, $R_0 = \emptyset$. The only default in $D$ is immediately inconsistent and so $A_{\preceq} = A_0$, $D_{\preceq} = \emptyset$, $I_{\preceq} = D$. ▲

**Example 3.103** Let $D = \{r_1, r_2, r_3\}$, $W = \{q\}$ where:

$$r_1 = \frac{q : M\neg p}{s} i, \quad r_2 = \frac{q :}{p}, \quad r_3 = \frac{p : M\neg r}{t}$$

Order $D$ as follows:

$$r_1 \preceq r_2 \preceq r_3$$

In this case $A_0 = Cn(\{q\})$, $R_0 = \emptyset$. In the next stage of our construction we get: $A_1 = Cn(\{q, s\})$, $R_1 = \{p\}$. Here the construction stops. The default $r_2$ is inconsistent, and the default $r_3$ is not applicable. Hence, $A_{\preceq} = A_1$. $I_{\preceq} = \{r_2\}$.

Now, consider the ordering $\preceq'$, where

$$r_2 \preceq' r_1 \preceq' r_3$$

We now get a different result: $A_0 = Cn(\{q\})$, $R_0 = \emptyset$. $A_1 = Cn(\{q, p\})$, $R_1 = \emptyset$. Finally, $A_2 = Cn(\{q, p, t\})$, $R_2 = \{r\}$. Here the construction stops. The rule $r_1$ is not applied but is not inconsistent. $A_{\preceq'} = A_2$, $I_{\preceq'} = \emptyset$. The partial extension constructed along $\preceq'$ is an extension of $(D, W)$. ▲

As before, for denumerable $D$, the well-orderings of type $\leq \omega$ are sufficient to construct all extensions of $(D, W)$ as partial extensions of $(D, W)$.

**Theorem 3.104** *If $|D| \leq \omega$ and $S$ is an extension of $(D, W)$ then there exists a well-ordering $\preceq$ in type $\leq \omega$ such that $A_{\preceq} = S$. In fact, the well-ordering constructed in the proof of Theorem 3.73 has this property.* ▲

It turns out that for the general case of partial extensions the analogous property fails. Our next example shows that there are partial extensions which require well-orderings of the type greater than $\omega$. The same example shows that, in opposition to extensions, partial extensions do *not* form an antichain.

**Example 3.105** Let $\{a_i : i < \omega\}$, $\{b_i : i < \omega\}$, $\{c_i : i < \omega\}$ be pairwise disjoint, infinite, sets of atoms. Consider a default theory $(D, W)$ where $D$ is the union of two infinite sets of defaults $\{r_i : i \geq 0\}$ and $\{s_i : i \geq 1\}$.

$$r_0 = \frac{: M\neg b_0}{a_0}, \quad r_1 = \frac{: M\neg b_1}{a_1}$$

$$r_i = \frac{a_1 \wedge \ldots \wedge a_{i-1} : M\neg b_i}{a_i}, \quad i \geq 2$$

$$s_i = \frac{a_i : Mc_i}{b_i}$$

Finally, define $W = \{a_0 \supset a_i : i < \omega\}$. Observe that $Cn(\{a_i : i \geq 0\} \cup \{b_i : i \geq 1\})$ is an extension of $(D, W)$.

Now, consider the following well-ordering $\preceq$ of defaults in $D$:

$$r_1 \preceq r_2 \preceq r_3 \ldots \preceq r_0 \preceq s_0 \preceq s_1 \preceq s_2 \preceq \ldots$$

The well-ordering $\preceq$ orders $D$ into the type $\omega + \omega$. Notice that $r_0$ follows in the well-ordering $\preceq$ all defaults $r_i$ for $i \geq 1$.

Let us compute now the sets $A_n$ for the ordering $\preceq$. It is an easy observation that for all $n < \omega$, $A_n = Cn(W \cup \{a_1, \ldots a_n\})$. $R_n = \{b_1, \ldots b_n\}$. Consequently,

$$\bigcup_{n<\omega} A_n = Cn(W \cup \{a_1, a_2, \ldots\}) \text{ and } \bigcup_{n<\omega} R_n = \{b_1, b_2, \ldots\}.$$

Next, $A_\omega = Cn(\{a_0, a_1, a_2 \ldots\})$. This is because in the step $\omega$ we use the default $r_0$ and then we have all the atoms $\{a_0, a_1, a_2, \ldots\}$. The theory $W$ can be derived from these atoms. $R_\omega = \{b_0, b_1, \ldots\}$. The construction stops at the step $\omega$; nothing more is computed.

Now, consider any well-ordering $\preceq'$ of $D$ in type $\omega$. The default $r_0$ must precede some of the rules $r_i$ for $i > 1$ (and in fact infinitely many of them). Since $r_0$ is preceded by finitely many defaults, and because $r_0$ has no prerequisite, there is finite stage $j$ in which $r_0$ becomes the first applicable rule. The atom $b_0$ appears only in the default $r_0$. The theories we construct are consistent and therefore the applicability of $r_0$ cannot be blocked. Now, at the stage $j$, $a_0$ is added to the set of accepted formulas. Using $W$ we derive all the atoms $a_i, i < \omega$ at once. Therefore in the further stages of the construction no other default of the form $a_i$ will ever be applied. Now, the set $R_j$ is finite and so it contains only a finite numbers of $b'$s. In the further stages of the construction *no* atom of the form $b_k$ will be added to the set of rejected formulas. This implies that in the subsequent stages of our construction after $j$ we will be able to use infinitely many of the rules $s_k$. Therefore the set of accepted formulas contains infinitely many atoms from of the form $b_k$. Hence $A_\preceq \neq A_{\preceq'}$. It should be also clear that $A_\preceq \subset A_{\preceq'}$.

Hence our example shows that some partial extensions cannot be computed along the orderings of type $\leq \omega$ and that the partial extensions can be included one in another. ▲

Notice that the partial extension construction controls not only the formulas accepted but also the way how those accepted formulas are actually proved.

It is easy to modify the algorithm in Figure 3.3 to construct a partial extension given a well-ordering of defaults. As before, we will need to construct a basis for a partial extension rather than the partial extension itself. We give the modified version in Figure 3.5. Given a set of defaults $D$, a set of formulas $A$ and another set of formulas $R$, By $CD(D, A, R)$ we denote the set of consistent, applicable defaults from the set of defaults $D$. That is a default $r$ belongs to $CD(D, A, R)$ is the following conditions are met:

**Compute_Partial_Extension**$(D, W, \preceq, A_{\preceq})$
*Input:* A finite default theory $(D, W)$, a total ordering $\preceq$ of the defaults in $D$;
*Output:* A set $A_{\preceq}$ of generators of partial extension generated by $\preceq$.

$A := W$;
$R := \emptyset$;
**while** $CD(D, A, R) \neq \emptyset$ **do**
    $d :=\preceq$-least default in $CD(D, A, R)$;
    $A := A \cup \{c(d)\}$;
    $R := R \cup J(d)$;
**od**
$A_{\preceq} := A$

**Fig. 3.5.** Algorithm for computing a partial extension of a default theory given an ordering of defaults

1. $c(r) \notin Cn(A), p(r) \in Cn(A)$.

2. $Cn(A \cup \{c(r)\}) \cap (R \cup J(r)) = \emptyset$.

We can estimate the number of calls to the satisfiability oracle made by this algorithm. Let $n = |D|$ and let $k$ be the maximum of the number of justifications in a default in $D$. The condition (1) calls the oracle twice. The condition (2) calls the oracle $|R| + |J(d)|$ times. This number can be estimated by $n \cdot k$. The **while** loop is executed at most $n$ times. Therefore there is $O(n^2 k)$ calls to the oracle altogether. The same number of calls is necessary to establish if $I_{\preceq} = \emptyset$, that is if the partial extension constructed along the well-ordering $\preceq$ is an extension of $(D, W)$.

Similarly as in Section 3.9, the algorithm in Figure 3.5 can be used to produce all partial extensions of a finite default theory.

## 3.12 Comments and remarks

The theory presented in this chapter is, in large part, due to Reiter, who introduced default logic in his seminal paper [Rei80]. The original presentation of Reiter was based on the operator $\Gamma$. A number of fundamental results are first proved in [Rei80].

The proof-theoretical approach based on the concept of an $S$-derivation was developed by the authors in [MT89b]. The idea that a set $D$ of defaults determines not one but many consequence relations, parameterized by a context $S$ which is used uniformly throughout the derivations, comes from that paper.

The notion of the reduct is due to Lifschitz and Gelfond [GL88] and has been developed in their investigations of stable semantics for logic programs. Our original development of default logic was based on the operator $R^{D,S}$ rather than $R^{D_S}$. The discovery of close connections between the Gelfond-Lifschitz approach to logic programming and default logic by Bidoit and Froidevaux [BF91b, BF91a] and the authors [MT89c] (see Chapter 6) led to our study of monotonic consequence operators $R^D$ and, subsequently, to the development of default logic based on the operators $R^{D_S}$ and $Cn^{D_S}$. This approach allows us to regard default logic as a study of properties of monotone, finitizable, and progressive operators and leads to further generalizations such as nonmonotonic rule systems studied by Marek, Nerode and Remmel [MNR90] and Reinfrank, Dressler and Brewka [RDB89]. In this context let us recall that the "operator-based" approach has proved very fruitful in the development of logic programming (see Apt [Apt90]).

The semantic approach presented in Section 3.5 is due to Lifschitz [Lif90], who credits Guerreiro and Casanova [GC90] for an inspiration. Lifschitz studies a semantical operator $\Sigma$ using the approach described in Section 3.5 to get a version of predicate default logic based on a fixed domain assumption.

The methods of computing extensions described in Section 3.9 are, to the best of our knowledge, original. The procedure **Derive_Candidate** was discovered independently by Remmel. Methods to compute extensions have also been studied by Ben-Eliyahu and Dechter [BED91].

The complexity of default reasoning has been studied by Kautz and Selman [KS89], Stillman [Sti90] and the authors [MT91]. The definitive result has been obtained by Gottlob [Got92]. We will discuss the complexity issues in nonmonotonic reasoning in Chapter 13.

The notion of a weak extension was introduced by the authors [MT89b] in an attempt to identify default logic structures related to autoepistemic logic (see Chapter 12 for details). Minimal sets closed under defaults were introduced under the name of extensions by Hanks and McDermott [HM86]. The notion of partial extension is due to Remmel who proved Theorems 3.99 and 3.101.

An interesting approach to default logic, not discussed in our book, is that of assigning to a default theory a *well-founded* semantics. This notion generalizes the well-founded semantics for logic programs and has been introduced and studied by Baral and Subrahmanian [BS91]. The approach is based on the observation that the operator $\Gamma^2$ is monotonic and, hence, possesses the least and the greatest fixpoints, say $X$ and $Y$. Then, the well-founded semantics is given by $(X, \mathcal{L} \setminus Y)$. If $\Gamma$ has a fixpoint, that is, if a default theory has an extension, it is also a fixpoint of $\Gamma^2$. However, $\Gamma^2$ has always fixpoints independently of whether extensions exist.

In this book we considered only propositional default logic. In his original paper [Rei80], Reiter proposed an approach to the case when defaults are built of formulas of predicate logic. The idea is, speaking informally, to treat a default as the abbreviation for the family of all *ground* defaults that can be obtained by replacing variables by elements of the Herbrand universe of the default theory. In such a treatment, a predicate default theory is, essentially, a propositional default theory in the language generated by the Herbrand base. A competing proposal for predicate default logic has been recently introduced and studied by Lifschitz [Lif90].

Reiter's approach to default logic, although elegant and intuitive, is not the only formal system for default reasoning. There have been several other approaches to default logic. Two of them, one by Łukaszewicz [Łuk84] and the other one by Brewka [Bre91a] are particularly worth noting.

Several books containing a presentation of default logic have been published recently. Of these, the monographs of Etherington [Eth88] and Besnard [Bes89] do not go beyond Reiter's original presentation. The most recent book by Brewka [Bre91b] presents extensions of the basic theory but provides no proofs.

# 4. Normal default theories

A common form of default reasoning involves default rules of the following type: ***under normal circumstances, if*** $\varphi$ ***then conclude*** $\psi$. Such rules allow us to derive $\psi$ from $\varphi$ unless the situation is not *normal*, that is, we know facts that contradict $\psi$. In other words, we conclude $\psi$ if $\varphi$ is known and $\psi$ is consistent. The intuitive meaning of the rule ***under normal circumstances, if*** $\varphi$ ***then conclude*** $\psi$ is captured well by the default

$$\frac{\varphi: M\psi}{\psi}.$$

Defaults whose only justification coincides with the consequent are called *normal*. Default theories all of whose defaults are normal are the main topic of this chapter. They have several useful properties, for instance, they always have at least one extension. Careful analysis of normal default logic provides us also with insights into two other nonmonotonic modes of reasoning: the Closed World Assumption (*CWA*) and the Optimistic Reasoner Assumption (*ORA*).

The technique we employ to investigate normal default logic is that of constructing extensions out of a well-ordering of defaults, developed for the case of general default logic in Section 3.7.

## 4.1 Normal default theories

One of the properties of default logic is that some default theories have no extensions. In this chapter we will study a class of default theories for which the existence of extensions is guaranteed.

**Definition 4.1** A default of the form

$$\frac{\varphi: M\psi}{\psi}$$

is called *normal*. A default theory $(D, W)$ is *normal* if every default in $D$ is normal. ▲

In our discussion of normal default logic we will use the terminology and some of the results from Section 3.7. In particular, $\eta_{\preceq}$ (also called $\eta$), $d_\xi$, $AD_\xi$, $AD_{\preceq}$ have the same meaning as in Definition 3.61. We start with a simple auxiliary fact.

**Lemma 4.2** *Let $(D, W)$ be a normal default theory and let $\preceq$ be a well-ordering of $D$. Then:*

1. *If $W$ is inconsistent then $\eta = 0$, that is, $AD_{\preceq} = \emptyset$.*
2. *If $W$ is consistent then $W \cup c(AD_{\preceq})$ is consistent.*

Proof: Let us assume that $W$ is inconsistent. Then $Cn(W \cup c(AD_{<0}))$ is inconsistent. Since every normal default has a justification, $GD(D, Cn(W \cup c(AD_{<0}))) = \emptyset$. Hence, $\eta = 0$.

Assume next that $W$ is consistent. We prove, by induction on $\xi$ that $W \cup c(AD_\xi)$ is consistent.

To this end consider an ordinal $\xi < \eta$ and assume that $W \cup c(AD_\alpha)$ is consistent for every $\alpha < \xi$. Since the sequence $AD_0, AD_1, \ldots$ is increasing, the compactness theorem (Theorem 2.34) implies that $W \cup c(AD_{<\xi})$ is consistent. By Definition 3.61, since $\xi < \eta$, $d_\xi$ is defined. Let $\beta$ be the only justification of $d_\xi$. It follows that $\beta$ is consistent with $W \cup c(AD_{<\xi})$. Consequently, $c(d_\xi)$ $(= \beta)$ is consistent with $W \cup c(AD_{<\xi})$. Hence, $W \cup c(AD_\xi)$ is consistent.

We have just proved that $W \cup c(AD_\xi)$ is consistent for every $\xi < \eta$. Reasoning as in the inductive step one can prove that $W \cup c(AD_{\preceq}) = W \cup c(AD_{<\eta})$ is consistent. ■

**Theorem 4.3** *Let $(D, W)$ be a normal default theory. For every well-ordering $\preceq$ of $D$, the theory generated by $\preceq$, $Cn(W \cup c(AD_{\preceq}))$, is an extension for $(D, W)$ and $AD_{\preceq} = GD(D, Cn(W \cup c(AD_{\preceq})))$. Moreover, every extension of $(D, W)$ is of this form.*

Proof: Our theorem is obvious if $W$ is inconsistent. Hence, assume that $W$ is consistent. Since $D$ consists of normal defaults only, $J(AD_{\preceq}) = c(AD_{\preceq})$. Let $\beta \in J(AD_{\preceq})$. Then $\beta \in W \cup c(AD_{\preceq})$. By Lemma 4.2, $W \cup c(AD_{\preceq})$ is consistent. Consequently, $\beta$ is consistent with $W \cup c(AD_{\preceq})$ and the condition (3.23) is satisfied. By Theorem 3.65, the first part of the assertion follows.

The second part of the assertion follows from Corollary 3.68. ■

**Corollary 4.4** *Every normal default theory has at least one extension.* ▲

Since every normal default has a justification, Corollary 3.60 applies to normal default theories. It implies that a normal default theory has an inconsistent extension if and only if $W$ is inconsistent.

**Corollary 4.5** *If $(D, W)$ is a normal default theory, then $(D, W)$ has an inconsistent extension if and only if $W$ is inconsistent.* ▲

We will now describe an algorithm to generate all extensions of a finite normal default theory $(D, W)$. Let us analyze again the algorithm **Derive_Candidate** described in Section 3.9. Let us assume that $|D| = n$. The results presented in Section 3.7 imply that if **Derive_Candidate** is run for all $n!$ orderings of a set $D$ of defaults then all extensions of $(D, W)$ are generated. In the case of general default theories, some orderings may generate theories which are not extensions. Hence, an additional criterion (the condition (3.23)) has to be used to eliminate them. In the case of normal default theories, Theorem 4.3 implies that no additional check is needed. Every ordering of $D$ generates a theory which is an extension. Thus, the algorithm **All_Extensions_2**, in the case of normal default theories, can be simplified. In Figure 4.1, we provide a detailed description of the version of the algorithm **All_Extensions_2** tailored to the class of normal default theories. The algorithm is referred to as **NDL_All_Extensions**.

---

**NDL_All_Extensions**$(D, W)$
*Input:* A finite normal default theory $(D, W)$;
*Output:* The list of all extensions of $(D, W)$;

**for** all well-orderings $\preceq$ of $D$ **do**
  **Derive_Candidate**$(\mathbf{D}, \mathbf{W}, \preceq, \mathbf{AD}_{\preceq})$;
  **output**$(W \cup c(AD_{\preceq}))$
**rof**

---

**Fig. 4.1.** Algorithm for computing all extensions of a normal default theory

Consider a normal default theory $(D, W)$. If there are $n$ defaults in $D$, then the total number of justifications in all defaults in $D$ is also $n$. Hence, the algorithm **Derive_Candidate** requires $O(n^2)$ calls to a propositional provability testing procedure. Since there are $n!$ orderings of $D$, the algorithm **NDL_All_Extensions** requires $O(n!n^2)$ calls to a propositional provability testing procedure. This number can be lowered to $O(2^n n^2)$ if the class of orderings is restricted. Namely, as argued at the end of Section 3.9, it is enough to consider only $2^n$ orderings $\preceq_U$, one for each subset $U$ of $D$, where $\preceq_U$ is any ordering in which the defaults in $U$ precede the defaults in $D \setminus U$. Finally, let us observe that the algorithm **NDL_All_Extensions** may generate the same extension several times, for different orderings of $D$.

We will now consider two examples.

**Example 4.6** Consider the normal default theory $(D, W)$, where

$$W = \{p\}, \quad \text{and} \quad D = \left\{ \frac{p\colon Mq}{q}, \frac{p\colon M\neg q}{\neg q}, \frac{q\colon Mr}{r} \right\}.$$

Consider the ordering of the defaults consistent with the way they are listed above. Then, $W \cup c(AD_{<0}) = \{p\}$. The default $\frac{p:Mq}{q}$ is applicable and so, $d_0 = \frac{p:Mq}{q}$ and $W \cup c(AD_0) = \{p, q\}$. The second default is not applicable with respect to $Cn(W \cup c(AD_{<1})) = Cn(W \cup c(AD_0))$ but the third one is. So, $d_1 = \frac{q:Mr}{r}$ and $W \cup c(AD_1) = \{p, q, r\}$. Only the second default has not been considered yet. Since it is inapplicable with respect to $Cn(W \cup c(AD_{<2})) = Cn(W \cup c(AD_1))$, the construction terminates and yields the theory $Cn(\{p, q, r\})$ which is an extension for $(D, W)$.

If we change the order of the last two defaults, then the same theory will be produced.

If the order of the first two defaults is changed, then we get $W \cup c(AD_{<0}) = \{p\}$ and $W \cup c(AD_{<1}) = \{p, \neg q\}$. Since neither $\frac{p:Mq}{q}$ nor $\frac{q:Mr}{r}$ is applicable with respect to $Cn(W \cup c(AD_{<1}))$, the construction terminates and produces the extension $Cn(\{p, \neg q\})$. ▲

**Example 4.7** Consider the default theory $(D, W)$ where

$$W = \{p\}, \quad \text{and} \quad D = \left\{ \frac{p\colon Mq}{q}, \frac{q\colon}{\neg p} \right\}.$$

This default theory is not normal, since the second default is not normal. It is a particularly simple default with no justification. Yet the algorithm **NDL_All_Extensions** does not work correctly in this case. The theory $(D, W)$ has no extensions. On the other hand, the algorithm **NDL_All_Extensions** will produce at least one theory. ▲

## 4.2 Properties of normal default theories

If $(D, W)$ is a default theory that possesses at least two different extensions then all the extensions of $(D, W)$ are consistent. Theorem 3.52 states that extensions of a default theory form an antichain. In the case of normal default theories a much stronger property holds. The union of any two different extensions of a normal default theory is inconsistent.

**Theorem 4.8** *Let $(D, W)$ be a normal default theory. Let $S_1$ and $S_2$ be two extensions of $(D, W)$. If $S_1 \neq S_2$, then $S_1 \cup S_2$ is inconsistent.*

Proof: Let $\preceq$ be a well-ordering of the defaults from $D$ in which the defaults from $GD(D, S_2)$ form an initial segment. Let $\eta$, $AD_\xi$, $\xi < \eta$, and $d_\xi$, $\xi < \eta$ be as described in Definition 3.61. Then, by Theorem 3.70, $S_2$ is generated by $\preceq$ and $GD(D, S_2) = AD_\preceq = \{d_\xi : \xi < \eta\}$.

By Theorem 3.57, $S_i = Cn(W \cup c(GD(D, S_i)))$, $i = 1, 2$. Since extensions form an antichain, $GD(D, S_2) \setminus GD(D, S_1) \neq \emptyset$. Let $\xi$ be the least ordinal such that $d_\xi \notin GD(D, S_1)$. Say, $d_\xi = \frac{\alpha : M\beta}{\beta}$. It follows that

$$\bigcup_{\zeta < \xi} AD_\zeta \subseteq GD(D, S_1).$$

By the definition of $d_\xi$,

$$\alpha \in Cn(W \cup c(\bigcup_{\zeta < \xi} AD_\zeta)) \subseteq Cn(W \cup c(GD(D, S_1))) = S_1.$$

Since $d_\xi \notin GD(D, S_1)$, $\neg\beta \in S_1$. On the other hand, since $d_\xi \in GD(D, S_2)$, $\beta \in S_2$. Hence, $S_1 \cup S_2$ is inconsistent. ■

The requirement that a theory $(D, W)$ be normal is essential in Theorem 4.8. In Example 3.35, we studied a default theory $(D, W)$, where

$$W = \{p\} \text{ and } D = \left\{ \frac{p : M\neg q}{r}, \frac{p : M\neg r}{q} \right\}.$$

We saw that this theory possesses two extensions $Cn(\{p, r\})$ and $Cn(\{p, q\})$. Clearly, the union of these two theories is consistent.

One of the properties of monotone proof systems is that if new inference rules are added then, for every theory $W$, the consequences of $W$ in the original system are also the consequences of $W$ in the enlarged system. A natural generalization of this property to the case of default theories is:

if $(D, W)$ and $(D', W)$ are two default theories and $D \subseteq D'$, then for every extension $E$ for $(D, W)$ there exists an extension $E'$ for $(D', W)$ such that $E \subseteq E'$.

This property is commonly referred to as *semi-monotonicity*. One of the demonstrations of nonmonotonicity of default reasoning is that the property of semi-monotonicity fails for general default theories. This is illustrated in Example 3.39. The situation is different for normal default theories. We will prove now that normal default theories have the property of semi-monotonicity.

**Theorem 4.9** (Semi-monotonicity of normal default logic) *Let $D_1, D_2$ be two disjoint sets of normal default rules. Let $W \subseteq \mathcal{L}$. Then, for every extension $S_1$ of $(D_1, W)$, there exists an extension $S_2$ of $(D_1 \cup D_2, W)$ such that $S_1 \subseteq S_2$.*

Proof: Since $S_1$ is an extension for $(D_1, W)$, there is an ordering $\preceq_1$ of the defaults in $D_1$ such that $S_1$ is generated by $\preceq_1$. Let $\eta_1$, $AD^1_\xi$, $\xi < \eta_1$ and $AD_{\preceq_1}$ be the ordinal and the sets of defaults as specified in Definition 3.61 for the ordering $\preceq_1$. Then, $S_1 = Cn(W \cup c(AD_{\preceq_1}))$.

Consider any ordering $\preceq$ of the defaults in $D_1 \cup D_2$ such that all the defaults in $D_1$ precede the defaults in $D_2$ and, among themselves, the defaults in $D_1$ are ordered according to the ordering $\preceq_1$. Let $\eta$, $AD_\xi$, $\xi < \eta$ and $AD_\preceq$ be the ordinal and the sets of defaults as specified in Definition 3.61 for the ordering $\preceq$. Define $S_2 = Cn(W \cup c(AD_\preceq))$. By Theorem 4.3, $S_2$ is an extension for $(D_1 \cup D_2, W)$. Furthermore, it follows from the definition of $\preceq$ that $\eta \geq \eta_1$ and that for every $\xi < \eta_1$, $AD^1_\xi = AD_\xi$. Consequently, $S_1 \subseteq S_2$. ■

Next, we will study the problem of determining whether a normal default theory has a unique extension. We will first give a simple sufficient condition for this to be the case.

**Corollary 4.10** *If $(D, W)$ is a normal default theory and $W \cup c(D)$ is consistent, then $(D, W)$ possesses a unique extension.*

Proof: Let $S_1$ and $S_2$ be extensions for $(D, W)$. It follows from Theorem 3.57 that $S_1 \cup S_2 \subseteq Cn(W \cup c(D))$. Since $W \cup c(D)$ is consistent, $S_1 \cup S_2$ is consistent. By Theorem 4.8, $S_1 = S_2$. ■

To conclude this section, we present a simple property of normal prerequisite-free default theories.

**Corollary 4.11** *If $(D, W)$ is a normal, prerequisite-free default theory and $S$ is an extension of $(D, W)$ then $S$ is complete with respect to all the formulas in $c(D)$. That is, if $\psi \in c(D)$ then either $\psi \in S$ or $\neg\psi \in S$.*

Proof: Since $S$ is an extension for $(D, W)$, $S$ is closed under all defaults in $D$. Let $\psi \in c(D)$. There exists a default $d \in D$ such that $\psi = c(d)$. Moreover, $J(d) = \{\psi\}$. Assume that $\neg\psi \notin S$. Since $S$ is closed under all defaults in $D$, $\psi \in S$. ■

## 4.3 The Closed World Assumption

The Closed World Assumption is one of the earliest nonmonotonic reasoning schemes. It was proposed by Reiter [Rei78] as a knowledge representation technique. In order to represent a knowledge set $T$ we need a simpler "basis" $W$ and a reasoning mechanism such that every formula in $T$ can be derived from $W$ and formulas not in $T$ cannot be so derived. If a classical logic is to be used as a reasoning mechanism, even a simple knowledge set $T$ might require a large theory $W$ as its representation. For example, it might be necessary to explicitly represent in $W$ a large amount of negative atomic information. The idea of the Closed World Assumption is to strengthen a reasoning mechanism by allowing an additional, extralogical, inference rule designed specifically for the purpose of deriving negative atomic information from $W$.

To better motivate the Closed World Assumption let us consider a simple example. Let $W$ be a database which is supposed to provide information whether there is a flight departing at a given time from New York to San Francisco. One approach to build $W$ could be to store, for every time $T$ of day, the answer YES or NO, depending on whether there is a flight from New York to San Francisco at time $T$ or not. But the size of $W$ would be very large then. Instead, one might store in $W$ only the YES facts. If we ask now whether at time $T$ there is a flight to San Francisco then, if $T$ is found in the database, the answer YES is returned. If, on the other hand, $T$ is not found, the answer NO is given.

When we analyze the mechanism for providing answers we see that in the first case the answer YES is given because $W$ logically implies the proposition "There is a plane to San Francisco at time $T$" (it is simply a member of $W$). On the other hand, the answer NO is given not because we are able to deduce a proposition "no plane to San Francisco at time $T$" from $W$ by means of classical logic. The answer NO was given because the information "There is a plane to San Francisco at time $T$" is missing. Hence, there is an extralogical inference mechanism used in the derivation of a NO answer.

This example motivates an inference rule which derives an atomic fact $p$ from a database $W$ if $p$ is logically derivable from $W$ and derives $\neg p$ if $p$ is not derivable from $W$ in classical (propositional) logic. If this additional inference rule is used, a database $W$ may imply more statements than in the case when only the classical provability operator is used. In some cases, facts inconsistent with $W$ can be derived. The discussion of the properties of theories that are implied by $W$ under the mechanism of the Closed World Assumption is the subject of this section.

We will now formally introduce the Closed World Assumption and study its properties. In particular, we will characterize situations when application of $CWA$ to a theory $W$ does not result in inconsistency.

**Definition 4.12** By the *Closed World Assumption* or, $CWA$, we mean the following inference scheme parameterized with a theory $W \subseteq \mathcal{L}$: *Whenever $p$ is an atom and $W \nvdash p$, conclude $\neg p$.* We will write $CWA$ as

$$\frac{W \nvdash p}{\neg p}.$$

We say that a theory $W \subseteq \mathcal{L}$ is *CWA-consistent*, if after adding to $W$ the atoms that can be derived by $CWA$, the resulting theory, $W \cup \{\neg p : p \in At \text{ and } W \nvdash p\}$, is consistent.

We denote $CWA(W) = Cn(W \cup \{\neg p : p \in At \text{ and } W \nvdash p\})$. ▲

The concepts of $CWA$ and $CWA$-consistency are illustrated in the following example.

**Example 4.13** Let $W_1 = \{p \vee q\}$. Then $W_1 \nvdash p$, and $W_1 \nvdash q$. Therefore $CWA(W_1) = Cn(\{p \vee q, \neg p, \neg q\})$. This theory is inconsistent. Therefore $W_1$ is not $CWA$-consistent.

Consider now the theory $W_2 = \{q \supset p, s \supset r, s\}$. Clearly, $W_2 \vdash s$ and $W_2 \vdash r$. On the other hand, $W_2 \nvdash q$ and $W_2 \nvdash p$. Therefore $CWA(W_2) = Cn(W_2 \cup \{\neg q, \neg p\}) = Cn(\{\neg p, \neg q, r, s\})$. This theory is consistent. Consequently, $W_2$ is $CWA$-consistent. ▲

There is a natural extension of $CWA$ in which the scope of applicability of $CWA$ is restricted to a prespecified subset of the set of atoms. In a special case, when this set of atoms coincides with the set of all atoms of the language, we get the usual $CWA$. We will now formally introduce $CWA$ with restricted scope and develop its theory. We will present the results on full $CWA$ as corollaries.

**Definition 4.14** Let $P$ be a subset of the set $At$ of atoms of the language $\mathcal{L}$. By the *Closed World Assumption with respect to $P$*, we mean the inference scheme of $CWA$ with scope restricted to the atoms in $P$. We denote it by $CWA^P$.

A theory $W$ is called *$CWA^P$-consistent* if the theory $W \cup \{\neg p : p \in P \text{ and } W \nvdash p\}$ is consistent. The theory $Cn(W \cup \{\neg p : p \in P \text{ and } W \nvdash p\})$ is denoted by $CWA^P(W)$. ▲

We will now present several results describing properties of the theories $CWA^P(W)$ and $CWA(W)$. In particular, we will establish connections between $CWA^P$ and normal default logic. We start with a simple result stating that $CWA^P$ produces theories which are complete with respect to the language $\mathcal{L}_P$ generated by $P$ (see Definition 2.17). This last property, according to Theorem 2.18, is equivalent to completeness with respect to the set $P$.

**Theorem 4.15** *For every $W \subseteq \mathcal{L}$ and for every $P \subseteq At$, $CWA^P(W)$ is complete with respect to $\mathcal{L}_P$ or, equivalently, with respect to $P$. In particular, $CWA(W)$ is complete.*

Proof: By Theorem 2.18, it suffices to show that for every $p \in P$ at least one of $p$ and $\neg p$ is in $CWA^P(W)$. Assume that $\neg p \notin CWA^P(W)$. By the definition, it follows that $W \vdash p$. Since $W \subseteq CWA^P(W)$ and $CWA^P(W)$ is closed under propositional consequence, $p$ belongs to $CWA^P(W)$. ■

The concept of $CWA$ historically precedes that of default logic. It turns out, however, that $CWA$ and $CWA^P$ are very closely related to an especially simple fragment of normal default logic. We will show now how the basic concept of $CWA$, namely $CWA$-consistency ($CWA^P$-consistency) can be expressed in the language of extensions of normal default theories.

Each instance

$$\frac{W \nvdash p}{\neg p}$$

of $CWA$ has a flavor of a normal default. Indeed, its meaning is: if $\neg p$ is consistent with $W$, then conclude $\neg p$. The normal default

$$\frac{: M\neg p}{\neg p}$$

has a similar intuitive content: if $\neg p$ is consistent (possible) conclude $\neg p$. Motivated by this similarity we will study now $CWA$ from the standpoint of normal default logic.

Consider a set of atoms $P \subseteq At$. Define a collection of defaults

$$D^{CWA^P} = \left\{ \frac{:M\neg p}{\neg p} : p \in P \right\}.$$

In the case when $P = At$, we will abbreviate $D^{CWA^{At}}$ to $D^{CWA}$.

In what follows we will characterize $CWA^P$-consistency of a theory $W$ in terms of some properties of the default theory $(D^{CWA^P}, W)$. We first prove a simple characterization of minimal sets for $(D^{CWA^P}, W)$, that is, inclusion-minimal sets containing $W$ and closed under propositional provability and the defaults in $D^{CWA^P}$

**Theorem 4.16** *Let $P$ be a nonempty set of atoms.*

1. *A theory $S \subseteq \mathcal{L}$ is closed under propositional provability and under defaults in $D^{CWA^P}$ if and only if $S$ is complete with respect to $P$.*
2. *A theory $S \subseteq \mathcal{L}$ is a minimal set for a default theory $(D^{CWA^P}, W)$, where $W \subseteq \mathcal{L}$, if and only if $S$ is an inclusion-minimal theory containing $W$ and complete with respect to $P$.*

Proof: (1) Consider an atom $p \in P$. Since $S$ is closed under the default $\frac{:M\neg p}{\neg p}$, $\neg\neg p \in S$ or $\neg p \in S$. Since $S$ is closed under propositional provability, $p \in S$ or $\neg p \in S$.

Conversely, if $S$ is complete with respect to $P$ then, by the definition of completeness, $S$ is closed under propositional provability. Consider a default $\frac{:M\neg p}{\neg p}$ from $D^{CWA^P}$. Assume that $\neg p \notin S$. By the completeness of $S$, $p \in S$. Hence, $S$ is closed under $D^{CWA^P}$.

Part (2) of the assertion follows directly from (1). ■

In Chapter 3 we proved that every extension for a default theory is a minimal set for this default theory (Theorem 3.92). Hence, we obtain the following corollary to Theorem 4.16.

**Corollary 4.17** *Let $P$ be a set of atoms. If a theory $S$ is an extension for a default theory $(D^{CWA^P}, W)$ then $S$ is an inclusion-minimal theory containing $W$ and is complete with respect to $P$.* ▲

To characterize extensions for $(D^{CWA^P}, W)$ we need one more notion. For any two theories $S_1$ and $S_2$ define

$$S_1 \sqsubseteq_P S_2 \text{ if } S_1 \cap P \subseteq S_2 \cap P.$$

The relation $\sqsubseteq_P$ is reflexive and transitive. Hence, it is a *preordering* of the class of all theories contained in $\mathcal{L}$. The relation $\sqsubseteq_P$ is not antisymmetric. It is easy to construct two theories $S_1$ and $S_2$ such that $S_1 \sqsubseteq_P S_2$ and $S_2 \sqsubseteq_P S_1$ but $S_1$ and $S_2$ are different. For example, if $q \in At \setminus P$, $Cn(P \cup \{q\}) \sqsubseteq_P Cn(P)$, $Cn(P) \sqsubseteq_P Cn(P \cup \{q\})$ and $Cn(P \cup \{q\}) \neq Cn(P)$. Hence, in general, $\sqsubseteq_P$ is not a partial ordering.

In the case when $P = At$ we will write $\sqsubseteq$ rather than $\sqsubseteq_{At}$. In this special case, it is easy to see that $\sqsubseteq$ is a partial ordering of complete theories. The example considered above shows that if $P \neq At$ then, even under the restriction to theories complete with respect to $P$, $\sqsubseteq_P$ is not a partial ordering.

Let $\mathcal{C}$ be a class of theories, that is, $\mathcal{C} \subseteq \mathcal{P}(\mathcal{L})$. We say that a theory $S \in \mathcal{C}$ is $\sqsubseteq_P$-*minimal* in $\mathcal{C}$ if there is no theory $S' \in \mathcal{C}$ such that $S' \sqsubseteq_P S$ and $S' \cap P \neq S \cap P$. We say that a theory $S \in \mathcal{C}$ is $\sqsubseteq_P$-*least* in the class $\mathcal{C}$ if for every theory $S' \in \mathcal{C}$, $S \sqsubseteq_P S'$. Since $\sqsubseteq_P$ is only a preordering such $\sqsubseteq_P$-least theory does not have to be unique.

These two notions coincide with the standard notions of a minimal element and a least element with respect to $\sqsubseteq_P$ in the case when $\sqsubseteq_P$ partially orders $\mathcal{C}$. In particular, if $\sqsubseteq_P$ partially orders $\mathcal{C}$, then:

1. There is at most one $\sqsubseteq_P$-least theory in $\mathcal{C}$.

2. If $S_1, S_2 \in \mathcal{C}$, $S_1$ is $\sqsubseteq_P$-minimal in $\mathcal{C}$ and $S_2 \sqsubseteq_P S_1$, then $S_1 = S_2$.

We will now exhibit a class of theories which is partially ordered by $\sqsubseteq_P$.

**Theorem 4.18** *Let $P$ be a set of atoms. For every theory $W \subseteq \mathcal{L}$, the class of all minimal sets for the default theory $(D^{CWA^P}, W)$ is partially ordered by $\sqsubseteq_P$. Equivalently, the class of all inclusion-minimal theories containing $W$ and complete with respect to $P$ is partially ordered by $\sqsubseteq_P$.*

Proof: Every minimal set for $(D^{CWA^P}, W)$ is of the form

$$Cn(W \cup U \cup \{\neg p : p \in P \setminus U\}),$$

for some $U \subseteq P$ (Corollary 3.90). Consider two minimal sets $S_1$ and $S_2$ for $(D^{CWA^P}, W)$, say

$$S_i = Cn(W \cup U_i \cup \{\neg p : p \in P \setminus U_i\}),$$

for some $U_i \subseteq P$, $i = 1, 2$. Assume that $S_1 \sqsubseteq_P S_2$ and $S_2 \sqsubseteq_P S_1$. Assume that $S_1 \neq S_2$. Then $S_1$ and $S_2$ are inclusion-incomparable. Hence, it follows that neither $S_1$ nor $S_2$ is inconsistent. Consequently, $U_i = S_i \cap P$, for $i = 1, 2$.

Since $S_1 \sqsubseteq_P S_2$ and $S_2 \sqsubseteq_P S_1$, $U_1 = U_2$. Hence, $S_1 = S_2$, a contradiction. Thus, $S_1 = S_2$ and $\sqsubseteq_P$ is antisymmetric. We have already observed that $\sqsubseteq_P$ is reflexive and transitive. Hence, $\sqsubseteq_P$ is a partial ordering of the class of minimal sets for $(D^{CWA^P}, W)$. The second part of the assertion follows from Theorem 4.16. ∎

We proved in the previous chapter that extensions form an antichain with respect to the relation of inclusion (Theorem 3.52). Now we will prove a similar property of extensions for $(D^{CWA^P}, W)$ with respect to the relation $\sqsubseteq_P$.

**Theorem 4.19** *Let $P$ be a set of atoms. For every theory $W \subseteq \mathcal{L}$ and for every two extensions $S_1$ and $S_2$ for $(D^{CWA^P}, W)$, if $S_1 \sqsubseteq_P S_2$ then $S_1 = S_2$.*

Proof: It follows from Theorem 3.57 that every extension can be described by means of its generating defaults. Consequently, for $i = 1, 2$,

$$S_i = Cn(W \cup c(GD(D^{CWA^P}, S_i))).$$

Hence, for $i = 1, 2$,

$$S_i = Cn(W \cup \{\neg p : p \in P \setminus S_i\}).$$

Since $S_1 \sqsubseteq_P S_2$, $P \setminus S_2 \subseteq P \setminus S_1$. Hence, $S_2 \subseteq S_1$. By Theorem 3.52, $S_1 = S_2$. ∎

Before we continue, we need an auxiliary lemma.

**Lemma 4.20** *Let $P$ be a set of atoms and let $W \subseteq \mathcal{L}$. For every theory $S$ containing $W$ and complete with respect to $P$ there is an extension $S'$ for $(D^{CWA^P}, W)$ such that $S' \sqsubseteq_P S$.*

Proof: The assertion is evident in the case when $S$ is inconsistent, because $(D^{CWA^P}, W)$, being a normal default theory, has at least one extension. So, assume that $S$ is consistent. In particular, it follows that $W$ is consistent and, consequently, that every extension for $(D^{CWA^P}, W)$ is consistent.

Define $D' = GD(D^{CWA^P}, S)$ and $D'' = D^{CWA^P} \setminus D'$. Let $\preceq$ be any well-ordering in which all the defaults in $D'$ precede the defaults in $D''$. By Theorem 4.3, the ordering $\preceq$ generates an extension $S'$ for $(D^{CWA^P}, W)$. Let $\eta$, $d_\xi$, $\xi < \eta$, and $AD_\preceq$ be the ordinal, the defaults and the set of defaults constructed for the ordering $\preceq$ according to Definition 3.61. Assume that $D'' \cap AD_\preceq \neq \emptyset$ (that is, that at some stage in the construction the default selected is not in $D'$). Let $\zeta$ be the smallest ordinal such that $d_\zeta \notin D'$. Let

$AD_{<\zeta}$ denote, as usual, the set of defaults selected prior to the stage $\zeta$ and let $T_{<\zeta} = Cn(W \cup c(AD_{<\zeta}))$ be the context used to select a default in the stage $\zeta$. Consider an ordinal $\xi < \zeta$. Let $p$ be an atom in $P$ such that $d_\xi = \frac{:M\neg p}{\neg p}$. Then, since $d_\xi \in D' = GD(D^{CWA^P}, S)$, $p \notin S$. By the completeness of $S$, $\neg p \in S$. Hence, $c(AD_{<\zeta}) \subseteq S$. Since $W \subseteq S$, we obtain that $T_{<\zeta} \subseteq S$. Since each default in $D^{CWA^P}$ is prerequisite-free, it follows that

$$D' \setminus AD_{<\zeta} = GD(D^{CWA^P} \setminus AD_{<\zeta}, S) \subseteq GD(D^{CWA^P} \setminus AD_{<\zeta}, T_{<\zeta}).$$

Hence, each of the defaults in $D' \setminus AD_{<\zeta}$ is a candidate for selection at stage $\zeta$. Since a default not in $D'$ has been selected and since each of the defaults in $D'$ precedes all defaults in $D''$, it follows that $D' \setminus AD_{<\zeta} = \emptyset$. In other words, $D' \subseteq AD_{<\zeta}$. Hence, $D' \subseteq AD_{\preceq} = GD(D^{CWA^P}, S')$.

Let $p \in P \cap S'$. Since $S'$ is consistent, $\frac{:M\neg p}{\neg p} \notin AD_{\preceq}$. Hence, $\frac{:M\neg p}{\neg p} \notin D'$. That is, $p \in S$. Hence, $S' \sqsubseteq_P S$. ∎

We will now characterize extensions for $(D^{CWA^P}, W)$ in terms of theories which are complete with respect to $P$, contain $W$, and are minimal with respect to inclusion and the relation $\sqsubseteq_P$.

**Theorem 4.21** *Let $P$ be a set of atoms and let $W \subseteq \mathcal{L}$. A theory $S \subseteq \mathcal{L}$ is an extension for $(D^{CWA^P}, W)$ if and only if $S$ is both inclusion-minimal and $\sqsubseteq_P$-minimal in the class of all theories containing $W$ and complete with respect to $P$.*

Proof: Assume that $S$ is an extension for $(D^{CWA^P}, W)$. Then $S$ contains $W$ and, by Corollary 4.17, $S$ is complete with respect to $P$. Since $S$ is a minimal set for $(D^{CWA^P}, W)$, it follows from Theorem 4.16 that $S$ is inclusion-minimal in the class of all theories containing $W$ and complete with respect to $P$.

Assume that $S$ is not $\sqsubseteq_P$-minimal in the class of theories containing $W$ and complete with respect to $P$. Let $S'$ be a theory containing $W$, complete with respect to $P$ and such that $S' \sqsubseteq_P S$ and $S' \cap P \neq S \cap P$ (see the definition of $\sqsubseteq_P$-minimality). By Lemma 4.20, there is an extension $S''$ for $(D^{CWA^P}, W)$ such that $S'' \sqsubseteq_P S'$. By the transitivity of $\sqsubseteq_P$, $S'' \sqsubseteq_P S$. Since both $S''$ and $S$ are extensions for $(D^{CWA^P}, W)$, by Theorem 4.19 it follows that $S'' = S$. Hence, $S = S'' \sqsubseteq_P S'$. Consequently, $S \cap P \subseteq S' \cap P$. Since $S' \sqsubseteq_P S$, $S' \cap P \subseteq S \cap P$. Hence, $S \cap P = S' \cap P$, a contradiction. It follows that $S$ is both inclusion-minimal and $\sqsubseteq_P$-minimal in the class of theories containing $W$ and complete with respect to $P$.

Conversely, let $S$ be inclusion-minimal and $\sqsubseteq_P$-minimal in the class of all theories containing $W$ and complete with respect to $P$. By Lemma 4.20, there is an extension $S'$ for $(D^{CWA^P}, W)$ such that $S' \sqsubseteq_P S$. Since $S'$ is an extension for $(D^{CWA^P}, W)$, $S'$ is a minimal set for $(D^{CWA^P}, W)$ and, by Theorem 4.16, an inclusion-minimal theory containing $W$ and complete with respect to $P$. We have:

1. Under the restriction to the inclusion-minimal theories which contain $W$ and are complete with respect to $P$ the relation $\sqsubseteq_P$ is a partial ordering (Theorem 4.18).

2. $S$ is $\sqsubseteq_P$-minimal in the class of all inclusion-minimal theories containing $W$ and complete with respect to $P$.

3. $S'$ is an inclusion-minimal theory containing $W$, complete with respect to $P$ and such that $S' \sqsubseteq_P S$.

Hence, the equality $S = S'$ follows. That is, the theory $S$ is an extension for $(D^{CWA^P}, W)$. ■

Let us now adapt the characterization given in Theorem 4.21 to the case when $P = At$.

**Corollary 4.22** *Let $W$ be consistent. A theory $S$ is an extension for $(D^{CWA}, W)$ if and only if $S$ is a consistent $\sqsubseteq$-minimal complete theory containing $W$.*

Proof: Since $W$ is consistent and each default in $D^{CWA}$ has a justification, extensions of $(D^{CWA}, W)$ are consistent. In addition, consistent and complete theories are inclusion-incomparable. Hence, they are all inclusion-minimal. Consequently, the result follows from Theorem 4.21. ■

We are now in a position to describe the property of $CWA^P$-consistency in the terminology of normal default logic.

**Theorem 4.23** *Let $P$ be a set of atoms. For every theory $W \subseteq \mathcal{L}$ the following conditions are equivalent:*

1. *$W$ is $CWA^P$-consistent.*

2. *The class of all consistent, inclusion-minimal theories containing $W$ and complete with respect to $P$ contains the $\sqsubseteq_P$-least theory.*

3. *The default theory $(D^{CWA^P}, W)$ has a unique extension and this extension is consistent.*

Proof: ((1) $\Rightarrow$ (2)) Let us assume that $W$ is $CWA^P$-consistent. Let

$$S = Cn(W \cup \{\neg p : p \in P \text{ and } W \not\vdash p\}) \; (= CWA^P(W)).$$

Obviously, $W \subseteq S$ and $S$ is complete with respect to $P$. Let $S'$ be any theory containing $W$ and complete with respect to $P$. Assume that $S' \subseteq S$. Consider an atom $p \in P$ such that $W \not\vdash p$. Since $S$ is consistent, $p \notin S$. Hence, $p \notin S'$. By the completeness of $S'$ with respect to $P$, $\neg p \in S'$. Since $S'$ is closed under propositional provability (as a theory complete with respect to $P$) and $W \subseteq S'$, $S \subseteq S'$. Consequently, $S = S'$ and $S$ is an inclusion-minimal theory containing $W$ and complete with respect to $P$.

Consider now an inclusion-minimal theory $S'$ containing $W$ and complete with respect to $P$. Let $p \in S \cap P$. By the definition of $CWA^P(W)$, $W \vdash p$. Since $W \subseteq S'$, $p \in S'$. Consequently, $S \sqsubseteq_P S'$. By (1), $S$ is consistent. Hence, (2) follows. In fact, our argument shows that it is $CWA^P(W)$ that is $\sqsubseteq_P$-least in the class of all consistent, inclusion-minimal theories containing $W$ and complete with respect to $P$.

((2) $\Rightarrow$ (3)) Assume that $S$ is a $\sqsubseteq_P$-least theory in the class of all consistent, inclusion-minimal theories containing $W$ and complete with respect to $P$. Then $S$ is in the same time $\sqsubseteq_P$-minimal and inclusion-minimal among the theories containing $W$ and complete with respect to $P$. By Theorem 4.21, $S$ is an extension for $(D^{CWA^P}, W)$. Consider an extension $S'$ for $(D^{CWA^P}, W)$. Then $W \subseteq S'$ and $S'$ is complete with respect to $P$. Since $S$ is $\sqsubseteq_P$-least among all theories containing $W$ and complete with respect to $P$, $S \sqsubseteq_P S'$. By Theorem 4.19, $S = S'$. Hence, $S$ is a unique extension for $(D^{CWA^P}, W)$. Since $S$ is consistent, (3) follows.

((3) $\Rightarrow$ (1)) Let $S$ be the unique extension of $(D^{CWA^P}, W)$. Assume also that $S$ is consistent. Let us define

$$D' = \left\{ \frac{: M\neg p}{\neg p} \in D^{CWA^P} : p \in S \right\}.$$

Clearly, $D' = D^{CWA^P} \setminus GD(D^{CWA^P}, S)$. Let $\preceq$ be any well-ordering in which all the defaults in $D'$ precede the defaults in $GD(D^{CWA^P}, S)$. By Theorem 4.3, the ordering $\preceq$ generates an extension $S'$ for $(D^{CWA^P}, W)$. Since $S$ is the unique extension for $(D^{CWA^P}, W)$, $S = S'$. Consequently, in the first step of the construction, none of the defaults in $D'$ is applicable. In other words, for every $p \in P$ such that $p \in S$, $W \vdash p$. That is, if $W \not\vdash p$, $p \notin S$. Since extensions for $(D^{CWA^P}, W)$ are complete with respect to $P$, $\neg p \in S$. Hence,

$$\begin{aligned} CWA^P(W) &= Cn(W \cup \{\neg p : p \in P \text{ and } W \not\vdash p\}) \\ &\subseteq Cn(W \cup \{\neg p : p \in P \setminus S\}) \\ &= Cn(W \cup c(GD(D^{CWA^P}, S))). \end{aligned}$$

Since $S$ is an extension for $(D^{CWA^P}, W)$, it follows by Theorem 3.57 that

$$S = Cn(W \cup c(GD(D^{CWA^P}, S))).$$

Hence, $CWA^P(W) \subseteq S$. Since $S$ is consistent, $CWA^P(W)$ is consistent. (In fact, $CWA^P(W) = S$. Indeed, since $W \subseteq S$,

$$\begin{aligned} S &= Cn(W \cup c(GD(D^{CWA^P}, S))) = Cn(W \cup \{\neg p : p \in P \setminus S\}) \\ &\subseteq Cn(W \cup \{\neg p : p \in P \text{ and } W \not\vdash p\}) = CWA^P(W). \end{aligned}$$

Hence, $S = CWA^P(W)$. ■

The last fragment of the argument used to prove Theorem 4.23 yields the following result.

**Theorem 4.24** *Let $P$ be a set of atoms and $W \subseteq \mathcal{L}$. If $W$ is $CWA^P$-consistent then $CWA^P(W)$ is a unique extension for $(D^{CWA^P}, W)$.* ▲

Since consistent and complete theories are inclusion-incomparable (and so, inclusion-minimal) we obtain a characterization of $CWA$-consistency as a corollary to Theorem 4.23.

**Corollary 4.25** *For every theory $W \subseteq \mathcal{L}$ the following conditions are equivalent:*

1. *$W$ is $CWA$-consistent.*
2. *the class of all consistent complete theories containing $W$ contains the $\sqsubseteq$-least theory.*
3. *the default theory $(D^{CWA}, W)$ has a unique extension and this extension is consistent.* ▲

As a byproduct of our considerations we obtain a particularly simple algorithm for testing the uniqueness of an extension for a default theory of the form $(D^{CWA^P}, W)$. Clearly, if $W$ is inconsistent, $\mathcal{L}$ is a unique extension for $(D^{CWA^P}, W)$. If $W$ is consistent, then all that we need to do is to check whether $CWA^P(W)$ is consistent and, if so, output $CWA^P(W)$ as the unique extension (Theorem 4.24).

We will now describe a class of theories which are $CWA$-consistent. First, let us recall (Example 4.13) that a theory $\{p \vee q\}$ is not $CWA$-consistent. The reason is that $\{p \vee q\}$ implies neither $p$ nor $q$. Consequently, both $\neg p$ and

$\neg q$ are added to the theory and the result is inconsistent. It turns out that if a theory consists of clauses, none of which contains two different atoms as disjuncts, then it is *CWA*-consistent. Such clauses are known as *Horn* clauses (see Definition 2.3). Theories consisting of Horn clauses are called *Horn* theories.

**Theorem 4.26** *If $W$ is a consistent Horn theory, then the class of complete consistent theories containing $W$ has a $\sqsubseteq$-least theory.*

Proof: Define inductively the following sequence of sets of atoms:

$$A_0 = At \cap W,$$

$$\begin{aligned} A_{n+1} &= A_n \cup \{p : \neg q_1 \vee \ldots \vee \neg q_m \vee p \vee \neg r_1 \vee \ldots \vee \neg r_t \in W \\ &\quad \text{and } q_1, \ldots, q_m, r_1, \ldots, r_t \in A_n\}. \end{aligned}$$

Finally, define

$$A_\omega = \bigcup_{n=0}^{\infty} A_n.$$

By an easy induction on $n$, it can be proved that $A_n \subseteq Cn(W)$. Consequently, $A_\omega \subseteq Cn(W)$.

Let $V$ be a valuation of $\mathcal{L}$ such that $V(p) = \mathbf{t}$ if and only if $p \in A_\omega$. We claim that $V$ is a model of $W$. Consider a clause $\varphi \in W$.

Case 1. The clause $\varphi$ is an atom. Then $\varphi \in A_\omega$ and $V(\varphi) = \mathbf{t}$, by the definition of $V$.

Case 2. The clause $\varphi$ is of the form

$$\neg q_1 \vee \ldots \vee \neg q_m \vee p \vee \neg r_1 \vee \ldots \vee \neg r_t.$$

If for some negative literal $\neg s$ of $\varphi$, $s \notin A_\omega$, then $V(s) = \mathbf{f}$, $V(\neg s) = \mathbf{t}$ and, consequently, $V(\varphi) = \mathbf{t}$. Hence, assume that for every negative literal $\neg s$ of $\varphi$, $s \in A_\omega$. Since the number of negative literals in $\varphi$ is finite, there is an integer $n$ such that for every negative literal $\neg s$ of $\varphi$, $s \in A_n$. Consequently, $p \in A_{n+1}$. Therefore, $p \in A_\omega$, $V(p) = \mathbf{t}$ and, consequently, $V(\varphi) = \mathbf{t}$.

Case 3. The clause $\varphi$ contains only negative literals. If for every negative literal $\neg s$ of $\varphi$, $W \vdash s$, then $W$ is inconsistent, a contradiction. Hence, there is a negative literal $\neg s$ of $\varphi$ such that $W \nvdash s$. Since $A_\omega \subseteq Cn(W)$, it follows that $A_\omega \nvdash s$. Since $A_\omega$ is a set of atoms, it follows that $s \notin A_\omega$. Consequently, $V(s) = \mathbf{f}$, $V(\neg s) = \mathbf{t}$ and $V(\varphi) = \mathbf{t}$.

Thus $V$ is a model of $W$. Let $T = Th(V)$, that is $T = \{\varphi : V(\varphi) = \mathbf{t}\}$. Then $T$ is a consistent and complete theory containing $W$ and $A_\omega = At \cap T$.

Consider any consistent and complete theory $T_1$ containing $W$. Then $Cn(W) \subseteq T_1$. So, $A_\omega \subseteq T_1$. Hence, $A_\omega \subseteq At \cap T_1$ and, consequently, $T \sqsubseteq T_1$. Thus $T$ is the $\sqsubseteq$-least complete and consistent theory containing $W$. ■

**Corollary 4.27** *If $W$ is a consistent Horn theory then $W$ is $CWA$-consistent.* ▲

The effect of $CWA$ is to conclude $\neg p$ from the inability to derive $p$ from the initial theory $W$. The scope of $CWA$ is restricted to atoms: if an atom is not provable, its negation is derived. This restriction is naturally motivated by a common practice that negative elementary facts are represented in databases implicitly by the fact of their absence. However, at least from the formal point of view, there are other classes of formulas to which one might want to apply the $CWA$ paradigm. It turns out that when $CWA$ is used for a class of formulas other than atoms, a reasoning system can be obtained which models entirely different intuitions. To illustrate how $CWA$ changes when its scope is modified and to show how normal default logic can be used to study the resulting reasoning systems we will briefly discuss one of them below.

By the *Optimistic Reasoner Assumption* ($ORA$ for short) we mean a reasoning scheme which allows us to conclude $p$ whenever $\neg p$ does not follow from the theory $W$ describing our knowledge base. That is, $ORA$ is a variant of $CWA$ where the scope of applicability of $CWA$ is the class of negated atoms rather than atoms as in the standard $CWA$. The Closed World Assumption is pessimistic in character — if an atomic fact cannot be proved from our knowledge base, it is false and its negation is accepted. The spirit of the Optimistic Reasoner Assumption is quite different. If an atomic fact is not explicitly falsified (that is, its negation cannot be derived), it is accepted as true.

Let $W \subseteq \mathcal{L}$ and let $P$ be a set of atoms. Define

$$ORA^P(W) = Cn(W \cup \{p : p \in P \text{ and } W \not\vdash \neg p\}).$$

If $ORA^P(W)$ is consistent, $W$ is said to be $ORA^P$*-consistent.* If $P = At$, we write $ORA(W)$ rather than $ORA^{At}(W)$ and $ORA$-consistent rather than $ORA^{At}$-consistent.

As in the case of $CWA$ we define a set of defaults $D^{ORA^P}$, where $P$ is a set of atoms, as follows:

$$D^{ORA^P} = \left\{ \frac{:Mp}{p} : p \in P \right\}.$$

In the case when $P = At$, we will abbreviate $D^{ORA^{At}}$ to $D^{ORA}$.

The properties of $ORA$ are dual to the properties of $CWA$. Below, we list without proof four results on $ORA$, corresponding to Theorem 4.21, Corollary 4.22, Theorem 4.23 and Corollary 4.25, respectively.

**Theorem 4.28** *Let $P$ be a set of atoms and let $W \subseteq \mathcal{L}$. A theory $S \subseteq \mathcal{L}$ is an extension for $(D^{ORA^P}, W)$ if and only if $S$ is both inclusion-maximal and $\sqsubseteq_P$-maximal in the class of all theories containing $W$ and complete with respect to $P$.* ▲

**Corollary 4.29** *Let $W$ be consistent. A theory $S$ is an extension for $(D^{ORA}, W)$ if and only if $S$ is a consistent $\sqsubseteq$-maximal complete theory containing $W$.* ▲

**Theorem 4.30** *Let $P$ be a set of atoms. For every theory $W \subseteq \mathcal{L}$ the following conditions are equivalent:*

1. *$W$ is $ORA^P$-consistent.*
2. *The class of all consistent, inclusion-maximal theories containing $W$ and complete with respect to $P$ contains the $\sqsubseteq_P$-greatest theory.*
3. *the default theory $(D^{ORA^P}, W)$ has a unique extension and this extension is consistent.* ▲

**Corollary 4.31** *For every theory $W \subseteq \mathcal{L}$ the following conditions are equivalent:*

1. *$W$ is $ORA$-consistent.*
2. *The class of all consistent complete theories containing $W$ contains the $\sqsubseteq$-greatest theory.*
3. *The default theory $(D^{ORA}, W)$ has a unique extension and this extension is consistent.* ▲

## 4.4 Comments and remarks

The theory presented in this chapter is essentially due to Reiter. He introduced $CWA$ in [Rei78] and normal defaults in [Rei80]. The relation between normal defaults and $CWA$ and the method of studying these concepts by means of well-orderings were first presented by the authors in [MT89a].

# 5. Representation theory for default logic

In propositional and predicate logics we often represent formulas and theories by means of formulas or theories of a simpler syntactic form but with the same logical properties. For instance, it is well known that for every propositional formula $\varphi$ there exists a formula $\psi$ in the conjunctive normal form such that $\varphi$ and $\psi$ have the same models (are logically equivalent). In fact, there is a simple algorithm to construct $\psi$ out of $\varphi$. In this chapter we study the *representability* issue for default logic. We provide several results showing that default theories can be represented by syntactically simpler theories in such a way that extensions are preserved. We show that, in some cases, these "normal form" default theories can easily be constructed. One of the highlights of this chapter is the result showing that, at the cost of adding new atoms to the language, every default theory can be assigned an equivalent default theory with all rules either monotonic, that is, justification-free, or semi-normal. We also show that prerequisite-free normal default theories can be represented within the framework of *CWA*. Finally, we discuss the representability of families of sets as collections of weak extensions of default theories.

## 5.1 Representability of default theories

A common way to simplify investigations of logics is to reduce the general case of arbitrary formulas to the case of formulas in a syntactically simple form, which makes them easy to study. That is, we replace each formula by its *normal form*. For this method to work, the formulas used as normal forms have to be logically equivalent to the formulas they replace. This requirement means that formulas and their normal forms have to have the same models. Secondly, if we are interested in algorithmic issues such as automated theorem proving, it is necessary that normal forms be easily constructed.

In this chapter we develop the *normal-form* (or, as we will usually say, *representation*) theory for default logic. Our motivations are similar as in the case of classical logics. The main difference is that we use the notion of an extension to define *equivalence* of default theories.

**Definition 5.1** Let $(D, W)$, $(D', W')$ be default theories. We say that $(D, W)$ is *equivalent* to $(D', W')$ (in symbols, $(D, W) \approx (D', W')$) if $(D, W)$ and $(D', W')$ have exactly the same extensions. ▲

It is obvious that $\approx$ is an equivalence relation.

**Example 5.2** Consider the default theory $(D, W)$, where

$$D = \left\{ \frac{p : M \neg r}{q}, \frac{q : M s}{t} \right\} \quad \text{and} \quad W = \{p\}.$$

Clearly, $(D, W)$ has exactly one extension, $Cn(\{p, q, t\})$.

Consider the default theory $(D', W')$, where

$$D' = \left\{ \frac{:}{p}, \frac{p : M \neg r}{q}, \frac{q : M s}{t} \right\} \quad \text{and} \quad W' = \emptyset.$$

The theory $Cn(\{p, q, t\})$ is the only extension for $(D', W')$. Hence, the theories $(D, W)$ and $(D', W')$ are equivalent. ▲

Our notion of equivalence generalizes standard logical equivalence. If $D = \emptyset$ then for every theory $W \subseteq \mathcal{L}$, $(D, W)$ has exactly one extension, $Cn(W)$. Consequently, in such case $(D, W)$ is equivalent to $(D, W')$ if and only if $W$ is logically equivalent to $W'$. Several other simple properties of the relation $\approx$ are gathered in the following theorems.

**Theorem 5.3**

1. *If* $Cn(W_1) = Cn(W_2)$ *then for every set* $D$ *of defaults,* $(D, W_1) \approx (D, W_2)$.
2. *If* $W \vdash (\alpha \equiv \alpha')$, $W \vdash (\psi_1 \equiv \psi_1'), \ldots, W \vdash (\psi_k \equiv \psi_k')$, $W \vdash (\gamma \equiv \gamma')$, *then for every* $D$,

$$(D \cup \left\{ \frac{\alpha : M\psi_1, \ldots, M\psi_k}{\gamma} \right\}, W) \approx (D \cup \left\{ \frac{\alpha' : M\psi_1', \ldots, M\psi_k'}{\gamma'} \right\}, W).$$

Proof: In both (1) and (2) one can prove that for every $S \subseteq \mathcal{L}$, both default theories $S$-prove the same formulas. Consequently, both theories have precisely the same extensions. ■

**Theorem 5.4** *Let*

$$d_0 = \frac{\varphi\colon M\beta_1,\ldots,M\beta_k}{\gamma_1 \wedge \gamma_2},\quad d_1 = \frac{\varphi\colon M\beta_1,\ldots,M\beta_k}{\gamma_1} \quad \textit{and}$$

$$d_2 = \frac{\varphi\colon M\beta_1,\ldots,M\beta_k}{\gamma_2}.$$

*Then, for every set of defaults $D$ and for every theory $W \subseteq \mathcal{L}$, $(D \cup \{d_0\}, W) \approx (D \cup \{d_1, d_2\}, W)$.*

Proof: Since all three defaults have the same prerequisite and the same set of justifications, it follows that for every theory $S \subseteq \mathcal{L}$,

$$Cn^{D\cup\{d_0\},S}(W) = Cn^{D\cup\{d_1,d_2\},S}(W).$$

Consequently, the assertion follows. ∎

**Theorem 5.5** *Let*

$$d_0 = \frac{\varphi\colon M(\delta_1 \vee \delta_2),\ldots,M\beta_k}{\gamma},\quad d_1 = \frac{\varphi\colon M\delta_1,\ldots,M\beta_k}{\gamma} \quad \textit{and}$$

$$d_2 = \frac{\varphi\colon M\delta_2,\ldots,M\beta_k}{\gamma}.$$

*Then, for every set of rules $D$ and every $W \subseteq \mathcal{L}$, $(D \cup \{d_0\}, W) \approx (D \cup \{d_1, d_2\}, W)$.*

Proof: Let $S$ be a theory closed under propositional consequence. Then, $\neg(\delta_1 \vee \delta_2) \notin S$ if and only if $\neg\delta_1 \notin S$ or $\neg\delta_2 \notin S$. Consequently, $d_0$ is applicable with respect to $S$ if and only if at least one of $d_1$ and $d_2$ is. Since all three rules $d_0$, $d_1$ and $d_2$ have the same prerequisite and the same consequent it follows that a formula $\psi$ has an $S$-proof from $(D \cup \{d_0\}, W)$ if and only if $\psi$ has an $S$-proof from $(D \cup \{d_1, d_2\}, W)$. Hence, as in the proof of the previous result,

$$Cn^{D\cup\{d_0\},S}(W) = Cn^{D\cup\{d_1,d_2\},S}(W)$$

and the result follows. ∎

We use the notion of equivalence to define when a default theory is representable by a class of default theories.

**Definition 5.6** Let $\mathcal{R}$ be a class of default theories. We say that a default theory $(D, W)$ is *representable* in $\mathcal{R}$ if there is a default theory $(D', W') \in \mathcal{R}$ such that $(D, W)$ is equivalent to $(D', W')$.

A class $\mathcal{Q}$ of default theories is *representable* in a class $\mathcal{R}$ of default theories if if every default theory in $\mathcal{Q}$ is representable in $\mathcal{R}$. Theories in $\mathcal{R}$ are *normal forms* for theories in $\mathcal{Q}$. ▲

We will first prove a negative result. Let $\mathcal{N}$ denote the class of all normal default theories. Normal default theories are syntactically simpler than general default theories and have several useful properties. For example, every normal default theory has at least one extension, which is not always the case for arbitrary default theories. In addition, let us recall that different extensions of normal theories not only are inclusion-incompatible, as in the case of general default theories, but their union is inconsistent. We will use this last property to show that the class of all default theories possessing extensions is *not* representable by the class $\mathcal{N}$ of normal default theories.

**Theorem 5.7** *The class $\mathcal{N}$ of all normal default theories does not represent the class of all default theories possessing extensions.*

Proof: Let $p, q \in At$, $p \neq q$. Consider the default theory $(D, W)$, where $D = \{\frac{:M\neg p}{q}, \frac{:M\neg q}{p}\}$ and $W = \emptyset$. Theory $(D, W)$ possesses two extensions $Cn(\{p\})$ and $Cn(\{q\})$. Since their union is consistent, there is no normal default theory with theories $Cn(\{p\})$ and $Cn(\{q\})$ as its extensions (see also Theorem 4.8). ■

In several cases, the two properties mentioned above, existence of extensions and inconsistency of the union of two different extensions, are essentially the only two features distinguishing normal default theories from unrestricted ones. We will make this statement formal in our next two results.

**Definition 5.8** A theory $T$ is *finitely generated* if there exists a finite set of formulas $\{\varphi_1, \ldots, \varphi_k\}$, called a set of *generators*, such that $Cn(T) = Cn(\{\varphi_1, \ldots, \varphi_k\})$. ▲

It turns out that collections of finitely generated, consistent and pairwise inconsistent theories can be represented as families of extensions of normal default theories.

**Theorem 5.9** *Let $\{T_\alpha : \alpha < \gamma\}$ be a nonempty collection of finitely generated, consistent, pairwise inconsistent theories ($\gamma$ is an ordinal, possibly infinite). Then there exists a normal default theory $(D, W)$ such that $T_\alpha$, $\alpha < \gamma$, are precisely the extensions of $(D, W)$.*

Proof: Let $\{\varphi_{\alpha,0}, \ldots, \varphi_{\alpha,k_\alpha}\}$ be a set of generators of the theory $T_\alpha$. Let $\psi_\alpha = \varphi_{\alpha,0} \wedge \ldots \wedge \varphi_{\alpha,k_\alpha}$. Then $T_\alpha = Cn(\{\psi_\alpha\})$. By the assumption, for $\alpha \neq \beta$, $T_\alpha \cup T_\beta$ is inconsistent, that is, $\{\psi_\alpha \wedge \psi_\beta\}$ is inconsistent.

For every $\alpha < \gamma$ define

$$d_\alpha = \frac{: M\psi_\alpha}{\psi_\alpha}.$$

Let $(D, W)$ be a normal default theory, where $D = \{d_\alpha : \alpha < \gamma\}$ and $W = \emptyset$. We claim that $\{T_\alpha : \alpha < \gamma\}$ is precisely the family of extensions of $(D, W)$. By Theorem 3.57, the only candidates for extensions of $(D, W)$ are of the form $T = Cn(\{\psi_{\alpha_\xi} : \xi < \eta\})$. If $\eta > 1$, that is, if $T$ is generated by more than one generator, then such a theory $T$ is inconsistent. Since extensions of normal default theories with nonempty set of defaults and empty $W$ are consistent, it follows that if $T$ is an extension for $(D, W)$, then $T = Cn(\{\psi_\alpha\}) = T_\alpha$, for some $\alpha < \gamma$.

Consider now a theory $T_\alpha$, $\alpha < \gamma$. Let $\beta < \gamma$ be such that $\beta \neq \alpha$. Then, $\psi_\alpha$ and $\psi_\beta$ are inconsistent. Hence, $\neg\psi_\beta \in T_\alpha$. Therefore for every $\alpha < \gamma$, the only rule applicable with respect to $T_\alpha$ is $d_\alpha$. That is, $D_{T_\alpha} = \{\frac{:}{\psi_\alpha}\}$. Hence,

$$Cn^{D, T_\alpha}(W) = Cn(\{\psi_\alpha\}) = T_\alpha.$$

Thus, $T_\alpha$ is an extension for $(D, W)$. ■

**Corollary 5.10** *If a default theory $(D, W)$ has at least one extension and the extensions for $(D, W)$ are finitely generated and pairwise inconsistent, then $(D, W)$ is representable in $\mathcal{N}$.*

Proof: If $\mathcal{L}$ is an extension for $(D, W)$ then it is the only extension for $(D, W)$. Define $D' = \emptyset$ and let $W'$ be an arbitrary inconsistent theory. Clearly, $(D', W')$ is a normal default theory and its only extension is $\mathcal{L}$. Hence, assume that all extensions of $(D, W)$ are consistent. In this case, the assertion follows directly from Theorem 5.9. ■

If we assume that a default theory $(D, W)$ has at least one and at most countably many pairwise inconsistent extensions then the assertion of Corollary 5.10 holds even without the assumption that extensions of $(D, W)$ are finitely generated. A default theory in the language $\mathcal{L}_{At}$ with $|At| = \omega$ can have a finite number, denumerably many or $2^\omega$ extensions. That is, metamathematical assumptions on the size of the continuum have no influence on the number of extensions. In particular, it is easy to construct a default theory whose family of extensions has cardinality continuum. Here is an example. Consider $W = \emptyset$ and $D = \{\frac{:Mp}{p} : p \in At\} \cup \{\frac{:M\neg p}{\neg p} : p \in At\}$. Then it is easy to see that the extensions of $(D, W)$ are precisely the complete theories in $\mathcal{L}_{At}$. There are precisely $2^\omega$ of them. We still do not know whether the assertion of Corollary 5.10 holds without the assumption that the extensions are finitely generated.

The representability result of Corollary 5.10 is relatively weak. Given a default theory $(D, W)$ satisfying its assumptions, in order to construct the corresponding normal default theory we have to find the extensions of $(D, W)$ first. Ideally, a representability result should provide a syntactic method of converting a given default theory into its "normal" form without the need to compute its extensions first. We will now present several such representability results.

For a theory $W \subseteq \mathcal{L}$ define

$$D(W) = \left\{ \frac{:}{\varphi} : \varphi \in W \right\}$$

That is, for every formula $\varphi \in W$ we include in $D(W)$ the default $\frac{:}{\varphi}$ which has no prerequisite and no justifications and, from the point of view of $S$-provability, is equivalent to $\varphi$. It is easy to see that for every $S \subseteq \mathcal{L}$,

$$Cn^{D_S}(W) = Cn^{(D \cup D(W))_S}(\emptyset).$$

Hence, we obtain the following two straightforward results.

**Theorem 5.11** *Let $S \subseteq \mathcal{L}$. Then $S$ is an extension of $(D, W)$ if and only if $S$ is an extension of $(D \cup D(W), \emptyset)$.* ▲

**Corollary 5.12** *Every default theory is representable in the class of default theories $(D, W)$ with $W = \emptyset$.* ▲

Corollary 5.12 is quite useful in technical considerations about default logic. In practical applications its role is, limited, however. The reason is that we want to separate an inference mechanism (defaults), which does not change or changes rarely, from the database of facts from which we reason (theory $W$), which changes frequently. This separation is easily expressible in the language of arbitrary default theories but vanishes completely when only theories with $W = \emptyset$ are allowed.

We will now prove a much deeper representability result. Recall that every propositional theory $S$ has precisely the same consequences as the theory $S' = Cn(S) \cap Clause$, that is, the theory consisting of those clauses which are provable from $S$. Hence, every propositional theory is faithfully represented by a set of clauses. We will now derive an analogous result for default logic.

**Definition 5.13** (Clausal defaults and default theories)

1. A default $d = \frac{\varphi:M\beta_1,\ldots,M\beta_k}{\gamma}$ is called *clausal* if:

   (a) $\varphi$ is a conjunction of clauses.

   (b) Each $\beta_j$ is the negation of a clause.

   (c) $\gamma$ is a clause.

2. A default theory $(D, W)$ is *clausal* if $W$ consists of clauses, and every rule in $D$ is clausal. ▲

We will show that every default theory is representable by a clausal default theory.

**Theorem 5.14** *For every default theory $(D, W)$ there exists a clausal default theory $(D', W')$ such that $(D, W) \approx (D', W')$.*

Proof: First observe that by Theorem 5.3 it suffices to take for $W'$ the set of clauses provable from $W$.

To construct a desired set of defaults $D'$ we first enforce the condition (a) of the definition of clausal defaults. Consider a default rule $d = \frac{\varphi:M\beta_1,\ldots,M\beta_k}{\gamma}$ in $D$. Let $\varphi'$ be a conjunctive normal form of $\varphi$ and let $d' = \frac{\varphi':M\beta_1,\ldots,M\beta_k}{\gamma}$. A direct application of Theorem 5.3 shows that the resulting theory $(D_1, W') = (\{d' : d \in D\}, W')$ has precisely the same extensions as $(D, W')$.

Next, we enforce the condition (c). By Theorem 5.3, we can assume that every consequent $\gamma$ of a default rule in $D_1$ is in a conjunctive normal form. The repeated application of Theorem 5.4 yields a theory $(D_2, W)$ such that $(D_2, W) \approx (D_1, W)$ and the consequents of rules in $D_2$ are clauses.

Finally, to satisfy the requirement (b) we will use Theorems 2.20 and 5.3. They imply that without loss of generality we may assume that each $\beta_i$ is in a double disjunctive normal form, that is, $\beta_i$ is a disjunction of negated clauses. Now we repeatedly apply Theorem 5.5 to the defaults in $D_2$ and produce a set $D'$ of clausal defaults such that $(D', W') \approx (D_2, W')$. Since $(D_2, W') \approx (D_1, W') \approx (D, W') \approx (D, W)$ and since $\approx$ is transitive, $(D', W') \approx (D, W)$ follows. ■

The size of the set of defaults $D'$ can be very large relative to the size of $D$. Let us consider the last step of the construction. Assume that each justification $\beta_i$ of a default $d \in D_2$ is a disjunction of $s_i$ negated clauses. Then the default $d$ contributes $s_1 \times \ldots \times s_k$ clausal defaults to $D'$.

**Example 5.15** Let $W = \{p \wedge q\}$ and

$$D = \left\{ \frac{p : M(r \vee s), M(\neg u \vee w)}{q}, \frac{p \wedge q : Mr, M(w \vee t)}{s} \right\}.$$

Applying the method described in the proof of Theorem 5.14 we obtain $W' = \{p, q\}$, and

$$D' = \left\{ \frac{p: Mr, M\neg u}{q}, \frac{p: Mr, Mw}{q}, \frac{p: Ms, M\neg u}{q}, \frac{p: Ms, Mw}{q}, \frac{p \wedge q: Mr, Mw}{s}, \frac{p \wedge q: Mr, Mt}{s} \right\}.$$

Theory $(D', W')$ is a clausal default theory and has exactly the same extensions as $(D, W)$. ▲

## 5.2 Semi-representability of default theories

We have already seen that the class of normal default theories is not wide enough to represent arbitrary default theories. In this section we will generalize the notion of a normal default by introducing the concept of a semi-normal default. We will define the notions of a semi-normal default theory and a weak semi-normal default theory. We will show that when new atoms are added to the language $\mathcal{L}$ any default theory is representable by a weak semi-normal one.

**Definition 5.16** A default of the form

$$\frac{\alpha: M(\beta \wedge \gamma)}{\gamma} \quad \text{or} \quad \frac{: M(\beta \wedge \gamma)}{\gamma}$$

is called *semi-normal.* A default theory $(D, W)$ is called *semi-normal* if all defaults in $D$ are semi-normal. A default theory $(D, W)$ is called *weak semi-normal* if every default in $D$ is semi-normal or has no justifications at all. ▲

The idea behind semi-normal defaults is quite simple. They generalize normal defaults. Indeed, if $\beta$ is a tautology then a semi-normal default $\frac{\alpha:M(\beta\wedge\gamma)}{\gamma}$ (or $\frac{:M(\beta\wedge\gamma)}{\gamma}$) is equivalent to the normal default $\frac{\alpha:M\gamma}{\gamma}$ (or $\frac{:M\gamma}{\gamma}$). Informally, one might say that derivations by means of semi-normal defaults are weaker than those by normal defaults. Instead of deriving the justification $\beta \wedge \gamma$ from $\alpha$ and the consistency of $\beta \wedge \gamma$ we derive only $\gamma$. In general, this mode of reasoning does not guarantee the existence of extensions.

**Example 5.17** Let $W = \emptyset$ and

$$D = \left\{ \frac{: M(p \wedge \neg q)}{p}, \frac{: M(q \wedge \neg r)}{q}, \frac{: M(r \wedge \neg p)}{r} \right\},$$

where $p$, $q$ and $r$ are distinct atoms. Then $(D, W)$ is a semi-normal default theory without extensions. ▲

Weak semi-normal theories are very close to semi-normal. In addition to semi-normal defaults they contain defaults without any justifications, which can be regarded as standard, monotone inference rules, and are applicable with respect to any context.

An arbitrary default theory can be represented by a weak semi-normal default theory provided the language $\mathcal{L}$ has been extended by new atoms. A precise description of the notion of representability used in this section is given below.

**Definition 5.18** A default theory $(D, W)$ is *semi-representable* in the class of default theories $\mathcal{R}$ if there is an extension $\mathcal{L}'$ of the language $\mathcal{L}$ and a theory $(D', W') \in \mathcal{R}$ such that $(D', W')$ is a default theory in the language $\mathcal{L}'$, and for every theory $S \subseteq \mathcal{L}$, $S$ is an extension of $(D, W)$ if and only if there is an extension $S'$ of $(D', W')$ such that $S = S' \cap \mathcal{L}$. ▲

**Theorem 5.19** *Every default theory is semi-representable in the class of weak semi-normal default theories. Moreover, one can require a one-to-one correspondence between extensions of the corresponding default theories.*

Proof: Let $(D, W)$ be a default theory. Let us suppose that every default $d \in D$ is of the form

$$d = \frac{\alpha^d : M\beta_1^d, \ldots, M\beta_{k_d}^d}{\gamma^d}.$$

For every such default $d$ let us add to $\mathcal{L}$ new atoms $c_1^d, \ldots, c_{k_d}^d$. Denote the resulting language by $\mathcal{L}'$.

For every default $d$ define $k_d + 1$ new rules. The first $k_d$ rules are of the form:

$$r_{d,i} = \frac{\alpha^d : M(\beta_i^d \wedge c_i^d)}{c_i^d},$$

$i = 1, \ldots, k_d$. In the case where $d$ has at least one justification, the last rule is of the form

$$s_d = \frac{c_1^d \wedge \ldots \wedge c_{k_d}^d :}{\gamma^d}.$$

Otherwise, $s_d = d$ (and there are no rules of the form $r_{d,i}$ at all). Next, define

$$R_d = \{r_{d,1}, \ldots, r_{d,k_d}, s_d\}$$

and, finally,

$$D' = \bigcup_{d \in D} R_d.$$

Clearly, $(D', W)$ is a weak seminormal theory.

We will prove that:

**(a)** For every extension $S$ of $(D, W)$ there exists an extension $S'$ of $(D', W)$ such that $S = S' \cap \mathcal{L}$.

**(b)** For every extension $S'$ of $(D', W)$, the theory $S' \cap \mathcal{L}$ is an extension of $(D, W)$.

**(c)** If $S', S''$ are extensions of $(D', W)$ and $S' \cap \mathcal{L} = S'' \cap \mathcal{L}$ then $S' = S''$.

These three facts together imply our theorem.

(a) For every default $d \in D$ define

$$C_d = \begin{cases} \emptyset & \text{if } S \not\vdash \alpha^d, \\ \{c_i^d : \neg\beta_i^d \notin S\} & \text{if } S \vdash \alpha^d. \end{cases}$$

Then define

$$C = \bigcup_{d \in D} C_d$$

and

$$S' = Cn(S \cup C).$$

Since $S$ and $S'$ are closed under propositional consequence, by Theorem 2.24 it follows that $S = S' \cap \mathcal{L}$. To complete our proof of (a) we need to show that $S'$ is an extension for $(D', W)$.

To this end we will prove several auxiliary facts. First, we will show that if a formula $\varphi \in \mathcal{L}$ has an $S$-proof from $W$ in $PC + D$ then $\varphi$ has an $S'$-proof from $W$ in $PC + D'$. We proceed by induction on the length of an $S$-proof of $\varphi$. Consider a formula $\varphi$ with an $S$-proof $P$ from $W$ in $PC + D$ of length $n$ and assume that the claim holds for every formula with an $S$-proof from $W$ in $PC + D$ of length less than $n$. If $\varphi$ is a tautology or belongs to $W$ then, clearly, $\varphi$ has an $S'$-proof from $W$ in $PC + D'$. If $\varphi$ is derived in $P$ by means of modus ponens applied to formulas $\psi$ and $\psi \supset \varphi$ appearing in $P$ before $\varphi$ then, by the induction hypothesis, $\psi$ and $\psi \supset \varphi$ are $S'$-provable from $W$ in $PC + D'$. Consequently, $\varphi$ is $S'$-provable from $W$ in $PC + D'$. The last possibility is that $\varphi$ is derived in $P$ by means of a default

$$d = \frac{\alpha^d : M\beta_1^d, \ldots, M\beta_{k_d}^d}{\gamma^d},$$

where $\gamma^d = \varphi$. It follows that $\alpha^d$ has a shorter $S$-proof from $W$ in $PC + D$ that $\varphi$. Consequently, $\alpha^d$ is $S'$-provable from $W$ in $PC + D'$. If $d$ has no justifications, then $d \in D'$ and $\varphi$ is $S'$-provable from $W$ in $PC + D'$. Hence, assume that $d$ has at least one justification. Then, for every justification $\beta_i^d$ of $d$, $\neg\beta_i^d \notin S$. By the definition of $C$ it follows that $\neg(\beta_i^d \wedge c_i^d) \notin S'$, for every justification $\beta_i^d$ of $d$. Consequently, for every $i$, $1 \leq i \leq k_d$, the atom $c_i^d$ is $S'$-provable from $W$ in $PC + D'$. Hence, we can use the default $s_d$ to obtain that $\varphi$ is $S'$-provable from $W$ in $PC + D'$.

In a similar way, also by induction on the length of a proof, one can show that if $\varphi \in \mathcal{L}$ and $\varphi$ has an $S'$-proof from $W$ in $PC + D'$ then $\varphi$ has an $S$-proof from $W$ in $PC + D$. We omit the details.

In the next step we will show that every atom $c_i^d \in C$ has an $S'$-proof from $W$ in $PC + D'$. By the definition of $C$ it follows that $S \vdash \alpha^d$. Consequently, since $S$ is an extension for $(D, W)$, $\alpha^d$ is $S$-provable from $W$ in $PC + D$. Hence, $\alpha^d$ is $S'$-provable from $W$ in $PC + D'$. By the definition of $C$, it follows that $\neg\beta_i^d \notin S$. Since $S' = Cn(S \cup C)$ and $C$ consists of new atoms, $\neg\beta_i^d \notin S'$. Again by the form of $S'$, it follows that $\neg(\beta_i^d \wedge c_i^d) \notin S'$. Thus, $c_i^d$ is $S'$-provable from $W$ in $PC + D'$.

The properties we have just proved imply that

$$S' \subseteq Cn_{\mathcal{L}'}^{D',S'}(W).$$

To prove the converse inclusion we need to show that every formula which has an $S'$-proof from $W$ in $PC+D'$ belongs to $S'$. This can be achieved again by induction on the length of an $S'$-proof. The proof is similar to the one provided above and we omit the details.

Summarizing, we obtain that

$$S' = Cn_{\mathcal{L}'}^{D',S'}(W).$$

Consequently, $S'$ is an extension for $(D', W)$.

(b) Consider the collection $U$ of all defaults occurring in $S'$-proofs from $W$ in $PC + D'$. These defaults are of three types:

**(i)** $\frac{\alpha^d :}{\gamma^d}$, where $\alpha^d, \gamma^d \in \mathcal{L}$.

**(ii)** $\frac{c_1^d \wedge \ldots \wedge c_k^d :}{\gamma^d}$, where $\gamma^d \in \mathcal{L}$ and $c_i^d$ are new atoms.

**(iii)** $\frac{\alpha^d : M(\beta_i^d \wedge c_i^d)}{c_i^d}$, where $\alpha^d, \beta_i^d \in \mathcal{L}$ and $c_i^d$ is a new atom.

It is now clear that a formula $\varphi \in \mathcal{L}$ is the consequent of a default in $U$ if and only if $\varphi = \gamma^d$ for some default $d \in D$ satisfying one of the following two conditions:

- $d$ is justification-free and $\alpha^d \in S' \cap \mathcal{L}$.

- $d$ has at least one justification and the atoms $c_1^d, \ldots, c_k^d$ all belong to $S'$.

The latter possibility occurs precisely when each of $\beta_1^d, \ldots, \beta_k^d$ is consistent with $S'$. Since all $\beta_i^d$ belong to $\mathcal{L}$, the consistency of $\beta_i^d$ with $S'$ is equivalent to the consistency of $\beta_i^d$ with $S$. Thus, we obtain that all the consequents of the defaults from $U$ are $S$-provable from $W$ in $PC + D$. Hence, it follows that $S = S' \cap \mathcal{L} \subseteq Cn_{\mathcal{L}}^{D,S}(W)$. The converse inclusion can be proved by a simple induction on the length of an $S$-proof in a way similar to that used in (a). Consequently, $S$ is an extension for $(D, W)$.

(c) Let $S'$ and $S''$ be extensions for $(D', W)$ such that $S' \cap \mathcal{L} = S'' \cap \mathcal{L}$. Denote $S' \cap \mathcal{L}$ by $S$. Both $S'$ and $S''$ are generated by some formulas from $\mathcal{L}$ and by some new atoms $c_i^d$. Since $S'$ and $S''$ coincide on $\mathcal{L}$, to prove that $S' = S''$ we need to show that they contain precisely the same new atoms. To this end, let us consider a new atom $c_i^d \in S'$. The only way to $S'$-prove $c_i^d$ is by means of the default $r_{d,i} = \frac{\alpha^d : M(\beta_i^d \wedge c_i^d)}{c_i^d}$. Thus, it follows that $\alpha^d \in S'$ and $\neg\beta_i^d \vee \neg c_i^d \notin S'$. Consequently, $\alpha^d \in S''$. In addition, since $\beta_i^d \in \mathcal{L}$, $\neg\beta_i^d \notin S''$. Clearly, $S'' = Cn(S \cup A)$, for some set $A$ of new atoms. Since $\neg\beta_i^d \in \mathcal{L}$, it follows that $\neg(\beta_i^d \wedge c_i^d) \notin S''$. Thus, $c_i^d \in S''$ (since $S''$ is closed under defaults in $D'$). Similarly, every new atom contained in $S''$ belongs to $S'$, too. Hence, (c) follows. ■

Let us call a default *unitary* if it has at most one justification. A default theory is *unitary* if all its defaults are unitary. The following result is an easy consequence of Theorem 5.19.

**Corollary 5.20** *Every default theory is semi-representable in the class of unitary default theories.* ▲

Next, we will briefly mention another semi-representability result. We will consider the class of normal default theories with all defaults prerequisite-free. To represent prerequisite-free normal theories we will use a subclass of the class of all normal default theories. A prerequisite-free normal default is called *simple* if it is of the form

$$\frac{: M\neg p}{\neg p},$$

where $p$ is an atom. A normal default theory is called *simple* if all its defaults are simple. We will show that every normal default theory with all defaults prerequisite-free is semi-representable by a simple normal default theory.

We have already encountered simple normal default theories earlier. In Section 4.3 we showed that the Closed World Assumption and its variants can be studied by means of such theories. Some of the results from Section 4.3 will be used here to derive properties of prerequisite-free normal default theories.

Consider a prerequisite-free normal default theory $(D, W)$. Extend the language $\mathcal{L}$ by adding for each default $d = \frac{:M\psi}{\psi}$ in $D$ a new atom $c_d$. Denote the new language $\mathcal{L}'$. Next, for each $d \in D$, define $\varphi_d = (\psi \equiv \neg c_d)$. Finally, define

$$W' = W \cup \{\varphi_d : d \in D\} \text{ and } D' = \left\{\frac{:M\neg c_d}{\neg c_d} : d \in D\right\}.$$

Clearly, $(D', W')$ is a simple normal default theory. It is also easy to see that

$$D' = D^{CWA^P},$$

where $P = \{c_d : d \in D\}$ (see Section 4.3).

We have the following result.

**Theorem 5.21**

1. *For every extension $S$ of $(D, W)$ there is a unique extension $S'$ of $(D', W')$ such that $S = S' \cap \mathcal{L}$.*
2. *Conversely, if $S'$ is an extension of $(D', W')$ then $S' \cap \mathcal{L}$ is an extension of $(D, W)$.* ▲

As a corollary to Theorem 5.21 we get the following result.

**Corollary 5.22** *Every prerequisite-free normal default theory is semi-representable in the class of simple normal default theories.* ▲

## 5.3 A representation result for weak extensions

In this section we discuss the issue of representability of collections of theories as families of weak extensions of default theories. We will show that every finite collection of finitely generated theories closed under propositional consequence is the family of weak extensions of a suitably chosen default theory.

**Theorem 5.23** *Let $T_1, \ldots, T_k$ be pairwise distinct, finitely generated and closed under propositional consequence theories contained in $\mathcal{L}$. Then, there exists a finite default theory $(D, W)$ such that $\{T_1, \ldots, T_k\}$ is precisely the family of weak extensions of $(D, W)$.*

Proof: We can assume that for every $i$, $1 \leq i \leq k$, $T_i = Cn(\{\varphi_i\})$, where $\varphi_i$ is a formula from $\mathcal{L}$. In addition, for all $i, j$ such that $i \neq j$, the formula $\varphi_i \equiv \varphi_j$ is not a tautology. Define a preordering (that is a reflexive and transitive relation) in the set $\{\varphi_i : i \leq k\}$ as follows:

$$\varphi_j \prec \varphi_i \text{ if and only if } \varphi_i \supset \varphi_j \text{ is a tautology.}$$

Clearly, the deduction theorem (Theorem 2.29) implies that $\varphi_j \prec \varphi_i$ if and only if $T_j \subseteq T_i$.

Define $W = \emptyset$. To define a set $D$ of defaults let us define first $Z_i = \{\varphi_j : (\varphi_j \not\prec \varphi_i)\}$. Thus, $\varphi_j \in Z_i$ if and only if $\varphi_i \supset \varphi_j$ is not a tautology. For each $i$, $1 \leq i \leq k$, define a default $d_i$ as follows. Let $Z_i = \{\beta_1, \ldots, \beta_n\}$. Then:

**(i)** If $\varphi_i$ is minimal in the preordering $\prec$, that is, if for all $j \neq i$, $\varphi_i \supset \varphi_j$ is not a tautology, then set

$$d_i = \frac{: M\neg\beta_1, \ldots, M\neg\beta_n}{\varphi_i}.$$

**(ii)** If $\varphi_i$ is not minimal in the preordering $\prec$ then set

$$d_i = \frac{\varphi_i : M\neg\beta_1, \ldots, M\neg\beta_n}{\varphi_i}.$$

Finally, define $D = \{d_i : 1 \leq i \leq k\}$.

Clearly, $(D, W)$ is a finite default theory. First, we claim that every $T_i$ is a weak extension of $(D, W)$. Indeed, one can verify that $d_i$ is the only applicable rule with respect to $T_i$. Since the consequent of $d_i$ is $\varphi_i$ and $T_i = Cn(\{\varphi_i\})$, $T_i$ is a weak extension for $(D, W)$.

Next, let us consider a weak extension $T$ of $(D, W)$. Then $T$ is generated by $W$ and the consequents of the generating defaults. Let $d_i \in GD(D, T)$.

Case 1. $d_i = \frac{:M\neg\beta_1,\ldots,M\neg\beta_n}{\varphi_i}$. In this case, $Z_i = \{\varphi_j : 1 \leq j \leq k, \text{ and } j \neq i\}$. Since $d_i \in GD(D, T)$, for every $j$ such that $j \neq i$, $\varphi_j \notin T$. Since $\varphi_i \in T$, $T = Cn(\{\varphi_i\})$.

Case 2. $d_i = \frac{\varphi_i : M\neg\beta_1, \ldots, M\neg\beta_n}{\varphi_i}$. In this case, $Z_i = \{\varphi_j : \varphi_i \supset \varphi_j$ is not a tautology$\}$. Hence, all these formulas $\varphi_j$ such that $\varphi_i \supset \varphi_j$ is not a tautology are not in $T$. All other formulas $\varphi_j$ are provable from $\varphi_i$ in the propositional calculus. Hence, $T = Cn(\{\varphi_i\})$. ■

## 5.4 Comments and remarks

Most of the results in this chapter has not been reported in the literature, although some of them are folklore. Theorems 5.7 and 5.9 must have been known to Reiter. The notion of semi-normal default rule has been introduced by Reiter and Etherington in [ER83]. Books by Besnard [Bes89] and Etherington [Eth88] contain more information on semi-normal default theories. Theorem 5.19 (although not its proof) is due to Yang, Blair and Brown [YBB92]. Theorem 5.23 is due to Zhang [Zha90].

# 6. Logic programming and nonmonotonic reasoning

For several years default logic and logic programming have been developed by disjoint groups of researchers and there have been few interactions between these two fields. Only quite recently, in the late 1980s, it was realized that the way in which logic programming interprets negation leads to nonmonotonic systems. Subsequently, connections between "negation as failure to prove" principle of logic programming and default and autoepistemic logics were established.

In the present chapter we discuss this fragment of logic programming which is directly related to default logic. We focus on the concepts of stable and supported models as they have very clear and natural default logic interpretations. We show that these notions can be naturally defined by means of extensions and weak extensions of some default theories.

We discuss two generalizations of logic programming: logic programming with classical negation and clausal logic programming. We show how the default logic interpretation of programs allows us to extend the notion of a stable model to these generalizations.

We also introduce and discuss the notions of stratification, local stratification and perfect model. We show that each locally stratified logic program has a unique perfect model and a unique stable model, and that these models coincide.

## 6.1 Logic programs — basic definitions

The syntax of logic programs is based on the language $\mathcal{L}_{pred}$ of predicate calculus, discussed earlier in Section 2.6. Throughout this chapter the notions of the Herbrand base and a Herbrand model are of key importance. Let us recall that the Herbrand base for the language $\mathcal{L}_{pred}$ is denoted by $B_H$. By $\mathcal{L}_H$

let us denote the language of propositional logic generated by the elements of $B_H$. Finally, by $Lit_H$ we will denote the literals of $\mathcal{L}_H$, that is, the elements of the Herbrand base and their negations.

In Section 2.6 we showed that a Herbrand interpretation $\mathcal{M}$ is uniquely determined by a subset of a Herbrand base consisting of those ground atomic formulas which are true in $\mathcal{M}$. Consequently, we have adopted the convention that whenever we talk about Herbrand interpretations or models, we have in mind the corresponding subset of the Herbrand base. Slightly abusing the notation, we will use the same symbol to denote both a subset of the Herbrand base and the corresponding Herbrand interpretation.

Clauses of logic programs are built of formulas of the language $\mathcal{L}_{pred}$ (usually, of atomic formulas but we will investigate here other possibilities, too). In addition, three more symbols are needed: ',', **not** and $\leftarrow$.

**Definition 6.1** A *logic program clause* is an expression of the form

$$p \leftarrow q_1, \ldots, q_m, \mathbf{not}(r_1), \ldots, \mathbf{not}(r_n), \tag{6.1}$$

where $p$, $q_i$ and $r_i$ are atomic formulas of $\mathcal{L}_{pred}$ different from $\bot, \top$. The expressions $q_1, \ldots, q_m$ and $\mathbf{not}(r_1), \ldots, \mathbf{not}(r_n)$ form the *body* of the clause and $p$ is the *head* of the clause. If $p$, all $q_i$ and all $r_i$ are ground atomic formulas then the clause is called *ground*. A *logic program* is a set of logic program clauses. ▲

The intuitive meaning of the logic program clause (6.1) is: *if all $q_i$ are true (derived) and none of $r_i$ is true (derivable), then $p$ is true (can be derived)*. There are many ways in which this intuition can be made precise. Some of them will be discussed in this chapter.

An important element of Definition 6.1 is that it uses new operators **not**, $\leftarrow$ and ','. The operator ',' is the least controversial among them. It is commonly interpreted as the conjunction operator. The operator **not** is meant to serve as a form of negation. One way to interpret **not** is to treat it simply as the standard negation operator from $\mathcal{L}_{pred}$. What distinguishes logic programming from classical logic is that **not** is often given other interpretations. Similarly, it is clear that $\leftarrow$ has some flavor of the implication: "*if* all the elements in the body, *then* the head". However, other interpretations of $\leftarrow$ are also possible. For example, we may treat it as an operator which defines an inference rule *if all $q_i$ and all* $\mathbf{not}(r_i)$ *are established, then conclude $p$* or, in other words, *to establish $p$, it is enough to establish all $q_i$ and all* $\mathbf{not}(r_i)$.

In this chapter we will focus our attention on two different interpretations of logic programs. In the first, the *declarative* interpretation, a logic program is regarded as a theory in $\mathcal{L}_{pred}$ and some classes of models of this theory are used to assign a meaning to the program. Our second approach, more *procedural* in nature, is to regard logic program clauses as inference rules, more precisely, as defaults. Such concepts as extensions or weak extensions, associated with the default theory corresponding to a program, are used to provide its meaning.

**Definition 6.2** The *predicate-calculus interpretation* of a clause (6.1) is the formula

$$(\forall x_1, \ldots, x_k)(q_1 \wedge \ldots \wedge q_m \wedge \neg r_1 \wedge \ldots \wedge \neg r_n \supset p),$$

where $x_1, \ldots, x_k$ are all free variables occurring in $q_i$, $1 \leq i \leq m$, $r_i$, $1 \leq i \leq n$, and $p$. The *predicate-calculus interpretation* of a logic program $P$ is the theory consisting of predicate-calculus interpretations of all clauses in $P$. The predicate-calculus interpretation of $P$ is denoted by $pc(P)$. ▲

Notice that the predicate-calculus interpretation of a logic program is a satisfiable universal theory.

The predicate-calculus interpretation of logic programs treats **not** as the standard negation, $\leftarrow$ as the implication, and assigns to a logic program a universal theory. We will use now the predicate-calculus interpretation of a logic program to define the notion of a model of a program $P$.

**Definition 6.3** Let $P$ be a logic program. Every model of the theory $pc(P)$ is called a *model* of the program $P$. Every Herbrand model of $pc(P)$ is called a *Herbrand model* of $P$. ▲

We have already observed that the theory $pc(P)$ is universal. Hence, Theorem 2.56 and Corollary 2.57 apply. Since the atom $\perp$ cannot appear in the head of any clause of a program, the logical interpretation of every program is consistent and we have the following simple result.

**Theorem 6.4** *Every logic program possesses a minimal Herbrand model.* ▲

To define the default interpretation of a program, we proceed in two steps. In the first step, we replace the program by its propositional grounding. In the second step, the grounding of the program is assigned a default theory.

**Definition 6.5** The *propositional grounding* of a logic program $P$, denoted *ground*($P$), is obtained from $P$ by replacing each clause (6.1) of $P$ by the set of all propositional clauses

$$p\Theta \leftarrow q_1\Theta, \ldots, q_m\Theta, \mathbf{not}(r_1\Theta), \ldots, \mathbf{not}(r_n\Theta),$$

where $\Theta$ ranges over the set of all ground substitutions whose domains include all variables occurring in the atomic formulas $p$, $q_i$ and $r_i$. ▲

It follows directly from the definition that if $P$ consists only of clauses without variables, then $ground(P)$ coincides with $P$.

The fundamental property of the propositional grounding of $P$ is that its propositional models are precisely the Herbrand models of $P$.

**Theorem 6.6** *Let $P$ be a logic program and let $M$ be a subset of the Herbrand base $B_H$. Then $M$ is a Herbrand model of $P$ if and only if $M$ is a propositional model of $pc(ground(P))$.*

Proof: By definition, $M \subseteq B_H$ is a Herbrand model of a program $P$ if and only if $M$ is a Herbrand model of the universal theory $pc(P)$. Since $ground(pc(P)) = pc(ground(P))$, the assertion is a simple consequence of Theorem 2.54. ■

The program $ground(P)$ can be regarded as a logic program in the language of propositional logic generated by the ground atomic formulas of $\mathcal{L}_{pred}$. Let us consider a clause

$$p \leftarrow q_1, \ldots, q_m, \mathbf{not}(r_1), \ldots, \mathbf{not}(r_n), \tag{6.2}$$

where $p$, $q_i$ and $r_i$ stand for ground atomic formulas. We have already mentioned that this clause could be interpreted as an inference rule saying roughly that *if all the premises $q_i$, $1 \leq i \leq m$, and $\mathbf{not}(r_i)$, $1 \leq i \leq n$, are established, then $p$ should be concluded.* While it is clear what is meant by establishing $q_i$ — some kind of a proof of $q_i$ should be provided — it is much less clear how to establish a premise $\mathbf{not}(r_i)$. The intuitive meaning of **not** is negation. But how could we prove the negation of an atomic formula when the consequents of all the inference rules in a program are (positive) atoms? To find a way out, we treat $\mathbf{not}(r_i)$ as the *inability* to prove $r_i$ rather than the (classical) negation of $r_i$. Such a treatment of negation is commonly known under the general term of *negation as failure to prove.*

Now, the question is to make precise what it means to say that $r_i$ cannot be proved. One possibility, and the one we take up in this book, is to use the *default* interpretation of negation as failure to prove. That is, we interpret $\mathbf{not}(r_i)$ as "$\neg r_i$ is possible" or "$\neg r_i$ is consistent" and make it precise by means of default extensions. Formally, we have the following definition.

**Definition 6.7** A *default interpretation* of a ground clause $C$ of the form (6.2) is the default

$$dl(C) = \frac{q_1 \wedge \ldots \wedge q_m : M\neg r_1, \ldots, M\neg r_n}{p}.$$

By a *default interpretation* of a logic program $P$ we mean the default theory $(dl(P), \emptyset)$, where

$$dl(P) = \{dl(C) : C \in ground(P)\}.$$

▲

We have used the predicate-calculus interpretation of logic programs to define the notion of a model of a program. We will now use the default interpretation of logic programs to assign to them some other classes of objects. First, we will prove an auxiliary result. Here and throughout this chapter, $Cn$ stands for the propositional consequence operator in the propositional language $\mathcal{L}_H$ generated by the Herbrand base.

**Theorem 6.8** *Let $P$ be a logic program in the language $\mathcal{L}_{pred}$. Let $S \subseteq \mathcal{L}$ be a theory closed under propositional consequence and the defaults from $dl(P)$. Then the set of ground atomic formulas in $S$, $S \cap B_H$, is a Herbrand model of $P$. Conversely, if $M \subseteq B_H$ is a Herbrand model of $P$ then $Cn(M)$ is closed under propositional consequence and the defaults in $dl(P)$.*

Proof: First, assume that $S$ is closed under propositional consequence and the defaults in $dl(P)$. Define $M = S \cap B_H$. To show that $M$ is a Herbrand model of $P$ we need to show that $M$ is a propositional model of $pc(ground(P))$.

Let us consider a formula

$$q_1 \wedge \ldots \wedge q_m \wedge \neg r_1 \wedge \ldots \wedge \neg r_n \supset p \tag{6.3}$$

from $pc(ground(P))$. Then the default

$$\frac{q_1 \wedge \ldots \wedge q_m : M\neg r_1, \ldots, M\neg r_n}{p} \tag{6.4}$$

belongs to $dl(P)$. Assume that for every $i$, $1 \leq i \leq m$, $M \models q_i$ and that for every $i$, $1 \leq i \leq n$, $M \models \neg r_i$. Since all $q_i$ and all $r_i$ are atoms, $q_i \in M$, $1 \leq i \leq m$, and $r_i \notin M$, $1 \leq i \leq n$. Hence, $q_1 \wedge \ldots \wedge q_m \in Cn(M)$ and for every $i$, $1 \leq i \leq n$, $r_i \notin Cn(M)$. Since $Cn(M)$ is closed under the default (6.4), $p \in Cn(M)$. Consequently, $p \in M$, that is, $M \models p$ (since $p$ is an atom). Thus, also the formula (6.3) is true in $M$ and it follows that $M$ is a model of $pc(ground(P))$. The remaining part of the assertion can be proved in a similar fashion. ■

**Corollary 6.9** *Let $P$ be a logic program. For every subset $M$ of the Herbrand base $B_H$, $M$ is a minimal Herbrand model of $P$ if and only if $Cn(M)$ is a minimal set for the default theory $(dl(P), \emptyset)$.* ▲

The next two corollaries show that the sets of all ground atomic formulas contained in extensions and weak extensions of $(dl(P), \emptyset)$ are models of a program $P$.

**Corollary 6.10** *Let $P$ be a logic program in the language $\mathcal{L}_{pred}$. Let $S \subseteq \mathcal{L}_H$ be an extension of $(dl(P), \emptyset)$. Then the set of ground atomic formulas in $S$, $S \cap B_H$, is a Herbrand model of $P$.*

Proof: Clearly, $S$ is closed under propositional consequence. In addition, by Theorem 3.49, $S$ is closed under the defaults in $dl(P)$. Hence, the assertion follows from Theorem 6.8. ■

In the same manner, by using Theorem 3.85 instead of Corollary 3.49, one can prove a similar statement for weak extensions.

**Corollary 6.11** *Let $P$ be a logic program in the language $\mathcal{L}_{pred}$. Let $S \subseteq \mathcal{L}_H$ be a weak extension of $(dl(P), \emptyset)$. Then the set of ground atomic formulas in $S$, $S \cap B_H$, is a Herbrand model of $P$.* ▲

The classes of Herbrand models of a program $P$, determined by extensions and weak extensions of $(dl(P), \emptyset)$, respectively, play an important role in the theory of logic programs.

**Definition 6.12** Let $P$ be a logic program and let $M$ be a subset of the Herbrand base $B_H$.

1. The set $M$ is a *stable* model of $P$ if $Cn(M)$ is an extension of $(dl(P), \emptyset)$.

2. The set $M$ is a *supported* model of $P$ if $Cn(M)$ is a weak extension of $(dl(P), \emptyset)$. ▲

**Remark 6.13** Our definitions of stable and supported models are different from the standard ones. We have chosen to define these notion by means of extensions and weak extensions to make as explicit as possible the default logic connection. We will show that our definitions are equivalent to the standard ones in the following sections of this chapter. ▲

Clearly, $B_H$ is a Herbrand model of every logic program. On the other hand, there are logic programs without stable and supported models.

**Example 6.14** Consider the program $P = \{p \leftarrow \mathbf{not}(p)\}$, where $p$ stands for some ground atomic formula of the language. Clearly, $pc(P) = \{\neg p \supset p\}$ and it has Herbrand models. It is also easy to see that $dl(P) = \{\frac{:M\neg p}{p}\}$. Since the default theory $(dl(P), \emptyset)$ has neither extensions nor weak extensions, $P$ has no stable models and no supported models. ▲

Using properties of extensions and weak extensions we will now derive some simple properties of stable and supported models.

**Theorem 6.15** *Let $P$ be a logic program and let $M$ be a stable model of $P$. Then:*

1. *$M$ is a supported model of $P$.*
2. *$M$ is a minimal Herbrand model of $P$.*

Proof: By Theorem 3.82, every extension of a default theory is a weak extension of this default theory. Hence, (1) follows. Part (2) of the assertion follows from Corollary 6.9 and Theorem 3.92, which states that every extension of a default theory is a minimal set for that default theory. ■

Even a cursory analysis shows that the predicate calculus and default interpretations of logic programs are different. For instance, every logic program has Herbrand models and minimal Herbrand models, which might be used to define the meaning of the program. On the other hand, as Example 6.14 shows, there are logic programs without stable and supported models. Thus, while it is always the case that stable and supported models of a program (if they exist) are indeed models of the program, there are programs whose models are neither stable nor supported. In the next several sections of this chapter we will investigate the properties of models and minimal models of programs as well as the properties of stable and supported models. We will also discuss their mutual relationships.

## 6.2 Horn programs

In Chapter 2 we defined the notions of a Horn clause and a program Horn clause in propositional logic. The definition can naturally be extended to the case of predicate calculus: a *Horn clause* in predicate logic is a formula of the form

$$r_1(\vec{x_1}) \vee \ldots \vee r_n(\vec{x_n}),$$

such that for some $i$, $1 \leq i \leq n$, $r_i(\vec{x_i})$ is an atomic formula or the negation of an atomic formula and, for all $j \neq i$, $r_j(\vec{x_j})$ is the negation of an atomic formula. If for some $i$, $1 \leq i \leq n$, $r_i(\vec{x_i})$ is an atomic formula and, for all $j \neq i$, $r_j(\vec{x_j})$ is the negation of an atomic formula then such a Horn clause is called a *program Horn clause*.

A logic program clause will be called *Horn* if its predicate-calculus interpretation is logically equivalent to a program Horn clause. Formally, we have the following definition.

**Definition 6.16** A logic program clause

$$p(\vec{x}) \leftarrow q_1(\vec{y_1}), \ldots, q_m(\vec{y_m})$$

is called a *Horn clause*. A logic program consisting of Horn clauses is called a *Horn program*. ▲

We have already seen that every logic program has a minimal Herbrand model. In particular, every Horn program has a minimal Herbrand model. However, a much stronger statement holds. A crucial property of Horn programs is that each such program $P$ possesses a *least* Herbrand model, that is, there is a Herbrand model of $P$ contained in every other Herbrand model of $P$.

**Theorem 6.17** *Every Horn program $P$ has a least Herbrand model.*

Proof: Consider the intersection of the family of all Herbrand models of $P$ (that is, the subsets of the Herbrand base that encode these models). This family is nonempty — $B_H$ is a Herbrand model of $P$. Denote the intersection by $M$. By definition, $M$ is contained in every Herbrand model of $P$. Hence, to prove the assertion, it suffices to show that $M$ is a model of $P$ or, equivalently, that $M$ is a (propositional) model of $pc(ground(P))$.

Let us consider a clause

$$q_1 \wedge \ldots \wedge q_m \supset p \tag{6.5}$$

from $pc(ground(P))$. Assume that for every $i$, $1 \leq i \leq m$, $M \models q_i$ (that is, $q_i \in M$). Let $H \subseteq B_H$ be any Herbrand model of $P$. Then $M \subseteq H$ and, consequently, for every $i$, $1 \leq i \leq m$, $q_i \in H$. Since $H$ is a (propositional) model of $pc(ground(P))$, $p \in H$. Hence, $p$ belongs to every Herbrand model of $P$. By the definition of $M$, $p \in M$. Thus, $M \models p$ and, consequently, $M \models q_1 \wedge \ldots \wedge q_m \supset p$. ■

Let us now consider a default interpretation of a Horn program $P$. Clearly, all defaults in $dl(P)$ are justification-free. It follows from Theorem 3.23 that the theory $(dl(P), \emptyset)$ has a unique extension. That is, the program $P$ has a unique stable model. In addition, since every stable model of a program is also a minimal Herbrand model, it follows from Theorem 6.17 that the unique stable model of $P$ is exactly the least Herbrand model of $P$. Hence, we obtain the following result.

**Theorem 6.18** *Let $P$ be a Horn program. Then, $P$ has a unique stable model and it is the least Herbrand model of $P$.* ▲

Our results show that in the case of Horn programs it makes no difference whether we interpret their clauses as formulas in the language of predicate calculus or whether we treat them as inference rules.

Since every stable model is supported, from Theorem 6.18 we obtain the following corollary.

**Corollary 6.19** *Let $P$ be a Horn program. Then the least Herbrand model of $P$ is supported.* ▲

We have seen that each Horn program has a unique stable model. The situation is different for supported models. A Horn program may have several supported models. For example, the program $\{p \leftarrow p\}$, where $p$ is a ground atomic formula, has two supported models: $\{\emptyset\}$ and $\{p\}$. Finally, let us observe that the union of any chain (with respect to inclusion relation) of supported models of a Horn program $P$ is a supported model of $P$. Hence, every Horn program has a greatest supported model.

## 6.3 Stable models of general logic programs

In this section we will describe the original definition of stable models as introduced by Gelfond and Lifschitz. We will show that our definition of stable models is equivalent to the original one.

**Definition 6.20** Let $M$ be a subset of the Herbrand base $B_H$.

1. A clause
$$p \leftarrow q_1, \ldots, q_m, \mathbf{not}(r_1), \ldots, \mathbf{not}(r_n),$$
where $p$, $q_i$ and $r_i$ stand for ground atomic formulas, is *irrelevant* with respect to $M$ if at least one $r_i$ is in $M$.

2. Let $P$ be a logic program. The *reduct* of $P$ with respect to $M$, denoted by $P^M$, is obtained from *ground*$(P)$ by:

   (a) Removing all clauses that are irrelevant with respect to $M$.

   (b) Removing each premise $\mathbf{not}(r_i)$ from the remaining clauses. ▲

Clearly, the reduct $P^M$ consists of ground Horn clauses. Therefore, $P^M$ has a least Herbrand model. This property is the basis of the following definition of stable models due to Gelfond and Lifschitz.

**Definition 6.21** (Original definition of stable models; Gelfond and Lifschitz) Let $P$ be a logic program and let $M$ be a subset of the Herbrand base. The set $M$ is a stable model of $P$ if $M$ coincides with the least Herbrand model of the reduct $P^M$. ▲

We will now show the equivalence of both definitions. Speaking precisely, we will prove the following theorem.

**Theorem 6.22** *Let $P$ be a logic program. A subset $M$ of the Herbrand base is a stable model of $P$ (according to Definition 6.12) if and only if $M$ is the least Herbrand model of $P^M$ (that is, if $M$ is stable according to the definition of Gelfond and Lifschitz).*

Proof: Let $M$ be a subset of the Herbrand base. We will first prove that the reduct of the set of defaults $dl(P)$ with respect to the theory $Cn(M)$ is closely related to the set of defaults which correspond to the clauses in the reduct $P^M$. Speaking precisely, we will show that

$$dl(P)_{Cn(M)} = \left\{ \frac{q_1 \wedge \ldots \wedge q_m}{p} : \frac{q_1 \wedge \ldots \wedge q_m :}{p} \in dl(P^M) \right\}. \qquad (6.6)$$

Using the notation introduced just before Corollary 3.27, where for a collection $D$ of monotone inference rules *default*$(D)$ has been defined as the set of corresponding justification-free defaults, the identity (6.6) can be rewritten as

$$default(dl(P)_{Cn(M)}) = dl(P^M).$$

To prove (6.6), consider an inference rule

$$\frac{q_1 \wedge \ldots \wedge q_m}{p}$$

in $dl(P)_{Cn(M)}$. Then there is a default

$$\frac{q_1 \wedge \ldots \wedge q_m : M\neg r_1, \ldots, M\neg r_n}{p}$$

in $dl(P)$ such that $r_i \notin Cn(M)$, for every $i$, $1 \leq i \leq n$. Consequently, by the definition of $dl(P)$, the clause

$$p \leftarrow q_1, \ldots, q_m, \mathbf{not}(r_1), \ldots, \mathbf{not}(r_n), \tag{6.7}$$

belongs to $ground(P)$ and it satisfies $r_i \notin Cn(M)$, for every $i$, $1 \leq i \leq n$. Since each $r_i$ is a ground atomic formula and $M$ is a subset of the Herbrand base, $r_i \in Cn(M)$ if and only if $r_i \in M$. Hence, the clause (6.7) has the property that $r_i \notin M$, for every $i$, $1 \leq i \leq n$. Consequently, the Horn clause

$$p \leftarrow q_1, \ldots, q_m$$

belongs to $P^M$ and the default

$$\frac{q_1 \wedge \ldots \wedge q_m :}{p}$$

belongs to $dl(P^M)$. Thus, the left-hand side of (6.6) is contained in the right-hand side. The converse inclusion can be proved in a similar fashion.

We will now prove the assertion. Let us consider the following sequence of statements:

1. $M$ is a stable model of $P$.
2. $Cn(M)$ is an extension of $(dl(P), \emptyset)$.
3. $Cn(M)$ is an extension of $(default(dl(P)_{Cn(M)}), \emptyset)$.
4. $Cn(M)$ is an extension of $(dl(P^M), \emptyset)$.
5. $M$ is a stable model of $P^M$.
6. $M$ is the least Herbrand model of $P^M$.

Statements (1) and (2) are equivalent by the definition of stable models of programs. The equivalence of statements (2) and (3) follows from Theorem 3.27. The equivalence of statements (3) and (4) follows from (6.6). Next, statements (4) and (5) are also equivalent, again by the definition of stable models. Finally, statements (5) and (6) are equivalent by Theorem 6.18. Hence, the assertion follows. ■

## 6.4 Supported models of logic programs

Let us recall that a subset $M$ of the Herbrand base is a supported model of a logic program $P$ if $Cn(M)$ is a weak extension of the default theory $(dl(P), \emptyset)$. We will now present some more familiar characterizations of supported models.

**Theorem 6.23** *Let $P$ be a logic program. Let $M$ be a subset of the Herbrand base. Then $M$ is a supported model of $P$ if and only if $M$ is a model of $P$ and the following condition is satisfied:*

**(SM)** *For every $p \in M$ there is a clause*

$$p \leftarrow q_1, \ldots, q_m, \mathbf{not}(r_1), \ldots, \mathbf{not}(r_n)$$

*in $ground(P)$ such that for every $i$, $1 \leq i \leq m$, $q_i \in M$ and for every $i$, $1 \leq i \leq n$, $r_i \notin M$.*

Proof: Let us assume that $M$ is a supported model of $P$. Then, by Corollary 6.11 and Definition 6.12, $M$ is a model of $P$. We will prove now that the condition (SM) is satisfied. Since $M$ is a supported model of $P$, $Cn(M)$ is a weak extension of $(dl(P), \emptyset)$. Hence, by Theorem 3.80,

$$Cn(M) = Cn(c(GD(dl(P), Cn(M)))),$$

that is, $Cn(M)$ is the set of consequences of the theory consisting of the consequents of those defaults from $dl(P)$ which are generating with respect to $Cn(M)$. Since both $M$ and $c(GD(dl(P), Cn(M)))$ consist of elements of the Herbrand base, it follows that

$$M = c(GD(dl(P), Cn(M))).$$

Hence, if a ground atomic formula $p$ belongs to $M$ then $p$ is a consequent of a generating default in $GD(dl(P), Cn(M))$. That is, there is a default

$$\frac{q_1 \wedge \ldots \wedge q_m : M\neg r_1, \ldots, M\neg r_n}{p}$$

in $dl(P)$ such that for every $i$, $1 \leq i \leq m$, $q_i \in Cn(M)$ and, for every $i$, $1 \leq i \leq n$, $r_i \notin Cn(M)$. By the definition of $dl(P)$, the clause

$$p \leftarrow q_1, \ldots, q_m, \mathbf{not}(r_1), \ldots, \mathbf{not}(r_n)$$

belongs to $ground(P)$. Moreover, since $M$ is a subset of the Herbrand base and $q_i$, $1 \leq i \leq m$, and $r_i$, $1 \leq i \leq n$, are elements of the Herbrand base, for every $i$, $1 \le i \le m$, $q_i \in M$ and, for every $i$, $1 \le i \le n$, $r_i \notin M$. Hence, the condition (SM) is satisfied.

Conversely, let us assume that $M$ is a model of $P$ and that the condition (SM) holds. By Theorem 3.80, and by our previous remarks, it is enough to show that

$$M = c(GD(dl(P), Cn(M))).$$

Consider $p \in M$. By the condition (SM) there is a clause

$$p \leftarrow q_1, \ldots, q_m, \mathbf{not}(r_1), \ldots, \mathbf{not}(r_n)$$

in $ground(P)$ such that for every $i$, $1 \le i \le m$, $q_i \in M$ and for every $i$, $1 \le i \le n$, $r_i \notin M$. Since $q_i$ and $r_i$ are ground atomic formulas and $M \subseteq B_H$, it follows that for every $i$, $1 \le i \le m$, $q_i \in Cn(M)$ and for every $i$, $1 \le i \le n$, $r_i \notin Cn(M)$. Hence, the default

$$\frac{q_1 \wedge \ldots \wedge q_m : M\neg r_1, \ldots, M\neg r_n}{p},$$

which belongs to $dl(P)$, is generating with respect to $Cn(M)$. Consequently, $p \in c(GD(dl(P), Cn(M)))$. Thus, $M \subseteq c(GD(dl(P), Cn(M)))$.

To prove the converse inclusion, consider a ground atomic formula $p$ from $c(GD(dl(P), Cn(M)))$. It follows that there is a default

$$\frac{q_1 \wedge \ldots \wedge q_m : M\neg r_1, \ldots, M\neg r_n}{p}$$

in $dl(P)$ such that for every $i$, $1 \le i \le m$, $q_i \in Cn(M)$ and for every $i$, $1 \le i \le n$, $r_i \notin Cn(M)$. Consider the clause

$$p \leftarrow q_1, \ldots, q_m, \mathbf{not}(r_1), \ldots, \mathbf{not}(r_n).$$

It belongs to $ground(P)$. In addition, for every $i$, $1 \le i \le m$, $q_i \in M$ and for every $i$, $1 \le i \le n$, $r_i \notin M$. Since $M$ is a model of $P$, it is a model of $pc(ground(P))$, too. Thus, $M$ satisfies the implication

$$q_1 \wedge \ldots \wedge q_m \wedge \neg r_1 \wedge \ldots \wedge \neg r_n \supset p.$$

It follows that $p \in M$ and, consequently, that $c(GD(dl(P), Cn(M))) \subseteq M$. ■

Next, we will characterize supported models as fixpoints of a certain operator.

**Definition 6.24** Let $P$ be a logic program. An operator $T_P: \mathcal{P}(B_H) \longrightarrow \mathcal{P}(B_H)$ is defined as follows: for a subset $S$ of $B_H$ and for a ground atomic formula $p \in B_H$, $p \in T_P(S)$ if and only if there is a clause

$$p \leftarrow q_1, \ldots, q_m, \mathbf{not}(r_1), \ldots, \mathbf{not}(r_n)$$

in *ground*$(P)$ such that for every $i$, $1 \leq i \leq m$, $q_i \in S$ and for every $i$, $1 \leq i \leq n$, $r_i \notin S$. ▲

It turns out that prefixpoints and fixpoints of the operator $T_P$ describe models and supported models of $P$. In addition, as we will see shortly, the operator $T_P$ can also be used to characterize stable models.

Let us recall that given an operator $T : \mathcal{P}(X) \longrightarrow \mathcal{P}(X)$, a subset $M$ of $X$ is called a *prefixpoint* of $T$ if $T(M) \subseteq M$ (Definition 2.66). Directly from the definition of the operator $T_P$ it follows that Herbrand models of a program $P$ are precisely the prefixpoints of $T_P$. Hence, we have the following result.

**Theorem 6.25** *Let $P$ be a logic program and let $M$ be a subset of the Herbrand base $B_H$. Then $M$ is a model of $P$ if and only if $M$ is a prefixpoint of $T_P$, that is, $T_P(M) \subseteq M$.* ▲

Our next theorem states that the fixpoints of $T_P$ are exactly the supported models of $P$.

**Theorem 6.26** *Let $P$ be a logic program and let $M \subseteq B_H$. Then $M$ is a supported model of $P$ if and only if $M$ is a fixpoint of $T_P$, that is, $T_P(M) = M$.*

Proof: By Theorem 6.23, it is enough to show that $T_P(M) = M$ if and only if $M$ is a model of $P$ and satisfies the condition (SM). Clearly, $T_P(M) \subseteq M$ is equivalent to $M$ being a model of $P$. In addition, it is easy to see that $M \subseteq T_P(M)$ is equivalent to the condition (SM). Hence, the assertion follows. ■

Let us return for a moment to the case of stable models. We will consider first the class of Horn programs. Several properties of the operator $T_P$, when $P$ is a Horn program, are gathered below.

**Theorem 6.27** *Let $P$ be a Horn program.*

1. *The operator $T_P$ is monotone and finitizable.*
2. *The operator $T_P$ possesses a least fixpoint.*
3. *The least fixpoint of $T_P$ is $T_P{\uparrow}\omega(\emptyset)$.*

4. *The least fixpoint of $T_P$, $T_P{\uparrow}\omega(\emptyset)$, is the least Herbrand model of $P$.*

Proof: Assertion (1) is evident and we omit the details. Assertions (2) and (3) follow from Knaster-Tarski Lemma (Theorem 2.69). To prove (4) observe that since $T_P{\uparrow}\omega(\emptyset)$ is a fixpoint of $T_P$, $T_P{\uparrow}\omega(\emptyset)$ is a Herbrand model of $P$ (Theorem 6.25). Consider now an arbitrary Herbrand model $M$ of $P$. Then, by the monotonicity of $T_P$, for every set $S \subseteq M$,

$$T_P(S) \subseteq T_P(M) \subseteq M.$$

Since $\emptyset \subseteq M$, it follows by an easy induction that for every non-negative integer $n$

$$T_P{\uparrow}n(\emptyset) \subseteq M.$$

Consequently, $T_P{\uparrow}\omega(\emptyset) \subseteq M$. Hence, $T_P{\uparrow}\omega(\emptyset)$ is the least Herbrand model of $P$. ∎

We will now describe stable models of an arbitrary logic program $P$ as the least fixpoints of the operators $T_Q$, for some suitably defined Horn programs $Q$.

**Theorem 6.28** *Let $P$ be a logic program and let $M$ be a subset of the Herbrand base. Then $M$ is a stable model of $P$ if and only if $M$ is the least fixpoint of the operator $T_{P^M}$, or equivalently, if and only if $M = T_{P^M}{\uparrow}\omega(\emptyset)$.*

Proof: By Theorem 6.22, $M$ is a stable model of $P$ if and only if $M$ is the least Herbrand model of $P^M$. By Theorem 6.27, this latter statement is equivalent to the statement that $M = T_{P^M}{\uparrow}\omega(\emptyset)$. ∎

**Remark 6.29** Not surprisingly, in the case when $P$ is a Horn program, there is a close relationship between the operator $T_P$ and the operator $B^D$ discussed in Section 3.1 and used in our discussion of general default theories (Chapter 3). Namely, it is easy to see that if $P$ is a Horn program and $S \subseteq B_H$ then

$$B^D(S) = S \cup T_P(S),$$

where $D$ consists of the monotone inference rules corresponding to the defaults in $dl(P)$ (observe that all these defaults are justification-free). ▲

We now resume our discussion of supported models. In the remainder of this section we will provide an elegant declarative characterization of supported models. That is, we will associate with a logic program a theory called the *ground completion* of the program, with the property that its Herbrand models correspond precisely to the supported models of the program. In order to do so, we need to allow for infinitary disjunctions in the formulas of the ground completion.

**Definition 6.30** Let $P$ be a propositional program. We construct the *ground completion* of $P$, denoted $gcomp(P)$, by considering all ground atomic formulas from $B_H$.

1. If a ground atomic formula $p$ is not the head of any clause of $ground(P)$ then define $gcomp_p = \neg p$.

2. Otherwise, for every clause $C$ in $ground(P)$ of the form

$$p \leftarrow q_1, \ldots, q_m, \mathbf{not}(r_1), \ldots, \mathbf{not}(r_n),$$

   where $q_i$ and $r_i$ stand for ground atomic formulas, define a formula

$$\varphi_C = (q_1 \wedge \ldots \wedge q_m \wedge \neg r_1 \wedge \ldots \wedge \neg r_n).$$

   Then define

$$gcomp_p = (p \equiv \bigvee_C \varphi_C),$$

   where $C$ ranges over all clauses of $P$ with $p$ as the head. (This is precisely the place where formulas with infinitary disjunctions are introduced.)

3. Define $gcomp(P) = \{gcomp_p : p \in B_H\}$. ▲

In the example that follows we illustrate the concept of the ground completion. In particular, we show that even though a program is finite, its ground completion may be infinite.

**Example 6.31** Let $P$ be the program consisting of the following clauses:

$nat(0) \leftarrow$
$nat(s(y)) \leftarrow nat(y)$
$q(0, s(0)) \leftarrow$
$p(x) \leftarrow q(x, y).$

Observe that $ground(P)$ contains, among others, the following formulas:

$p(0) \leftarrow q(0,0)$
$p(0) \leftarrow q(0, s(0))$

$$\vdots$$

$$p(0) \leftarrow q(0, \underbrace{s \ldots s}_{n}(0))$$

Moreover, no other clause in *ground*($P$) has $p(0)$ in the head. Hence, the ground completion of $p(0)$ is the formula

$$gcomp_{p(0)} = (p(0) \equiv \bigvee_{n<\omega} q(0, \underbrace{s \ldots s}_{n}(0)))$$

Note also that the formula $gcomp_{p(0)}$ is infinitary as it involves an infinite disjunction. ▲

We have not studied in this book the infinitary propositional logic. While there are problems with the proof theory for such a logic, the semantics of propositional calculus can easily be extended to the case when infinite disjunctions and conjunctions are allowed. A propositional valuation is said to satisfy an infinite disjunction if it satisfies at least one of its disjuncts. It is said to satisfy an infinite conjunction if it satisfies all of its conjuncts.

Herbrand models of *gcomp*($P$) are, in particular, models of $P$. Namely, we have the following result.

**Theorem 6.32** *If $M$ is a Herbrand model of gcomp($P$) then $M$ is a Herbrand model of $P$.*

Proof: Consider a clause

$$A = (p \leftarrow q_1, \ldots, q_m, \mathbf{not}(r_1), \ldots, \mathbf{not}(r_n)),$$

of *ground*($P$) ($p$, $q_i$ and $r_i$ stand for ground atomic formulas). We need to prove that $M$ is a model of a formula

$$q_1 \wedge \ldots \wedge q_m \wedge \neg r_1 \wedge \ldots \wedge \neg r_n \supset p.$$

Let us denote this formula by $\psi$. If $p \in M$ then $M \models \psi$. So, assume that $p \notin M$. Since $M$ is a model of *gcomp*($P$), $M \models gcomp_p$. Let us recall that since $p$ is the head of a clause in *ground*($P$),

$$gcomp_p = (p \equiv \bigvee_{C} \varphi_C),$$

where $C$ ranges over all clauses of $P$ with $p$ as the head and $\varphi_C$ is the formula specified in Definition 6.30. In particular,

$$\varphi_A = q_1 \wedge \ldots \wedge q_m \wedge \neg r_1 \wedge \ldots \wedge \neg r_n.$$

Since $p \notin M$, $M \not\models \bigvee_C \varphi_C$. In particular,

$$M \not\models \varphi_A.$$

Hence, $M \models \psi$ in this case, too. ■

In fact, a stronger result holds. Not only are the models of *gcomp*($P$) models of $P$. They are precisely the supported models of $P$.

**Theorem 6.33** *Let $P$ be a logic program and let $M$ be a subset of the Herbrand base. Then $M$ is a supported model of $P$ if and only if $M$ is a model of gcomp($P$).*

Proof: First, let us assume that $M$ is a model of *gcomp*($P$). Then $M$ is a model of $P$. Hence, it follows that $T_P(M) \subseteq M$ (Theorem 6.25). Next, we will prove that $M \subseteq T_P(M)$. Let $p \in M$. Since $M \models gcomp_p$ and since $p \in M$, there is a clause in *ground*($P$) with $p$ as the head, say

$$p \leftarrow q_1, \ldots, q_m, \mathbf{not}(r_1), \ldots, \mathbf{not}(r_n),$$

such that

$$M \models q_1 \wedge \ldots \wedge q_m \wedge \neg r_1 \wedge \ldots \wedge \neg r_n.$$

That is, for every $i$, $1 \leq i \leq m$, $q_i \in M$ and for every $i$, $1 \leq i \leq n$, $r_i \notin M$. Hence, $p \in T_P(M)$.

Consequently, $M = T_P(M)$ and, by Theorem 6.26, $M$ is a supported model of $P$.

Conversely, assume that $M$ is a supported model of $P$. Then $M = T_P(M)$. We will show that $M$ is a model of *gcomp*($P$). Let us consider an element of *gcomp*($P$), a formula $gcomp_p$, where $p$ is a ground atomic formula.

If $p$ is not the head of any clause in *ground*($P$), then $gcomp_p = \neg p$. In such case, by the definition of the operator $T_P$, $p \notin T_P(M)$. Consequently, $p \notin M$. Hence, $M \models \neg p$ and $M \models gcomp_p$.

The only other possibility is when $p$ is the head of a clause in *ground*($P$). In this case,

$$gcomp_p = (p \equiv \bigvee_C \varphi_C),$$

where $C$ ranges over all clauses of $P$ with $p$ as the head and $\varphi_C$ is the formula described in Definition 6.30.

Assume first that $M \models p$. Then, $p \in T_P(M)$. Hence, there is a clause $A \in$ *ground*($P$) of the form

$$p \leftarrow q_1, \ldots, q_m, \mathbf{not}(r_1), \ldots, \mathbf{not}(r_n) \tag{6.8}$$

such that for every $i$, $1 \le i \le m$, $q_i \in M$ and for every $i$, $1 \le i \le n$, $r_i \notin M$. Consequently,

$$M \models q_1 \wedge \ldots \wedge q_m \wedge \neg r_1 \wedge \ldots \wedge \neg r_n.$$

Since $\varphi_A = (q_1 \wedge \ldots \wedge q_m \wedge \neg r_1 \wedge \ldots \wedge \neg r_n)$, it follows that $M \models p \equiv \bigvee_C \varphi_C$.

Assume next that $M \not\models p$. Let $A$ be a clause in *ground*$(P)$ of the form (6.8). Since $p \notin T_P(M)$, there is $i$, $1 \le i \le m$, such that $q_i \notin M$, or there is $i$, $1 \le i \le n$, such that $r_i \in M$. Hence, it follows that $M \not\models \varphi_A$. Consequently, also in this case, $M \models p \equiv \bigvee_C \varphi_C$. Thus, $M \models gcomp(P)$, as claimed. ■

Theorem 6.33 provides a declarative characterization of the notion of a supported model, which has been defined procedurally through the notion of a weak extension of a certain default theory.

**Remark 6.34** Given a finite predicate logic program $P$ one can associate with $P$ a finite theory, *comp*$(P)$, called the *completion of P* which has similar properties to those of *gcomp*$(P)$. In particular, Herbrand models of *comp*$(P)$ are exactly the models of *gcomp*$(P)$ and Herbrand models of *comp*$(P)$ are the supported models of $P$. The construction of *comp*$(P)$ is quite complex and since we only very briefly touch in this book on the issues in predicate logic we do not give it here. For more details the reader is referred to the original paper by Clark, where this notion was first introduced, and to the survey paper by Apt (see the last section for the exact references). ▲

## 6.5 Logic programming with classical negation

The class of logic programs is sometimes too narrow to properly capture our knowledge pertinent to a particular situation. Gelfond and Lifschitz give the following example which they attribute to McCarthy: one of the rules governing the behavior of a school bus is that it may cross the railway tracks only under the condition that there is no train approaching. If by a propositional letter $c$ we denote the statement "cross the tracks" and by $t$ the statement "train is coming", a possible logic programming formalization of this rule might be

$$c \leftarrow \mathbf{not}(t).$$

But this formalization informally means (under default interpretation of logic programs and, more generally, under those interpretations which treat negation as "negation as failure to prove") that the bus should cross the tracks if only there is no evidence that the train is approaching. This is unacceptable, as the information about the train may not be available. For example, the driver's vision may be blocked. What we want is the classical negation $\neg$ and the rule

$$c \leftarrow \neg t$$

which says that the tracks should be crossed under the condition that there is evidence that no train is approaching.

Often, it is necessary to have both types of negation in one clause. For example when we want to formalize a situation when an action of getting more information is warranted in the case of absence of both positive and negative information. For example, a hiring committee before looking carefully at applications may want to narrow down the pool of candidates by verifying that they are eligible for the position (that is, satisfy all formal requirements for the job). In the case of incomplete applications they may not be able to decide whether a candidate is eligible or not. In such case they should conclude that a letter requesting more information be sent, which can naturally be expressed by the rule

$$\mathit{request\_more\_information} \leftarrow \mathbf{not}(\mathit{eligible}), \mathbf{not}(\neg \mathit{eligible}).$$

Our discussion implies that a version of logic programming where the basic building blocks are literals (atoms or negated atoms) may be more suitable in knowledge representation applications than the standard logic programming formalism. We will study such an extension now. We will refer to it as *logic programming with classical negation.*

**Definition 6.35** A *CN-clause* (*CN* stands for classical negation) is an expression of the form

$$p \leftarrow q_1, \ldots, q_m, \mathbf{not}(r_1), \ldots, \mathbf{not}(r_n), \tag{6.9}$$

where $p$, $q_i$ and $r_i$ are literals of $\mathcal{L}_{pred}$, that is, atomic formulas of $\mathcal{L}_{pred}$ or their negations.

A *CN-program* is a set of *CN*-clauses. ▲

As in the case of standard logic programming there are several possible ways in which *CN*-programs can be interpreted. The notions of the predicate-calculus interpretation, the propositional grounding and the default interpretation naturally generalize to the case of *CN*-programs. One simply has to replace in Definitions 6.2, and 6.5 the notions of logic programs and logic program clauses by the notions of *CN*-programs and *CN*-clauses, and to stipulate that $p$, $q_i$ and $r_i$ stand for atomic formulas of $\mathcal{L}_{pred}$ or their negations. To extend the notion of the default interpretation we first introduce some new notation. For every ground literal $a$ we define

$$\overline{a} = \begin{cases} \neg a & \text{if } a \text{ is an atom,} \\ c & \text{if } a = \neg c, \text{ where } c \text{ is an atom.} \end{cases}$$

**Definition 6.36** A *default interpretation* of a ground clause $C$ of the form

$$p \leftarrow q_1, \ldots, q_m, \mathbf{not}(r_1), \ldots, \mathbf{not}(r_n),$$

where $p$, $q_i$ and $r_i$ are ground literals, is the default

$$dl(C) = \frac{q_1 \wedge \ldots \wedge q_m : M\overline{r_1}, \ldots, M\overline{r_n}}{p}.$$

A *default interpretation* of a logic program $P$ is the default theory $(dl(P), \emptyset)$, where

$$dl(P) = \{dl(C) : C \in ground(P)\}.$$

▲

Clearly, for every ground literal $a$, $\overline{a} \equiv \neg a$ is a tautology and we could have used the default

$$\frac{q_1 \wedge \ldots \wedge q_m : M\neg r_1, \ldots, M\neg r_n}{p}$$

as the default interpretation of a *CN*-clause. Our choice is motivated by a simple aesthetic reason. Under our interpretation, defaults used to interpret *CN*-clauses are built of literals. Note also that both possibilities coincide in the case of standard program clauses (that is, those not involving $\neg$).

We will define now the notion of a *stable answer set* for a *CN*-program. Stable answer sets consist of literals of the language $\mathcal{L}_H$ generated by the Herbrand base $B_H$ (recall that we denote this set by $Lit_H$).

**Definition 6.37** Let $P$ be a *CN*-program and let $M$ be a subset of $Lit_H$. The set $M$ is a *stable answer set* for $P$ if $Cn(M)$ is an extension of $(dl(P), \emptyset)$.

▲

In the case of standard logic programs, we studied their predicate-calculus and default interpretations. While in general these interpretations have different properties, they coincide in the case of programs without the operator **not** (Horn programs) — the least Herbrand model of a Horn program is precisely its only stable model. That is, in the case of Horn programs it does not matter whether we regard clauses as formulas in the language $\mathcal{L}_{pred}$ or as inference rules. This is not so in the case of **not**-*free CN-programs*, that is, *CN*-programs without any occurrence of the operator **not**.

**Example 6.38** Consider the following **not**-free *CN*-program $P$:

$$\neg b \leftarrow$$
$$b \leftarrow \neg a,$$

where $a$ and $b$ are ground atomic formulas. The predicate-calculus interpretation of $P$ is given by

$$pc(P) = \{\neg b, \neg a \supset b\}.$$

Clearly, $a \in Cn(pc(P))$. In particular, $a$ belongs to every model of $P$. On the other hand,

$$dl(P) = \left\{\frac{\neg a:}{b}, \frac{:}{\neg b}\right\}.$$

The theory $(dl(P), \emptyset)$ has exactly one extension $Cn(\{\neg b\})$ and it does not contain $a$. ▲

It follows directly from the definitions of stable models and stable answer sets that under the restriction to standard logic programs (without classical negation), the notion of a stable answer set reduces to that of a stable model. A formal statement to this effect is contained in the following result.

**Theorem 6.39** *Let $P$ be a CN-program not involving the operator $\neg$ (that is, $P$ is a standard logic program).*

1. *If a set of literals $M \subseteq Lit_H$ is a stable answer set for $P$, then $M$ is a subset of $B_H$.*
2. *A subset $M$ of the Herbrand base $B_H$ is a stable answer set for $P$ if and only if $M$ is a stable model for $P$.*

Proof: Since the operator $\neg$ does not occur in $P$, the consequents of the defaults in $dl(P)$ are elements of $B_H$. Consequently, each extension of $(dl(P), \emptyset)$ is of the form $Cn(M)$ for some subset $M$ of $B_H$. Each stable answer set, being a set of literals of an extension of $(dl(P), \emptyset)$, is therefore a subset of $B_H$. Hence, (1) follows. Part (2) of the assertion follows from part (1) and the definitions of stable models and stable answer sets. ■

The original definition of stable answer sets by Gelfond and Lifschitz does not contain any references to the default logic. We have chosen to define stable answer sets as sets of literals of extensions of some default theory to stress the fact that logic programming with stable answer sets is essentially nothing else but a fragment of default logic. Now, we will provide the characterization of stable answer sets that was used as the original definition by Gelfond and Lifschitz. We start by extending the definition of the reduct to the case of *CN*-programs.

**Definition 6.40** Let $M$ be a subset of $Lit_H$.

1. A *CN*-clause

$$p \leftarrow q_1, \ldots, q_m, \mathbf{not}(r_1), \ldots, \mathbf{not}(r_n),$$

   where $p$, $q_i$ and $r_i$ stand for ground atomic formulas or their negations, is *irrelevant* with respect to $M$ if at least one $r_i$ is in $M$.

2. Let $P$ be a *CN*-program. The *reduct* of $P$ with respect to $M$, denoted by $P^M$, is obtained from *ground*($P$) by:

   (a) Removing all clauses that are irrelevant with respect to $M$.

   (b) Removing each premise $\mathbf{not}(r_i)$ from the remaining clauses. ▲

The reduct of a logic program is a Horn program. Consequently, it has the least Herbrand model. Here the situation is different. The reduct of a *CN*-program does not contain any occurrences of the **not** operator but may still contain the occurrences of the $\neg$ operator. Consequently, it is not a Horn program and it may have several different minimal Herbrand models. Despite that, **not**-free *CN*-programs are, in a sense, similar to Horn programs. Namely, there is a natural way of assigning to each **not**-free *CN*-program a set of literals. In the case of a Horn program, this set of literals consists of atoms only and coincides with the least Herbrand model of the program. In general, though, the set of literals we associate with a **not**-free *CN*-program, even if this set consists of atoms only, does not determine a model of the program (see Example 6.43 below).

**Definition 6.41** Let $M$ consist of ground atomic formulas and their negations (that is, $M \subseteq Lit_H$). The set $M$ is *closed* under a ground **not**-free *CN*-clause of the form

$$p \leftarrow q_1, \ldots, q_m,$$

precisely when one of the following two cases holds:

1. $M = Lit_H$.

2. $M$ is consistent (that is, does not contain a pair of contradictory literals from $Lit_H$) and whenever for all $i$, $1 \leq i \leq m$, $q_i \in M$, then $p \in M$.

The set $M$ is *closed* under a *CN*-program $P$, which does not contain any occurrence of **not**, if $M$ is closed under every *CN*-clause of $ground(P)$. ▲

**Theorem 6.42** *Let $P$ be a* **not**-*free CN-program. Then, the collection of those subsets of $Lit_H$ that are closed under $P$ is nonempty and contains the least element.*

Proof: Let us denote by $\mathcal{C}$ the collection of those subsets of $Lit_H$ that are closed under $P$. Clearly, $Lit_H$ is closed under $P$. So, $\mathcal{C}$ is nonempty. Define $M$ to be the intersection of all elements in $\mathcal{C}$. There are two possibilities. If $\mathcal{C} = \{Lit_H\}$ then $M = Lit_H$ and, by definition, it is closed under $P$. In the other case, $\mathcal{C}$ contains a consistent subset of $Lit_H$. In this case $M$ is consistent. Consider a clause $C$ from $ground(P)$. Say, $C$ is of the form

$$p \leftarrow q_1, \ldots, q_m.$$

If for some $i$, $1 \leq i \leq m$, $q_i \notin M$, then $M$ is closed under $C$. If for every $i$, $1 \leq i \leq m$, $q_i \in M$ then $q_i \in M'$, for every $M' \in \mathcal{C}$. Since each $M'$ is closed under $C$, it follows that $p \in M'$. Hence, $p \in \bigcap \mathcal{C} = M$. Thus, $M$ is closed under $C$ and, consequently, under $P$. It follows that $M \in \mathcal{C}$ and it is contained in every $M' \in \mathcal{C}$. Hence, the assertion follows. ■

Let us look briefly at the case of Horn programs. The least subset of $Lit_H$ closed under a Horn program $P$ clearly contains ground atomic formulas only. It is also easy to see that a set of ground atomic formulas is closed under a Horn program $P$ if and only if it is a model of $P$. Hence, in the case of Horn programs the concepts of the least Herbrand model and the least set of literals closed under $P$ coincide.

**Example 6.43** Consider the **not**-free *CN*-program $P$ consisting of two clauses:

$r \leftarrow$
$p \leftarrow \neg q$

The least set of literals closed under $P$ is equal to $\{r\}$. Clearly, $\{r\}$ is not a model of $pc(P)$. Hence, if a **not**-free *CN*-program $P$ is not a Horn program, the least set of literals closed under $P$, even if it consists of atoms only, does not have to be a model of $pc(P)$. ▲

The least subset of $Lit_H$ closed under a **not**-free *CN*-program $P$ will be denoted by $L(P)$. Using the notions of the reduct and the operator $L$ we get the following characterization of stable answer sets, which was originally introduced by Gelfond and Lifschitz as their definition.

**Theorem 6.44** *Let $P$ be a CN-program. A subset $M$ of $Lit_H$ is a stable answer set for $P$ if and only if $M$ is the least set closed under the reduct $P^M$, that is, $M = L(P^M)$.*

Proof: Let us observe that for every literal $p \in Lit_H$ and for every set of literals $M \subseteq Lit_H$, such that $M$ is consistent or $M = Lit_H$, we have $p \in M$ if and only if $p \in Cn(M)$. Once this observation is made, the proof of Theorem 6.22 can be repeated almost literally to prove the present assertion. We leave the details to the reader. ■

In the case of a Horn program $P$ the least Herbrand model of $P$ can be described as the least fixpoint of the operator $T_P$. Consequently, a characterization of stable models in terms of fixpoints of operators $T_{P^M}$ follows. The operator $T_P$ cannot be used to characterize stable answer sets in a straightforward manner. Let us consider the following example.

**Example 6.45** Let $P$ be the **not**-free *CN*-program consisting of the following clauses:

$$q \leftarrow$$
$$p \leftarrow q$$
$$\neg p \leftarrow q.$$

The operator $T_P$ (adapted to the case of literals; in the definition of $T_P$ one has to replace ground atomic formulas by ground literals) applied to this program would generate $q$ in the first step and would generate $p$ and $\neg p$ in the second. Then no other literals would be generated. Hence, $T_P \uparrow \omega(\emptyset) = \{q, p, \neg p\}$. On the other hand, the least set of literals closed under $P$ is $Lit_H$ and $Lit_H \neq \{q, p, \neg p\}$. ▲

The reason why the operator $T_P$ is inadequate in the case of *CN*-programs is simple. In the case of logic programs, we are dealing with sets of ground atomic formulas, and closure under logical consequence will not generate new ground atomic formulas. Hence, the operator $T_P$, which does not refer to any mechanism of predicate logic, suffices. On the other hand, *CN*-programs are built of literals. If a set of literals is inconsistent, then its closure under logical consequence will contain all literals. The operator $T_P$ does not contain any mechanism to simulate this behavior.

There are two possible approaches one might consider in order to develop a fixpoint characterization of stable answer sets. The first is to modify the operator $T_P$, the other is to encode the necessary fragment of logic into a program. We investigate both approaches below.

**Definition 6.46** Let $P$ be a *CN*-program. An operator $G_P: \mathcal{P}(Lit_H) \rightarrow \mathcal{P}(Lit_H)$ is defined as follows: for a subset $S$ of $Lit_H$ and for a ground literal $p \in Lit_H$, $p \in G_P(S)$ if and only if at least one of the following conditions holds:

1. There is a clause

$$p \leftarrow q_1, \ldots, q_m, \mathbf{not}(r_1), \ldots, \mathbf{not}(r_n)$$

   in *ground*($P$) such that for every $i$, $1 \leq i \leq m$, $q_i \in S$ and for every $i$, $1 \leq i \leq n$, $r_i \notin S$.

2. There are clauses

$$a \leftarrow q_1, \ldots, q_m, \mathbf{not}(r_1), \ldots, \mathbf{not}(r_n)$$

   and

$$\neg a \leftarrow q'_1, \ldots, q'_{m'}, \mathbf{not}(r'_1), \ldots, \mathbf{not}(r'_{n'})$$

   in *ground*($P$) such that for every $i$, $1 \leq i \leq m$, $q_i \in S$, for every $i$, $1 \leq i \leq m'$, $q'_i \in S$, for every $i$, $1 \leq i \leq n$, $r_i \notin S$ and for every $i$, $1 \leq i \leq n'$, $r'_i \notin S$. ▲

The following property is an immediate consequence of the definition.

**Theorem 6.47** *Let $P$ be a* **not**-*free CN-program. Then, for every non-negative integer $n$, either $G_P{\uparrow}n(\emptyset)$ is consistent or $G_P{\uparrow}n(\emptyset) = Lit_H$. Consequently, either $G_P{\uparrow}\omega(\emptyset)$ is consistent or $G_P{\uparrow}\omega(\emptyset) = Lit_H$.* ▲

The operator $G_P$ has similar properties to those of the operator $T_P$. In particular, if $P$ is a **not**-free *CN*-program we have the following result analogous to Theorem 6.27.

**Theorem 6.48** *Let $P$ be a* **not**-*free CN-program.*

1. *The operator $G_P$ is monotone and finitizable.*
2. *The operator $G_P$ possesses a least fixpoint.*
3. *The least fixpoint of $G_P$ is $G_P{\uparrow}\omega(\emptyset)$.*
4. *A set of literals $M$ is closed under $P$ if and only if $G_P(M) \subseteq M$.*
5. *The least fixpoint of $G_P$, $G_P{\uparrow}\omega(\emptyset)$, is the least set of ground literals closed under $P$.* ▲

Proof: The proof is essentially the same as that of Theorem 6.27. Assertion (1) follows directly from the definition of the operator $G_P$, and assertions (2) and (3) follow from the Knaster-Tarski lemma (Theorem 2.69).

(4) Let us consider a set $M$ of literals such that $G_P(M) \subseteq M$. Consider a clause $C$ from $ground(P)$, where $C$ is of the form

$$p \leftarrow q_1, \ldots, q_m.$$

If for some $i$, $1 \leq i \leq m$, $q_i \notin M$, then $M$ is closed under $C$. Hence, assume that for every $i$, $1 \leq i \leq m$, $q_i \in M$. Then, by the definition of $G_P$, $p \in G_P(M)$. Since $G_P(M) \subseteq M$, $p \in M$. Thus, $M$ is closed under $C$. Consequently, $M$ is closed under $P$. The converse implication can be proved similarly.

(5) Consider now an arbitrary set $M$ of ground literals, closed under $P$. By (4), $G_P(M) \subseteq M$. By the monotonicity of $G_P$, for every set $S \subseteq M$,

$$G_P(S) \subseteq G_P(M) \subseteq M.$$

Since $\emptyset \subseteq M$, it follows by an easy induction that for every non-negative integer $n$

$$G_P{\uparrow}n(\emptyset) \subseteq M.$$

Consequently, $G_P{\uparrow}\omega(\emptyset) \subseteq M$. Since $G_P{\uparrow}\omega(\emptyset)$ is a fixpoint of $G_P$ (assertion (3)), it follows by (4) that $G_P{\uparrow}\omega(\emptyset)$ is closed under $P$. Hence, $G_P{\uparrow}\omega(\emptyset)$ is the least set of ground literals closed under $P$. ■

**Corollary 6.49** *Let $P$ be a CN-program and let $M \subseteq Lit_H$ be a subset of the set of ground literals. Then $M$ is a stable answer set of $P$ if and only if $M$ is the least subset of $Lit_H$ closed under $P^M$, or equivalently, if and only if $M = G_{P^M}{\uparrow}\omega(\emptyset)$.* ▲

Next, we will show how stable answer sets of a *CN*-program $P$ can be characterized by means of fixpoints of the operator $T_Q$, for a suitably defined program $Q$. The price we pay in order to accomplish that goal is quite high. We have to add to $P$ infinitely many new clauses. These new clauses, called *built-in* clauses, allow us to generate all ground literals as soon as two contradictory ground literals are generated.

**Definition 6.50** A *built-in* clause is any clause of one of the following forms:

1. $q(x_1, \ldots, x_k) \leftarrow p(y_1, \ldots, y_l), \neg p(y_1, \ldots, y_l)$
2. $\neg q(x_1, \ldots, x_k) \leftarrow p(y_1, \ldots, y_l), \neg p(y_1, \ldots, y_l)$

For a *CN*-program $P$, by the *closure* of $P$, denoted $\overline{P}$, we mean the *CN*-program obtained by adding to $P$ all ground instances of built-in clauses. ▲

Let us assume that a *CN*-program $P$ does not contain **not**. Then, it is easy to prove by induction that for every non-negative integer $n$

$$G_P\uparrow n(\emptyset) \subseteq T_{\overline{P}}\uparrow\omega(\emptyset)$$

and

$$T_{\overline{P}}\uparrow n(\emptyset) \subseteq G_P\uparrow\omega(\emptyset)$$

Consequently,

$$G_P\uparrow\omega(\emptyset) = T_{\overline{P}}\uparrow\omega(\emptyset).$$

Hence, we obtain the following corollary.

**Corollary 6.51** *Let $P$ be a CN-program and let $M \subseteq Lit_H$ be a subset of the set of ground literals. Then $M$ is a stable answer set of $P$ if and only if $M = T_{(\overline{P})^M}\uparrow\omega(\emptyset)$.*

Proof: The set $M$ is a stable answer set for $P$ if and only if $M = G_{P^M}\uparrow\omega(\emptyset)$. This last identity is equivalent to the equality

$$M = T_{\overline{P^M}}\uparrow\omega(\emptyset). \tag{6.10}$$

Observe that built-in clauses do not contain any occurrence of the operator **not**. Hence, for every set $M$ of ground literals and for every *CN*-program $Q$ we have

$$(\overline{Q})^M = \overline{Q^M}.$$

Consequently, (6.10) is equivalent to

$$M = T_{(\overline{P})^M}\uparrow\omega(\emptyset).$$

Hence, the assertion follows. ■

We conclude this section by showing that *CN*-programs can be translated into standard logic programs so that there is a *one-to-one* and *onto* correspondence between stable answer sets and stable models. The technique we use is to introduce new predicate symbols. Namely, for every predicate symbol $p$ we introduce a new predicate symbol $p'$ of the same arity. We will denote the extended language by $\mathcal{L}'_{pred}$. Our intention is to use an atomic formula $p'(t_1,\ldots,t_k)$ instead of the formula $\neg p(t_1,\ldots,t_k)$. Speaking formally, for every atomic formula $p(t_1,\ldots,t_k)$ of $\mathcal{L}_{pred}$ we define

$$tr(p(t_1,\ldots,t_k)) = p(t_1,\ldots,t_k)$$

and

$$tr(\neg p(t_1,\ldots,t_k)) = p'(t_1,\ldots,t_k).$$

Next, for a clause $C$ of the form

$$p \leftarrow q_1,\ldots,q_m,\mathbf{not}(r_1),\ldots,\mathbf{not}(r_n),$$

where $p$, $q_i$ and $r_i$ are atomic formulas of $\mathcal{L}_{pred}$ or their negations, we define the clause $tr(C)$ (in the extended language) by

$$tr(C) = (tr(p) \leftarrow tr(q_1),\ldots,tr(q_m),\mathbf{not}(tr(r_1)),\ldots,\mathbf{not}(tr(r_n)))$$

Finally, for a *CN*-program $P$ we define

$$tr(P) = \{tr(C): C \in P\},$$

and, for a set of ground literals $M$, we set

$$tr(M) = \{tr(p): p \in M\}.$$

Under these definitions, we are able to establish the following relationship between *CN*-programs and ordinary logic programs.

**Theorem 6.52** *Let $P$ be a CN-program. A set $M$ of ground literals of $\mathcal{L}_{pred}$ is a stable answer set for $P$ if and only if $tr(M)$ is a stable model of the program $tr(\overline{P})$. Conversely, every stable model of $tr(\overline{P})$ is of the form $tr(M)$ for some stable answer set $M$ for $P$.*

Proof: First, it is easy to show by induction that for every non-negative integer $n$ we have

$$tr(T_{(\overline{P})^M}\uparrow n(\emptyset)) = T_{tr(\overline{P})^{tr(M)}}\uparrow n(\emptyset).$$

Consequently, by Theorem 6.28 and Corollary 6.51, $M$ is a stable answer set for $P$ if and only if $tr(M)$ is a stable model of the program $tr(\overline{P})$.

Conversely, let us consider a stable model $H$ of $tr(\overline{P})$. Since $tr(\overline{P})$ contains all the rules of the form

$$q(x_1,\ldots,x_k) \leftarrow p(y_1,\ldots,y_l),p'(y_1,\ldots,y_l),$$

or

$$q'(x_1,\ldots,x_k) \leftarrow p(y_1,\ldots,y_l),p'(y_1,\ldots,y_l),$$

it follows that either $H$ contains all ground atomic formulas of the extended language $\mathcal{L}'_{pred}$, or there is no ground atomic formula $p(t_1,\ldots,t_k)$ in $\mathcal{L}_{pred}$ such that both $p(t_1,\ldots,t_k)$ and $p'(t_1,\ldots,t_k)$ are in $H$. Hence, there is a set of ground literals $M$ (in the language $\mathcal{L}_{pred}$) such that $H = tr(M)$. Thus, the second part of the assertion follows from the first part. ■

Let us observe that it is necessary to translate the closure $\overline{P}$ of a *CN*-program $P$ in order to recover *all* answer sets for $P$. However, if we are only interested in consistent answer sets, it is enough to translate $P$ only. More, formally, we have the following result.

**Theorem 6.53** *Let $P$ be a CN-program and let $M$ be a consistent set of ground literals of $\mathcal{L}_{pred}$. The set $M$ is a stable answer set for $P$ if and only if $tr(M)$ is a stable model of the program $tr(P)$. Conversely, every stable model $M'$ of $tr(P)$ such that $M'$ does not contain any pair of atoms $p$, $p'$, is of the form $tr(M)$ for some consistent stable answer set $M$ for $P$.* ▲

## 6.6 Clausal logic programming

The class of logic programs can be extended even further. In this section we will consider logic program clauses built of clauses of the language $\mathcal{L}_{pred}$. Our approach is similar to that of the previous section and we will omit most of the details.

**Definition 6.54** A *clausal* logic program clause (*CL*-clause) is an expression of the form

$$C \leftarrow A_1, \ldots, A_m, \mathbf{not}(B_1), \ldots, \mathbf{not}(B_n), \tag{6.11}$$

where $C$, $A_i$ and $B_i$ are clauses of the language $\mathcal{L}_{pred}$, that is, disjunctions of atomic formulas and their negations. A *clausal logic program*, or a *CL-program*, is a set of logic program clauses. ▲

By $Clause_H$ we will denote the collection of all ground clauses of $\mathcal{L}_{pred}$, that is, disjunctions of ground literals from $\mathcal{L}_{pred}$. As in the case of ordinary logic programming and logic programming with classical negation, one can define, in a natural way, the notions of propositional grounding and default interpretation for a *CL*-program $P$. As before, we will denote them by $ground(P)$ and $dl(P)$, respectively.

Given a *CL*-program $P$, and using its default interpretation, one can now distinguish some sets of ground clauses as stable answer sets for $P$. Formally, we have the following definition.

**Definition 6.55** Let $P$ be a *CL*-program. A subset $M$ of $Clause_H$ is a *stable answer set* for $P$ if $Cn(M)$ is an extension of $(dl(P), \emptyset)$. ▲

It is easy to see that for a *CL*-program $P$, every default in $dl(P)$ is *clausal* in the sense of Definition 5.13. Consequently, it follows from Theorem 5.14 that clausal logic programs can represent arbitrary default theory. Namely, as an immediate corollary to Theorem 5.14 we obtain the following result.

**Corollary 6.56** *For every default theory $(D, W)$ there exists a (propositional) CL-program $P$ such that a theory $S$ is an extension of $(D, W)$ if and only if the set of clauses in $S$ is a stable answer set for $P$.* ▲

Perhaps the most important result on clausal logic programming is a characterization of stable answer sets of *CL*-programs in terms of fixpoints of the operator $T_Q$, where $T_Q$ is adapted to the case of clauses and $Q$ is obtained from $P$ by adding to it a clausal logic program clauses encoding inference rules of logic.

**Definition 6.57** In what follows, $p$ stands for an arbitrary atomic formula $p(y_1, \ldots, y_k)$, where $y_1, \ldots, y_k$ are variables, $C$ and $C'$ stand for arbitrary (possibly empty) disjunctions of atomic formulas and their negations, with each disjunct of the form $q(z_1, \ldots, z_n)$ or $\neg q(z_1, \ldots, z_n)$, for some predicate symbol $q$ and variables $z_1, \ldots, z_n$.

A *built-in* *CL*-clause is any *CL*-clause in one of the following forms:

1. $C \leftarrow p, \neg p$
2. $C \vee C' \leftarrow (C \vee p), (C' \vee \neg p)$
3. $C \vee p \vee \neg p \leftarrow$
4. $C \vee p \leftarrow C$
5. $C \vee \neg p \leftarrow C$

For a *CL*-program $P$, by the *closure* of $P$, denoted $\overline{P}$, we denote the *CL*-program obtained by adding to $P$ all built-in clauses. ▲

Next, we need to extend the notion of reduct to the case of *CL*-programs.

**Definition 6.58** Let $M$ be a subset of $Clause_H$.

1. A *CL*-clause

$$C \leftarrow A_1, \ldots, A_m, \mathbf{not}(B_1), \ldots, \mathbf{not}(B_n),$$

where $C$, $A_i$ and $B_i$ stand for ground clauses from $Clause_H$, is *irrelevant* with respect to $M$ if at least one $B_i$ is in $M$.

2. Let $P$ be a *CL*-program. The *reduct* of $P$ with respect to $M$, denoted by $P^M$, is obtained from *ground*($P$) by:

   (a) Removing all clauses that are irrelevant with respect to $M$.

   (b) Removing each premise $\mathbf{not}(B_i)$ from the remaining clauses. ▲

Finally, we obtain the following characterization of stable answer sets of *CL*-programs.

**Theorem 6.59** *Let $P$ be a CL-program and let $M \subseteq Clause_H$ be a subset of the set of ground clauses. Then $M$ is a stable answer set of $P$ if and only if $M = T_{(\overline{P})^M}\uparrow\omega(\emptyset)$.* ▲

## 6.7 Stratified and locally stratified logic programs

We will study now two classes of logic programs which have similar properties to those of Horn programs, but permit the operator **not** in the body. These programs, called *stratified* and *locally stratified*, have the desired feature that there is a natural and systematic way of selecting one of their minimal Herbrand models as the *intended* model. Let us consider some examples first.

**Example 6.60** Consider the following logic program $P$:

$$p \leftarrow q, \mathbf{not}(r)$$
$$w \leftarrow p$$
$$q \leftarrow,$$

where $p, q, r$ and $w$ are ground atomic formulas. The program $P$ possesses two minimal Herbrand models, $M_1 = \{q, p, w\}$, and $M_2 = \{q, r\}$. First, notice that the atomic formula $r$ is in $M_2$ "without any computational reason". It cannot be derived from $P$. Hence, $M_2$ does not properly capture the properties of the program $P$ (if we regard $P$ as a collection of inference rules). The first model, on the other hand, has much nicer properties. The formula $q$ is in $M_1$ since it must be in every model — the program affirms $q$. Since there is no way to compute $r$, we apply Closed World Assumption and conclude $r$ false. Then, *out of this data*, we compute $p$ and finally $w$. ▲

But the situation is not always as simple as in Example 6.60.

**Example 6.61** Consider the program $P$ consisting of the following ground clauses:

$$q \leftarrow \mathbf{not}(p)$$
$$p \leftarrow \mathbf{not}(q)$$

Here, the situation is completely symmetric. The two minimal models, $M_1 = \{p\}$ and $M_2 = \{q\}$ cannot be distinguished in the manner applied in Example 6.60. There is no way in which we can say which of $p$ and $q$ should fail. ▲

The difference between these two examples is that in the first case the program determined a natural hierarchy of (ground) atomic formulas. In the first case, we could decide the validity of $r$ and $q$ before we had to deal with $p$. In the second example we could not do this. These examples should serve as a motivation for the following definition.

**Definition 6.62** (Stratification, local stratification)

1. A program $P$ is *locally stratified* if there is a function *rank* assigning to every element of the Herbrand base $B_H$ an ordinal number such that for every clause

$$p \leftarrow q_1, \ldots, q_n, \mathbf{not}(r_1), \ldots, \mathbf{not}(r_m)$$

from *ground*($P$), $rank(q_i) \leq rank(p)$ for all $i \leq n$ and $rank(r_j) < rank(p)$ for all $j \leq m$.

2. A program $P$ is *stratified* if there is a function *rank* assigning to every predicate symbol of $\mathcal{L}_{pred}$ an ordinal number such that for every clause

$$p(\vec{x}) \leftarrow q_1(\vec{y_1}), \ldots, q_n(\vec{y_n}), \mathbf{not}(r_1(\vec{z_1})), \ldots, \mathbf{not}(r_m(\vec{z_m}))$$

of $P$, $rank(q_i) \leq rank(p)$ for all $i \leq n$ and $rank(r_j) < rank(p)$ for all $j \leq m$.

3. Function *rank* is called a *local-stratification* function, or a *stratification* function, respectively.

4. Finally, the precedence relation imposed on the elements of the Herbrand base $B_H$ by a (local-)stratification function *rank* is called the *(local) stratification* corresponding to *rank*. ▲

Without loss of generality we may restrict our attention to local-stratification functions whose ranges (sets of values) are ordinals. Indeed, it is easy to see that if *rank* is a local-stratification function for a program $P$, then there is a local-stratification function *rank'* for $P$ such that the range of *rank'* is an ordinal and for every two ground atomic formulas $p$ and $q$ from $B_H$, $rank(p) < rank(q)$ if and only if $rank'(p) < rank'(q)$. In other words, the function *rank'* defines precisely the same local-stratification of $P$ as *rank*. From now on, we always assume that the range of any local-stratification function *rank* is an ordinal. The same conventions will be adopted for stratification functions.

Suppose that $P$ is a stratified logic program. Let *rank* be a corresponding stratification function (there may be several different stratification functions under which $P$ is stratified). Consider a ground atomic formula $p(\vec{x})\Theta$, where $p$ is a predicate symbol and $\Theta$ a ground substitution. Define $rank'(p(\vec{x})\Theta) = rank(p)$. It is easy to see that *rank'* defines a local stratification of $P$. Hence, we get the following result.

**Theorem 6.63** *Let $P$ be a stratified logic program. Then, $P$ is locally stratified.* ▲

Consider next a Horn program $P$ and a function *rank* which assigns the ordinal 0 to every predicate symbol of the language. Clearly, *rank* is a stratification function for $P$. Consequently, $P$ is stratified and, by Theorem 6.63, locally stratified. These observations are gathered in the following theorem.

**Theorem 6.64** *Let $P$ be a Horn program. Then, $P$ is stratified and locally stratified.* ▲

Let us reconsider Examples 6.60 and 6.61. Clearly, the program from the first one of them is stratified (hence, also locally stratified). Just assign 0 to the predicate symbols of $q$ and $r$ and 1 to the predicate symbols of $p$ and $w$. It is quite obvious that there is no stratification nor local-stratification function in the second example.

The notion of stratification is strictly stronger than that of local stratification. To see this, let us consider another example.

**Example 6.65** Let $P$ be the program consisting of the following clauses:

$$p(0) \leftarrow \mathbf{not}(q(x))$$
$$q(s(x)) \leftarrow \mathbf{not}(q(x)).$$

Clearly, this program is not stratified. The requirement that the predicate symbols appearing under the scope of the operator **not** have the rank strictly smaller than the predicate symbol in the head cannot be satisfied in this case. On the other hand, the program $P$ is locally stratified. For instance, let *rank* be a function such that

$$rank(q(s^n(0))) = n$$
$$rank(p(0)) = \omega$$
$$rank(p(s^n(0))) = 0 \text{ for } n > 0.$$

We leave it for the reader to check that, indeed, *rank* is a local-stratification function for $P$. ▲

We will describe now a construction of a certain Herbrand model of a locally stratified logic program $P$. Next, we will examine its properties. This model, called *perfect*, will turn out to be a minimal model of the program. Consequently, in the case of Horn programs, it coincides with the least Herbrand model. All these results apply also to stratified programs.

In order to define the perfect model of a locally stratified program we need some definitions and simple facts. First, given a collection of sets $\{A_\xi : \xi < \alpha\}$, where $\alpha$ is an ordinal, we define, for each $\xi < \alpha$

$$A_{<\xi} = \bigcup_{\eta<\xi} A_\eta,$$

and

$$A_{\leq\xi} = \bigcup_{\eta\leq\xi} A_\eta.$$

**Definition 6.66** Let $\alpha$ be an ordinal, $rank: B_H \to \{\xi : \xi < \alpha\}$ a function, and $P$ a logic program. Define:

$$B_\xi = \{p \in B_H : \ rank(p) = \xi\},$$

and

$$P_\xi = \{C \in ground(P) : \ rank(head(C)) = \xi\},$$

where $head(C)$ stands for the head of a clause $C$. ▲

Let us consider a locally stratified program $P$. Let *rank* be a local-stratification function for $P$, with the range $\{\xi : \xi < \alpha\}$, for some ordinal $\alpha$. Then the sets $B_\xi$ form a partition of the Herbrand base $B_H$ into disjoint nonempty subsets. Similarly, the sets $P_\xi$ form a partition of $P$ into disjoint, but possibly empty, subsets. Sets $B_\xi$ and $P_\xi$ will be called *strata*.

To see that $P_\xi$ may be empty, let us look at Example 6.60. Consider the local-stratification function for $P$ defined as follows: $rank(q) = 0$, $rank(r) = 1$, and $rank(p) = rank(w) = 2$. It is easy to see that $P_1 = \emptyset$.

Several simple properties of the sets introduced in Definition 6.66 are gathered below.

**Theorem 6.67** *For every program $P$ and every function $rank: B_H \to \{\xi : \xi < \alpha\}$:*

1. $B_{<0} = \emptyset$ *and* $P_{<0} = \emptyset$.
2. $B_{<\xi} \cap B_\xi = \emptyset$.
3. $B_{<\xi} \cup B_\xi = B_{\leq\xi}$.
4. $B_{<\xi} = \bigcup_{\eta<\xi} B_{\leq\eta}$.
5. $P_{<\xi} \cap P_\xi = \emptyset$.
6. $P_{<\xi} \cup P_\xi = P_{\leq\xi}$.
7. $P_{<\xi} = \bigcup_{\eta<\xi} P_{\leq\eta}$. ▲

**Example 6.65 (continued)** For the locally stratified program $P$ described in Example 6.65 and the local-stratification function given there we have:

$$P_{\leq 0} = \emptyset$$
$$P_{\leq 1} = \{q(s(0)) \leftarrow \mathbf{not}(q(0))\}$$
$$P_{\leq 2} = P_{\leq 1} \cup \{q(s^2(0)) \leftarrow \mathbf{not}(q(s(0)))\}$$
$$\vdots$$
$$P_{<\omega} = \bigcup_{n<\omega} P_{\leq n}$$
$$P_{\leq\omega} = P_{<\omega} \cup \{p(0) \leftarrow \mathbf{not}(q(s^n(0))) : n < \omega\} = ground(P).$$

We also have $P_0 = \emptyset$. ▲

Let $P$ be a locally stratified program and let $rank: B_H \to \{\xi : \xi < \alpha\}$ be its local-stratification function. Observe that $P_0$ is a Horn program (although it may be empty!). Indeed if $\mathbf{not}(q)$ were an element of the body of a ground clause from $P_0$ with $p$ as the head, we would have that $rank(q) < rank(p) = 0$, a contradiction.

Since $P_0$ is a Horn program, it always possesses a least model. In higher strata $P_\xi$, $\xi > 0$, the ground atomic formulas from $B_0$ are never recomputed (redefined), as they do not occur in the heads of the clauses in $P_\xi$, $\xi > 0$. This property serves as the main motivation for the crucial idea of stratification, which is to use the least model of $P_0$ to provide answers to queries concerning the ground atomic formulas from $B_0$. Now, we can move to the next strata $B_1$. Only the ground atomic formulas from $B_0$ may occur under the scope of the operator **not** in the clauses from $P_1$. The truth values of these atoms are already decided. We can use this knowledge to transform $P_1$ into a Horn program and use its least model to decide truth values of ground atomic formulas from $B_1$. Again, due to stratification, these values are not affected by the higher strata and will remain unchanged throughout the construction. This process is repeated until all strata are exhausted.

At each stage of the construction we are transforming a program $P_\xi$ into a Horn program using information gathered earlier about the truth values of ground atomic formulas belonging to lower strata. To this end we use the notion of the reduct introduced by Gelfond and Lifschitz and studied in the context of stable models.

We will now recursively define a collection $\{M_\xi\}_{\xi<\alpha}$ of subsets of the Herbrand base $B_H$. The union of these subsets will be called a perfect Herbrand structure for a locally stratified program $P$.

**Definition 6.68** Let $P$ be a locally stratified program and let *rank* be its local-stratification function with the range $\{\xi : \xi < \alpha\}$.

1. For every ordinal $\xi$, $0 \le \xi \le \alpha$, we define $M_\xi$ to be the least Herbrand model of the Horn program $\{p \leftarrow : p \in M_{<\xi}\} \cup Q_\xi$, where $Q_\xi$ stands for the reduct of $P_\xi$ with respect to the set of ground atomic formulas $M_{<\xi} = \bigcup_{\eta<\xi} M_\eta$, that is, $Q_\xi = P_\xi^{M_{<\xi}}$.
2. The set $M = \bigcup_{\xi<\alpha} M_\xi$ is called a *perfect structure* for the program $P$. ▲

Let us look at what happens when $\xi = 0$. In this case, $M_{<0} = \emptyset$ and $Q_0$ is the reduct of $P_0$ with respect to the empty set of ground atomic formulas. Since $P_0$ is a Horn program, $Q_0 = P_0$. Hence, $M_0$ is the least Herbrand model of $P_0$, which agrees with our informal discussion before.

We will prove now some auxiliary results which exhibit several useful properties of the sets $M_\xi$ and programs $Q_\xi$.

**Lemma 6.69** *Under the notation from Definition 6.68 the following statements hold:*

*1. For all ordinals $\xi_1 < \xi_2 < \alpha$, $M_{\xi_1} \subseteq M_{\xi_2}$.*

2. *For every $\xi < \alpha$, $M_\xi \subseteq B_{\leq\xi}$.*

3. *For every $\xi < \alpha$, $M_\xi \setminus M_{<\xi} \subseteq B_\xi$.*

4. *For every $\xi < \alpha$, for every ground atomic formula $p \in M_\xi \setminus M_{<\xi}$ and for every ground atomic formula $q \in M_{<\xi}$, $rank(p) > rank(q)$.*

5. *For all $\eta < \xi < \alpha$, for every ground atomic formula $p \in M_\xi \setminus M_\eta$ and for every ground atomic formula $q \in M_\eta$, $rank(p) > rank(q)$.*

Proof: Part (1) is evident, we omit its proof. To prove part (2) one needs to note that the least Herbrand model of a ground Horn program is contained in the set of the heads of all clauses of the program. Then the assertion follows by a straightforward induction on $\xi$. To prove part (3), let us observe that every element of $M_\xi$ is either an element of $M_{<\xi}$ or is the head of a clause from $Q_\xi$. Since the heads of clauses from $Q_\xi$ are in $B_\xi$, the assertion follows. Assertion (4) follows from assertions (3) and (2) and assertion (5) is a simple corollary of assertion (4). ∎

Let us recall that we have $Q_{<\xi} = \bigcup_{\eta<\xi} Q_\eta$, $Q_{\leq\xi} = \bigcup_{\eta\leq\xi} Q_\eta$.

**Lemma 6.70** *Under the notation of Definition 6.68, $M_\xi$ is the least Herbrand model of $Q_{\leq\xi}$, for every $\xi < \alpha$.*

Proof: We proceed by induction on $\xi$. Observe that $Q_{\leq 0} = Q_0$ and $M_{<0} = \emptyset$. By Definition 6.68, $M_0$ is the least Herbrand model of $M_{<0} \cup Q_0 = Q_0 = Q_{\leq 0}$. Hence, the basis for the induction is established.

Consider now an ordinal $\xi$ such that $0 < \xi < \alpha$. We will first prove that $M_\xi$ is a model of $Q_{\leq\xi}$. Clearly, by Definition 6.68, $M_\xi$ is a model of $Q_\xi$. Consider then an arbitrary clause $C$ of the form

$$r \leftarrow q_1, \ldots, q_m$$

from $Q_{<\xi}$. (Let us recall that each $Q_\eta$ is a ground program. Consequently, $C$ is a ground clause.) There is $\eta < \xi$ such that $C \in Q_\eta$. By Definition 6.68, $M_\eta$ is a model of $C$. It also follows that $rank(r) = \eta$, and, by stratification, it follows that $rank(q_i) \leq \eta$ for every $i$, $1 \leq i \leq m$. Since every element in $M_\xi \setminus M_\eta$ has rank greater than $\eta$, it follows that $M_\xi$ is a model of $C$. Hence, $M_\xi$ is a model of $Q_{<\xi}$. We have already seen that $M_\xi$ is a model of $Q_\xi$. Consequently, $M_\xi$ is a model of $Q_{\leq\xi}$.

Next, we will prove that $M_\xi$ is the least Herbrand model of $Q_{\leq\xi}$. Let $M$ be any Herbrand model of $Q_{\leq\xi}$. Then $M$ is a Herbrand model of $Q_{\leq\eta}$, for every $\eta < \xi$. By the induction hypothesis, $M_\eta \subseteq M$. Consequently, $M_{<\xi} \subseteq M$.

Hence, $M$ is a Herbrand model of $M_{<\xi} \cup Q_\xi$. By Definition 6.68, $M_\xi \subseteq M$. ■

**Corollary 6.71** *Under the notation of Definition 6.68, $M_{<\alpha}$ is the least Herbrand model of $Q_{<\alpha}$.*

Proof: Let $C$ be a clause from $Q_\xi$, where $\xi < \alpha$. By Definition 6.68, $M_\xi$ is a model of $C$. By Lemma 6.69, every element in $M_{<\alpha} \setminus M_\xi$ has rank greater than $\xi$. By stratification, every ground atomic formula occurring in $C$ has rank at most $\xi$. Since $M_\xi \subseteq M_{<\alpha}$, it follows that $M_{<\alpha}$ is a model of $C$. Consequently, $M_{<\alpha}$ is a model of $Q_{<\alpha}$.

Next, consider any Herbrand model $M$ of $Q_{<\alpha}$. Then $M$ is a Herbrand model of $Q_{\leq\xi}$, for every $\xi < \alpha$. By Lemma 6.70, $M_\xi \subseteq M$. Consequently, $M_{<\alpha} \subseteq M$. ■

**Lemma 6.72** *Under the notation of Definition 6.68, $P^{M_{<\alpha}} = Q_{<\alpha}$.*

Proof: We will first show that for $\xi < \alpha$

$$P_\xi^{M_{<\alpha}} = P_\xi^{M_{<\xi}}. \tag{6.12}$$

Since $M_{<\xi} \subseteq M_{<\alpha}$, the inclusion

$$P_\xi^{M_{<\alpha}} \subseteq P_\xi^{M_{<\xi}}$$

follows directly from the definition of the reduct. To prove the converse inclusion, let us consider a clause $C$ of the form

$$r \leftarrow q_1, \ldots, q_m$$

from $P_\xi^{M_{<\xi}}$. Then there exists a clause $C'$ of the form

$$r \leftarrow q_1, \ldots, q_m, \mathbf{not}(r_1), \ldots, \mathbf{not}(r_n)$$

in $P_\xi$ such that $r_i \notin M_{<\xi}$, for every $i$, $1 \leq i \leq n$. By stratification, $rank(r_i) < rank(r) = \xi$. On the other hand, every element in $M_{<\alpha} \setminus M_{<\xi}$ has rank at least $\xi$ (Lemma 6.69). Hence, $r_i \notin M_{<\alpha}$, for every $i$, $1 \leq i \leq n$. Consequently, $C \in P_\xi^{M_{<\alpha}}$.

Next, observe that

$$P^{M_{<\alpha}} = (\bigcup_{\xi<\alpha} P_\xi)^{M_{<\alpha}} = \bigcup_{\xi<\alpha} P_\xi^{M_{<\alpha}}.$$

Hence, by (6.12),

$$P^{M_{<\alpha}} = \bigcup_{\xi<\alpha} P_\xi^{M_{<\xi}} = \bigcup_{\xi<\alpha} Q_\xi = Q_{<\alpha},$$

as required. ■

Now, we are ready to show that perfect structures for a locally stratified program are, in fact, stable models of the program. Thus, in particular, perfect structures for a locally stratified program are models of the program and models of the ground completion of the program.

**Theorem 6.73** *Let $P$ be a locally stratified logic program. Let rank be its local-stratification function with the range $\{\xi : \xi < \alpha\}$ and let $M$ be the corresponding perfect structure for $P$. Then:*

1. *$M$ is a stable model for $P$.*
2. *$M$ is a minimal Herbrand model for $P$.*
3. *$M$ is a model of the ground completion of $P$.*

Proof: From Corollary 6.71 it follows that $M$ is the least Herbrand model of $Q_{<\alpha}$. Hence, by Lemma 6.72, $M$ is the least Herbrand model of $P^M$. Consequently, $M$ is a stable model of $P$ (Theorem 6.22). Parts (2) and (3) follow from Theorem 6.15 and Theorem 6.33. ■

Since every stratified program is locally stratified, we obtain the following corollary.

**Corollary 6.74** *Let $P$ be a stratified logic program. Let rank be its stratification function and let $M$ be the corresponding perfect structure for $P$. Then:*

1. *$M$ is a stable model for $P$.*
2. *$M$ is a minimal Herbrand model for $P$.*
3. *$M$ is a model of the ground completion of $P$.* ▲

In view of Theorem 6.73, from now on we will use the term *perfect model* rather than *perfect structure* for $P$.

A local-stratification function for a program (if exists) is not necessarily unique. Thus we can have different stratifications of a program. It turns out that the corresponding perfect models coincide. In other words, every locally stratified program has a unique perfect model. To prove this fact, we need one more lemma.

**Lemma 6.75** *Let $P$ be a locally stratified logic program and let $M_1$ and $M_2$ be stable models of $P$. Then $M_1 = M_2$.*

Proof: Let *rank* be a local-stratification function for $P$ with the range $\{\xi : \xi < \alpha\}$ for some ordinal $\alpha$. By $H^i_\xi$, $\xi < \alpha$, we will denote the least Herbrand model of $P^{M_i}_{<\xi}$, $i = 1, 2$. Let us also note that by the stability of $M_i$, $i = 1, 2$, $M_i$ is the least Herbrand model of $P^{M_i}$. We will prove first that, for $i = 1, 2$, if $p$ is a ground atomic formula and $rank(p) < \xi$, then

$$p \in H^i_\xi \text{ if and only if } p \in M_i. \tag{6.13}$$

Clearly, it is sufficient to prove this equivalence only for $i = 1$.

Since $P^{M_1}_{<\xi} \subseteq P^{M_1}$, it follows that $H^1_\xi \subseteq M_1$. Hence, if a ground atomic formula $p$, with rank less than $\xi$, is in $H^1_\xi$ then $p$ is in $M_1$.

To prove the converse implication, let us consider a ground atomic formula $p$ such that $rank(p) < \xi$ and $p \notin H^1_\xi$. Let $M$ be the subset of the Herbrand base obtained from $H^1_\xi$ by adding to it all ground atomic formulas of rank at least $\xi$.

Clearly, $p \notin M$. In addition, let us observe that the head of every clause in $P^{M_1} \setminus P^{M_1}_{<\xi}$ has rank at least $\xi$. Hence, $M$ is a model of $P^{M_1} \setminus P^{M_1}_{<\xi}$. Let us consider a clause

$$r \leftarrow q_1, \ldots, q_m$$

from $P^{M_1}_{<\xi}$. By stratification, $rank(q_i) \leq rank(r) < \xi$. Assume that $q_1, \ldots, q_m \in M$. Since every element in $M \setminus H^1_\xi$ has rank at least $\xi$, it follows that $q_1, \ldots, q_m \in H^1_\xi$. Since $H^1_\xi$ is a model of $P^{M_1}_{<\xi}$, $r \in H^1_\xi$. Consequently, $r \in M$. Thus, $M$ is a Herbrand model of $P^{M_1}_{<\xi}$. It follows that $M$ is a Herbrand model of $P^{M_1}$. Since $M_1$ is the least Herbrand model of $P^{M_1}$, we have $M_1 \subseteq M$. Recall that $p \notin M$. Therefore, $p \notin M_1$.

Let us consider an ordinal $\xi < \alpha$ and assume that for every $\eta < \xi$, $P^{M_1}_\eta = P^{M_2}_\eta$. Consequently, $P^{M_1}_{<\xi} = P^{M_2}_{<\xi}$ and $H^1_\xi = H^2_\xi$. It follows from (6.13) that

$$M_1 \cap B_{<\xi} = M_2 \cap B_{<\xi}. \tag{6.14}$$

Let us consider now a clause $C$ from $P_\xi$ of the form

$$r \leftarrow q_1, \ldots, q_m, \mathbf{not}(r_1), \ldots, \mathbf{not}(r_n).$$

By stratification, $rank(r_j) < \xi$, for every $j$, $1 \leq j \leq n$, and, by (6.14), $r_j \in M_1$ if and only if $r_j \in M_2$. Consequently, $P^{M_1}_\xi = P^{M_2}_\xi$.

Thus, by induction, $P^{M_1} = P^{M_2}$. Since $M_1$ is the least Herbrand model of $P^{M_1}$ and $M_2$ is the least Herbrand model of $P^{M_2}$, $M_1 = M_2$. ■

**Theorem 6.76** *Let $P$ be a locally stratified logic program. Then its perfect model is unique. That is, it does not depend on a particular local-stratification function used in its construction. Moreover, $P$ possesses a unique stable model which coincides with the perfect model of $P$.*

Proof: The assertion follows from Theorem 6.73 and Lemma 6.75. ■

We proved that a locally stratified program possesses a unique stable model. One can ask if the converse is true, that is, if a program $P$ with a unique stable model must be stratified. This is not the case.

**Example 6.77** Consider the following program $P$:

$$p \leftarrow \mathbf{not}(q)$$
$$q \leftarrow r, \mathbf{not}(p)$$

Clearly, $P$ is not locally stratified. If *rank* were a local-stratification function, the first clause would imply $rank(q) < rank(p)$ and the second one would imply $rank(p) < rank(q)$, a contradiction. It is easy to check, however, that $P$ possesses the unique stable model $M = \{p\}$. ▲

## 6.8 Proof schemes and stable models

In this section we provide one more characterization of stable models. It is in the same spirit as the characterization of supported models of a program as models of its ground completion. Given a logic program $P$ we will construct a theory in the language of propositional calculus generated by the elements of the Herbrand base $B_H$ such that its models are exactly the stable models of $P$. In general, our theory will involve infinitary disjunctions but in the case when $ground(P)$ is finite no infinitary disjunctions will be necessary.

The results of this section are obtained by means of a technique similar to the method of context-dependent proofs used in Chapter 3.

**Definition 6.78** Let $P$ be a logic program and let $M \subseteq B_H$. A sequence $p_1, \ldots, p_n$ is an $M$-proof of $p$ from the program $P$ (or simply an *$M$-proof* if $P$ is given) if these conditions are met:

1. $p_n = p$.

2. For every $1 \leq i \leq n$, there is a ground clause

$$C_i = (p_i \leftarrow q_1, \ldots, q_t, \mathbf{not}(r_1), \ldots, \mathbf{not}(r_m))$$

such that:

(a) $\{q_1, \ldots, q_t\} \subseteq \{p_1, \ldots, p_{i-1}\}$.
(b) $\{r_1, \ldots, r_m\} \cap M = \emptyset$. ▲

The notion of an $M$-proof does not make explicit which clauses were used in the process of $M$-derivation. In addition, an $M$-proof uses the complement of $M$ for testing the applicability of clauses. Since $M$-proofs are finite and since each clause is finite, knowledge of only a finite part of that complement is actually needed to verify that a sequence of atoms forms an $M$-proof. The following definition introduces a concept of a proof scheme which addresses these issues.

**Definition 6.79** Let $P$ be a logic program. A *proof scheme* with respect to $P$ (or simply a *proof scheme* if $P$ is understood) is a sequence $s$ of triples $\langle C_1, p_1, U_1\rangle, \ldots \langle C_n, p_n, U_n\rangle$ such that for every $i$, $1 \leq i \leq n$,

1. $C_i \in ground(P)$.

2. $p_i$ is the head of $C_i$.

3. If $C_i = p_i \leftarrow q_1, \ldots, q_t, \mathbf{not}(r_1), \ldots, \mathbf{not}(r_m)$ then:

(a) $\{q_1, \ldots, q_t\} \subseteq \{p_1, \ldots, p_{i-1}\}$.
(b) $U_i = U_{i-1} \cup \{r_1, \ldots, r_m\}$.

If $s = \langle C_1, p_1, U_1\rangle, \ldots \langle C_n, p_n, U_n\rangle$ is a proof scheme then $p_n$ is called the *conclusion* of $s$ and is denoted by $con(s)$, and $U_n$ is called the *support* of $s$ and is denoted by $supp(s)$. ▲

We have the following simple result relating the notions of a proof scheme and an $M$-proof. It is a straightforward consequence of Definitions 6.79 and 6.78, and we omit its proof.

**Theorem 6.80** *Let $P$ be a logic program. For every subset $M$ of the Herbrand base $B_H$ and for every proof scheme $s = \langle C_1, p_1, U_1\rangle, \ldots \langle C_n, p_n, U_n\rangle$, if $M \cap supp(s) = \emptyset$ then $p_1, \ldots, p_n$ is an $M$-proof of $p_n$. Conversely, if $\langle p_1, \ldots, p_n\rangle$ is* *an $M$-proof of $p_n$, then there exists a proof scheme $s$ such that $M \cap supp(s) = \emptyset$ and $con(s) = p_n$.* ▲

Let $s$ be a proof scheme. We say that a set $M \subseteq B_H$ *admits* $s$ if $M \cap supp(s) = \emptyset$. By Theorem 6.80, if $M$ admits a proof scheme $s$, then the sequence of ground atomic formulas induced by $s$ is an $M$-proof.

Stable models of a logic program can be characterized in terms of proof schemes and $M$-proofs. We gather these characterizations in the result that follows.

**Theorem 6.81** *Let $P$ be a logic program and let $M$ be a subset of $B_H$. The following statements are equivalent:*

1. *$M$ is a stable model of $P$.*
2. *For every $p \in B_H$, $p \in M$ if and only if there is a proof scheme $s$ such that $M$ admits $s$ and $con(s) = p$.*
3. *For every $p \in B_H$, $p \in M$ if and only if there is an $M$-proof of $p$.*

Proof: Assume that $M$ is a stable model of $P$. Then, $M$ is precisely the set of ground atomic formulas in $T_{P^M}\uparrow\omega(\emptyset)$. It is easy to see by induction that a ground atomic formula is in $T_{P^M}\uparrow\omega(\emptyset)$ if and only if it has an $M$-proof. Hence, the equivalence of (1) and (3) follows. The equivalence of (2) and (3) follows from Theorem 6.80. ■

Some proof schemes are redundant in the sense that they perform derivations irrelevant for the purpose of deriving their conclusions. We will define now the notion of a minimal proof scheme.

**Definition 6.82** A proof scheme $s$ is *minimal* if there is no proof scheme $s'$ such that:

1. $con(s) = con(s')$.
2. The set of clauses of $s'$ is a *proper* subset of the set of clauses of $s$. ▲

Since proof schemes are finite, the following observation is evident.

**Theorem 6.83** *For every proof scheme $s$ there exists a minimal proof scheme $s'$ such that $con(s) = con(s')$ and $supp(s') \subseteq supp(s)$.* ▲

Theorem 6.83 allows us to characterize stable models in terms of minimal proof schemes.

**Corollary 6.84** *Let $P$ be a logic program and let $M$ be a subset of $B_H$. The following statements are equivalent:*

1. *$M$ is a stable model of $P$.*
2. *For every $p \in B_H$, $p \in M$ if and only if there is a minimal proof scheme $s$ such that $M$ admits $s$ and $con(s) = p$.* ▲

We will use now Corollary 6.84 to construct, given a logic program $P$, a theory whose models are exactly the stable models of $P$. We will adopt several notational conventions. For a set $\Phi$ of formulas, we define $\bigwedge \Phi$ to be the conjunction of the formulas in $\Phi$ and $\bigvee \Phi$ to be the disjunction of the formulas in $\Phi$. In addition, we define $\neg\Phi = \{\neg\varphi : \varphi \in \Phi\}$.

**Definition 6.85** Let $P$ be a logic program. Let $p \in B_H$. The *defining boolean equation* for $p$, denoted $def_p$, is the formula:

$$p \equiv \bigvee \left\{ \bigwedge \neg supp(s) : \ s \ \text{ is a minimal proof scheme for } \ p \right\}.$$

The collection of defining boolean equations for all elements of $B_H$, $\{def_p : p \in B_H\}$, is denoted by $Def(P)$. ▲

In the case of infinite propositional logic programs (in particular, in the case of finite predicate programs with function symbols) there may be infinitely many minimal proof schemes for $p$. Consequently, our defining equations may involve infinitary disjunctions.

Before we state the characterization result for stable models of logic programs let us consider the following example.

**Example 6.86** Let $P$ be the program consisting of the clauses:

$$\begin{array}{l} p \leftarrow \mathbf{not}(q) \\ r \leftarrow p, \mathbf{not}(u) \\ p \leftarrow \mathbf{not}(w) \\ w \leftarrow \mathbf{not}(p) \end{array}$$

Since there is no proof scheme with conclusion $q$ we have $def_q = (q \equiv \bot)$. Similarly, $def_u = (u \equiv \bot)$. The remaining three equations are: $def_p = (p \equiv \neg q \vee \neg w)$, $def_r = (r \equiv (\neg q \wedge \neg u) \vee (\neg w \wedge \neg u))$ and $def_w = (w \equiv \neg p)$. ▲

**Theorem 6.87** *The following statements are equivalent:*

1. *$M$ is a stable model for $P$.*

2. *The valuation $V$ of $B_H$, determined by $M$, is a solution to the set $Def(P)$ of boolean equations.*

Proof: The equivalence follows immediately from Corollary 6.84 and Definition 6.85. ■

## 6.9 Comments and remarks

Our presentation of logic programming stresses essentially only one aspect of this field — its connections with default logic. Hence, by no means can it be treated as a comprehensive overview of the area. For excellent texts on logic programming the reader is referred to Apt [Apt90] and Lloyd [Llo84].

Our definitions of stable and supported models stress the connections of these notions with default logic and the notions of extensions and weak extensions. The connections between stable models and extensions were discovered by Bidoit and Froidevaux [BF91a, BF91b] and, independently, by the authors [MT89c].

The notion of the operator $T_P$ and the duality between "logic" and "default" interpretations of Horn programs is due to van Emden and Kowalski [vEK76], and Apt and van Emden [AvE82]. The use of operators in the studies of semantics of logic programs, now a common practice in the area, comes from these papers.

The completion of a program was introduced by Clark [Cla78] in an attempt to describe a semantics for general logic programs. Apt and van Emden [AvE82] characterized the completion in terms of the fixpoints the operator $T_P$.

Apt, Blair and Walker [ABW88] defined and studied the class of stratified programs. Locally stratified programs and perfect models were introduced by Przymusinski [Prz88]. Expressibility of stratified and locally stratified logic programs was studied in Apt and Blair [AB90] and Blair, Marek and Schlipf [BMS91].

Stable models of a logic program were defined by Gelfond and Lifschitz [GL88]. The relationship between perfect and stable models was announced by Gelfond and Lifschitz [GL88] and proved by Marek and Subrahmanian

[MS89]. Proof schemes and the boolean-equation representation of stable models was introduced by Marek, Nerode and Remmel [MNR92c].

Logic programs with classical negation were introduced and studied by Gelfond and Lifschitz [GL90]. Logic programming with clauses was studied by Yang, Blair and Brown [YBB92].

Current implementations of logic programming (in particular PROLOG interpreters) do not treat negation in a manner entirely consistent with the default interpretation of negation, **not**. The complexity results which we will show in Chapter 13 imply that checking if a ground atom $p$ belongs to at least one stable model of the program is an *NP*-complete task. Thus any real system interpreting programs with negation can only approximate stable models.

# 7. Modal logic

The aim of this chapter is to provide an overview of selected topics in modal logic to the extent necessary for the development of the theory of nonmonotonic modal logics later in the book. An exhaustive presentation of modal logics can be found in several good monographs on the subject, for example, Chellas[Che80], Fitting [Fit83], and Hughes and Cresswell [HC84]. A detailed study of index logics can be found in the dissertation of Segerberg [Seg71].

There are two aspects of our presentation that require underlining. The notion of provability we study allows for the necessitation rule to be applied to any formula in the proof and not only to the theorems of the logic, as is usually the case. Secondly, our discussion is not limited to normal modal logics only. We also study several subnormal modal logics, as some of them are important from the nonmonotonic reasoning standpoint. Some subnormal modal logics are discussed in Segerberg [Seg71] and Fitting [Fit83], but those that naturally arise in the area of nonmonotonic reasoning are not treated there.

In order for our book to be self-contained, we introduce here all the necessary definitions and provide proofs for most of the key results.

## 7.1 Modal logics — syntax and basic properties

To study formal properties of concepts such as *knowledge* or *belief* we need to extend our language of propositional calculus by a modal operator whose intended meaning will be *is known* or *believed.*

**Definition 7.1**

1. The language $\mathcal{L}_L$ of propositional modal logic is defined as the least set $\mathcal{U}$ of strings over the alphabet $At \cup \{\neg, \vee, \wedge, \supset, L, (, )\}$ such that:

(a) $At \subseteq \mathcal{U}$.

(b) If $\varphi \in \mathcal{U}$ then $\neg\varphi \in \mathcal{U}$.

(c) If $\varphi_1, \varphi_2 \in \mathcal{U}$ then $(\varphi_1 \vee \varphi_2) \in \mathcal{U}$.

(d) If $\varphi_1, \varphi_2 \in \mathcal{U}$ then $(\varphi_1 \wedge \varphi_2) \in \mathcal{U}$.

(e) If $\varphi_1, \varphi_2 \in \mathcal{U}$ then $(\varphi_1 \supset \varphi_2) \in \mathcal{U}$.

(f) If $\varphi \in \mathcal{U}$ then $L\varphi \in \mathcal{U}$.

When writing formulas, the outermost pair of parenthesis enclosing the formula will usually be omitted.

2. The elements of $\mathcal{L}_L$ are called *well-formed formulas* or simply *formulas* of the language of modal logic.

3. We will use special symbols to abbreviate some formulas. Similarly as in Chapter 2, we write $\varphi \equiv \psi$ for $(\varphi \supset \psi) \wedge (\psi \supset \varphi)$. In addition, we abbreviate $\neg L \neg\varphi$ as $M\varphi$.

4. Every formula of the form $L\varphi$, where $\varphi \in \mathcal{L}_L$, will be called a *modal atom*. ▲

Modal logics which we consider here interpret connectives $\neg, \vee, \wedge$ and $\supset$ in the standard way as described in Chapter 2. Consequently, one can consider the logic of propositional calculus in the language $\mathcal{L}_L$. Namely, $\mathcal{L}_L$ can be looked at as the language of propositional calculus with the set of atoms $At \cup \{L\varphi : \varphi \in \mathcal{L}_L\}$. In other words, modal atoms can be treated as "regular" propositional variables. Tautologies of the propositional calculus in the language $\mathcal{L}_L$ are precisely the results of uniform substitutions of modal atoms for some propositional variables in the tautologies in $\mathcal{L}$. For example, the formulas $Lp \vee \neg Lp$ and $Lp \wedge (Lp \vee Lq) \equiv Lp$ are propositional tautologies of $\mathcal{L}_L$ but $L(p \vee q) \equiv L(q \equiv p)$ is not (we assume here that $p, q \in At$). The logic of propositional calculus over the language $\mathcal{L}_L$, as described above, will be denoted by $PC_L$. The provability relation in the logic $PC_L$ will be denoted by $\vdash$ and the consequence operator by $Cn$. Even though we use the same symbols as in Chapter 2, it will always be clear from the context whether they refer to the logic of propositional calculus in the language $\mathcal{L}$ or $\mathcal{L}_L$.

**Definition 7.2** For a formula $\varphi \in \mathcal{L}_L$ we recursively define the notion of the *L-depth* of $\varphi$, denoted by $d_L(\varphi)$, as follows:

1. If $\varphi \in At$, then $d_L(\varphi) = 0$.

2. If $\varphi = \neg\psi$, then $d_L(\varphi) = d_L(\psi)$.

3. If $\varphi = \psi_1 \vee \psi_2$, $\varphi = \psi_1 \wedge \psi_2$ or if $\varphi = \psi_1 \supset \psi_2$ then $d_L(\varphi) = \max\{d_L(\psi_1), d_L(\psi_2)\}$.
4. If $\varphi = L\psi$, then $d_L(\varphi) = d_L(\psi) + 1$.

The set of all formulas of $\mathcal{L}_L$ with $L$-depth at most $n$ will be denoted by $\mathcal{L}_{L,n}$. That is,

$$\mathcal{L}_{L,n} = \{\varphi \in \mathcal{L}_L : d_L(\varphi) \leq n\}.$$ ▲

Applying the results on normal forms (Theorem 2.20) to the logic $PC_L$ immediately yields normal form results for formulas from the modal language $\mathcal{L}_L$.

**Theorem 7.3** *Let $\varphi \in \mathcal{L}_L$. There are formulas $\psi_1, \psi_2, \psi_3 \in \mathcal{L}_L$ such that:*

1. *All atoms and modal atoms occurring in $\psi_i$, $i = 1, 2, 3$, occur in $\varphi$.*
2. $\vdash \varphi \equiv \psi_i$, $i = 1, 2, 3$.
3. $d_L(\psi_i) \leq d_L(\varphi)$, $i = 1, 2, 3$.

*In addition, $\psi_1, \psi_2$ and $\psi_3$ are of the form:*

1. (Modal Disjunctive Normal Form) $\psi_1 = \Theta_1 \vee \ldots \vee \Theta_k$ *in $\mathcal{L}_L$ and each $\Theta_i$, $1 \leq i \leq k$, is a conjunction*

$$L\alpha_1 \wedge \ldots \wedge L\alpha_m \wedge \neg L\beta_1 \wedge \ldots \wedge \neg L\beta_n \wedge \gamma,$$

*where $\gamma \in \mathcal{L}$ (we allow that one or both of m, n are equal to 0).*

2. (Modal Conjunctive Normal Form) $\psi_2 = \Theta_1 \wedge \ldots \wedge \Theta_k$ *in $\mathcal{L}_L$, and each $\Theta_i$, $1 \leq i \leq k$, is a disjunction*

$$L\alpha_1 \vee \ldots \vee L\alpha_m \vee \neg L\beta_1 \vee \ldots \vee \neg L\beta_n \vee \gamma,$$

*where $\gamma \in \mathcal{L}$ (we allow that one or both of m, n are equal to 0).*

3. $\psi_3 = \Theta_1 \wedge \ldots \wedge \Theta_k$ *in $\mathcal{L}_L$ and each $\Theta_i$, $1 \leq i \leq k$, is an implication*

$$L\alpha_1 \wedge \ldots \wedge L\alpha_m \wedge \neg L\beta_1 \wedge \ldots \wedge \neg L\beta_n \supset \gamma,$$

*where $\gamma \in \mathcal{L}$ (we allow that one or both of m, n are equal to 0).* ▲

The modal operator $L$ can be interpreted in a variety of ways. An intended meaning of $L$ in this book is that of *knowledge* or *belief*. That is, $L\varphi$ will usually mean $\varphi$ *is known* or *believed*. We postpone a presentation of a model theory for modal logics, which will make precise the meaning of the operator $L$, until the next section. We start with a description of proof theory.

Logics allow us to derive formulas from premises by means of *inference rules*. Let us recall that an *inference rule* can formally be defined as a subset of the cartesian product $\mathcal{L}_L \times \ldots \times \mathcal{L}_L$. The *modus ponens* inference rule was discussed in Chapter 2 in the context of the logic of propositional calculus. For the needs of modal logics over the language $\mathcal{L}_L$ it is defined as $\{(\varphi \supset \psi, \varphi, \psi) \colon \varphi, \psi \in \mathcal{L}_L\}$ and is usually written as

$$\frac{\varphi \supset \psi, \varphi}{\psi}.$$

In addition to modus ponens, we will require one more inference rule. The *necessitation* rule is defined as $\{(\varphi, L\varphi) \colon \varphi \in \mathcal{L}_L\}$. Customarily, it is written as

$$\frac{\varphi}{L\varphi}.$$

**Definition 7.4** Given a formula $\varphi \in \mathcal{L}_L$, propositional variables $p_1, \ldots, p_k \in At$ and formulas $\psi_1, \ldots, \psi_k \in \mathcal{L}_L$, the formula $\varphi(p_1/\psi_1, \ldots, p_k/\psi_k)$ — a *substitution instance* of $\varphi$ — is defined recursively as follows:

1. In the case when $\varphi \in At$, $\varphi(p_1/\psi_1, \ldots, p_k/\psi_k) = \psi_i$ if $\varphi = p_i$, for some $i$, $1 \le i \le k$, and $\varphi(p_1/\psi_1, \ldots, p_k/\psi_k) = \varphi$, otherwise.
2. If $\varphi = \neg\beta$, then $\varphi(p_1/\psi_1, \ldots, p_k/\psi_k) = \neg\beta(p_1/\psi_1, \ldots, p_k/\psi_k)$.
3. If $\varphi = \beta_1 \vee \beta_2$, then $\varphi(p_1/\psi_1, \ldots, p_k/\psi_k) = \beta_1(p_1/\psi_1, \ldots, p_k/\psi_k) \vee \beta_2(p_1/\psi_1, \ldots, p_k/\psi_k)$.
4. If $\varphi = \beta_1 \wedge \beta_2$, then $\varphi(p_1/\psi_1, \ldots, p_k/\psi_k) = \beta_1(p_1/\psi_1, \ldots, p_k/\psi_k) \wedge \beta_2(p_1/\psi_1, \ldots, p_k/\psi_k)$.
5. If $\varphi = \beta_1 \supset \beta_2$, then $\varphi(p_1/\psi_1, \ldots, p_k/\psi_k) = \beta_1(p_1/\psi_1, \ldots, p_k/\psi_k) \supset \beta_2(p_1/\psi_1, \ldots, p_k/\psi_k)$.
6. If $\varphi = L\beta$, then $\varphi(p_1/\psi_1, \ldots, p_k/\psi_k) = L\beta(p_1/\psi_1, \ldots, p_k/\psi_k)$.

A theory $T \subseteq \mathcal{L}_L$ is said to be closed under the *rule of uniform substitution* if for every formula $\varphi$ from $T$, every substitution instance of $\varphi$ is in $T$.

▲

Speaking informally, $\varphi(p_1/\psi_1, \ldots, p_k/\psi_k)$ is obtained from $\varphi$ by simultaneously substituting $\psi_i$ for $p_i$ in $\varphi$, for every $i$, $1 \le i \le k$.

We are now ready to formally define the notions of a modal logic and a proof (or derivation) in a modal logic.

**Definition 7.5**

1. By a *modal logic over the language* $\mathcal{L}_L$ we mean any set $\mathcal{S}$ of formulas from $\mathcal{L}_L$ such that:

   **(ML1)** $\mathcal{S}$ contains all propositional tautologies in the language $\mathcal{L}_L$.

   **(ML2)** $\mathcal{S}$ is closed under uniform substitution, modus ponens and necessitation.

   The elements of $\mathcal{S}$ are called *theorems* of the logic $\mathcal{S}$.

2. Let $\mathcal{S}$ be a modal logic and let $AX \subseteq \mathcal{S}$. If $\mathcal{S}$ is the least set of formulas from $\mathcal{L}_L$ which contains $AX$ and satisfies conditions (ML1) and (ML2), then we say that $\mathcal{S}$ is *axiomatized by* $AX$, and we call the members of $AX$ *axioms* of $\mathcal{S}$. ▲

If a modal logic $\mathcal{S}$ is axiomatized by the set of axioms $AX$, then $\mathcal{S}$ can be described procedurally in terms of the notion of provability.

**Definition 7.6** Let $AX \subseteq \mathcal{L}_L$ and let $\varphi \in \mathcal{L}_L$. A sequence $\psi_1, \ldots, \psi_n$ of formulas from $\mathcal{L}_L$ is called a *proof of* $\varphi$ from axioms $AX$ if $\psi_n = \varphi$ and for every $k$, $1 \leq k \leq n$, at least one of the following conditions holds:

1. $\psi_k$ is a substitution instance of a tautology in $\mathcal{L}$ (or, equivalently, a tautology in $\mathcal{L}_L$).
2. $\psi_k$ is a substitution instance of an axiom from $AX$.
3. $\psi_k$ is the result of applying modus ponens to formulas $\psi_i$ and $\psi_j$, for some $i, j < k$.
4. $\psi_k$ is the result of applying necessitation to a formula $\psi_i$, for some $i < k$.

▲

We will now show that the set of theorems of a modal logic $\mathcal{S}$ can be described by means of the notion of proof.

**Theorem 7.7** *Let $\mathcal{S}$ be a modal logic and let $AX$ be its axiomatization. Then the set of all theorems of $\mathcal{S}$ exactly coincides with the set of those formulas in $\mathcal{L}_L$ that have a proof from $AX$.*

Proof: It follows directly from Definition 7.6 that the set of all formulas possessing a proof from axioms $AX$ contains $AX$.

Moreover, it is easy to see that this set also contains all tautologies in $\mathcal{L}_L$. Hence, it satisfies condition (ML1).

Suppose that $\varphi$ and $\varphi \supset \psi$ have proofs. Then concatenating these proofs and appending $\psi$ yields a proof of $\psi$. Consequently, the set of all formulas possessing a proof from $AX$ is closed under modus ponens. Similarly, it is closed under necessitation.

We will now show that the set of all formulas possessing a proof from $AX$ is closed under uniform substitution. Let $\varphi \in \mathcal{L}_L$, $p_1, \ldots, p_k \in At$ and formulas $\psi_1, \ldots, \psi_k \in \mathcal{L}_L$. We will show that if $\varphi$ has a proof from $AX$, then $\varphi(p_1/\psi_1, \ldots, p_k/\psi_k)$ has a proof from $AX$. In the argument, for every formula $\alpha \in \mathcal{L}_L$, we will abbreviate $\alpha(p_1/\psi_1, \ldots, p_k/\psi_k)$ by $\alpha(P/\Psi)$.

Assume that $\varphi$ has a proof from $AX$, say $\varphi_1, \ldots, \varphi_n$. Consider a sequence of formulas $\varphi_1(P/\Psi), \ldots, \varphi_n(P/\Psi)$. It is easy to see that for each $i = 1, \ldots, n$:

1. If $\varphi_i$ is a substitution instance of a tautology in $\mathcal{L}$, so is $\varphi_i(P/\Psi)$.
2. If $\varphi_i$ is a substitution instance of an axiom from $AX$, so is $\varphi_i(P/\Psi)$.
3. If $\varphi_i$ is the result of applying modus ponens to formulas $\varphi_r$ and $\varphi_s$, for some $r, s < i$, then $\varphi_i(P/\Psi)$ is the result of applying modus ponens to formulas $\varphi_r(P/\Psi)$ and $\varphi_s(P/\Psi)$.
4. If $\varphi_i$ is the result of applying necessitation to a formula $\varphi_j$, for some $j < i$, then $\varphi_i(P/\Psi)$ is the result of applying necessitation to a formula $\varphi_j(P/\Psi)$.

It follows from (1) – (4) that $\varphi_1(P/\Psi), \ldots, \varphi_n(P/\Psi)$ is a proof of $\varphi(P/\Psi)$ from $AX$.

Hence, the set of all formulas provable from $AX$ satisfies condition (ML2).

To complete the argument for the theorem we need to show that the set of formulas provable from $AX$ is contained in every set of formulas containing $AX$ and satisfying (ML1) and (ML2). This can be done by induction on the length of a proof of a formula. We omit the details. ■

**Corollary 7.8** *If modal logics $\mathcal{S}$ and $\mathcal{T}$ are axiomatized by sets $AX_\mathcal{S}$ and $AX_\mathcal{T}$, respectively, and $AX_\mathcal{S} \subseteq AX_\mathcal{T}$, then $\mathcal{S} \subseteq \mathcal{T}$.* ▲

The notion of a proof, as introduced in Definition 7.6 can be generalized.

**Definition 7.9**

1. Let $\mathcal{S}$ be a modal logic and let $AX$ be a set of axioms of $\mathcal{S}$. Given a formula $\varphi \in \mathcal{L}_L$ and a subset $I \subseteq \mathcal{L}_L$, a sequence $\psi_1, \ldots, \psi_n$ of formulas from $\mathcal{L}_L$ is called a *proof of $\varphi$ from premises $I$ in logic $\mathcal{S}$* if $\psi_n = \varphi$ and for every $k$, $1 \leq k \leq n$, at least one of the following conditions holds:

   (a) $\psi_k$ is a substitution instance of a tautology in $\mathcal{L}$.

   (b) $\psi_k$ is a substitution instance of an axiom from $AX$.

   (c) $\psi_k \in I$.

   (d) $\psi_k$ is the result of applying modus ponens to formulas $\psi_i$ and $\psi_j$, for some $i, j < k$.

   (e) $\psi_k$ is the result of applying necessitation to a formula $\psi_i$, for some $i < k$.

   We write $I \vdash_{\mathcal{S}} \varphi$ to denote that there is a proof of $\varphi$ from premises $I$ in logic $\mathcal{S}$ and we define

$$Cn_{\mathcal{S}}(I) = \{\varphi \in \mathcal{L}_L : I \vdash_{\mathcal{S}} \varphi\}.$$

2. A theory $I \subseteq \mathcal{L}_L$ is *consistent with a modal logic $\mathcal{S}$* if $Cn_{\mathcal{S}}(I) \neq \mathcal{L}_L$, that is, if for no formula $\varphi \in \mathcal{L}_L$, both $I \vdash_{\mathcal{S}} \varphi$ and $I \vdash_{\mathcal{S}} \neg\varphi$. A theory which is not consistent with $\mathcal{S}$ is called *inconsistent* with $\mathcal{S}$. ▲

**Remark 7.10** The notion of provability defined by conditions (a), (c) and (d) of Definition 7.9 coincides with the notion of provability in the logic $PC_L$. ▲

**Remark 7.11** Another notion of provability from a set of premises is usually considered in papers and monographs on modal logics. This other relation, which we will denote by $\vdash^{\mathcal{S}}$, can be defined in terms of our notion of provability as follows: $I \vdash^{\mathcal{S}} \varphi$ if for some finite subset $\{\varphi_1, \ldots, \varphi_k\}$ of $I$, $\vdash_{\mathcal{S}} \varphi_1 \wedge \ldots \wedge \varphi_k \supset \varphi$. Clearly, for every $\varphi \in \mathcal{L}_L$,

$$\vdash^{\mathcal{S}} \varphi \text{ if and only if } \vdash_{\mathcal{S}} \varphi$$

and

$$\text{if } I \vdash^{\mathcal{S}} \varphi \text{ then } I \vdash_{\mathcal{S}} \varphi.$$

But this last implication cannot, in general, be reversed. For example, if $p$ is a propositional variable, $\{p\} \vdash_{\mathcal{S}} Lp$. On the other hand, if $p \supset Lp$ is not a theorem of $\mathcal{S}$, then $\{p\} \not\vdash^{\mathcal{S}} Lp$. Thus, the relations $\vdash_{\mathcal{S}}$ and $\vdash^{\mathcal{S}}$ are, in general, different. We will return to this issue at the end of this section. ▲

Now, we will present some simple properties of the consequence operator for modal logics. The first two of them, gathered in Theorem 7.12, follow directly from the definition of provability. Their proofs are omitted.

**Theorem 7.12** *Let $\mathcal{S}$ and $\mathcal{T}$ be modal logics such that $\mathcal{S} \subseteq \mathcal{T}$. Then:*

1. *For every theory $I \subseteq \mathcal{L}_L$, $Cn(I) \subseteq Cn_{\mathcal{S}}(I) \subseteq Cn_{\mathcal{T}}(I)$.*

2. *For all theories $I, I' \subseteq \mathcal{L}_L$, if $Cn_{\mathcal{S}}(I) = Cn_{\mathcal{S}}(I')$ then $Cn_{\mathcal{T}}(I) = Cn_{\mathcal{T}}(I')$. This statement remains true also when $\mathcal{S}$ stands for the logic $PC_L$.* ▲

Theorem 7.7 yields the following properties of the set $Cn_{\mathcal{S}}(\emptyset)$.

**Corollary 7.13** *For every modal logic $\mathcal{S}$, $Cn_{\mathcal{S}}(\emptyset)$ is the set of theorems of $\mathcal{S}$, that is, the set $Cn_{\mathcal{S}}(\emptyset)$ is the least subset of $\mathcal{L}_L$ satisfying conditions* (ML1) *and* (ML2) *of Definition 7.5.* ▲

To specify a modal logic one has to present a set of axioms of the logic. All logics considered in this book can be described by means of a finite number of axioms. However, it is well-known that not every modal logic admits such a description (see [HC84]).

All modal logics, which are referred to by name later in the book, are specified by some selection of axioms from the following list:

| | |
|---|---|
| k: | $L(p \supset q) \supset (Lp \supset Lq)$ |
| t: | $Lp \supset p$ |
| d: | $\neg Lp \vee \neg L \neg p$ |
| 4: | $Lp \supset LLp$ |
| 5: | $\neg L \neg Lp \supset Lp$ |
| b: | $\neg L \neg Lp \supset p$ |
| w5: | $\neg L \neg Lp \supset (p \supset Lp)$ |
| f: | $(p \wedge MLq) \supset L(Mp \vee q)$ |

The names of some modal logics that are studied in the book, together with their axioms, are given in Table 7.1. With only a few exceptions, the name of the logic is the boldface string of the names of its axioms.

| Name of logic | Axioms |
|---|---|
| **N** | $\emptyset$ (no axioms) |
| **t** | t |
| **b** | b |
| **5** | 5 |
| **K** | k |
| **T** | k, t |
| **S4** | k, t, 4 |
| **S4f** | k, t, 4, f |
| **kd45** | k, d, 4, 5 |
| **Sw5** | k, t, 4, w5 |
| **S5** | k, t, 4, 5 |

**Fig. 7.1.** Axiomatizations of some modal logics

Clearly, the logic **N** is the weakest modal logic. Yet, at the price of reasoning from infinitely many premises, for every modal logic $\mathcal{S}$ the consequence operator $Cn_{\mathcal{S}}$ can naturally be expressed in terms of the consequence operator $Cn_{\mathbf{N}}$. The following theorem is an immediate consequence of Definition 7.5(2).

**Theorem 7.14** *Let $\mathcal{S}$ be a modal logic specified by the set of axioms $AX$. Let $\overline{AX}$ be the closure of $AX$ under the rule of uniform substitution. Then for every theory $I \subseteq \mathcal{L}_L$,*

$$Cn_{\mathcal{S}}(I) = Cn_{\mathbf{N}}(I \cup \overline{AX}).$$ ▲

Perhaps the most important class of modal logics is the class of *normal* modal logics.

**Definition 7.15** A modal logic is *normal* if it contains axiom k. All other modal logics are called *subnormal.* ▲

**Remark 7.16** In this book we require a modal logic to satisfy the conditions (ML1) and (ML2) of Definition 7.5. Thus, according to our definition, subnormal modal logics are those that satisfy the conditions (ML1) and (ML2), but do not contain the axiom k. In the literature, the term *subnormal* modal logic has often a different meaning. Modal formal systems in the language $\mathcal{L}_L$ have been studied for which weaker conditions than (ML1) or (ML2) of Definition 7.5 are postulated. For example, sometimes it is not required that a system be closed under necessitation or that it contain all propositional tautologies. Such modal systems, even if they do contain the axiom k, are also called *subnormal* in the literature. ▲

Subnormal modal logics, in the sense used in this book, so far have received little attention. Such modalities as *knowledge* or *belief*, as commonly understood, satisfy the axiom k and, thus, it is only natural that the logics introduced to formally model these modalities were designed to satisfy it, too. The case of nonmonotonic modal reasoning about knowledge or belief is different. Roughly speaking, belief sets that are constructed by nonmonotonic modal logics are always closed under provability in the logic **S5**, no matter whether $\mathcal{S}$ contains k or not. So, the stipulation that $\mathcal{S}$ contains k is no longer essential.

Normal modal logics have several important properties that fail, in general, for subnormal logics. Some of them are gathered in our next theorem.

**Theorem 7.17** *Let $\mathcal{S}$ be a normal modal logic. Then for every theory $I \subseteq \mathcal{L}_L$ and for all formulas $\alpha, \beta \in \mathcal{L}_L$, the following properties hold:*

1. *If $I \vdash_{\mathcal{S}} \alpha \supset \beta$, then $I \vdash_{\mathcal{S}} L\alpha \supset L\beta$ and $I \vdash_{\mathcal{S}} M\alpha \supset M\beta$.*
2. *If $I \vdash_{\mathcal{S}} \alpha \equiv \beta$, then $I \vdash_{\mathcal{S}} L\alpha \equiv L\beta$ and $I \vdash_{\mathcal{S}} M\alpha \equiv M\beta$.*
3. *$\vdash_{\mathcal{S}} L(\alpha \wedge \beta) \equiv L\alpha \wedge L\beta$.*
4. *$\vdash_{\mathcal{S}} M\alpha \wedge L\beta \supset M(\alpha \wedge \beta)$.*
5. *For every $\varphi \in \mathcal{L}_L$ and for every propositional variable $p \in At$, if $I \vdash_{\mathcal{S}} \alpha \equiv \beta$, then $I \vdash_{\mathcal{S}} \varphi(p/\alpha) \equiv \varphi(p/\beta)$.*

Proof: (1) We first prove that if $I \vdash_{\mathcal{S}} \alpha \supset \beta$, then $I \vdash_{\mathcal{S}} L\alpha \supset L\beta$. Clearly, applying the necessitation rule, $I \vdash_{\mathcal{S}} L(\alpha \supset \beta)$. Since k is an axiom of $\mathcal{S}$, we also have $I \vdash_{\mathcal{S}} L(\alpha \supset \beta) \supset (L\alpha \supset L\beta)$. Hence, by modus ponens, $I \vdash_{\mathcal{S}} L\alpha \supset L\beta$.

Next, observe that if $I \vdash_{\mathcal{S}} \alpha \supset \beta$, then $I \vdash_{\mathcal{S}} \neg\beta \supset \neg\alpha$ (since $\mathcal{S}$ contains the logic of propositional calculus $PC_L$). Consequently, $I \vdash_{\mathcal{S}} L\neg\beta \supset L\neg\alpha$. Again by the properties of propositional provability, we obtain that $I \vdash_{\mathcal{S}} \neg L\neg\alpha \supset \neg L\neg\beta$. Thus, $I \vdash_{\mathcal{S}} M\alpha \supset M\beta$ follows.

(2) This assertion is a direct consequence of (1).

(3) Since $\vdash (\alpha \wedge \beta) \supset \alpha$, by (1) it follows that $\vdash_{\mathcal{S}} L(\alpha \wedge \beta) \supset L\alpha$. Similarly, we have $\vdash_{\mathcal{S}} L(\alpha \wedge \beta) \supset L\beta$. Thus, $\vdash_{\mathcal{S}} L(\alpha \wedge \beta) \supset (L\alpha \wedge L\beta)$.

To show that the converse implication is a theorem of $\mathcal{S}$, we proceed as follows. First, we observe that by the axiom k, $\vdash_{\mathcal{S}} L(\beta \supset (\alpha \wedge \beta)) \supset (L\beta \supset L(\alpha \wedge \beta))$. Since $\alpha \supset (\beta \supset (\alpha \wedge \beta))$ is a tautology, it follows that $\vdash_{\mathcal{S}} L\alpha \supset L(\beta \supset (\alpha \wedge \beta))$. Thus, using properties of propositional provability we get $L\alpha \supset (L\beta \supset L(\alpha \wedge \beta))$ and, next, $(L\alpha \wedge L\beta) \supset L(\alpha \wedge \beta)$.

(4) Since $M$ abbreviates $\neg L \neg$, it follows that $M\alpha \wedge L\beta \supset M(\alpha \wedge \beta)$ is propositionally equivalent to $L\neg(\alpha \wedge \beta) \supset (L\beta \supset L\neg\alpha)$. Since $\neg(\alpha \wedge \beta) \equiv (\beta \supset \neg\alpha)$ is a tautology, the assertion follows from (2).

(5) We proceed by induction on the length of $\varphi$. First suppose that $\varphi \in At$. Then, there are two possibilities.

1. The atom $p$ does not occur in $\varphi$. Then $\varphi(p/\alpha) = \varphi(p/\beta)$ and $I \vdash_{\mathcal{S}} \varphi(p/\alpha) \equiv \varphi(p/\beta)$ is evident.

2. $\varphi = p$. In this case, $\varphi(p/\alpha) = \alpha$ and $\varphi(p/\beta) = \beta$. Since $I \vdash_{\mathcal{S}} \alpha \equiv \beta$, $I \vdash_{\mathcal{S}} \varphi(p/\alpha) \equiv \varphi(p/\beta)$ follows.

Next, consider a formula $\varphi$ of length greater than 1 and assume that the claim holds for all shorter formulas. If $\varphi = \neg\psi$ then, by induction hypothesis, $I \vdash_{\mathcal{S}} \psi(p/\alpha) \equiv \psi(p/\beta)$. Consequently, $I \vdash_{\mathcal{S}} \neg\psi(p/\alpha) \equiv \neg\psi(p/\beta) (= \varphi(p/\alpha) \equiv \varphi(p/\beta))$. The cases when $\varphi = \psi_1 \vee \psi_2$, $\varphi = \psi_1 \wedge \psi_2$ and $\varphi = \psi_1 \supset \psi_2$ can be dealt with in the same manner. The case when $\varphi = L\psi$ follows from (2). ■

**Remark 7.11 (continued)** We have already observed earlier that the two relations of provability, $\vdash_{\mathcal{S}}$ and $\vdash^{\mathcal{S}}$, are different. Here, we will discuss the question of the mutual relationship between these two notions. For a theory $I \subseteq \mathcal{L}_L$, define

$$L^k I = \{L^k\varphi : \varphi \in I\},\ k = 0, 1, \ldots, \quad \text{and}$$

$$L^\infty I = \bigcup_{k=0}^{\infty} L^k I,$$

where $L^0\varphi = \varphi$ and $L^{k+1}\varphi = L(L^k\varphi)$. It is easy to see that

$$\text{if } L^\infty I \vdash^{\mathcal{S}} \varphi \text{ then } I \vdash_{\mathcal{S}} \varphi.$$

If $\mathcal{S}$ is a normal modal logic, then one can show that also conversely

$$\text{if } I \vdash_{\mathcal{S}} \varphi \text{ then } L^\infty I \vdash^{\mathcal{S}} \varphi. \tag{7.1}$$

If, in addition, $\mathcal{S}$ contains the axiom 4, then we have an even stronger result

$$LI \cup I \vdash^{S} \varphi \text{ if and only if } I \vdash_{S} \varphi.$$

It follows that, for normal systems, both notions are closely connected. However, for subnormal logics, one can show that (7.1) fails. ▲

## 7.2 General Kripke semantics

In the previous section we have described a proof theory for modal logics. Now, we will develop a model theory and prove corresponding soundness and completeness results.

**Definition 7.18** A *Kripke structure* is a triple

$$\mathcal{M} = \langle M, \{R_\psi\}_{\psi \in \mathcal{L}_L}, V\rangle,$$

where:

1. $M$ is a nonempty set; its elements are referred to as *possible worlds* or, simply, *worlds*.
2. $V: M \to \{0,1\}^{At}$ is a function assigning a propositional valuation to each world $a \in M$.
3. For every $\psi \in \mathcal{L}_L$, $R_\psi$ is a binary relation on $M$, referred to as the *accessibility* relation for $\psi$. ▲

Let us note that while "standard" Kripke models contain just one accessibility relation, our Kripke structures contain infinitely many of them, one for each formula in the language. However, both concepts are closely related. The connection will be made clear in the next section.

**Definition 7.19** Given a Kripke structure $\mathcal{M} = \langle M, \{R_\psi\}_{\psi \in \mathcal{L}_L}, V\rangle$, we define the *satisfiability relation* between formulas $\varphi$ of $\mathcal{L}_L$ and worlds $a \in M$. We denote it by $\langle \mathcal{M}, a\rangle \models \varphi$. This relation is defined for $a \in M$ by induction on the complexity of formula $\varphi \in \mathcal{L}_L$. In other words, we specify what it means for $\varphi$ to be *satisfied* (or *valid*) in the world $a \in M$.

1. If $\varphi$ is an atom, then $\langle \mathcal{M}, a\rangle \models \varphi$ if and only if $V(a)(\varphi) = \mathbf{t}$.
2. If $\varphi = \neg\psi$, then $\langle \mathcal{M}, a\rangle \models \varphi$ if and only if it is not true that $\langle \mathcal{M}, a\rangle \models \psi$.
3. If $\varphi = \psi_1 \vee \psi_2$ then $\langle \mathcal{M}, a\rangle \models \varphi$ if and only if $\langle \mathcal{M}, a\rangle \models \psi_1$ or $\langle \mathcal{M}, a\rangle \models \psi_2$.

4. If $\varphi = \psi_1 \wedge \psi_2$ then $\langle \mathcal{M}, a \rangle \models \varphi$ if and only if $\langle \mathcal{M}, a \rangle \models \psi_1$ and $\langle \mathcal{M}, a \rangle \models \psi_2$.

5. If $\varphi = \psi_1 \supset \psi_2$ then $\langle \mathcal{M}, a \rangle \models \varphi$ if and only if it is not true that $\langle \mathcal{M}, a \rangle \models \psi_1$, or if $\langle \mathcal{M}, a \rangle \models \psi_2$.

6. If $\varphi = L\psi$, then $\langle \mathcal{M}, a \rangle \models \varphi$ if and only if for every $a'$ such that $(a, a') \in R_\psi$, $\langle \mathcal{M}, a' \rangle \models \psi$.

We write $\langle \mathcal{M}, a \rangle \not\models \varphi$ to denote that it is not true that $\langle \mathcal{M}, a \rangle \models \varphi$.

We say that a Kripke structure $\mathcal{M} = \langle M, \{R_\psi\}_{\psi \in \mathcal{L}_L}, V \rangle$ *satisfies* $\varphi$ (or that $\varphi$ is *valid* in $\mathcal{M}$) if for all $a \in M$, $\langle \mathcal{M}, a \rangle \models \varphi$. We write $\mathcal{M} \models \varphi$ if $\varphi$ is valid in $\mathcal{M}$. We say that $\mathcal{M}$ *satisfies* a theory $I$ (or that $I$ is *valid* in $\mathcal{M}$) if $\mathcal{M} \models \varphi$ for every $\varphi \in I$. We write $\mathcal{M} \models I$ if $I$ is valid in $\mathcal{M}$.

For a Kripke structure $\mathcal{M}$, we define the *theory of the structure* $\mathcal{M}$ to be the set of all formulas valid in the structure and denote this set by $Th(\mathcal{M})$. That is, $Th(\mathcal{M}) = \{\varphi \in \mathcal{L}_L : \mathcal{M} \models \varphi\}$. ▲

Notice that this definition generalizes usual definition of satisfaction for valuations. If $\mathcal{M}$ consists of a single world then for modal-free formulas satisfaction in $\mathcal{M}$ and in this unique world coincide. For general $\mathcal{M}$ there is an important difference: whereas for a valuation $v$, $\{\varphi \in \mathcal{L} : \models_v \varphi\}$ is a complete theory (that is, for every $\psi$ contains either $\psi$ or $\neg\psi$), $Th(\mathcal{M})$ does not have to be complete.

**Theorem 7.20** *Let $\mathcal{M}$ be a Kripke structure.*

1. *For every formula $\varphi \in \mathcal{L}_L$, if $\mathcal{M} \models \varphi$ then $\mathcal{M} \models L\varphi$. In other words, $Th(\mathcal{M})$ is closed under the necessitation rule.*

2. *For all formulas $\varphi, \psi \in \mathcal{L}_L$, if $\mathcal{M} \models \varphi$ and $\mathcal{M} \models \varphi \supset \psi$, then $\mathcal{M} \models \psi$. In other words, $Th(\mathcal{M})$ is closed under the modus ponens rule.*

Proof: Let $\mathcal{M} = \langle M, \{R_\psi\}_{\psi \in \mathcal{L}_L}, V \rangle$. Since $\mathcal{M} \models \varphi$, we have $\langle \mathcal{M}, a \rangle \models \varphi$, for every $a \in M$. Consider a world $b \in M$. Then, in particular, $\langle \mathcal{M}, b' \rangle \models \varphi$, for every $b' \in M$ such that $(b, b') \in R_\varphi$. Hence, $\langle \mathcal{M}, b \rangle \models L\varphi$ and, consequently, $\mathcal{M} \models L\varphi$. Part (2) of the theorem can be proved similarly. ■

**Definition 7.21** Let $\mathcal{C}$ be a class of Kripke structures and let $\mathcal{S}$ be a modal logic.

1. We say that logic $\mathcal{S}$ is *sound with respect to* class $\mathcal{C}$ if for every theory $I \subseteq \mathcal{L}_L$, for every formula $\varphi \in \mathcal{L}_L$ and for every structure $\mathcal{M} \in \mathcal{C}$, $I \vdash_{\mathcal{S}} \varphi$ and $\mathcal{M} \models I$ imply $\mathcal{M} \models \varphi$.

2. We say that logic $\mathcal{S}$ is *complete with respect to* class $\mathcal{C}$ if for every theory $I \subseteq \mathcal{L}_L$ and for every formula $\varphi \in \mathcal{L}_L$ the following happens: If for every structure $\mathcal{M} \in \mathcal{C}$ such that $\mathcal{M} \models I$ we have $\mathcal{M} \models \varphi$, then $I \vdash_{\mathcal{S}} \varphi$.

3. If a modal logic $\mathcal{S}$ is both sound and complete with respect to a class of structures $\mathcal{C}$ then we say that $\mathcal{S}$ is *characterized* by the class $\mathcal{C}$. ▲

In the next result we establish sufficient conditions that guarantee that a modal logic $\mathcal{S}$ is sound with respect to a class $\mathcal{C}$ of Kripke structures.

**Theorem 7.22** *Let $\mathcal{S}$ be a modal logic. Let $\mathcal{C}$ be a class of Kripke structures such that for every $\mathcal{M} \in \mathcal{C}$ and for every axiom $\psi$ of $\mathcal{S}$, all substitution instances of $\psi$ are valid in $\mathcal{M}$. Then for every theory $I \subseteq \mathcal{L}_L$, for every formula $\varphi \in \mathcal{L}_L$ and for every Kripke structure $\mathcal{M} \in \mathcal{C}$, if $I \vdash_{\mathcal{S}} \varphi$ and $\mathcal{M} \models I$ then $\mathcal{M} \models \varphi$. In other words, $\mathcal{S}$ is sound with respect to $\mathcal{C}$.*

Proof: We proceed by induction on the length of a proof of a formula. Consider a formula $\varphi$ with a derivation from $I$ in $\mathcal{S}$ of length $n$. Assume also, as an induction hypothesis, that the assertion holds for every formula with a derivation from $I$ in $\mathcal{S}$ of length less than $n$.

There are several possibilities. If $\varphi_n$ is (1) a substitution instance of a tautology from $\mathcal{L}$, or (2) a substitution instance of an axiom of $\mathcal{S}$, or (3) $\varphi_n \in I$, then the assertion is evident. These three cases establish also the basis of induction.

Another possibility is that $\varphi_n$ is derived from $\varphi_i$ and $\varphi_j$, where $i, j < n$, by means of modus ponens. Since $I \vdash_{\mathcal{S}} \varphi_i$ and $I \vdash_{\mathcal{S}} \varphi_j$, by the induction hypotheses it follows that $\mathcal{M} \models \varphi_i$ and $\mathcal{M} \models \varphi_j$. Consequently, $\mathcal{M} \models \varphi_n$ follows by Theorem 7.20(2).

Finally, if $\varphi_n$ follows by necessitation from a formula $\varphi_i$, where $i < n$, then $I \vdash_{\mathcal{S}} \varphi_i$ and, by the induction hypothesis, $\mathcal{M} \models \varphi_i$. Consequently, $\mathcal{M} \models L\varphi_i (= \varphi_n)$ follows by Theorem 7.20(1). ■

**Corollary 7.23** *The logic* **N** *is sound with respect to the class of all Kripke structures.* ▲

Now, we will consider the more difficult problem of completeness. We will describe a method of proving completeness results known as the method of *canonical models*. This method is well known and well explained in many monographs on modal logic [Che80, Fit83, HC84]. Our discussion follows the standard scheme for the canonical model argument. Two points that need to be stressed here are that our completeness results provide a semantic characterization for the provability relation $I \vdash_{\mathcal{S}} \varphi$ and that they apply to normal as well as subnormal modal logics.

**Definition 7.24** Let $\mathcal{S}$ be a modal logic and let $I \subseteq \mathcal{L}_L$. A theory $T \subseteq \mathcal{L}_L$ is called *I-inconsistent with* $\mathcal{S}$ if there is a nonempty, finite set $\{\varphi_1, \ldots, \varphi_n\} \subseteq T$, such that $I \vdash_{\mathcal{S}} \neg\varphi_1 \vee \ldots \vee \neg\varphi_n$. We say that $T$ is *I-consistent with* $\mathcal{S}$ if $T$ is not $I$-inconsistent with $\mathcal{S}$. ▲

Several results that follow gather useful properties of sets that are $I$-consistent with $\mathcal{S}$.

**Lemma 7.25** *Let* $\mathcal{S}$ *be a modal logic.*

1. *For every* $I \subseteq \mathcal{L}_L$, $\emptyset$ *is* $I$*-consistent with* $\mathcal{S}$.
2. *A theory* $I$ *is consistent with* $\mathcal{S}$ *if and only if there exist nonempty theories* $I$*-consistent with* $\mathcal{S}$.

Proof: (1) follows directly from Definition 7.24.

(2) Assume first that $I$ is consistent with $\mathcal{S}$. Then, there is a formula $\varphi \in \mathcal{L}_L$ such that $I \not\vdash_{\mathcal{S}} \varphi$. Clearly, the theory $\{\neg\varphi\}$ is $I$-consistent with $\mathcal{S}$. Conversely, let $T$ be a nonempty theory $I$-consistent with $\mathcal{S}$. Let $\varphi \in T$. Then $I \not\vdash_{\mathcal{S}} \neg\varphi$. Hence, $I$ is consistent with $\mathcal{S}$. ■

**Lemma 7.26** *If* $T$ *is* $I$*-consistent with* $\mathcal{S}$ *then* $T$ *is consistent with the logic* $PC_L$ *(that is* $T$ *is propositionally consistent).*

Proof: Suppose that $T$ is propositionally inconsistent. Then there is a formula $\varphi \in T$ such that $T \vdash \neg\varphi$. Consequently, there is a finite subset $\{\varphi_1, \ldots, \varphi_n\} \subseteq T$ such that $\{\varphi_1, \ldots, \varphi_n\} \vdash \neg\varphi$. By the deduction theorem for propositional logic, we get $\vdash \neg\varphi_1 \vee \ldots \vee \neg\varphi_n \vee \neg\varphi$. Therefore, $I \vdash_{\mathcal{S}} \neg\varphi_1 \vee \ldots \vee \neg\varphi_n \vee \neg\varphi$. ■

The following lemma is the key element of every canonical-model argument.

**Lemma 7.27** *If $T \subseteq \mathcal{L}_L$ is $I$-consistent with $\mathcal{S}$ and $\neg L\varphi \in T$, then $\{\neg\varphi\}$ is $I$-consistent with $\mathcal{S}$.*

Proof: Assume that $\{\neg\varphi\}$ is $I$-inconsistent with $\mathcal{S}$. Then $I \vdash_{\mathcal{S}} \neg\neg\varphi$. Consequently, $I \vdash_{\mathcal{S}} \varphi$. Hence, $I \vdash_{\mathcal{S}} L\varphi$. This, of course, implies that $I \vdash_{\mathcal{S}} \neg\neg L\varphi$, which means that $T$ is $I$-inconsistent with $\mathcal{S}$. ■

**Definition 7.28** An $I$-consistent with $\mathcal{S}$ theory $T$ is called *maximal* if for every $I$-consistent with $\mathcal{S}$ theory $U$ such that $T \subseteq U$, $T = U$. ▲

**Lemma 7.29** *Let $I \subseteq \mathcal{L}_L$ and let $\mathcal{S}$ be a modal logic. If $T \subseteq \mathcal{L}_L$ is $I$-consistent with $\mathcal{S}$ then there is a maximal $I$-consistent with $\mathcal{S}$ theory $U$ such that $T \subseteq U$.*

Proof: The family of all theories which are $I$-consistent with $\mathcal{S}$ is nonempty (Lemma 7.25). The union of every chain (with respect to inclusion) of $I$-consistent with $\mathcal{S}$ theories containing $T$ is $I$-consistent with $\mathcal{S}$ and contains $T$. Consequently the Kuratowski-Zorn lemma (Theorem 2.37) is applicable and so there exists a maximal $I$-consistent with $\mathcal{S}$ theory that contains $T$. ■

The next lemma collects properties of maximal $I$-consistent with $\mathcal{S}$ theories.

**Lemma 7.30** *Let $I \subseteq \mathcal{L}_L$ be consistent with a modal logic $\mathcal{S}$. Let $T$ be a maximal $I$-consistent with $\mathcal{S}$ theory.*

1. *For every $\varphi \in \mathcal{L}_L$, $T$ contains exactly one of the formulas $\varphi$, $\neg\varphi$.*
2. *For every $\varphi$, $\psi \in \mathcal{L}_L$, $\varphi \wedge \psi \in T$ if and only if $\varphi \in T$ and $\psi \in T$.*
3. *For every $\varphi$, $\psi \in \mathcal{L}_L$, $\varphi \vee \psi \in T$ if and only if $\varphi \in T$ or $\psi \in T$.*
4. *For every $\varphi \in \mathcal{L}_L$, if $I \vdash_{\mathcal{S}} \varphi$ then $\varphi \in T$. In particular, $I \subseteq T$ and for every substitution instance $\varphi$ of an axiom of $\mathcal{S}$, $\varphi \in T$.*
5. *For every $\varphi, \psi \in \mathcal{L}_L$, if $\varphi \in T$ and $\varphi \supset \psi \in T$, then $\psi \in T$.*

Proof: (1) By Lemma 7.26, $T$ is propositionally consistent. Consequently, at most one of $\varphi$, $\neg\varphi$ belongs to $T$. Assume that neither $\varphi \notin T$ nor $\neg\varphi \notin T$. By maximality of $T$ there exist formulas $\alpha_1, \ldots, \alpha_k \in T$ and $\beta_1, \ldots, \beta_m \in T$ such that $I \vdash_{\mathcal{S}} \neg\alpha_1 \vee \ldots \vee \neg\alpha_k \vee \neg\varphi$ and $I \vdash_{\mathcal{S}} \neg\beta_1 \vee \ldots \vee \neg\beta_m \vee \neg\neg\varphi$. If $k = 0$ and $m = 0$ then $I$ is inconsistent with $\mathcal{S}$, contradicting our assumption. Hence, $k+m \geq 1$. Using propositional calculus, $I \vdash_{\mathcal{S}} \neg\alpha_1 \vee \ldots \vee \neg\alpha_k \vee \neg\beta_1 \vee \ldots \vee \neg\beta_m$. Since $k + m \geq 1$, $T$ is $I$-inconsistent with $\mathcal{S}$, a contradiction.

(2) If $\varphi \wedge \psi \in T$ and $\varphi \notin T$ then, by (1), $\neg\varphi \in T$. Since $\neg\neg\varphi \vee \neg(\varphi \wedge \psi)$ is a tautology, $I \vdash_{\mathcal{S}} \neg\neg\varphi \vee \neg(\varphi \wedge \psi)$. Hence, $T$ is $I$-inconsistent with $\mathcal{S}$, a contradiction. Hence, $\varphi \in T$ and, by the symmetry argument, $\psi \in T$, as well.

Conversely, assume that $\varphi \in T$, $\psi \in T$ and $\varphi \wedge \psi \notin T$. Then, by (1), $\neg(\varphi \wedge \psi) \in T$. But then $T$ is propositionally inconsistent, a contradiction with Lemma 7.26.

(3) This claim can be proved in a similar way as (2). The details are omitted.

(4) If $\varphi \notin T$ then, by (1), $\neg\varphi \in T$. Since $T$ is $I$-consistent with $\mathcal{S}$, $I \not\vdash_{\mathcal{S}} \neg\neg\varphi$. So, $I \not\vdash_{\mathcal{S}} \varphi$.

(5) Assume that $\varphi \in T$, $\varphi \supset \psi \in T$ and $\psi \notin T$. Then $\{\varphi, \varphi \supset \psi, \neg\psi\}$ is a subset of $T$. Consequently, $T$ is propositionally inconsistent, a contradiction with Lemma 7.26. ■

Now, for a logic $\mathcal{S}$ and a consistent with $\mathcal{S}$ theory $I \subseteq \mathcal{L}_L$, we will define the notion of a canonical structure and then show that formulas valid in this structure are exactly those that can be proved in $\mathcal{S}$ from $I$.

**Definition 7.31** Let $\mathcal{S}$ be a modal logic and let $I \subseteq \mathcal{L}_L$ be consistent with $\mathcal{S}$. Define $M$ to be the set of all maximal $I$-consistent with $\mathcal{S}$ theories $T \subseteq \mathcal{L}_L$. Next, for every formula $\varphi \in \mathcal{L}_L$, define a relation $R_\varphi \subseteq M \times M$ by

$$(a, b) \in R_\varphi \quad \text{if} \quad L\varphi \notin a \quad \text{or} \quad \varphi \in b.$$

Finally, for every $a \in M$ and for every propositional variable $p$, define $V(a)(p) = \mathbf{t}$ if $p \in a$ and $V(a)(p) = \mathbf{f}$, otherwise.

The Kripke structure $\langle M, \{R_\varphi\}_{\varphi \in \mathcal{L}_L}, V\rangle$ (note that by Lemma 7.25 we have $M \neq \emptyset$ and, therefore, the structure $\langle M, \{R_\varphi\}_{\varphi \in \mathcal{L}_L}, V\rangle$ is well-defined) is called the *$\mathcal{S}$-canonical* structure for $I$ and is denoted by $\mathcal{M}^{can}_{\mathcal{S}}(I)$. ▲

Now, we are ready to prove the *fundamental theorem for canonical structures*.

**Theorem 7.32** *Let $I \subseteq \mathcal{L}_L$ be a theory consistent with $\mathcal{S}$. For every formula $\varphi \in \mathcal{L}_L$, and for every world $a$ of the $\mathcal{S}$-canonical structure $\mathcal{M}_{\mathcal{S}}^{can}(I)$ for $I$, $\langle \mathcal{M}_{\mathcal{S}}^{can}(I), a\rangle \models \varphi$ if and only if $\varphi \in a$.*

Proof: In the proof we abbreviate $\mathcal{M}_{\mathcal{S}}^{can}(I)$ by $\mathcal{M}$. We proceed by induction on the length of $\varphi$.

1. If $\varphi$ is atomic, then the definition of $V$ implies the assertion.

2. If $\varphi = \neg\vartheta$ then $\langle \mathcal{M}, a\rangle \models \varphi$ precisely when $\langle \mathcal{M}, a\rangle \not\models \vartheta$. By the induction hypothesis, this latter statement is equivalent to $\vartheta \notin a$. Since the theory $a$ is maximal $I$-consistent with $\mathcal{S}$, therefore, by Lemma 7.30(1), it follows that this last fact is equivalent to $\varphi \in a$.

3. If $\varphi = \vartheta_1 \wedge \vartheta_2$ or $\varphi = \vartheta_1 \vee \vartheta_2$, then we reason as in (2), using Lemma 7.30(2) and (3), respectively.

4. If $\varphi = L\vartheta$, the argument is a bit more complicated. First, assume that $\langle \mathcal{M}, a\rangle \not\models L\vartheta$. Then for some $a'$ such that $(a, a') \in R_\vartheta$, $\langle \mathcal{M}, a'\rangle \not\models \vartheta$. By the induction hypothesis, $\vartheta \notin a'$. Since $(a, a') \in R_\vartheta$, $L\vartheta \notin a$.

Next, suppose that $L\vartheta \notin a$. Then by Lemma 7.30(1), $\neg L\vartheta \in a$. Consequently, by Lemma 7.27, the set $\{\neg\vartheta\}$ is $I$-consistent with $\mathcal{S}$. Let $a'$ be a maximal $I$-consistent with $\mathcal{S}$ set containing $\{\neg\vartheta\}$ (the existence of such a set follows from Lemma 7.29). By Lemma 7.30(1), $\vartheta \notin a'$. By the induction hypothesis, $\langle \mathcal{M}, a'\rangle \not\models \vartheta$. It is easy to see that, by the definition of $R_\vartheta$, $(a, a') \in R_\vartheta$. Thus, $\langle \mathcal{M}, a\rangle \models \neg L\vartheta$. Consequently, $\langle \mathcal{M}, a\rangle \not\models L\vartheta$. ■

**Corollary 7.33** *For every modal logic $\mathcal{S}$ and for every consistent with $\mathcal{S}$ theory $I \subseteq \mathcal{L}_L$*

$$\mathcal{M}_{\mathcal{S}}^{can}(I) \models I$$

Proof: By Lemma 7.30(4), for every world $a$ of the $\mathcal{S}$-canonical structure for $I$ we have $I \subseteq a$. Thus, the assertion follows from Theorem 7.32. ■

Our next result shows that each modal logic $\mathcal{S}$ is complete with respect to the class of all $\mathcal{S}$-canonical structures.

**Corollary 7.34** *Let $I \subseteq \mathcal{L}_L$ be a theory consistent with $\mathcal{S}$. For every formula $\varphi \in \mathcal{L}_L$, if $\mathcal{M}_{\mathcal{S}}^{can}(I) \models \varphi$ then $I \vdash_{\mathcal{S}} \varphi$.*

Proof: Assume that $I \not\vdash_{\mathcal{S}} \varphi$. Then the set $\{\neg\varphi\}$ is $I$-consistent with $\mathcal{S}$ and, by Lemma 7.29, there is a world $a$ in $\mathcal{M}_{\mathcal{S}}^{can}(I)$ such that $\neg\varphi \in a$. By Theorem 7.32, $\langle \mathcal{M}_{\mathcal{S}}^{can}(I), a\rangle \models \neg\varphi$. Consequently, $\mathcal{M}_{\mathcal{S}}^{can}(I) \not\models \varphi$. ■

Theorem 7.22, Corollary 7.33 and Corollary 7.34 together give us the following completeness result.

**Corollary 7.35** *Let $\mathcal{S}$ be a modal logic and let $I \subseteq \mathcal{L}_L$ be a theory consistent with $\mathcal{S}$. Then*

$$\mathcal{M}_{\mathcal{S}}^{can}(I) \models \varphi \textit{ if and only if } I \vdash_{\mathcal{S}} \varphi.$$

▲

As an immediate corollary, we obtain that each modal logic $\mathcal{S}$ is complete with respect to any class of Kripke structures that contains the class of all $\mathcal{S}$-canonical structures.

**Corollary 7.36** *Let $\mathcal{S}$ be a modal logic and $\mathcal{C}$ be a class of Kripke structures. If for every consistent with $\mathcal{S}$ theory $I \subseteq \mathcal{L}_L$, $\mathcal{M}_{\mathcal{S}}^{can}(I) \in \mathcal{C}$, then the logic $\mathcal{S}$ is complete with respect to $\mathcal{C}$. In particular, $\mathcal{S}$ is characterized by the class of all canonical structures $\mathcal{M}_{\mathcal{S}}^{can}(I)$, where $I$ ranges over all theories consistent with $\mathcal{S}$.* ▲

We will now apply our general results, Theorem 7.22 and Corollary 7.36, to obtain soundness and completeness results for some specific modal logics. The corresponding classes of structures are described in terms of properties of accessibility relations.

**Definition 7.37** Let $R \subseteq M \times M$.

1. If for every $a \in M$, $(a,a) \in R$, then $R$ is called *reflexive*.
2. If for all $a, b \in M$ such that $(a,b) \in R$, $(b,a) \in R$, then $R$ is called *symmetric*.
3. If for all $a, b, c \in M$ such that $(a,b) \in R$ and $(b,c) \in R$, $(a,c) \in R$, then $R$ is called *transitive*.
4. If $R$ is reflexive, symmetric and transitive, then $R$ is called an *equivalence* relation.
5. If $R = M \times M$, then $R$ is called *universal*.
6. If for all $a, b, c \in R$ such that $(a,b) \in R$ and $(a,c) \in R$ we have also $(b,c) \in R$, then $R$ is called *euclidean*.
7. If $a \in M$ and for every $b \in M$, $(a,b) \notin R$, then $a$ is called a *dead end* of $R$. ▲

**Lemma 7.38** *Let $\mathcal{S}$ be a modal logic and let $I \subseteq \mathcal{L}_L$ be a theory consistent with $\mathcal{S}$. Let $M$ be the set of worlds and let $\{R_\psi\}_{\psi \in \mathcal{L}_L}$ be the accessibility relations of the $\mathcal{S}$-canonical structure for $I$.*

1. *If* t *is an axiom of $\mathcal{S}$, then for every $\psi \in \mathcal{L}_L$, $R_\psi$ is reflexive.*
2. *If* 4 *is an axiom of $\mathcal{S}$, then for every formula $\psi \in \mathcal{L}_L$,*
$$R_{L\psi} \circ R_\psi \subseteq R_\psi \tag{7.2}$$
*(here $\circ$ is the operation of composition introduced in Definition 2.39).*
3. *If* b *is an axiom of $\mathcal{S}$, then for every formula $\psi \in \mathcal{L}_L$ and for all worlds $a, a'$,*
$$\textit{if } (a, a') \in R_{\neg L\psi} \textit{ then } (a', a) \in R_\psi. \tag{7.3}$$
4. *If* 5 *is an axiom of $\mathcal{S}$, then for every formula $\psi \in \mathcal{L}_L$, and for all worlds $a, a', a''$,*
$$(a, a') \in R_\psi \textit{ and } (a, a'') \in R_{\neg L\psi} \textit{ implies } (a'', a') \in R_\psi. \tag{7.4}$$

Proof: (1) Let $a \in M$. Assume that $L\psi \in a$. By Lemma 7.30(4) and (5), it follows that $\psi \in a$. Thus, by the definition of the accessibility relations in the canonical structures, $(a, a) \in R_\psi$, that is, $R_\psi$ is reflexive.

(2) Let $a, a'' \in M$ be such that $(a, a'') \in R_{L\psi} \circ R_\psi$. Then, there is a world $a' \in M$ such that $(a, a') \in R_{L\psi}$ and $(a', a'') \in R_\psi$. Assume $L\psi \in a$. Since $L\psi \supset LL\psi \in a$ (Lemma 7.30(4) and (5)), $LL\psi \in a$. Since $(a, a') \in R_{L\psi}$, $L\psi \in a'$. Hence, since $(a', a'') \in R_\psi$, $\psi \in a''$. Consequently, $(a, a'') \in R_\psi$.

(3) Let $\psi \in \mathcal{L}_L$ and let $a, a' \in M$. Assume that $(a, a') \in R_{\neg L\psi}$. To prove that $(a', a) \in R_\psi$, assume $L\psi \in a'$. Then $\neg L\psi \notin a'$. Since $(a, a') \in R_{\neg L\psi}$, $L\neg L\psi \notin a$. Consequently, $\neg L\neg L\psi \in a$. Since every substitution instance of the axiom b belongs to $a$, it follows that $\psi \in a$. Hence, $(a', a) \in R_\psi$.

(4) Consider a formula $\psi \in \mathcal{L}_L$ and worlds $a, a', a'' \in M$ such that $(a, a') \in R_\psi$ and $(a, a'') \in R_{\neg L\psi}$. Assume that $\psi \notin a'$. Then, by the definition of the accessibility relations in canonical structures, $L\psi \notin a$. Consequently, by Lemma 7.30(1), $\neg L\psi \in a$. By Lemma 7.30(4), $\neg L\psi \supset L\neg L\psi \in a$ and, by Lemma 7.30(5), $L\neg L\psi \in a$. Again by the definition of the accessibility relations in canonical structures, $\neg L\psi \in a''$. Consequently, $L\psi \notin a''$ and $(a'', a') \in R_\psi$. ∎

In general there is no connection between different accessibility relations of a Kripke structure $\mathcal{M}$. If we impose conditions like those of Lemma 7.38 then some sentences become automatically true in $\mathcal{M}$.

**Corollary 7.39**

1. *Logic* **N** *is characterized by the class of all Kripke structures.*
2. *Logic* **t** *is characterized by the class of all Kripke structures whose all accessibility relations are reflexive.*
3. *Logic* **4** *is characterized by the class of all Kripke structures whose accessibility relations* $R_\psi$ *satisfy (7.2).*
4. *Logic* **b** *is characterized by the class of all Kripke structures whose accessibility relations* $R_\psi$ *are such that for every formula* $\psi \in \mathcal{L}_L$ *and for all worlds* $a, a'$ *(7.3) holds.*
5. *Logic* **5** *is characterized by the class of all Kripke structures whose accessibility relations* $R_\psi$ *are such that for every formula* $\psi \in \mathcal{L}_L$*, and for all worlds* $a, a', a''$*, (7.4) holds.*

Proof: (1) Soundness was established in Corollary 7.23. Completeness follows directly from Corollary 7.36.

The "completeness" part of the statements (2) - (5) follows from Lemma 7.38 and Corollary 7.36. To prove the "soundness" part, we use Theorem 7.22. We give the details only for the case when $\mathcal{S}$ contains the axiom b (statement (4)). Other cases are dealt with in a similar manner.

(4) Consider an arbitrary substitution instance $\neg L \neg L\varphi \supset \varphi$ of the axiom b. Let $\mathcal{M} = \langle M, \{R_\psi\}_{\psi \in \mathcal{L}_L}, V\rangle$ be a Kripke structure such that for every formula $\psi \in \mathcal{L}_L$ and for all worlds $a, a' \in M$ condition (7.3) holds. Assume that $\langle \mathcal{M}, a\rangle \models \neg L \neg L\varphi$. Then there is $a' \in M$ such that $(a, a') \in R_{\neg L\varphi}$ and $\langle \mathcal{M}, a'\rangle \models L\varphi$. By (7.3), it follows that $(a', a) \in R_\varphi$. Hence, $\langle \mathcal{M}, a\rangle \models \varphi$. Consequently, every instance of the axiom b is valid in $\mathcal{M}$. Hence, soundness follows by Theorem 7.22. ■

Kripke structures will also be referred to as **N**-*structures* and Kripke structures in which all accessibility relations are reflexive will be called **t**-*structures*. Similarly, a Kripke structure is called a **4**-*structure* if for every $\psi \in \mathcal{L}_L$ its accessibility relations $R_\psi$, satisfy the condition (7.2). The concepts of **b**-structures and **5**-structures are defined analogously.

We conclude this section with more terminology and some useful auxiliary results.

**Definition 7.40** Let $M$ be a nonempty set and let $R \subseteq M \times M$. If $M' \subseteq M$ is a minimal nonempty subset of $M$ such that for every $a, a' \in M$ if $(a, a') \in R$

then either $a, a' \in M'$ or $a, a' \in M \setminus M'$, then the relation $\{(a, a') : a, a' \in M', (a, a') \in R\}$ is called a *component* of $R$. ▲

**Definition 7.41** If $\mathcal{M} = \langle M, \{R_\psi\}_{\psi \in \mathcal{L}_L}, V \rangle$ is a Kripke structure and $R_\psi|_{M'}$ is a component of the accessibility relation $R_\psi$ for every $\psi \in \mathcal{L}_L$, then $\langle M', \{R_\psi|_{M'}\}_{\psi \in \mathcal{L}_L}, V|_{M'} \rangle$ is a component of $\mathcal{M}$. ▲

**Definition 7.42** Given two Kripke structures $\mathcal{M} = \langle M, \{R_\psi\}_{\psi \in \mathcal{L}_L}, V \rangle$ and $\mathcal{M}' = \langle M', \{R'_\psi\}_{\psi \in \mathcal{L}_L}, V' \rangle$, where $M \cap M' = \emptyset$, the Kripke structure $\langle M \cup M', \{R_\psi \odot R'_\psi\}_{\psi \in \mathcal{L}_L}, V \cup V' \rangle$ is called the *concatenation* of $\mathcal{M}$ and $\mathcal{M}'$, and is denoted by $\mathcal{M} \odot \mathcal{M}'$. (Recall that the operation of concatenation of relations, also denoted by $\odot$, has been introduced in Definition 2.39). ▲

The following lemma is a direct consequence of the definitions of satisfiability relation, component of a structure and concatenation of structures.

**Lemma 7.43** *Let $\mathcal{M}$ be a Kripke structure and let $\{\mathcal{M}_i\}_{i \in I}$ be the family of all components of $\mathcal{M}$. Then:*

1. *For every formula $\psi \in \mathcal{L}_L$,*

$$\mathcal{M} \models \psi \text{ if and only if for every } i \in I,\ \mathcal{M}_i \models \psi.$$

2. *If $\mathcal{N}$ is a Kripke structure with the set of worlds disjoint from the set of worlds of $\mathcal{M}$, then*

$$\mathcal{M} \odot \mathcal{N} \models \psi \text{ if and only if for every } i \in I,\ \mathcal{M}_i \odot \mathcal{N} \models \psi.$$

3. *If $\mathcal{N}$ is a Kripke structure with the set of worlds disjoint from the set of worlds of $\mathcal{M}$, then for every world $a$ of $\mathcal{N}$,*

$$\langle \mathcal{M} \odot \mathcal{N}, a \rangle \models \psi \text{ if and only if } \langle \mathcal{N}, a \rangle \models \psi.$$ ▲

## 7.3 Kripke semantics for normal logics

The results of the previous section apply also to normal modal logics. But, as is well known, in the case of normal modal logics simpler structures, with only one accessibility relation, can be used to semantically characterize the provability relation. In the previous section, we used the term Kripke *structure* to refer to models with separate accessibility relations for different formulas.

In this section, we will use the traditional term *Kripke model* when referring to models with one accessibility relation.

The theory of Kripke models and the corresponding semantic characterization of the provability relation in normal modal logics is well understood and is presented in several monograph texts on modal logics [Che80, Fit83, HC84]. In this section, we will derive the fundamental theorem for canonical models of *normal* modal logics from the general theory developed in the previous section. Then for various normal modal logics $\mathcal{S}$, we will present classes $\mathcal{C}$ of Kripke models such that $\mathcal{S}$ is sound and complete with respect to $\mathcal{C}$.

**Definition 7.44** A *Kripke structure* $\langle M, \{R_\psi\}_{\psi \in \mathcal{L}_L}, V\rangle$ such that for some $R \subseteq M \times M$, $R_\psi = R$, for every $\psi \in \mathcal{L}_L$, is called a *Kripke model* and is denoted by $\langle M, R, V\rangle$. ▲

According to our definition, Kripke models are special Kripke structures. In particular, the definition of the satisfiability relation $\models$ given in Section 7.2 applies to Kripke models. It is easy to see that our definitions of Kripke models and satisfiability are equivalent to the standard definitions of these notions.

To every Kripke *structure* we will assign now a Kripke *model*. Namely, for a Kripke structure $\mathcal{M} = \langle M, \{R_\psi\}_{\psi \in \mathcal{L}_L}, V\rangle$ we define

$$\mathcal{M}^\cap = \langle M, \bigcap_{\psi \in \mathcal{L}_L} R_\psi, V\rangle.$$

**Theorem 7.45** *Let $\mathcal{S}$ be a normal modal logic. For every consistent with $\mathcal{S}$ theory $I \subseteq \mathcal{L}_L$, for every formula $\varphi \in \mathcal{L}_L$, and for every world $a$ of the $\mathcal{S}$-canonical structure $\mathcal{M}_{\mathcal{S}}^{can}(I)$ of $I$,*

$$\langle \mathcal{M}_{\mathcal{S}}^{can}(I), a\rangle \models \varphi \quad \textit{if and only if} \quad \langle (\mathcal{M}_{\mathcal{S}}^{can}(I))^\cap, a\rangle \models \varphi.$$

In the proof we use an auxiliary lemma. This lemma, in one form or another, plays a key role in the "standard" proof of the fundamental theorem for canonical models. See, for example, the proof in [HC84], or our proof in Section 7.2 (Lemma 7.27).

**Lemma 7.46** *Let $\mathcal{S}$ be a normal modal logic and let $I \subseteq \mathcal{L}_L$ be a theory consistent with $\mathcal{S}$. For every $I$-consistent with $\mathcal{S}$ theory $T \subseteq \mathcal{L}_L$ and for every formula $\varphi \in \mathcal{L}_L$, if $\neg L\varphi \in T$ then $\{\psi \colon L\psi \in T\} \cup \{\neg\varphi\}$ is $I$-consistent with $\mathcal{S}$.*

Proof of the lemma: Since $\neg L\varphi \in T$, by the $I$-consistency of $T$ it follows that $I \not\vdash_S \neg\neg L\varphi$. Hence, by propositional calculus, $I \not\vdash_S L\varphi$. Consequently,

$$I \not\vdash_S \varphi. \tag{7.5}$$

Assume that $\{\psi : L\psi \in T\} \cup \{\neg\varphi\}$ is not $I$-consistent with $\mathcal{S}$. It follows from the definition of an $I$-inconsistent set and from (7.5) that there is a nonempty finite subset $\{\psi_1, \ldots, \psi_n\}$ of $\{\psi : L\psi \in T\}$ such that

$$I \vdash_S \neg\psi_1 \vee \ldots \vee \neg\psi_n \vee \neg\neg\varphi.$$

Consequently,

$$I \vdash_S \psi_1 \wedge \ldots \wedge \psi_n \supset \varphi.$$

Applying part (1) and then part (3) of Theorem 7.17 yields

$$I \vdash_S L\psi_1 \wedge \ldots \wedge L\psi_n \supset L\varphi.$$

Hence,

$$I \vdash_S \neg L\psi_1 \vee \ldots \vee \neg L\psi_n \vee \neg\neg L\varphi.$$

Since $\{L\psi_1, \ldots, L\psi_n, \neg L\varphi\} \subseteq T$, therefore $T$ is $I$-inconsistent with $\mathcal{S}$, a contradiction. ■

**Proof of Theorem 7.45**: We will use the following notation: $\mathcal{M} = \mathcal{M}_{\mathcal{S}}^{can}(I)$ and $R = \bigcap_{\psi \in \mathcal{L}_L} R_\psi$. We proceed by induction on the length of a formula. In the case of the basis of induction and in the induction step when $\varphi = \neg\vartheta$, $\vartheta_1 \wedge \vartheta_2$, $\vartheta_1 \vee \vartheta_2$ and $\vartheta_1 \supset \vartheta_2$, the argument is easy and is omitted. The only remaining case is when $\varphi = L\vartheta$.

Assume first that $\langle \mathcal{M}, a \rangle \models L\vartheta$. Consider a world $b$ such that $(a, b) \in R$. Then, $(a, b) \in R_\vartheta$. Hence, $\langle \mathcal{M}, b \rangle \models \vartheta$. By the induction hypothesis, $\langle \mathcal{M}^\cap, b \rangle \models \vartheta$ and, consequently, $\langle \mathcal{M}^\cap, a \rangle \models L\vartheta$.

Next, assume that $\langle \mathcal{M}, a \rangle \not\models L\vartheta$. Recall that worlds of the canonical model are maximal $I$-consistent with $\mathcal{S}$ theories. Hence, by Theorem 7.32 and Lemma 7.30(1), $\neg L\vartheta \in a$. By Lemma 7.46, there is a world $b$ such that $\{\psi : L\psi \in a\} \cup \{\neg\vartheta\} \subseteq b$. It follows that for every formula $\psi \in \mathcal{L}_L$, if $L\psi \in a$ then $\psi \in b$. Consequently, for every formula $\psi \in \mathcal{L}_L$, $(a, b) \in R_\psi$. Hence, $(a, b) \in R$. Since $\neg\vartheta \in b$, it follows by Theorem 7.32 that $\langle \mathcal{M}, b \rangle \models \neg\vartheta$ or, equivalently, that $\langle \mathcal{M}, b \rangle \not\models \vartheta$. By the induction hypothesis, $\langle \mathcal{M}^\cap, b \rangle \not\models \vartheta$. Since $(a, b) \in R$, $\langle \mathcal{M}^\cap, a \rangle \not\models L\vartheta$. ■

Corollary 7.35 and Theorem 7.45 imply the following crucial property of models $(\mathcal{M}_{\mathcal{S}}^{can}(I))^\cap$.

**Corollary 7.47** *Let $\mathcal{S}$ be a normal modal logic. For every theory $I \subseteq \mathcal{L}_L$ which is consistent with $\mathcal{S}$, and for every formula $\varphi \in \mathcal{L}_L$,* *if* $(\mathcal{M}_S^{can}(I))^\cap \models \varphi$ *if and only if* $I \vdash_S \varphi$. ▲

**Remark 7.48** It is easy to see that the Kripke model $(\mathcal{M}_S^{can}(I))^\cap$ coincides with the "standard" canonical model for normal modal logics. In particular, if $R$ is the accessibility relation in $(\mathcal{M}_S^{can}(I))^\cap$ then, for every two worlds $a, a'$ of $(\mathcal{M}_S^{can}(I))^\cap$, $(a, a') \in R$ if and only if $\{\varphi : L\varphi \in a\} \subseteq a'$, which is precisely the standard definition of the accessibility relation in the canonical model. Therefore we will call the model $(\mathcal{M}_S^{can}(I))^\cap$ the *$\mathcal{S}$-canonical model for $I$*. ▲

Corollary 7.47 implies that a normal modal logic $\mathcal{S}$ is complete with respect to any class of Kripke models that contains the class of all Kripke models of the form $(\mathcal{M}_S^{can}(I))^\cap$, where $I \subseteq \mathcal{L}_L$ is consistent with $\mathcal{S}$.

**Corollary 7.49** *Let $\mathcal{S}$ be a normal modal logic and $\mathcal{C}$ be a class of Kripke models. If for every consistent with $\mathcal{S}$ theory $I \subseteq \mathcal{L}_L$, $(\mathcal{M}_S^{can}(I))^\cap \in \mathcal{C}$, then the logic $\mathcal{S}$ is complete with respect to $\mathcal{C}$.* ▲

As in Section 7.2, we will now derive soundness and completeness results for several specific normal modal logics.

**Lemma 7.50** *Let $\mathcal{S}$ be a normal modal logic and let $I$ be a theory consistent with $\mathcal{S}$. Denote $(\mathcal{M}_S^{can}(I))^\cap = \langle M, R, V\rangle$.*

1. *If* t *is a theorem of $\mathcal{S}$ then $R$ is reflexive.*
2. *If* 4 *is a theorem of $\mathcal{S}$ then $R$ is transitive.*
3. *If* b *is a theorem of $\mathcal{S}$ then $R$ is symmetric.*
4. *If* d *is a theorem of $\mathcal{S}$ then $R$ has no dead ends.*
5. *If* 5 *is a theorem of $\mathcal{S}$ then $R$ is euclidean.*

Proof: Let $R_\psi$, $\psi \in \mathcal{L}_L$, be the accessibility relations of the canonical structure $\mathcal{M}_S^{can}(I)$. By the definition of $(\mathcal{M}_S^{can}(I))^\cap$,

$$R = \bigcap_{\psi \in \mathcal{L}_L} R_\psi .$$

(1) By Lemma 7.38(1), for every $\psi \in \mathcal{L}_L$, $R_\psi$ is reflexive. Consequently, $R$ is reflexive.

(2) Assume that for some worlds $a, a', a'' \in M$, $(a, a') \in R$ and $(a', a'') \in R$. Let $\psi \in \mathcal{L}_L$. Then $(a, a') \in R_{L\psi}$ and $(a', a'') \in R_\psi$. By Lemma 7.38(2), $(a, a'') \in R_\psi$. Consequently, $(a, a'') \in R$ and $R$ is transitive.

(3) Assume that $(a, a') \in R$. Consider a formula $\psi \in \mathcal{L}_L$. Then, $(a, a') \in R_{\neg L\psi}$. By Lemma 7.38(3), $(a', a) \in R_\psi$. Since $\psi$ stands for an arbitrary formula in $\mathcal{L}_L$, $(a', a) \in R$.

(4) Consider a world $a$. It easily follows from the definition of the accessibility relations $R_\psi$ in the canonical structure that

$$\{a' : (a, a') \in \bigcap_{\psi \in \mathcal{L}_L} R_\psi\} = \{a' : (a, a') \in \bigcap_{\{\psi \in \mathcal{L}_L : L\psi \in a\}} R_\psi\}$$

If $\{\psi \in \mathcal{L}_L : L\psi \in a\} = \emptyset$, then $\bigcap_{\{\psi \in \mathcal{L}_L : L\psi \in a\}} R_\psi = M \times M$. In particular, $a$ is not a dead end. So, assume that $\{\psi \in \mathcal{L}_L : L\psi \in a\} \neq \emptyset$. Let $\varphi$ be such that $L\varphi \in a$. Since d is an axiom of $\mathcal{S}$, $\neg L \neg \varphi \in a$. By Lemma 7.46, it follows that $\{\psi \in \mathcal{L}_L : L\psi \in a\} \cup \{\varphi\}$ is $I$-consistent with $\mathcal{S}$. Thus, there is a world $a' \in M$ such that $\{\psi \in \mathcal{L}_L : L\psi \in a\} \subseteq a'$. Thus, $(a, a') \in \bigcap_{\psi \in \mathcal{L}_L} R_\psi$ and $a$ is not a dead end.

(5) Assume that for some worlds $a, a', a'' \in M$, $(a, a') \in R$ and $(a, a'') \in R$. Let $\psi \in \mathcal{L}_L$. Then $(a, a') \in R_\psi$ and $(a, a'') \in R_{\neg L\psi}$. By Lemma 7.38(4), $(a'', a') \in R_\psi$. Consequently, $(a'', a') \in R$ and $R$ is euclidean. ■

Now we are ready to present soundness and completeness results for several normal modal logics.

**Corollary 7.51**

1. *Logic* **K** *is characterized by the class of all Kripke models.*
2. *Logic* **T** *is characterized by the class of all reflexive Kripke models.*
3. *Logic* **S4** *is characterized by the class of all reflexive and transitive Kripke models.*
4. *Logic* **S5** *is characterized by the class of all Kripke models whose accessibility relations are equivalence relations.*
5. *Logic* **kd45** *is characterized by the class of all Kripke models that are transitive, euclidean and have no dead ends.*
6. *Logic* **S4f** *is characterized by the class of all reflexive and transitive Kripke models $\langle M, R, V \rangle$ satisfying this condition:*

$$\textit{if } (a, a') \in R, (a', a) \notin R \textit{ and } (a, a'') \in R, \quad \textit{then } (a'', a') \in R. \tag{7.6}$$

7. *Logic* **Sw5** *is characterized by the class of all Kripke models whose accessibility relation $R$ is reflexive, transitive and satisfies the following condition:*

$$\begin{array}{l} \textit{for all worlds } a, a', a'' \\ \textit{if } a \neq a', a \neq a'', (a, a') \in R \textit{ and } (a, a'') \in R \\ \textit{then } (a', a'') \in R \textit{ and } (a'', a') \in R. \end{array} \tag{7.7}$$

Proof: In each of the cases (1) — (7), it is straightforward to verify that all substitution instances of the axioms of the corresponding logic are valid in every model from the specified class. Thus, the "soundness" part follows in each case from Theorem 7.22. The "completeness" part will follow from Corollary 7.49 once it is established that the class of models specified in each of the assertions contains all models of the form $(\mathcal{M}_{\mathcal{S}}^{can}(I))^{\cap}$, where $\mathcal{S}$ stands for the corresponding logic and $I$ is consistent with $\mathcal{S}$.

Thus, the "completeness" part of (1) is evident and, in the case of (2), (3) and (5) follows directly from Lemma 7.50. Throughout the rest of the proof the accessibility relation of $(\mathcal{M}_{\mathcal{S}}^{can}(I))^{\cap}$ will be denoted by $R$.

(4) First, observe that the axiom b ($\neg L \neg Lp \supset p$) is a theorem of **S5**. Also, t and 4 are axioms of **S5**. Thus, by Lemma 7.50, $R$ is reflexive, transitive and symmetric. That is, $R$ is an equivalence relation.

(6) By Lemma 7.50(1) and (2), $R$ is reflexive and transitive. To complete the proof, consider worlds $a, a', a''$ of the canonical model such that $(a, a') \in R$, $(a', a) \notin R$ and $(a, a'') \in R$. Since $(a', a) \notin R$, there is a formula $\varphi$ such that $L\varphi \in a'$ but $\varphi \notin a$. Consider an arbitrary formula $L\psi \in a''$. Then $\neg L\psi \notin a''$. Consequently, $L\neg L\psi \notin a$. Thus, $ML\psi \in a$. Also $\neg\varphi \in a$. Since every instance of the axiom f is in $a$, $L(M\neg\varphi \vee \psi) \in a$. Consequently, $M\neg\varphi \vee \psi \in a'$. On the other hand, $L\varphi \in a'$. Thus, it follows that $L\neg\neg\varphi \in a'$. Hence, $\psi \in a'$, which proves that $(a'', a') \in R$.

(7) First, let us observe that for every $\varphi, \psi \in \mathcal{L}_L$ the formula $ML\varphi \supset (\psi \supset L(\varphi \vee \psi))$ is a theorem of **Sw5**. Indeed, by Theorem 7.17, $\vdash_{\mathbf{K}} ML\varphi \supset ML(\varphi \vee \psi)$. We also have $\vdash ((\varphi \vee \psi) \supset L(\varphi \vee \psi)) \supset (\psi \supset L(\varphi \vee \psi))$. Consequently, $\vdash_{\mathbf{Sw5}} ML\varphi \supset (\psi \supset L(\varphi \vee \psi))$ follows.

Reflexivity and transitivity of $R$ follow from Lemma 7.50(1) and (2). To complete the proof, consider worlds $a$, $a'$ and $a''$ such that $a \neq a'$, $a \neq a''$, $(a, a') \in R$ and $(a, a'') \in R$. Since $a \neq a''$, there is a formula $\psi$ such that $\psi \in a$ and $\psi \notin a''$. Let $L\varphi \in a'$. Then, $\neg L\varphi \notin a'$. Consequently, $L\neg L\varphi \notin a$ and $ML\varphi \in a$. Thus, $\psi \supset L(\varphi \vee \psi) \in a$. Since $\psi \in a$, therefore $L(\varphi \vee \psi) \in a$. Since $(a, a'') \in R$, $\varphi \vee \psi \in a''$. Since $\psi \notin a''$, $\neg\psi \in a''$ and, consequently, $\varphi \in a''$. Thus, $(a', a'') \in R$. The argument that $(a'', a') \in R$ is analogous. ■

We will now simplify the characterizations of logics **S5, Sw5, S4f** and **kd45** given in Corollary 7.51.

**Theorem 7.52**

1. *Logic* **S5** *is characterized by the class of all universal Kripke models.*
2. *Logic* **kd45** *is characterized by the class of all Kripke models of the form* $\langle \{a\} \cup M, (\{a\} \cup M) \times M, V \rangle$, *where* $M \neq \emptyset$.
3. *Logic* **S4f** *is characterized by the class of all Kripke models of the form* $\langle M_1 \cup M_2, ((M_1 \cup M_2) \times M_2) \cup (M_1 \times M_1), V \rangle$, *where* $M_1 \cap M_2 = \emptyset$ *and* $M_1, M_2 \neq \emptyset$.
4. *Logic* **Sw5** *is characterized by the class of all Kripke models of the form* $\langle \{a\} \cup M, \{(a,a)\} \cup ((\{a\} \cup M) \times M), V \rangle$.

Proof: Let $\mathcal{S}$ stand for **S5, S4f, kd45** and **Sw5**. Denote by $\mathcal{C}$ the class of Kripke models described in Corollary 7.51 as characterizing $\mathcal{S}$.

Define $\mathcal{C}_1$ to be the subclass of $\mathcal{C}$ consisting of those Kripke models in $\mathcal{C}$ which have only one component. Since $\mathcal{S}$ is sound with respect to $\mathcal{C}$, the soundness with respect to $\mathcal{C}_1$ follows.

Observe next that for every model $\mathcal{M} \in \mathcal{C}$ and for every formula $\varphi \in \mathcal{L}_L$, $\mathcal{M} \models \varphi$ if and only if for every component $\mathcal{M}'$ of $\mathcal{M}$, $\mathcal{M}' \models \varphi$. In each of the four cases under consideration, every component of a model from $\mathcal{C}$ belongs to $\mathcal{C}_1$. Hence, the completeness of $\mathcal{S}$ with respect to $\mathcal{C}_1$ follows. Summarizing, $\mathcal{S}$ is characterized by the class $\mathcal{C}_1$ of Kripke models.

(1) $\mathcal{S} =$ **S5**. An equivalence relation with exactly one component is a universal relation. Hence, $\mathcal{C}_1$ is the class of all universal Kripke models and the assertion (1) follows.

(2) $\mathcal{S} =$ **kd45**. In this case, one can show that the class $\mathcal{C}_1$ consists of Kripke models of the form

$$\langle M_1 \cup M_2, (M_1 \cup M_2) \times M_2, V \rangle, \tag{7.8}$$

where $M_2 \neq \emptyset$ and $M_1 \cap M_2 = \emptyset$.

Let $\mathcal{C}_2$ be the class consisting of all Kripke models of the form (7.8) with $M_1 \neq \emptyset$. Since $\mathcal{C}_2 \subseteq \mathcal{C}_1$, it follows that **kd45** is sound with respect to $\mathcal{C}_2$.

Consider a universal Kripke model $\mathcal{M} = \langle M_2, M_2 \times M_2, V \rangle$ (a model of the form (7.8) with $M_1 = \emptyset$). Let $a$ be any element not in $M_2$. Let $\mathcal{M}' = \langle \{a\} \cup M_2, (\{a\} \cup M_2) \times M_2, W \rangle$, where $W|_{M_2} = V$ and $W(a) = V(m)$, for some $m \in M_2$. For every formula $\varphi \in \mathcal{L}_L$ we have $\mathcal{M} \models \varphi$ if and only if $\mathcal{M}' \models \varphi$. Since $\mathcal{M}' \in \mathcal{C}_2$, the completeness of **kd45** with respect to $\mathcal{C}_2$ follows.

Consider a Kripke model $\mathcal{M} \in \mathcal{C}_2$, say

$$\mathcal{M} = \langle M_1 \cup M_2, (M_1 \cup M_2) \times M_2, V\rangle,$$

where $M_1 \neq \emptyset$. For every $a \in M_1$ define

$$\mathcal{M}_a = \langle \{a\} \cup M_2, (\{a\} \cup M_2) \times M_2, V|_{(\{a\} \cup M_2)}\rangle.$$

It is easy to see that $\mathcal{M} \models \varphi$ if and only if for every $a \in M$, $\mathcal{M}_a \models \varphi$. Consequently, **kd45** is complete with respect to the class of all models of the form $\langle \{a\} \cup M, (\{a\} \cup M) \times M, V\rangle$, where $M \neq \emptyset$. Soundness follows from the fact that every such Kripke model belongs to $\mathcal{C}_1$. Hence, logic **kd45** is characterized by the class of Kripke models of the form $\langle \{a\} \cup M, (\{a\} \cup M) \times M, V\rangle$, where $M \neq \emptyset$.

(3) $\mathcal{S} =$ **S4f**. It is easy to see that a relation $R$ on $M$, which is reflexive and transitive, has one connected component, and satisfies condition (7.6) is of the form $((M_1 \cup M_2) \times M_2) \cup (M_1 \times M_1)$, where $M_1 \cap M_2 = \emptyset$ and $M_2 \neq \emptyset$. Reasoning as in the proof of (2), one can now show that **S4f** is characterized by the class of Kripke models of the form $((M_1 \cup M_2) \times M_2) \cup (M_1 \times M_1)$, where $M_1 \cap M_2 = \emptyset$ and $M_1, M_2 \neq \emptyset$.

(4) The proof is similar to that for **kd45** and is omitted. ■

Kripke models in each of the classes described in Corollary 7.51(1) – (3) and in Theorem 7.52 are called *$\mathcal{S}$-models*, where $\mathcal{S}$ is the name of the corresponding logic. The class of all $\mathcal{S}$-models is denoted by $\mathcal{C}_\mathcal{S}$. For example, models with universal accessibility relations are called **S5**-models and the class of all **S5**-models is denoted by $\mathcal{C}_{\mathbf{S5}}$.

An **S5**-model $\mathcal{M} = \langle M, M \times M, V\rangle$ will be written as $\langle M, V\rangle$. An **S4f**-model $\langle M_1 \cup M_2, ((M_1 \cup M_2) \times M_2) \cup (M_1 \times M_1), V\rangle$, will be written as $\langle M_1, M_2, V\rangle$. An **Sw5**-model $\langle \{a\} \cup M, \{(a,a)\} \cup ((\{a\} \cup M) \times M), V\rangle$ will be denoted by $\langle a, M, V\rangle$ and the same symbol will be used to denote a **kd45**-model $\langle \{a\} \cup M, (\{a\} \cup M) \times M, V\rangle$.

We conclude this section with a theorem which states the so called *finite model property* for normal logics mentioned in Corollary 7.51. This result is important for establishing decidability of these logics.

**Theorem 7.53** *Let $\mathcal{S}$ be one of the modal logics* **K**, **T**, **S4**, **S4f**, **Sw5**, **kd45** *and* **S5**. *Let $I \subseteq \mathcal{L}_L$ be finite and let $\varphi \in \mathcal{L}_L$. Then $I \vdash_\mathcal{S} \varphi$ if and only if for every finite $\mathcal{S}$-model $\mathcal{M}$ of $I$, $\mathcal{M} \models \varphi$.* ▲

## 7.4 Index logics

In this section, we discuss basic properties of index logics and use them to establish the hierarchy of modal logics between $\mathcal{S}$ and **S5**, where $\mathcal{S}$ stands for **kd45**, **Sw5** and **S4f**.

**Definition 7.54** We say that a modal logic $\mathcal{S}$ is *determined* by the class of Kripke models $\mathcal{C}$ if for every formula $\varphi \in \mathcal{L}_L$, $\vdash_{\mathcal{S}} \varphi$ precisely when $\mathcal{M} \models \varphi$, for every $\mathcal{M} \in \mathcal{C}$. ▲

Let us observe that the notion of being determined by a class of Kripke models is related to that of being characterized by a class of models. Namely, if a logic $\mathcal{S}$ is characterized by a class of Kripke models $\mathcal{C}$, then $\mathcal{S}$ is also determined by the class $\mathcal{C}$. The converse implication does not hold in general. For example, it is well known that logic **S4** is determined but not characterized by the class of all Kripke models with finite sets of worlds and with reflexive and transitive accessibility relations.

**Definition 7.55** A finite sequence $\Lambda = \langle \lambda_1, \ldots, \lambda_n \rangle$ of integers, where $n > 0$, and for each $i$, $1 \leq i \leq n$, $\lambda_i$ is a non-negative integer or $\omega$ (the first infinite ordinal), is called an *index*. The number of terms in the index $\langle \lambda_1, \ldots, \lambda_n \rangle$, that is, the number $n$, is called the *length* of the index.

We say that a Kripke model $\langle M, R, V \rangle$ *has index* $\Lambda = \langle \lambda_1, \ldots, \lambda_n \rangle$ if the following two conditions hold:

1. $M$ is a disjoint union of sets $M_1, \ldots, M_n$ such that for each $i$, $1 \leq i \leq n$, if $\lambda_i = 0$ or $\lambda_i = 1$ then $|M_i| = 1$, and if $\lambda_i > 1$ then $\lambda_i$ is the cardinality of $M_i$.
2. $R = R_1 \odot \ldots \odot R_n$, where $R_i = M_i \times M_i$ if $\lambda_i > 0$ and $R_i = \emptyset$, otherwise.

If a modal logic $\mathcal{S}$ is determined by the class of all Kripke models with an index $\Lambda$ then $\mathcal{S}$ is called the *index logic with index* $\Lambda$ and is denoted by $\mathcal{S}_\Lambda$. ▲

**Theorem 7.56** *Let $\mathcal{M}$ be a Kripke model of the form $\mathcal{M}_1 \odot \ldots \odot \mathcal{M}_n$, where each model $\mathcal{M}_i$ is universal or consists of just one world not related to itself. There exists a Kripke model $\mathcal{M}'$ of the form $\mathcal{M}'_1 \odot \ldots \odot \mathcal{M}'_n$ such that:*

1. *If $\mathcal{M}_i$ is universal then $\mathcal{M}'_i$ is universal.*
2. *If $\mathcal{M}_i$ consists of exactly one world not related to itself then $\mathcal{M}'_i = \mathcal{M}_i$.*

*3. $\mathcal{M}'$ is denumerable.*

*4. $Th(\mathcal{M}) = Th(\mathcal{M}')$.*

Proof: Let $\mathcal{M} = \langle M, R, V\rangle$. For every $i$ let $a_i$ be an arbitrary world in the model $\mathcal{M}_i$. Next, for every formula $\varphi \in \mathcal{L}_L$ such that $\mathcal{M} \not\models \varphi$, let $a_\varphi$ be a world in $M$ such that:

**(a)** $\langle \mathcal{M}, a_\varphi\rangle \not\models \varphi$.

**(b)** If $a_\varphi \in M_i$ and for some $b \in M_j$ we have $\langle \mathcal{M}, b\rangle \not\models \varphi$, then $j \leq i$.

Define $M' = \{a_1, \ldots, a_n\} \cup \{a_\varphi : \mathcal{M} \not\models \varphi\}$ and $\mathcal{M}' = \langle M', R', V'\rangle$, where $R' = R|_{M'}$ and $V' = V|_{M'}$. It is clear that $\mathcal{M}'$ satisfies conditions (1), (2) and (3) of the assertion.

To show that (4) holds, we will prove that for every formula $\varphi \in \mathcal{L}_L$ and for every $a \in M'$,

$$\langle \mathcal{M}, a\rangle \models \varphi \text{ if and only if } \langle \mathcal{M}', a\rangle \models \varphi. \tag{7.9}$$

The proof is by induction on the complexity of a formula. If $\varphi$ is an atom, claim (7.9) is obvious. If $\varphi = \neg\psi$ then we reason as follows: $\langle \mathcal{M}, a\rangle \models \varphi$ if and only if $\langle \mathcal{M}, a\rangle \not\models \psi$. By the induction hypothesis, this latter statement is equivalent to $\langle \mathcal{M}', a\rangle \not\models \psi$ which, in turn, is equivalent to $\langle \mathcal{M}', a\rangle \models \varphi$. The cases when the main connective in $\varphi$ is $\vee, \wedge$ and $\supset$ can be dealt with similarly; we omit the details.

So, assume that $\varphi = L\psi$. If $\langle \mathcal{M}, a\rangle \models \varphi$ then $\langle \mathcal{M}, b\rangle \models \psi$ for every $b \in M$ such that $(a, b) \in R$. In particular, $\langle \mathcal{M}, b\rangle \models \psi$ for every $b \in M'$ such that $(a, b) \in R'$. By the induction hypothesis, $\langle \mathcal{M}', b\rangle \models \psi$ for every $b \in M'$ such that $(a, b) \in R'$. Thus, $\langle \mathcal{M}', a\rangle \models \varphi$.

To prove the converse implication, assume that $\langle \mathcal{M}, a\rangle \not\models \varphi$. Then, there exists $b \in M$ such that $(a, b) \in R$ and $\langle \mathcal{M}, b\rangle \not\models \psi$. Clearly, $\mathcal{M} \not\models \psi$. Thus, $a_\psi \in M'$. Moreover, by the definition of $a_\psi$, it follows that $(a, a_\psi) \in R$. Consequently, $(a, a_\psi) \in R'$. Since, by the induction hypothesis, $\langle \mathcal{M}', a_\psi\rangle \not\models \psi$, it follows that $\langle \mathcal{M}', a\rangle \not\models \varphi$.

Equation (7.9) directly implies that if $\mathcal{M} \models \varphi$ then $\mathcal{M}' \models \varphi$. In addition, if $\mathcal{M} \not\models \varphi$ then $\langle \mathcal{M}, a_\varphi\rangle \not\models \varphi$. Again by (7.9), $\langle \mathcal{M}', a_\varphi\rangle \not\models \varphi$ and, consequently, $\mathcal{M}' \not\models \varphi$. Thus, $Th(\mathcal{M}) = Th(\mathcal{M}')$.

Since for each model $\mathcal{M} \in \mathcal{C}$, the set of worlds in $\mathcal{M}'$ is denumerable, the assertion follows. ■

**Corollary 7.57**

1. $\mathbf{S5} = S_{\langle\omega\rangle}$.
2. $\mathbf{kd45} = S_{\langle 0,\omega\rangle}$.
3. $\mathbf{Sw5} = S_{\langle 1,\omega\rangle}$.
4. $\mathbf{S4f} = S_{\langle\omega,\omega\rangle}$.

Proof: The proofs are similar in all four cases. We will give only the proof of (4). It follows from Theorem 7.52(3) and Theorem 7.56 that the logic **S4f** is characterized, and therefore determined, by the class of all models with indices of the form $\langle m, n\rangle$, where $m$ and $n$ are positive integers or $\omega$. Each such model can be regarded as an **S4f**-model $(M_1, M_2, V)$, where $|M_1| = m$ and $|M_2| = n$. By duplicating worlds in $M_1$ and in $M_2$ one can construct an **S4f**-model $\mathcal{M}' = (M_1', M_2', V')$, such that $|M_1'| = |M_2'| = \omega$ and $Th(\mathcal{M}') = Th(\mathcal{M})$. Consequently, logic **S4f** is determined by the class of all Kripke models with index $\langle\omega,\omega\rangle$. ■

The following deep result was obtained by Segerberg [Seg71]. We give it here without proof.

**Theorem 7.58**

1. *Let $\mathcal{S}$ be a modal logic* properly *containing the intersection of all index logics with indices of length at most n and with all components positive. Then $\mathcal{S}$ is the intersection of a finite number of index logics with indices of length at most n and with all components positive.*
2. *Let $\mathcal{S}$ be a modal logic* properly *containing the intersection of all index logics with indices of length at most n. Then $\mathcal{S}$ is the intersection of a finite number of index logics with indices of length at most n.* ▲

**Theorem 7.59** *Let $\mathcal{S}$ be a modal logic* properly *contained in* **S5**.

1. *If $\mathcal{S}$* properly *contains* **Sw5**, *then $\mathcal{S} = S_{\langle\omega\rangle} \cap S_{\langle 1,m\rangle}$, for some integer $m \geq 1$.*
2. *If $\mathcal{S}$* properly *contains* **kd45**, *then at least one of the following conditions holds:*
   (a) $\mathcal{S} = S_{\langle\omega\rangle} \cap S_{\langle 0\rangle}$.

*(b)* $\mathcal{S} = S_{\langle\omega\rangle} \cap S_{\langle 0\rangle} \cap S_{\langle 0,m\rangle}$, *for some integer* $m \geq 1$.

*(c)* $\mathcal{S} = S_{\langle\omega\rangle} \cap S_{\langle 0,m\rangle}$, *for some integer* $m \geq 1$.

Proof: Both claims can be proved in a similar fashion. We will prove only (2) here. By Corollary 7.57(2), logic **kd45** contains the intersection of all index logics with indices of length at most 2. Since **kd45** is properly contained in $\mathcal{S}$, by Theorem 7.58(2) it follows that there are indices $\Lambda_1, \ldots, \Lambda_n$, each of length at most 2, such that $\mathcal{S} = S_{\Lambda_1} \cap \ldots \cap S_{\Lambda_n}$. Since $\mathcal{S} \subseteq$ **S5**, without loss of generality we can assume that one of the indices, say $\Lambda_1$, is of the form $\langle\omega\rangle$. Also, since for any positive integer $m$, $S_{\langle\omega\rangle} \subseteq S_{\langle m\rangle}$, without loss of generality, we can assume that all other indices are of the form $\langle k, m\rangle$ or $\langle 0\rangle$. Suppose that for some $i$, $2 \leq i \leq n$, $\Lambda_i = \langle k, m\rangle$. If $k \geq 1$, then the axiom 5 does not belong to $S_{\langle k,m\rangle}$. Consequently, $k = 0$. Since axiom d does not belong to $S_{\langle 0,0\rangle}$, $m > 0$. Observe also that if $1 \leq r \leq s$, then $S_{\langle 0,s\rangle} \subseteq S_{\langle 0,r\rangle}$. Putting all these observations together implies that there are only four possibilities for $\mathcal{S}$ left:

1. $\mathcal{S} = S_{\langle\omega\rangle}$.
2. $\mathcal{S} = S_{\langle\omega\rangle} \cap S_{\langle 0\rangle}$.
3. $\mathcal{S} = S_{\langle\omega\rangle} \cap S_{\langle 0\rangle} \cap S_{\langle 0,m\rangle}$, for some integer $m \geq 1$.
4. $\mathcal{S} = S_{\langle\omega\rangle} \cap S_{\langle 0,m\rangle}$, for some integer $m \geq 1$.

The first possibility is excluded since $\mathcal{S}$ is *properly* contained in **S5** $= S_{\langle\omega\rangle}$. Hence, the assertion follows. ∎

**Theorem 7.60** *Let* $\mathcal{S}$ *be a modal logic contained in* **S5** *and properly containing* **S4f**. *Then there are positive integers m and n such that* $S_{\langle\omega,m\rangle} \cap S_{\langle n,\omega\rangle}$ *is contained in* $\mathcal{S}$.

Proof: By Corollary 7.57(4), logic **S4f** contains the intersection of all index logics with indices of length at most 2 and with each term positive. By Theorem 7.58(1), since $\mathcal{S}$ properly contains **S4f** it follows that there are indices $\Lambda_1, \ldots, \Lambda_n$, of length at most two and with all terms positive, such that $\mathcal{S} = S_{\Lambda_1} \cap \ldots \cap S_{\Lambda_n}$. Now, the assertion follows from these easy observations:

1. For every positive integer $n$, $S_{\langle\omega\rangle} \subseteq S_{\langle n\rangle}$.
2. For all $m, n$, where $m$ and $n$ stand either for a positive integer or $\omega$, $S_{\langle m,n\rangle} \subseteq S_{\langle n\rangle}$.

3. Let each of $m_1, m_2, n_1, n_2$ denote a positive integer or $\omega$. If $m_1 \leq m_2$ and $n_1 \leq n_2$ (here we assume that for any integer $n$, $n \leq \omega$), then $S_{(m_2,n_2)} \subseteq S_{(m_1,n_1)}$. ∎

## 7.5 Comments and remarks

The material presented in Section 7.1 is standard and can be found in any monograph on modal logics, for example, in Chellas [Che80], Fitting [Fit83], and Hughes and Cresswell [HC84].

Substantial fragment of this chapter deals with logics which are not normal such as **N** and **t**. Logic **N** was introduced by the authors in [MT90]. Other subnormal modal logics with applications to nonmonotonic reasoning can also be found in this paper. The semantics of **N**-structures was introduced by Fitting and the authors in [FMT92]. The generalization of this semantics to other modal logics as well as the results from Section 7.2 are due to Truszczyński [Tru92]. The possible world semantics for normal modal logics was introduced by Kripke [Kri63]. Our approach, deriving this semantics from more general results on possible world semantics for arbitrary logics, follows the presentation contained in Truszczyński [Tru92].

It is perhaps worth mentioning that logics which are not normal (but in a different sense than our subnormal logics) were studied in the literature. Historically first modal logics, S1, S2 and S3, are not normal and more non-normal logics are discussed in Kripke [Kri65], Segerberg [Seg71] and Fitting [Fit83].

The technique of using Theorem 7.58 by Segerberg to characterize logics contained in **S5** and properly containing **kd45** or **Sw5** is due to Schwarz [Sch91a]. An extension of this technique to the case of the logic **S4f** is described in Schwarz and Truszczyński [ST92a].

# 8. Stable theories

Stable theories were proposed as formal representations of belief sets of an ideally rational reasoning agent possessing a full power of introspection. A set of beliefs of such an agent should, of course, be closed under propositional consequence but, in addition, it should be closed under:

- Positive introspection — if $\varphi$ is an agent's belief, then the statement "I believe in $\varphi$" is also an agent's belief (in other words, an agent's beliefs are closed under necessitation).
- Negative introspection — if $\varphi$ is not an agent's belief, then the statement "I do not believe in $\varphi$" is an agent's belief.

Stable theories are of fundamental importance in nonmonotonic modal logics, which is why we study the concept in this book. This chapter gives a comprehensive exposition of the topic of stable theories. Section 8.1 provides a formal definition of a stable theory and several basic properties. Characterizations of stable theories are discussed in Section 8.2. Algorithms that, given a stable theory $T$ and a formula $\vartheta$, decide whether $\vartheta \in T$, are presented in Section 8.3.

## 8.1 Stable theories and their elementary properties

Our informal concepts of positive and negative introspection can be expressed in a natural way by means of modal formulas from the language $\mathcal{L}_L$, thus yielding a formal definition of a stable theory.

**Definition 8.1** A theory $T \subseteq \mathcal{L}_L$ is *stable* if $T$ satisfies the following three conditions:

**(St1)** $T$ is closed under propositional consequence.

**(St2)** (Positive introspection) For every $\varphi \in \mathcal{L}_L$, if $\varphi \in T$ then $L\varphi \in T$.

**(St3)** (Negative introspection) For every $\varphi \in \mathcal{L}_L$, if $\varphi \notin T$ then $\neg L\varphi \in T$. ▲

Note that the condition (St2) is equivalent to the fact that $T$ is closed under necessitation. Several other simple properties of stable theories that follow directly from the definition are given below.

We start with some notation. For a theory $T \subseteq \mathcal{L}_L$ we define

$$[T]_n = T \cap \mathcal{L}_{L,n}.$$

Recall that $\mathcal{L}_{L,n}$ denotes the collection of all formulas in $\mathcal{L}_L$ with $L$-depth at most $n$. Thus, $[T]_n$ consists of exactly those formulas of $T$ which have $L$-depth not exceeding $n$. In particular, $[T]_0$ consists of all formulas in $T$ that do not contain the modal operator $L$ at all. Such formulas will be called *modal-free* and $[T]_0$ will be referred to as the *modal-free part* of $T$.

Our first result gathers some simple properties of the theories $[T]_n$ in the case when a theory $T$ is stable.

**Theorem 8.2** *Let a theory $T \subseteq \mathcal{L}_L$ be stable.*

1. *For every integer $n \geq 0$, the theory $[T]_n$ is closed under propositional consequence in the language $\mathcal{L}_{L,n}$.*

2. *The theory $T$ is consistent if and only if $[T]_n$ is consistent, for some integer $n \geq 0$.*

3. *The theory $T$ is consistent if and only if $[T]_0$ is consistent.*

Proof: Every proof of a formula $\varphi$ from $[T]_n$ is a proof from $T$. Since both $T$ and $\mathcal{L}_{L,n}$ are closed under propositional consequence, so is $[T]_n$. Thus, part (1) follows. Parts (2) and (3) follow immediately. ■

The next result states that the implications in conditions (St2) and (St3) can be reversed, with an additional requirement of consistency needed in the case of (St3).

**Theorem 8.3** *Let a theory $T \subseteq \mathcal{L}_L$ be stable. Then:*

1. *If $L\varphi \in T$, then $\varphi \in T$.*

*2. If $T$ is consistent and $\neg L\varphi \in T$, then $\varphi \notin T$.*

Proof: To prove (1), assume that $L\varphi \in T$ but $\varphi \notin T$. Then, $T$ is consistent and, by (St3), $\neg L\varphi \in T$, a contradiction. To prove (2), assume that $\neg L\varphi \in T$. Since $T$ is consistent, $L\varphi \notin T$. Thus, by (St2), $\varphi \notin T$. ■

One might argue that the condition (St1) is too weak. Propositional closure treats all modal formulas $L\varphi$ as propositional variables and thus does not force any relationship between modal atoms $L\varphi$ for different but propositionally equivalent formulas $\varphi$. For example, a theory closed under propositional consequence may contain a formula $L\varphi$ but not $L(\varphi \vee \varphi)$. Nor does the property of propositional closure imply any connections between formulas $\varphi$ and $L\varphi$. For example, there are propositionally consistent theories containing both $\neg\varphi$ and $L\varphi$. Thus, one might expect that a more useful and intuitive concept could be obtained if the condition (St1) were replaced by the requirement that $T$ be closed under the consequence operator of some classical normal modal logic, like **S4** or **S5**. Our next result shows that it is not necessary to explicitly require that. In fact, the conditions (St2) and (St3) imply that each stable theory is closed under the consequence operator of the logic **S5**.

**Theorem 8.4** *If a theory $T \subseteq \mathcal{L}_L$ is stable, then $T$ is closed under* **S5**-*consequence. In particular $T$ contains all instances of the axiom schemata of* **S5**.

Proof: We need to check that every instance of the axiom schemata k, t, 4 and 5 is in $T$.
Axiom schema k. Consider a formula $L(\varphi \supset \psi) \supset (L\varphi \supset L\psi)$. Since $T$ is closed under propositional consequence, it is enough to prove that

$$\neg L(\varphi \supset \psi) \vee \neg L\varphi \vee L\psi \in T.$$

This is so if $\neg L(\varphi \supset \psi) \in T$ or $\neg L\varphi \in T$. So, assume that $\neg L(\varphi \supset \psi) \notin T$ and $\neg L\varphi \notin T$. Then, by (St3), $\varphi \supset \psi \in T$ and $\varphi \in T$. Since $T$ is closed under propositional consequence, $\psi \in T$. Hence, by (St2), $L\psi \in T$ and, consequently, $\neg L(\varphi \supset \psi) \vee \neg L\varphi \vee L\psi \in T$.
Axiom schema t. Consider an instance $\psi = L\varphi \supset \varphi$ of schema t. If $\varphi \in T$ then $\psi \in T$. If $\varphi \notin T$, then by (St3), $\neg L\varphi \in T$ and , consequently, $\psi \in T$.
Axiom schema 4. Consider a formula $\psi = L\varphi \supset LL\varphi$. If $\neg L\varphi \in T$, then $\psi \in T$. So, assume that $\neg L\varphi \notin T$. By (St3), it follows that $\varphi \in T$. Hence, by applying (St2) twice, we get $LL\varphi \in T$. Consequently, $\psi \in T$.
Axiom schema 5. Consider a formula $\psi = \neg L\varphi \supset L\neg L\varphi$. If $\varphi \in T$, then $L\varphi \in T$ and, consequently, $\psi \in T$. So, let $\varphi \notin T$. Then $\neg L\varphi \in T$ and, by (St2), $L\neg L\varphi \in T$. Hence, $\psi \in T$. ■

As a corollary, we obtain that a stable theory is closed under the consequence operator in any modal logic $\mathcal{S}$ contained in **S5**. Consequently, for any modal logic $\mathcal{S}$ contained in **S5**, a stable theory is consistent if and only if it is consistent with $\mathcal{S}$.

**Corollary 8.5** *Let $\mathcal{S}$ be any modal logic contained in* **S5** *and let a theory $T \subseteq \mathcal{L}_L$ be stable. Then:*

1. *The theory $T$ is closed under $\mathcal{S}$-consequence.*
2. *The theory $T$ is consistent if and only if $T$ is consistent with $\mathcal{S}$.* ▲

In several aspects stable theories are similar to ultrafilters in boolean algebras. We give two such properties below. The first states that consistent stable theories form an antichain. The second is a form of completeness. Both play a crucial role in our presentation of the properties of stability.

**Theorem 8.6** *Let $T_1, T_2$ be consistent stable theories. If $T_1 \subseteq T_2$, then $T_1 = T_2$.*

Proof: Assume that $T_2 \neq T_1$. Then there is $\varphi \in \mathcal{L}_L$ such that $\varphi \in T_2 \setminus T_1$. By (St2), we have $L\varphi \in T_2$ and, by (St3), $\neg L\varphi \in T_1$. But then $\neg L\varphi \in T_2$, contradicting the consistency of $T_2$. ■

The logical notion of completeness of a theory can be expressed as follows: for every $\varphi$, at least one of $\varphi$ and $\neg\varphi$ is in the theory. The completeness of a theory often simplifies proofs that involve analysis of the membership of formulas in the theory. Stable theories are not, in general, complete. But we have the following weaker property — completeness with respect to modal atoms.

**Theorem 8.7** *If $T$ is stable, then at least one of $L\varphi$, $\neg L\varphi$ belongs to $T$. If, in addition, $T$ is consistent, then exactly one of $L\varphi$, $\neg L\varphi$ belongs to $T$.*

Proof: It is enough to prove the second part of the claim. Consider a consistent and stable theory $T$. First, assume that $L\varphi \notin T$. Then, by (St2), $\varphi \notin T$. Consequently, $\neg L\varphi \in T$. Next, assume that $\neg L\varphi \notin T$. Then, by (St3), $\varphi \in T$. Hence, $L\varphi \in T$. ■

A theory $T$ has the *disjunctive property* if the following statement holds for $T$: $\varphi \vee \psi \in T$ if and only if $\varphi \in T$ or $\psi \in T$. Complete theories have the disjunctive property. Our weaker variant of the completeness property for stable theories implies a corresponding, weaker, variant of the disjunctive property. It states that a disjunction of modal literals and a formula from $\mathcal{L}_L$ belongs to a stable theory if and only if at least one of the disjuncts does.

**Theorem 8.8** *Let $T \subseteq \mathcal{L}_L$ be a stable theory. Then*

$$L\varphi_1 \vee \ldots \vee L\varphi_k \vee \neg L\psi_1 \vee \ldots \vee \neg L\psi_m \vee \vartheta \in T$$

*if and only if $L\varphi_i \in T$, for some $i$, $1 \leq i \leq k$, or $\neg L\psi_i \in T$, for some $i$, $1 \leq i \leq m$, or $\vartheta \in T$.*

Proof: The "if" part is evident, and follows from closure of $T$ under propositional consequence. We will prove the "only if" part here. To this end, assume that for every $i$, $1 \leq i \leq k$, $L\varphi_i \notin T$ and that for every $i$, $1 \leq i \leq m$, $\neg L\psi_i \notin T$. By Theorem 8.7, for every $i$, $1 \leq i \leq k$, $\neg L\varphi_i \in T$ and for every $i$, $1 \leq i \leq m$, $L\psi_i \in T$. Since $T$ is closed under propositional consequence and since

$$L\varphi_1 \vee \ldots \vee L\varphi_k \vee \neg L\psi_1 \vee \ldots \vee \neg L\psi_m \vee \vartheta \in T,$$

it follows that $\vartheta \in T$. ■

For consistent stable theories $T$, we have the following corollary to Theorem 8.8, that expresses the membership of a disjunction

$$L\varphi_1 \vee \ldots \vee L\varphi_k \vee \neg L\psi_1 \vee \ldots \vee \neg L\psi_m \vee \vartheta$$

in $T$ in terms of the membership in $T$ of the formulas $\varphi_i$, $\psi_i$ and $\vartheta$. Since this corollary reduces the question of membership testing to the same problem but for formulas of smaller $L$-depths, it plays a crucial role in algorithms that decide whether or not a formula belongs to a given stable and consistent theory.

**Corollary 8.9** *Let $T \subseteq \mathcal{L}_L$ be stable and consistent. Then*

$$L\varphi_1 \vee \ldots \vee L\varphi_k \vee \neg L\psi_1 \vee \ldots \vee \neg L\psi_m \vee \vartheta \in T,$$

*if and only if $\varphi_i \in T$, for some $i$, $1 \leq i \leq k$, or $\psi_i \notin T$, for some $i$, $1 \leq i \leq m$, or $\vartheta \in T$.*

Proof: The assertion follows directly from Theorems 8.8 and 8.3. ■

The simple facts gathered in this section form the basis for our study of more advanced properties of stable theories — their characterizations and algorithms for membership testing.

## 8.2 Characterizations of stable theories

Our goal in this section is to present several equivalent characterizations of stable theories. These characterizations are of three types:

- Semantic — stable theories are characterized as theories satisfied by some Kripke models.
- Syntactic — constructions are given that build a stable theory with a prescribed modal-free part.
- Fixpoint — stable theories are characterized as solutions to an equation involving a certain operator.

Our first results connect stable theories and Kripke model semantics. We have seen earlier that stable theories are closed under **S5**-consequence. In fact, a stronger statement holds. It turns out that stable and consistent theories are exactly the theories consisting of formulas true in some Kripke model with a universal accessibility relation. Since provability in **S5** is sound and complete with respect to universal Kripke models, closure of stable theories under **S5**-consequence is an easy corollary to this result.

We start with a theorem stating that for every universal Kripke model $\mathcal{M}$, all formulas true in $\mathcal{M}$ form a stable theory.

**Theorem 8.10** *Let $\mathcal{M}$ be a Kripke model with a universal accessibility relation. Then $Th(\mathcal{M})$ is stable.*

Proof: By Theorem 7.20, $Th(\mathcal{M})$ satisfies the conditions (St1) and (St2). To prove that (St3) holds as well, consider $\vartheta \notin Th(\mathcal{M})$. There is $m \in M$ such that $\langle \mathcal{M}, m \rangle \not\models \vartheta$. Since the accessibility relation of $\mathcal{M}$ is universal, $\langle \mathcal{M}, m' \rangle \models \neg L\vartheta$ holds for every $m' \in M$. Thus $\neg L\vartheta \in Th(\mathcal{M})$. This proves (St3) and establishes the stability of $Th(\mathcal{M})$. ■

This theorem has an important consequence. It implies that each set of formulas, which is consistent with **S5**, is contained in a theory that is consistent and stable.

**Corollary 8.11** *Let $U \subseteq \mathcal{L}_L$ be consistent with* **S5**. *Then there is a stable and consistent theory $T$ such that $U \subseteq T$.*

Proof: Let $\mathcal{M}$ be a Kripke model whose accessibility relation is an equivalence relation and such that $\mathcal{M} \models U$. Let $\mathcal{M}'$ be a component of $\mathcal{M}$. Then, $Th(\mathcal{M}')$

is consistent and, by Theorem 8.10, stable. Since $U \subseteq Th(\mathcal{M})$ and $Th(\mathcal{M}) \subseteq Th(\mathcal{M}')$, the assertion follows. ∎

Not only is every theory consisting of all formulas true in some universal Kripke model stable, but the converse holds, too. That is, every consistent stable theory can be represented as the theory of all formulas true in some such model.

**Theorem 8.12** *Let $T \subseteq \mathcal{L}_L$ be stable and consistent. Then $T = Th(\mathcal{M})$, for some Kripke model $\mathcal{M}$ with a universal accessibility relation.*

Proof: Since $T$ is consistent, by Corollary 8.5(2), $T$ is consistent with **S5**. Let $\mathcal{M} = (\mathcal{M}^{can}_{\mathbf{S5}}(T))^{\cap}$. By Corollary 7.47, $T = Th(\mathcal{M})$. By Lemma 7.50 the accessibility relation of $\mathcal{M}$ is an equivalence relation. Let $\mathcal{M}'$ be any component of $\mathcal{M}$. Then $\mathcal{M}'$ has a universal accessibility relation. Clearly, $Th(\mathcal{M}')$ is consistent and, by Theorem 8.10, stable. In addition, since $\mathcal{M}'$ is a component of $\mathcal{M}$, $Th(\mathcal{M}') \supseteq Th(\mathcal{M}) = T$. Hence, Theorem 8.6 implies that $Th(\mathcal{M}') = T$. ∎

Theorems 8.10 and 8.12 together provide us with a first complete characterization of consistent and stable theories — they are precisely the theories consisting of the formulas true in universal Kripke models.

**Theorem 8.13** *Let a theory $T \subseteq \mathcal{L}_L$ be consistent. Then $T$ is stable if and only if $T = Th(\mathcal{M})$, for some Kripke model $\mathcal{M}$ with a universal accessibility relation.* ▲

We have seen earlier that the modal-free part of any stable theory is closed under propositional consequence. Theorem 8.10 allows us to prove a version of a converse statement, namely that each theory consisting of modal-free formulas and closed under propositional consequence is the modal-free part of some stable theory.

**Theorem 8.14** *Let $S \subseteq \mathcal{L}$ be closed under propositional consequence. Then there is a stable theory $T \subseteq \mathcal{L}_L$ such that $[T]_0 = S$ (or, equivalently, $T \cap \mathcal{L} = S$).*

Proof: If $S$ is inconsistent, then $S = \mathcal{L}$. Clearly, $T = \mathcal{L}_L$ satisfies the assertion. So, assume that $S$ is consistent. Consider the **S5**-model $\mathcal{M} = \langle M, V \rangle$, where $M$ consists of all propositional valuations in which all formulas from $S$ are true and $V(w) = w$, for every $w \in M$. Since $S$ is closed under propositional provability, for a formula $\varphi \in \mathcal{L}$ we have that $\varphi \in S$ if and only if $\varphi$ is true in every valuation from $M$ (by the completeness theorem). Consequently, $Th(\mathcal{M}) \cap \mathcal{L} = S$. In addition, by Theorem 8.10, $Th(\mathcal{M})$ is stable. Hence, the assertion follows. ■

Similar methods imply yet another characterization of consistent stable theories. This time, the characterization has a syntactic flavor. Stable theories are shown to coincide with maximal theories consistent with **S5**.

**Theorem 8.15** *Let $T \subseteq \mathcal{L}_L$. Theory $T$ is stable and consistent if and only if $T$ is a maximal theory consistent with* **S5**.

Proof: The "only if" part follows by Corollaries 8.5 and 8.11, and from the fact that stable and consistent theories form an antichain (Theorem 8.6). To prove the "if" part, assume that $T$ is a maximal theory in $\mathcal{L}_L$ consistent with **S5**. Reasoning as in the proof of Theorem 8.12 there is a model $\mathcal{M}$, such that the accessibility relation of $\mathcal{M}$ is an equivalence relation and $T \subseteq Th(\mathcal{M})$. Let $\mathcal{M}'$ be a component of $\mathcal{M}$. Then $T \subseteq Th(\mathcal{M}) \subseteq Th(\mathcal{M}')$. Since $Th(\mathcal{M}')$ is consistent with **S5** and $T$ is maximal with this property, $Th(\mathcal{M}') = T$. Thus, $T$ is stable and consistent. ■

So far we have developed the theory of stability by exploiting connections between stable theories and logic **S5**. In particular, we obtained characterizations of stable theories in terms of maximal theories consistent with **S5** (Theorem 8.15) and as theories true in universal models (Theorem 8.13). Now we will investigate connections between stable theories and other normal modal systems. In particular, we will derive two different characterizations of stable theories, this time in terms of Kripke models with arbitrary accessibility relations, and in terms of provability in the logic **K**.

Theorem 8.12 states that every stable and consistent theory is precisely the set of all formulas true in some universal Kripke model. Our next result is of a similar type. Given a stable and consistent theory $T$, it specifies necessary and sufficient conditions for a Kripke model $\mathcal{M}$ to have the property that $T = Th(\mathcal{M})$.

**Theorem 8.16** *Let $T \subseteq \mathcal{L}_L$ be a stable theory and let $S = [T]_0$.*

1. *Let $\mathcal{M}$ be a Kripke model such that $\mathcal{M} \models S \cup \{\neg L\varphi : \varphi \in \mathcal{L} \setminus S\}$. Then:*
   (a) *For every $\varphi \in \mathcal{L}_L$, $\varphi \in T$ if and only if $\mathcal{M} \models \varphi$, in other words, $T = Th(\mathcal{M})$.*
   (b) *For every $\varphi \in \mathcal{L}_L$, there exists $\omega \in \mathcal{L}$ such that $\mathcal{M} \models (\varphi \equiv \omega)$.*
2. *Conversely, if $T = Th(\mathcal{M})$ for some Kripke model $\mathcal{M}$, then $\mathcal{M} \models S \cup \{\neg L\varphi : \varphi \in \mathcal{L} \setminus S\}$.*

Proof: (1) First observe that since $S$ is satisfied by a Kripke model, $S$ is consistent. Thus, by Theorem 8.2, $T$ is consistent. In the proof, we proceed by induction on the depth $n$ of nesting of $L$ in $\varphi$. Let us recall that by $\mathcal{L}_{L,n}$ we denote the set of all formulas from $\mathcal{L}_L$ with $L$-depth at most $n$. Consider $\varphi \in \mathcal{L}_{L,0}$ $(= \mathcal{L})$. If $\varphi \in S$, then $\mathcal{M} \models \varphi$, by the assumptions of the theorem. If $\varphi \notin S$, then $\mathcal{M} \models \neg L\varphi$. Thus, $\mathcal{M} \not\models \varphi$. Consequently, the condition (a) follows. For formulas of $L$-depth 0 the condition (b) is evident.

Consider now $n > 0$ and assume that the assertion holds for all formulas of $L$-depth at most $n-1$. Let $\varphi \in \mathcal{L}_{L,n}$. Then by Theorem 7.3, there is a formula $\psi \in \mathcal{L}_{L,n}$, propositionally equivalent to $\varphi$, and such that

$$\psi = \psi_1 \wedge \ldots \wedge \psi_k,$$

where each of $\psi_i$ is of the form

$$L\alpha_1 \vee \ldots \vee L\alpha_m \vee \neg L\beta_1 \vee \ldots \vee \neg L\beta_p \vee \gamma,$$

and $\gamma \in \mathcal{L}$. Since $T$ is closed under propositional consequence, $\varphi \in T$ if and only if $\psi \in T$. In addition, $\mathcal{M} \models (\varphi \equiv \psi)$. Thus, to prove the induction step, it is sufficient to prove that conditions (a) and (b) hold for $\psi$.

Assume that $\psi \in T$. Then each formula $\psi_i \in T$. By Corollary 8.9, one of the following three conditions holds for $\psi_i$:

**(i)** For some $j$, $1 \leq j \leq m$, $\alpha_j \in T$.

**(ii)** For some $j$, $1 \leq j \leq p$, $\beta_j \notin T$.

**(iii)** $\gamma \in T$.

If (i) holds, then it follows by the induction hypothesis that $\mathcal{M} \models \alpha_j$. Consequently, $\mathcal{M} \models L\alpha_j$ and $\mathcal{M} \models \psi_i$.

If (ii) holds, we reason as follows. By the induction hypothesis, $\mathcal{M} \not\models \beta_j$. Also by the induction hypothesis, there is a formula $\omega \in \mathcal{L}$ such that $\mathcal{M} \models (\beta_j \equiv \omega)$. Clearly, $\mathcal{M} \not\models \omega$. Since $\omega \in \mathcal{L}$, by the induction hypothesis, $\omega \notin T$. Since $S = [T]_0$, $\omega \notin S$. By the assumptions of the theorem, $\mathcal{M} \models \neg L\omega$. Consequently, by Theorem 7.17, $\mathcal{M} \models \neg L\beta_j$. Thus, $\mathcal{M} \models \psi_i$.

If (iii) holds, then $\gamma \in S$ and $\mathcal{M} \models \gamma$ follows from the assumptions of the theorem. Consequently, $\mathcal{M} \models \psi_i$.

Thus, in each case $\mathcal{M} \models \psi_i$. Consequently, $\mathcal{M} \models \psi$ follows.

Conversely, assume that $\psi \notin T$. Then, for some $i$, $\psi_i \notin T$. By Corollary 8.9, for this $\psi_i$ the following three conditions hold:

**(i)** For all $j$, $1 \leq j \leq m$, $\alpha_j \notin T$.

**(ii)** For all $j$, $1 \leq j \leq p$, $\beta_j \in T$.

**(iii)** $\gamma \notin T$.

By the induction hypothesis, for each $\alpha_j$ there is $\alpha'_j \in \mathcal{L}$ such that $\mathcal{M} \models (\alpha_j \equiv \alpha'_j)$. Also by the induction hypothesis, $\mathcal{M} \not\models \alpha_j$. Thus $\mathcal{M} \not\models \alpha'_j$. Since $\alpha'_j \in \mathcal{L}$, we obtain that $\alpha'_j \notin S$. Consequently, $\mathcal{M} \models \neg L\alpha'_j$. Thus, for all $\alpha_j$, $\mathcal{M} \models \neg L\alpha_j$.

Since all $\beta_j \in T$, by the induction hypothesis it follows that for all $\beta_j$, $\mathcal{M} \models L\beta_j$. Thus, for every world $m$ of the model $\mathcal{M}$, $(\mathcal{M}, m) \models \psi_i$ if and only if $(\mathcal{M}, m) \models \gamma$. By (iii), $\gamma \notin T$. Since $\gamma \in \mathcal{L}$, $\mathcal{M} \not\models \gamma$ follows. Consequently, $\mathcal{M} \not\models \psi_i$ and $\mathcal{M} \not\models \psi$.

It remains to prove condition (b) for $\psi$. Let us denote by $c(\psi_i)$ the modal-free disjunct of $\psi_i$. If this part is not present in $\psi_i$, take for $c(\psi_i)$ any contradiction from $\mathcal{L}$. Let $\Psi = \{\psi_i : \psi_i \notin T\}$. Define $\omega$ to be the conjunction of the formulas $c(\psi_i)$ for all $\psi_i \in \Psi$. If $\Psi = \emptyset$, take any tautology from $S$ as $\omega$. As we have just seen, for every $\psi_i \notin T$ (that is, $\psi_i \in \Psi$) and for every world $m$ of the model $\mathcal{M}$ we have $(\mathcal{M}, m) \models \psi_i$ if and only if $(\mathcal{M}, m) \models c(\psi_i)$. Since for all $\psi_i \notin \Psi$ we have $\mathcal{M} \models \psi_i$ (by (a)), we obtain that for every world $m$ of the model $\mathcal{M}$, $(\mathcal{M}, m) \models \psi$ if and only if $(\mathcal{M}, m) \models \omega$. Thus, $\mathcal{M} \models (\psi \equiv \omega)$ follows.

(2) This part is easy to prove. It follows from two facts: that $\mathcal{M} \models T$ and that $S \cup \{\neg L\varphi : \varphi \in \mathcal{L} \setminus S\} \subseteq T$. ■

The proof of Theorem 8.16 relies on an important assumption that $T$ is stable. If $T$ is stable and a Kripke model $\mathcal{M}$ satisfies $\mathcal{M} \models S \cup \{\neg L\varphi : \varphi \in \mathcal{L} \setminus S\}$, where $S = [T]_0$, then $Th(\mathcal{M}) = T$ which, among other things, implies that $Th(\mathcal{M})$ is stable. But Theorem 8.16 does not explicitly state that if $S$ is closed under propositional consequence and $\mathcal{M} \models S \cup \{\neg L\varphi : \varphi \in \mathcal{L} \setminus S\}$ then $Th(\mathcal{M})$ is stable. This statement, however, is true. To prove it one has to combine Theorem 8.16 with Theorem 8.14, which states that each modal-free theory closed under propositional consequence in $\mathcal{L}$ is a modal-free part of some stable theory.

**Theorem 8.17** *Let $S \subseteq \mathcal{L}$ be closed under propositional consequence and let $\mathcal{M}$ be a Kripke model such that $\mathcal{M} \models S \cup \{\neg L\varphi : \varphi \in \mathcal{L} \setminus S\}$. Then $Th(\mathcal{M})$ is stable and $Th(\mathcal{M}) \cap \mathcal{L} = S$.*

Proof: Let $T$ be a stable theory such that $[T]_0 = S$. Such a stable theory exists by Theorem 8.14. Clearly, $\mathcal{M} \models [T]_0 \cup \{\neg L\varphi : \varphi \in \mathcal{L} \setminus [T]_0\}$. Thus, by Theorem 8.16, $T = Th(\mathcal{M})$. In particular, $Th(\mathcal{M})$ is stable and $Th(\mathcal{M}) \cap \mathcal{L} = S$. ■

Now we can fully characterize stable theories in terms of arbitrary Kripke models. The characterization is given in the next theorem.

**Theorem 8.18** *Let a theory $T \subseteq \mathcal{L}_L$ be consistent. Then the theory $T$ is stable if and only if $T = Th(\mathcal{M})$ for some Kripke model $\mathcal{M}$ such that $\mathcal{M} \models [T]_0 \cup \{\neg L\varphi : \varphi \in \mathcal{L} \setminus [T]_0\}$.*

Proof: First, assume that $T$ is stable and consistent. Then, by Theorem 8.12, there exists a Kripke model $\mathcal{M}$ such that $T = Th(\mathcal{M})$. Since $[T]_0 \cup \{\neg L\varphi : \varphi \in \mathcal{L} \setminus [T]_0\} \subseteq T$, it follows that $\mathcal{M} \models [T]_0 \cup \{\neg L\varphi : \varphi \in \mathcal{L} \setminus [T]_0\}$.

The converse implication follows directly from Theorem 8.17. ■

Theorems 8.16 and 8.18 yield the following corollary, which states that there are no two different stable theories with the same modal-free part. In other words, the modal-free part uniquely determines a stable theory.

**Corollary 8.19** *If $T_1$ and $T_2$ are stable and $[T_1]_0 = [T_2]_0$ then $T_1 = T_2$.*

Proof: If $[T_1]_0$ is inconsistent, then $[T_2]_0$ is inconsistent, and we have $T_1 = T_2 = \mathcal{L}_L$. So, assume that $[T_1]_0$ is consistent. Then, by Theorem 8.18, $T_1 = Th(\mathcal{M})$, for some Kripke model $\mathcal{M}$ such that $\mathcal{M} \models [T_1]_0 \cup \{\neg L\varphi : \varphi \in \mathcal{L} \setminus [T_1]_0\}$. Since $[T_1]_0 = [T_2]_0$, $\mathcal{M} \models [T_2]_0 \cup \{\neg L\varphi : \varphi \in \mathcal{L} \setminus [T_2]_0\}$. Since $T_2$ is stable, Theorem 8.16 implies that $Th(\mathcal{M}) = T_2$. Hence, $T_1 = T_2$. ■

Consider a stable and consistent theory $T$. Let $S$ be a modal-free part of $T$. Let us recall that in the proof of Theorem 8.14 we defined an **S5**-model $\mathcal{M} = \langle M, V \rangle$, where $M$ consists of all propositional valuations in which all formulas from $S$ are true and $V(w) = w$, for every $w \in M$. This model has the property that $Th(\mathcal{M}) \cap \mathcal{L} = S$. Since $Th(\mathcal{M})$ is stable (Theorem 8.10), it follows from Corollary 8.19 that $Th(\mathcal{M}) = T$. We will call $\mathcal{M}$ the *canonical model for a stable theory* $T$ and we will denote it by $\mathcal{M}_T$. Observe that $\mathcal{M}_T$ is the **S5**-canonical model for $S$.

It follows that there is a one-to-one correspondence between subsets of $\mathcal{L}$ closed under propositional consequence and stable theories.

**Definition 8.20** By $ST(S)$ we denote the unique stable theory $T$ such that $T \cap \mathcal{L} = Cn(S)$. ▲

Another corollary of Theorem 8.17 gives a syntactic description of the operator $ST$ in terms of the provability operator of any normal modal logic $\mathcal{S}$ contained in **S5**.

**Corollary 8.21** *Let $S \subseteq \mathcal{L}$ be closed under propositional consequence.*

1. *The theory $T = Cn_{\mathbf{K}}(S \cup \{\neg L\varphi : \varphi \in \mathcal{L} \setminus S\})$ is stable and $T \cap \mathcal{L} = S$. Hence,*
$$ST(S) = Cn_{\mathbf{K}}(S \cup \{\neg L\varphi : \varphi \in \mathcal{L} \setminus S\}).$$

2. *For any normal modal logic $\mathcal{S}$ contained in* **S5**,
$$ST(S) = Cn_{\mathcal{S}}(S \cup \{\neg L\varphi : \varphi \in \mathcal{L} \setminus S\}).$$

Proof: (1) If $S$ is propositionally inconsistent then $Cn_{\mathbf{K}}(S \cup \{\neg L\varphi : \varphi \in \mathcal{L} \setminus S\}) = \mathcal{L}_L$ and the assertion follows. Hence, assume that $S$ is propositionally consistent. Then $ST(S)$ is consistent with **S5**. Since $S \cup \{\neg L\varphi : \varphi \in \mathcal{L} \setminus S\} \subseteq ST(S)$, it follows that $S \cup \{\neg L\varphi : \varphi \in \mathcal{L} \setminus S\}$ is consistent with **K**. By Corollary 7.47, there exists a model $\mathcal{M}$ such that $Th(\mathcal{M}) = Cn_{\mathbf{K}}(S \cup \{\neg L\varphi : \varphi \in \mathcal{L} \setminus S\})$. Thus, $\mathcal{M} \models S \cup \{\neg L\varphi : \varphi \in \mathcal{L} \setminus S\}$, and the assertion follows by Theorem 8.17.

(2) Since $ST(S)$ is stable, it is closed under provability in **S5**. Thus,
$$Cn_{\mathcal{S}}(S \cup \{\neg L\varphi : \varphi \in \mathcal{L} \setminus S\}) \subseteq ST(S).$$
The converse inclusion follows by (1). ■

It turns out that the operator $ST(S)$ can be described entirely in terms of the propositional provability operator. First, we have the following simple property of stable theories.

**Theorem 8.22** *Let $T \subseteq \mathcal{L}_L$ be stable. Then, for every integer $n \geq 0$,*
$$[T]_{n+1} = Cn([T]_n \cup \{L\varphi : \varphi \in [T]_n\} \cup \{\neg L\varphi : \varphi \in \mathcal{L}_{L,n} \setminus [T]_n\}) \cap \mathcal{L}_{L,n+1}.$$

Proof: Clearly, we have $[T]_n \subseteq [T]_{n+1}$. In addition, by stability of $T$ and the definition of $[T]_n$ and $[T]_{n+1}$, we have $L\varphi \in [T]_{n+1}$ for $\varphi \in [T]_n$, and $\neg L\varphi \in [T]_{n+1}$ for $\varphi \in \mathcal{L}_{L,n} \setminus [T]_n$. Thus, the inclusion
$$[T]_{n+1} \supseteq Cn([T]_n \cup \{L\varphi : \varphi \in [T]_n\} \cup \{\neg L\varphi : \varphi \in \mathcal{L}_{L,n} \setminus [T]_n\}) \cap \mathcal{L}_{L,n+1}$$
follows.

The converse inclusion is evident if $T$ is inconsistent (because $[T]_n$ is inconsistent in such a case, too). So, let us assume that $T$ is consistent. To prove the converse inclusion in this case, let us consider a formula $\omega \in [T]_{n+1}$. Let $\psi$ be a formula in the normal form, propositionally equivalent to $\omega$, whose existence is guaranteed by Theorem 7.3. We have then that

$$\psi = \vartheta_1 \wedge \ldots \wedge \vartheta_k,$$

where each $\vartheta_i$ is of the form

$$L\alpha_1 \vee \ldots \vee L\alpha_m \vee \neg L\beta_1 \vee \ldots \vee \neg L\beta_p \vee \gamma.$$

By Theorem 7.3 the $L$-depth of each $\alpha_j$ and $\beta_j$ is at most $n$, and $\gamma \in \mathcal{L}$. Since $\psi \in [T]_{n+1}$, then each $\vartheta_i \in [T]_{n+1}$ and, by Theorem 8.8, one of the three cases holds:

**(a)** $L\alpha_j \in [T]_{n+1}$, for some $j$, $1 \leq j \leq m$.

**(b)** $\neg L\beta_j \in [T]_{n+1}$, for some $j$, $1 \leq j \leq p$.

**(c)** $\gamma \in [T]_{n+1}$.

If (a) holds, $L\alpha_j \in \{L\varphi : \varphi \in [T]_n\}$ and $\vartheta_i \in Cn([T]_n \cup \{L\varphi : \varphi \in [T]_n\} \cup \{\neg L\varphi : \varphi \in \mathcal{L}_{L,n} \setminus [T]_n\}) \cap \mathcal{L}_{L,n+1}$.

If (b) holds then, since $T$ is consistent, $\neg L\beta_j \in \{\neg L\varphi : \varphi \in \mathcal{L}_{L,n} \setminus [T]_n\}$. Consequently, $\vartheta_i \in Cn([T]_n \cup \{L\varphi : \varphi \in [T]_n\} \cup \{\neg L\varphi : \varphi \in \mathcal{L}_{L,n} \setminus [T]_n\}) \cap \mathcal{L}_{L,n+1}$.

Finally, if (c) holds, $\gamma \in [T]_0 \subseteq [T]_n$.

Thus, in all three cases, $\vartheta_i \in Cn([T]_n \cup \{L\varphi : \varphi \in [T]_n\} \cup \{\neg L\varphi : \varphi \in \mathcal{L}_{L,n} \setminus [T]_n\}) \cap \mathcal{L}_{L,n+1}$. Hence, $\psi \in Cn([T]_n \cup \{L\varphi : \varphi \in [T]_n\} \cup \{\neg L\varphi : \varphi \in \mathcal{L}_{L,n} \setminus [T]_n\}) \cap \mathcal{L}_{L,n+1}$. Consequently, $\omega \in Cn([T]_n \cup \{L\varphi : \varphi \in [T]_n\} \cup \{\neg L\varphi : \varphi \in \mathcal{L}_{L,n} \setminus [T]_n\}) \cap \mathcal{L}_{L,n+1}$, and the converse inclusion follows. ■

For a theory $S \subseteq \mathcal{L}$, let us define

$$ST_0(S) = Cn(S),$$

and for $n \geq 0$,

$$ST_{n+1}(S) = Cn(ST_n(S) \cup \{L\varphi : \varphi \in ST_n(S)\} \cup \{\neg L\varphi : \varphi \in \mathcal{L}_{L,n} \setminus ST_n(S)\}) \cap \mathcal{L}_{L,n+1}.$$

We have the following theorem.

**Theorem 8.23** *For any theory $S \subseteq \mathcal{L}$,*

$$ST(S) = \bigcup_{n=0}^{\infty} ST_n(S).$$

Proof: Follows easily by induction on $n$ from Theorem 8.22. ■

We can modify the construction of the operator used to construct $ST(S)$ somewhat. First define, for a theory $T \subseteq \mathcal{L}_L$,

$$[T]_{=n} = \{\varphi \in T : d_L(\varphi) = n\}$$

Then define

$$st_0(S) = Cn(S) \cap \mathcal{L},$$

and for $n \geq 0$,

$$\begin{aligned} st_{n+1}(S) = Cn(st_n(S) \cup \{L\varphi : \varphi \in [st_n(S)]_{=n}\} \\ \cup \{\neg L\varphi : \varphi \in [\mathcal{L}_{L,n}]_{=n} \setminus st_n(S)\}) \cap \mathcal{L}_{L,n+1}. \end{aligned}$$

We have the following theorem.

**Theorem 8.24** *For any theory $S \subseteq \mathcal{L}$,*

$$ST(S) = \bigcup_{n=0}^{\infty} st_n(S).$$

▲

Together with Theorem 8.14, Corollary 8.19 implies that there is a one-to-one correspondence between stable theories and theories in $\mathcal{L}$ that are closed under propositional consequence. This observation allows us to find the cardinality of the collection of all stable theories in $\mathcal{L}_L$.

**Corollary 8.25**

1. *If the language $\mathcal{L}$ possesses a finite number of atoms, then there is a finite number of stable theories in $\mathcal{L}_L$.*
2. *If the set of atoms of the language $\mathcal{L}$ is denumerably infinite then there are exactly continuum many (that is, $2^{\aleph_0}$) stable theories.* ▲

Next, we will present two fixpoint characterizations of stable theories. The term *fixpoint* is used in the sense that stable theories are characterized as the solutions to the equation $T = F(T)$, where $F$ is an operator defined on the power set of $\mathcal{L}_L$. Usually, this operator $F$ involves the consequence operator of some logic $\mathcal{S}$. Our first approach utilizes the consequence operator of the propositional logic.

**Theorem 8.26** *A theory $T \subseteq \mathcal{L}_L$ is stable if and only if*

$$T = Cn([T]_0 \cup \{L\varphi : \varphi \in T\} \cup \{\neg L\varphi : \varphi \notin T\}).$$

Proof: Clearly, if

$$T = Cn([T]_0 \cup \{L\varphi : \varphi \in T\} \cup \{\neg L\varphi : \varphi \notin T\})$$

then $T$ satisfies conditions (St1) – (St3) required for stability.

To prove the converse implication, consider a stable theory $T$. The inclusion

$$Cn([T]_0 \cup \{L\varphi : \varphi \in T\} \cup \{\neg L\varphi : \varphi \notin T\}) \subseteq T$$

is straightforward. To prove the converse inclusion, first observe that

$$\{L\varphi : \varphi \in [T]_n\} \subseteq \{L\varphi : \varphi \in T\},$$

and

$$\{\neg L\varphi : \varphi \in \mathcal{L}_{L,n} \setminus [T]_n\} \subseteq \{\neg L\varphi : \varphi \notin T\}.$$

Next, using these inclusions and Theorem 8.22, one can easily show by induction that for every nonnegative integer $n$,

$$[T]_n \subseteq Cn([T]_0 \cup \{L\varphi : \varphi \in T\} \cup \{\neg L\varphi : \varphi \notin T\}).$$

Hence, the inclusion

$$T \subseteq Cn([T]_0 \cup \{L\varphi : \varphi \in T\} \cup \{\neg L\varphi : \varphi \notin T\})$$

follows. ■

The characterization of stable theories given in Theorem 8.26 will reappear in the context of autoepistemic logic in Chapter 10.

Our next fixpoint characterization is in terms of the provability operator for the logic **S5**.

**Theorem 8.27** *Let $T \subseteq \mathcal{L}_L$ be consistent. Then $T$ is stable if and only if*

$$T = Cn_{\mathbf{S5}}(\{\neg L\varphi : \varphi \notin T\}).$$

Proof: First, assume stability of $T$. Let $\mathcal{M}$ be any Kripke model with a universal accessibility relation and such that $\mathcal{M} \models \{\neg L\varphi : \varphi \notin T\}$. Let $\varphi \in T$. If $\mathcal{M} \not\models \varphi$ then, since the accessibility relation of $\mathcal{M}$ is universal, $\mathcal{M} \models \neg L\varphi$. Consequently, $\mathcal{M} \models L\neg L\varphi$. But, since $T$ is consistent, $\neg L\varphi \notin T$. Thus, $\mathcal{M} \models \neg L\neg L\varphi$, a contradiction. Hence, $\mathcal{M} \models T$. In other words, $T \subseteq Th(\mathcal{M})$. Since both $T$ and $Th(\mathcal{M})$ are stable, $T = Th(\mathcal{M})$ follows by Theorem 8.6. Consequently, $T = Cn_{\mathbf{S5}}(\{\neg L\varphi : \varphi \notin T\})$.

Next, assume that $T = Cn_{\mathbf{S5}}(\{\neg L\varphi : \varphi \notin T\})$. Then $T$ is closed under propositional consequence and necessitation. Thus $T$ satisfies (St1) and (St2). Since $\{\neg L\varphi : \varphi \notin T\} \subseteq T$, (St3) follows, too. ■

We conclude this section with a theorem which gathers all characterizations of stable theories we have obtained above.

**Theorem 8.28** *Let a theory $T \subseteq \mathcal{L}_L$ be consistent. Then the following conditions are equivalent:*

1. *$T$ is stable.*
2. *$T = Th(\mathcal{M})$, for some universal Kripke model $\mathcal{M}$.*
3. *$T = Th(\mathcal{M})$, for some Kripke model $\mathcal{M}$ such that $\mathcal{M} \models [T]_0 \cup \{\neg L\varphi : \varphi \in \mathcal{L} \setminus [T]_0\}$.*
4. *$T$ is a maximal theory consistent with* **S5**.
5. *$T = Cn([T]_0 \cup \{L\varphi : \varphi \in T\} \cup \{\neg L\varphi : \varphi \notin T\})$.*
6. *$T = Cn_{\mathbf{S5}}(\{\neg L\varphi : \varphi \notin T\})$.* ▲

Let us note here that the characterization given by the condition (5) applies also in the case of inconsistent theories.

## 8.3 Algorithms

In this section, we discuss algorithmic issues related to stable theories. Specifically, we concentrate on the problem of testing whether a formula belongs to a stable theory. As we have seen, stable theories are uniquely determined by their modal-free parts. Hence, the membership testing problem can be formalized as follows:

> Given a set $S \subseteq \mathcal{L}$ and a formula $\varphi \in \mathcal{L}_L$, decide whether $\varphi \in ST(S)$.

In what follows, we assume that $S$ is finite. Some of our algorithms can be modified to cover a more general case where $S$ is given by means of an *oracle* which, given a formula from $\mathcal{L}$, determines whether it is a consequence of $S$ or not.

We present four algorithms for membership testing in stable theories. They are based on some of the characterizations and properties of stable theories obtained in the previous sections of this chapter.

We start with two auxiliary, technical, results. When checking whether $\varphi \in ST(S)$, they allow us to replace the language $\mathcal{L}$, which may have infinitely many propositional variables, with a language whose set of propositional variables is finite.

**Theorem 8.29** *Let $\mathcal{K}$ and $\mathcal{L}$ be two languages of propositional calculus such that $\mathcal{K} \subseteq \mathcal{L}$. Let $\mathcal{K}_L$ and $\mathcal{L}_L$ be the corresponding modal languages. If $T \subseteq \mathcal{L}_L$ is stable (in $\mathcal{L}_L$), then $T \cap \mathcal{K}_L$ is stable in $\mathcal{K}_L$.*

Proof: Denote $T \cap \mathcal{K}_L$ by $T'$. Clearly, $T'$ is closed under propositional consequence (in $\mathcal{K}_L$). Let $\varphi \in T'$. Then $\varphi \in T$. By stability of $T$, $L\varphi \in T$. Since $L\varphi \in \mathcal{K}_L$, $L\varphi \in T'$. Finally, consider $\varphi \in \mathcal{K}_L$ such that $\varphi \notin T'$. Then $\varphi \notin T$. By stability of $T$, $\neg L\varphi \in T$. Consequently, $\neg L\varphi \in T'$. ■

In the next theorem and in the first of our algorithms we use the following notation. By $At$ we denote the set of all propositional variables occurring in $\varphi$ and in the formulas in $S$. The sublanguage of $\mathcal{L}$ generated by the propositional variables in $At$ is denoted by $\mathcal{K}$ and the corresponding modal language by $\mathcal{K}_L$. Finally, $ST_{\mathcal{K}}$ denotes the operator $ST$ defined for the language $\mathcal{K}_L$.

**Theorem 8.30** *Under the notation introduced above, if $\varphi \in \mathcal{K}_L$, then $\varphi \in ST(S)$ if and only if $\varphi \in ST_{\mathcal{K}}(S)$.*

Proof: It follows by Theorem 8.29 that $ST(S) \cap \mathcal{K}_L$ is stable in $\mathcal{K}_L$. Since

$$(ST(S) \cap \mathcal{K}_L) \cap \mathcal{K} = Cn_{\mathcal{K}}(S),$$

by Corollary 8.19 it follows that

$$ST_{\mathcal{K}}(S) = ST(S) \cap \mathcal{K}_L.$$

Since $\varphi \in \mathcal{K}_L$, the assertion follows. ■

This last theorem allows us to decide whether $\varphi \in ST(S)$ by deciding whether $\varphi \in ST_{\mathcal{K}}(S)$.

Our first algorithm makes use of the characterization of stable theories as theories of formulas true in some universal Kripke model. The idea of the algorithm can informally be described as follows. First, we construct a universal Kripke model $\mathcal{M}$ such that $Th(\mathcal{M}) \cap \mathcal{K} = Cn(S)$. Clearly, $Th(\mathcal{M}) = ST(S)$. Thus, to find out whether $\varphi \in ST(S)$, it suffices to determine whether $\mathcal{M} \models \varphi$. The algorithm is described in more detail in Figure 8.1.

Since $S$ is finite, the set $At$ is finite. Hence, the set $M$ is finite and the valuations $V(w)$ have finite representations. Consequently, following the

---

**InStabA**$(S, \varphi)$
*Input:* A finite set $S \subseteq \mathcal{L}$ and a formula $\varphi \in \mathcal{L}_L$;
*Output:* **InStabA** returns **true** if $\varphi \in ST(S)$, and **false**, otherwise.

Compute the set $At$ of all propositional variables occurring in $\varphi$ and in the formulas in $S$;
($\star$ All that follows below assumes $\mathcal{K}_L$ as the working modal language $\star$)
Define $M$ to consist of all valuations $w$ of $At$ such that $w \models S$ (in propositional logic);
$R := M \times M$;
$V(w) := w$, for each $w \in M$;
$\mathcal{M} := \langle M, R, V \rangle$;
**for** $w \in M$ **do**
    **if** $\langle \mathcal{M}, w \rangle \not\models \varphi$ **then return{false}**
**rof**
**return{true}**.

---

**Fig. 8.1.** Algorithm **InStabA**

inductive definition of the relation of entailment, it can be established in a finite number of steps whether $\langle \mathcal{M}, w \rangle \models \varphi$. This argument shows that the algorithm will always terminate. The correctness of the algorithm follows from Theorem 8.30 and the following fact.

**Theorem 8.31** *The universal model $\mathcal{M}$ constructed in the algorithm* **InStabA** *has the following property:*

$$Th(\mathcal{M}) \cap \mathcal{K} = Cn(S)$$

*(where $Cn$ is the propositional consequence operator in $\mathcal{K}$).*

Proof: By the definition of $M$, it follows that $Cn(S) \subseteq Th(\mathcal{M})$. Consider $\vartheta \in \mathcal{K} \setminus Cn(S)$. Then, there is a valuation $w$ of $At$ such that $w(S) = \mathbf{t}$ and $w(\vartheta) = \mathbf{f}$. Since $w \in M$, it follows that $\vartheta \notin Th(\mathcal{M})$. ∎

Our next algorithm is based on Corollary 8.9. This corollary allows us to reduce the question of whether a formula $\vartheta$ in the form

$$L\varphi_1 \vee \ldots \vee L\varphi_k \vee \neg L\psi_1 \vee \ldots \vee \neg L\psi_m \vee \gamma$$

is a member of a consistent stable theory $T$ to several problems of the same type but involving formulas with a smaller $L$-depth. By applying Corollary 8.9

**InStabB**$(S, \varphi)$
*Input:* A finite set $S \subseteq \mathcal{L}$ and a formula $\varphi \in \mathcal{L}_L$;
*Output:* **InStabB** returns **true** if $\varphi \in ST(S)$, and **false**, otherwise.

**if** $S$ is inconsistent (**InCn**$(S, \bot)$) **then return**{**true**}
**else if** $\varphi \in \mathcal{L}$ **then return**{**InCn**$(S, \varphi)$}
**else**
transform $\varphi$ to a normal form $\varphi' = \varphi_1 \wedge \ldots \wedge \varphi_n$,
where each $\varphi_i$ is of the form

$$L\alpha_1 \vee \ldots \vee L\alpha_k \vee \neg L\beta_1 \vee \ldots \vee \neg L\beta_m \vee \gamma;$$

**for** $i := 1$ **to** $n$ **do**
**if not InStabB**$(S, \alpha_j)$, $1 \leq j \leq k$, **and not InStabB**$(S, \gamma)$
**and InStabB**$(S, \beta_j)$, $1 \leq j \leq m$, **then return**{**false**}
**rof**
**return**{**true**}.

**Fig. 8.2.** Algorithm **InStabB**

repeatedly, we eventually reduce the problem to several membership testing problems, each involving a modal-free formula. The stable theory $T$ is represented as $ST(S)$, for some finite set $S \subseteq \mathcal{L}$. If $\alpha \in \mathcal{L}$, then $\alpha \in T = ST(S)$ if and only if $\alpha \in Cn(S)$. Thus, by repeated application of Corollary 8.9, we reduce the question whether $\vartheta \in T = ST(S)$ to several problems of the type: does a modal-free formula $\alpha$ belong to $Cn(S)$. In Figure 8.2, we give a detailed description of the algorithm. By **InCn**$(S, \vartheta)$, we denote a procedure deciding whether a given modal-free formula $\vartheta$ is a propositional consequence of a set of modal-free formulas $S$. Algorithm **InStabB** is recursive. Since all recursive calls are made for formulas of smaller $L$-depth than $\varphi$, the recursion is well-founded and the algorithm terminates. The correctness of the algorithm follows from Corollary 8.9, and from the fact that stable theories are closed under propositional consequence, which guarantees that $\varphi \in ST(S)$ if and only if $\varphi' \in ST(S)$.

Let us mention here that algorithm **InStabB** is applicable not only if $S$ is finite, but also when a stable theory $ST(S)$ is given by means of a procedure (oracle) for deciding the membership of modal-free formulas in $Cn(S)$. This observation implies, in particular, the following result.

**Corollary 8.32** *Let $S \subseteq \mathcal{L}$. If $Cn(S)$ is recursive (under some enumeration of $\mathcal{L}$), then $ST(S)$ is recursive.* ▲

Algorithm **InStabB** can be further refined. Assume that we have to check whether a given formula $\vartheta$ is a member of several stable theories $ST(S)$. Then the recursive process executed by **InStabB** of identifying modal-free formulas whose membership in $Cn(S)$ has to be checked will be repeated each time **InStabB** is called for $\vartheta$. Our next algorithm eliminates this inefficiency by first computing for a formula $\vartheta$ several sets of modal-free formulas. Then, based on the membership of modal-free formulas from these set in $Cn(S)$, the algorithm decides whether $\vartheta \in ST(S)$. To make this idea precise, we need more definitions.

**Definition 8.33** A pair $(U, V)$, where $U, V \subseteq \mathcal{L}$, is called a *condition*. A condition is *finite* if both $U$ and $V$ are finite. A stable theory $T$ *satisfies* a condition $(U, V)$ if $U \subseteq T$ and $V \cap T = \emptyset$. A condition $(U, V)$ *forces* a formula $\vartheta$ if and only if every stable theory $T$ satisfying $(U, V)$ contains $\vartheta$. A collection $\mathcal{F}$ of conditions is *complete for* $\vartheta$ if every condition of $\mathcal{F}$ forces $\vartheta$ and, for every consistent stable theory $T$, if $\vartheta \in T$ then $T$ satisfies some condition $(U, V)$ in $\mathcal{F}$. ▲

If we have a complete set of forcing conditions for $\vartheta$, deciding whether $\vartheta$ belongs to a stable theory $ST(S)$ can be reduced to the problem of deciding whether $ST(S)$ satisfies any of the conditions in $\mathcal{F}$. If so, $\vartheta \in ST(S)$, and otherwise $\vartheta \notin ST(S)$. Checking whether $ST(S)$ satisfies any of the conditions in $\mathcal{F}$ requires only repeated application of a procedure for deciding the membership of a modal-free formula in $Cn(S)$.

Next, we prove a series of theorems whose purpose is to provide us with a method to construct, for any $\vartheta \in \mathcal{L}$, a finite and complete set of finite forcing conditions. In what follows, the crucial role is played by the concept of a purely autoepistemic formula — a natural generalization of a modal atom $L\varphi$.

**Definition 8.34** A formula $\vartheta \in \mathcal{L}_L$ that is built from modal atoms by means of boolean connectives is called ***purely autoepistemic***. A theory consisting of purely autoepistemic formulas is called ***purely autoepistemic***. ▲

Stable theories are complete with respect to purely autoepistemic formulas. This property of stable theories, which we prove below, generalizes Theorem 8.7.

**Theorem 8.35** *If $\vartheta \in \mathcal{L}_L$ is a purely autoepistemic formula, then for every stable theory $T$, at least one of $\vartheta$, $\neg\vartheta$ belongs to $T$. If, in addition, $T$ is consistent, then exactly one of $\vartheta$, $\neg\vartheta$ belongs to $T$.*

Proof: First, assume that $\vartheta$ is of the form

$$L\varphi_1 \vee \ldots \vee L\varphi_k \vee \neg L\psi_1 \vee \ldots \vee \neg L\psi_m. \tag{8.1}$$

Assume that $\vartheta \notin T$. Since $T$ is closed under propositional consequence, it follows that $L\varphi_i \notin T$, $1 \leq i \leq k$, and $\neg L\psi_i \notin T$, $1 \leq i \leq m$. By Theorem 8.7, $\neg L\varphi_i \in T$, $1 \leq i \leq k$, and $L\psi_i \in T$, $1 \leq i \leq m$. Consequently, $\neg\vartheta \in T$ (again by closure of $T$ under propositional consequence).

Next, assume that $\vartheta$ is an arbitrary purely autoepistemic formula. Then, $\vartheta$ is equivalent to a formula $\vartheta' = \alpha_1 \wedge \ldots \wedge \alpha_n$, where each $\alpha_i$ is of the form (8.1). Since $T$ is closed under propositional consequence, it is enough to show that at least one of $\vartheta'$, $\neg\vartheta'$ is in $T$. Let us suppose that $\vartheta' \notin T$. Then for some $i$, $1 \leq i \leq n$, $\alpha_i \notin T$. Since $\alpha_i$ is of the form (8.1), $\neg\alpha_i \in T$. Consequently, $\neg\vartheta' \in T$. ∎

As an immediate corollary, we obtain the following counterpart to Theorem 8.8.

**Corollary 8.36** *Let a formula $\varphi \in \mathcal{L}_L$ be purely autoepistemic, and let a theory $T \subseteq \mathcal{L}_L$ be stable. Then $\varphi \vee \psi \in T$ if and only if $\varphi \in T$ or $\psi \in T$.* ▲

In our solution to the problem of constructing, for a given formula $\vartheta \in \mathcal{L}_L$, a complete set of forcing conditions, the key role is played by purely autoepistemic formulas of $L$-depth 1. We proceed in two steps. First, we show that the general problem can be reduced to the problem of finding a complete set of forcing conditions for a purely autoepistemic formula of $L$-depth 1. Next, we deal with this restricted form of the problem. It is not hard to see that a complete set of forcing conditions for a purely autoepistemic formula of $L$-depth 1 can easily be constructed out of the disjunctive normal form of the formula. Below, we present details of this approach.

**Theorem 8.37** *For every formula $\vartheta \in \mathcal{L}_L$ there exists a purely autoepistemic formula $\vartheta'$ such that $d_L(\vartheta') = 1$ and for every stable theory $T$, $\vartheta \in T$ if and only if $\vartheta' \in T$.*

Proof: We proceed by induction on the $L$-depth of $\vartheta$. First assume that $\vartheta \in \mathcal{L}$. Then, $\vartheta' = L\vartheta$ satisfies the assertion. Consider now a formula $\vartheta \in \mathcal{L}_L$ with the $L$-depth $n \geq 0$, and assume that the assertion holds for all formulas with $L$-depths smaller than $n$. Let a formula $\omega$ be a modal conjunctive normal form of $\vartheta$ (Theorem 7.3), that is,

$$\varphi_1 \wedge \ldots \wedge \varphi_p,$$

where each $\varphi_i$ is of the form

$$L\alpha_1 \vee \ldots \vee L\alpha_k \vee \neg L\beta_1 \vee \ldots \vee \neg L\beta_m \vee \gamma.$$

In this normal form, all formulas $\gamma$ are modal-free, and the $L$-depth of $\omega$ is at most $n$. Since $\vartheta$ and $\omega$ are propositionally equivalent and stable theories are closed under propositional consequence, we have that for every stable theory $T$, $\vartheta \in T$ if and only if $\omega \in T$.

The $L$-depths of the formulas $\alpha_j$ and $\beta_j$ are smaller than $n$. Thus, the induction hypothesis applies to each of them. Let $\alpha'_j$, $1 \leq j \leq k$, and $\beta'_j$, $1 \leq j \leq m$, be purely autoepistemic formulas of the $L$-depth 1 such that for every stable theory $T$, $\alpha_j \in T$ if and only if $\alpha'_j \in T$ and $\beta_j \in T$ if and only if $\beta'_j \in T$. Define

$$\varphi'_i = \alpha'_1 \vee \ldots \vee \alpha'_k \vee \neg\beta'_1 \vee \ldots \vee \neg\beta'_m \vee L\gamma.$$

Consider a stable theory $T$. We will show that $\varphi_i \in T$ if and only if $\varphi'_i \in T$. If $T$ is inconsistent ($T = \mathcal{L}_L$), the claim is obvious. So, assume that $T$ is consistent. First, assume that $\varphi_i \in T$. Then, by Corollary 8.9, one of the following three conditions holds:

**(i)** For some $j$, $1 \leq j \leq k$, $\alpha_j \in T$.

**(ii)** For some $j$, $1 \leq j \leq m$, $\beta_j \notin T$.

**(iii)** $\gamma \in T$.

If (i) holds, then $\alpha'_j \in T$ and so, $\varphi'_i \in T$. If (ii) holds, then $\beta'_j \notin T$. Since $\beta'_j$ is purely autoepistemic, by Theorem 8.35, $\neg\beta'_j \in T$. Hence, also $\varphi'_i \in T$. Finally, if (iii) holds, then $L\gamma \in T$, by stability of $T$. Thus, in all three cases $\varphi'_i \in T$.

The converse implication can be established in the same fashion, using Corollary 8.36 instead of Theorem 8.9.

We have just proved that for every stable theory $T$, $\varphi_i \in T$ if and only if $\varphi'_i \in T$. Define

$$\vartheta' = \varphi'_1 \wedge \ldots \wedge \varphi'_p.$$

Since each stable theory $T$ is closed under propositional consequence, it follows that $\omega = \varphi_1 \wedge \ldots \wedge \varphi_p \in T$ if and only if $\vartheta' \in T$. Formula $\vartheta'$ is purely autoepistemic and has the $L$-depth equal to 1. Thus, the proof of the induction step is complete. ■

The next result shows how to construct a finite complete set of finite forcing conditions for a purely autoepistemic formula of the $L$-depth 1, given in a disjunctive normal form:

$$\psi_1 \vee \ldots \vee \psi_n, \tag{8.2}$$

where

$$\psi_i = L\alpha_1 \wedge \ldots \wedge L\alpha_k \wedge \neg L\beta_1 \wedge \ldots \wedge \neg L\beta_m,$$

and $\alpha_i, \beta_i \in \mathcal{L}$.

**Theorem 8.38** *Let $\vartheta \in \mathcal{L}_L$ be a purely autoepistemic formula of the L-depth 1 given in its modal disjunctive normal form (8.2). For each $i$, $1 \leq i \leq n$, define*

$$U_i = \{\alpha_1, \ldots, \alpha_k\},$$

*and*

$$V_i = \{\beta_1, \ldots, \beta_m\}.$$

*Then $\mathcal{F} = \{(U_1, V_1), \ldots, (U_n, V_n)\}$ is a complete set of forcing conditions for $\vartheta$.*

Proof: Assume first that a stable theory $T$ satisfies condition $(U_i, V_i)$. Then each $\alpha_j \in T$ and each $\beta_j \notin T$. By stability, each $L\alpha_j \in T$ and each $\neg L\beta_j \in T$. Thus,

$$\psi_i = L\alpha_1 \wedge \ldots \wedge L\alpha_k \wedge \neg L\beta_1 \wedge \ldots \wedge \neg L\beta_m \in T,$$

and, consequently, $\vartheta \in T$.

Conversely, let a theory $T$ be stable and consistent and let $\vartheta \in T$. All $\psi_i$ are purely autoepistemic and, by Corollary 8.36, some $\psi_i \in T$. By the stability and consistency of $T$, each $\alpha_j \in T$ and each $\beta_j \notin T$. Consequently, $U_i \subseteq T$ and $V_i \cap T = \emptyset$ and $T$ satisfies $(U_i, V_i)$. ∎

Our results on purely autoepistemic formulas imply the correctness of yet another algorithm for checking whether a formula $\vartheta$ belongs to $ST(S)$, where $S$ is finite. The algorithm is described in Figure 8.3.

Algorithm **InStabC** has an important advantage over the two previous algorithms. Once we compute for $\varphi$ a complete set $\mathcal{F}$ of forcing conditions, we can save it, and use it to check whether $\varphi \in ST(S)$ for several different input stable theories $ST(S)$. This eliminates the need for repeated, and very costly, transformations of formulas to their normal forms, performed by **InStabB**, as well as by the first part of **InStabC**.

Let us mention that also algorithm **InStabC**, like **InStabB**, requires only a procedure that decides whether a modal-free formula is a member of $Cn(S)$, rather than an explicit knowledge of $S$.

We will look again at the membership problem for stable theories. For a formula $\varphi \in \mathcal{L}_L$ define $tma(\varphi)$ as the set $\{\psi : L\psi$ is a subformula of $\varphi$ and some occurrence of the string $L\psi$ in $\varphi$ is not within the scope of the operator $L\}$. Thus $tma(\varphi)$ is the set of formulas $\psi$ such that $L\psi$ is a "top-level modal atom" of $\varphi$.

---

**InStabC**$(S, \varphi)$
*Input:* A finite set $S \subseteq \mathcal{L}$ and a formula $\varphi \in \mathcal{L}_L$;
*Output:* **InStabC** returns **true** if $\varphi \in ST(S)$, and **false**, otherwise.

Compute a purely autoepistemic formula $\varphi'$ of $L$-depth equal to 1 and such that for every stable theory $T$, $\varphi \in T$ if and only if $\varphi' \in T$ (proceed in a recursive way as in the proof of Theorem 8.37);

Find a modal disjunctive normal form $\varphi''$ of $\varphi'$;

Construct out of $\varphi''$ a complete set $\mathcal{F}$ of forcing conditions for $\varphi$ (as in Theorem 8.38);
**for** all conditions $(U, V) \in \mathcal{F}$ **do**
  **if** $(U \subseteq Cn(S))$ **and** $(V \cap Cn(S) = \emptyset)$
  **then return**{**true**}
**rof**
**return**{**false**}.

---

**Fig. 8.3.** Algorithm **InStabC**

Next, given a stable theory $T$ define:

$$(L\psi)_T = \begin{cases} \top & \text{if } \psi \in T, \\ \bot & \text{otherwise.} \end{cases}$$

Now, for every $\psi \in tma(\varphi)$, substitute in $\varphi$ every top-level occurrence of a formula $L\psi$ by a formula $(L\psi)_T$. The resulting formula, $(\varphi)_T$ belongs to $\mathcal{L}$; all occurrences of the operator $L$ have been eliminated.

We now have the following result.

**Lemma 8.39** *Let $\varphi \in \mathcal{L}_L$ and let $T \subseteq \mathcal{L}_L$ be a stable theory. Then $\varphi \in T$ if and only if $(\varphi)_T \in T$.*

Proof: It is easy to check that for every valuation $V$ of $\mathcal{L}_L$, such that for every $\xi \in \mathcal{L}_L$, $V(L\xi) = \mathbf{t}$ if and only if $\xi \in T$, $V(\varphi) = V((\varphi)_T)$. This implies that $T \vdash \varphi \equiv (\varphi)_T$. Since $T$ is closed under propositional consequence, the desired equivalence follows. ■

Lemma 8.39 implies the correctness of the algorithm for testing the membership problem for stable theories presented in Figure 8.4.

---

**InStabD**$(S, \varphi)$
*Input:* A finite set $S \subseteq \mathcal{L}$ and a formula $\varphi \in \mathcal{L}_L$;
*Output:* **InStabD** returns **true** if $\varphi \in ST(S)$, and **false**, otherwise.

($\star$ in the algorithm, $T$ stands for $ST(S)$ $\star$)
$A := tma(\varphi)$;
**for** all $\psi \in A$ **do**
  **if InStabD**$(S, \psi)$ then $(L\psi)_T = \top$
  **else** $(L\psi)_T = \bot$;
**rof**
compute $(\varphi)_T$;
**if** $(\varphi)_T \in Cn(S)$ **return**{**true**}
**else return**{**false**}.

---

**Fig. 8.4.** Algorithm **InStabD**

Notice that if the length of $\varphi$ is $n$ then algorithm **InStabD** makes $O(n)$ calls to the propositional satisfiability oracle. Indeed, there is one call to the propositional satisfiability oracle per each recursive call to **InStabD** and the number of these calls is bounded by the number of occurrences of the operator $L$ in the formula $\varphi$. Consequently, it follows that the problem of testing whether a formula $\varphi$ belongs to a stable theory $ST(S)$ is in the class $\Delta_2^P$ (see Chapter 13 for more details).

## 8.4 Comments and remarks

The concept of a stable theory was introduced by Stalnaker in [Sta80]. Its clear mathematical form is due to Moore [Moo85]. Many facts proved in this chapter come from that paper, although not necessarily with the same proofs. Moore studied stable theories as an auxiliary tool to investigate autoepistemic expansions. These will be studied in Chapter 10 where we will use the theory outlined in this chapter. Theorems 8.4 and 8.8 appear in Halpern and Moses [HM85]. They attribute them to van Benthem, Fitting and others.

Characterizations of stable theories summarized in Theorem 8.28 come from the following sources: McDermott [McD82], Moore [Moo84, Moo85], Halpern and Moses [HM85] and Truszczyński [Tru91c]. The constructive characterization of the operator $ST$ was given by Marek [Mar89].

The algorithms for testing the membership in stable theories come from three sources. The algorithm **InStabA** is due to Moore, [Moo84]. The recursive algorithm **InStabB** is due to Marek [Mar89], and the algorithm **InStabC**, based on the transformation of a formula into a purely autoepistemic form comes from authors [MT91]. Algorithm **InStabD** has been introduced by Niemelä [Nie92] and Schwarz and Truszczyński [ST92a]. Importance of algorithm InStabD for complexity considerations was pointed out to us by Gottlob. Purely autoepistemic formulas are called "fully modalized" by Chellas ([Che80], p. 142).

# 9. Modal nonmonotonic logics

Monotonic modal logics have long been used as means to study formal properties of such concepts as knowledge and belief. But they are not quite adequate to study knowledge (or belief) sets an agent constructs when he is reasoning not only from what is known to him but also from what is *possible* or *consistent to assume*. Such reasonings are defeasible, that is, when the agent learns new facts some of the previous derivations are not valid any more. In order to capture the phenomenon of nonmonotonicity, classical provability operators have to be modified. We have already studied in detail one nonmonotonic formalism, namely the logic of defaults, defined by means of the notion of a context-dependent proof.

In this chapter, we describe a general technique which allows us to assign to each monotonic modal logic a nonmonotonic counterpart. We assume that an agent whose reasoning we are modeling is ideally rational and has a full power of introspection. We proceed similarly as in the case of default logic. For a modal logic $\mathcal{S}$, we introduce the notion of a context-dependent proof and use it to define $\mathcal{S}$-expansions for a theory describing an agent's initial assumptions. Like extensions for a default theory, $\mathcal{S}$-expansions serve as candidate theories for an agent's belief set.

This chapter contains an extensive study of modal nonmonotonic logics as introduced by McDermott in [McD82]. We try, however, to model the presentation (at least the initial part) on a paradigm similar to one used in our presentation of default logic. Hence we start with the concept of a context-dependent proof in modal logic. From this concept we develop the formal definition of $\mathcal{S}$-expansion. In Section 9.3, we introduce a semantic description of $\mathcal{S}$-expansions, by means of a minimal model semantics for modal logics. We study the issues of consistency with respect to introspection in Section 9.4. Several characterization of expansions are provided in Section 9.5.

## 9.1 Context-dependent proofs in a modal logic

The goal we set for ourselves in this chapter is to introduce and then study a modal system for describing the reasoning of an agent satisfying the following properties:

**(A1)** The agent possesses full power of introspection, that is, if $\varphi$ is an agent's belief then so is $L\varphi$, and if $\varphi$ is not an agent's belief, then $\neg L\varphi$ is.

**(A2)** The agent reasons, using a modal logic $\mathcal{S}$, from his initial assumptions and from modal formulas expressing what is, for him, possible or consistent to assume.

**(A3)** The agent assigns belief sets only to theories consistent with $\mathcal{S}$.

Condition (A3) states that only a theory consistent with $\mathcal{S}$ can be the basis of a belief set. Another possibility is to assign to a theory which is inconsistent with $\mathcal{S}$ the whole language $\mathcal{L}_L$ as the corresponding belief set. We decided to adopt condition (A3) as it will lead us to a more uniform theory of nonmonotonic modal logics.

Condition (A2) needs to be made more precise. In particular, we have to specify which formulas are possible or consistent for the agent to assume and which modal formulas might be used to express these notions. As agent builds up his belief set, the notion of consistency with respect to his beliefs could change (and, in general, will). Formulas previously consistent with agent's beliefs may not be consistent anymore as additional beliefs are derived. We overcome this problem using the same technique as in the case of default logic. Specifically, we introduce the notion of a *context*. This is a theory with respect to which consistency is checked throughout the whole reasoning process. Let $T$ be a context, that is, a theory in the modal language $\mathcal{L}_L$. Since the agent reasons by means of the logic $\mathcal{S}$, a formula $\neg\varphi$ is consistent with $T$ if $T \not\vdash_{\mathcal{S}} \varphi$. The fact that the formula $\neg\varphi$ is consistent can be expressed by means of the modal formula $\neg L\varphi$. This is the approach we will adopt here. There are also other possibilities. One of them is to use the formula $M\neg\varphi$. It turns out that both approaches coincide if $\mathcal{S}$ is normal. For subnormal modal logics, the approaches differ. We will return to this issue after Theorem 9.4.

To conclude this part of our discussion, the intended meaning of condition (A2) is that the agent reasons, using a modal logic $\mathcal{S}$, from the set of initial assumptions $I$ and from all formulas of the form $\neg L\varphi$, where $T \not\vdash_{\mathcal{S}} \varphi$. This brings us to the notion of a $T$-proof.

**Definition 9.1** Let $\mathcal{S}$ be a modal logic and let $T \subseteq \mathcal{L}_L$. A sequence $\varphi_1, \ldots, \varphi_n$ of formulas from $\mathcal{L}_L$ is a *proof with respect to* $T$ (or simply, a $T$*-proof*) *from a theory* $I \subseteq \mathcal{L}_L$ *in the logic* $\mathcal{S}$ if, for each $i = 1, \ldots, n$, at least one of the following six conditions holds:

1. $\varphi_i \in I$.
2. $\varphi_i = \neg L\psi$, for some formula $\psi$ such that $T \not\vdash_{\mathcal{S}} \psi$.
3. $\varphi_i$ is a propositional tautology in $\mathcal{L}_L$.
4. $\varphi_i$ is an instance of an axiom schema of $\mathcal{S}$.
5. $\varphi_i$ is the effect of applying modus ponens to $\varphi_j$, $\varphi_k$, where $j, k < i$.
6. $\varphi_i$ is the effect of applying necessitation to $\varphi_j$, where $j < i$.

The theory $T$ is referred to as the *context* of a $T$-proof. If $\sigma = \varphi_1, \ldots, \varphi_n$ is a $T$-proof from $I$ in the logic $\mathcal{S}$ and $\varphi_n = \varphi$, then we say that $\sigma$ is a $T$-proof of $\varphi$ from $I$ in the logic $\mathcal{S}$, and denote this fact by $I \vdash_{\mathcal{S},T} \varphi$. ▲

Let us note the close analogy which exists between the notion of an $S$-proof in default logic and the notion of a $T$-proof introduced above.

It follows directly from Definition 9.1 that a formula $\varphi$ has a $T$-proof from $I$ in logic $\mathcal{S}$ if and only if $\varphi$ has a proof in $\mathcal{S}$ from $I \cup \{\neg L\varphi : T \not\vdash_{\mathcal{S}} \varphi\}$. In other words,

$$\{\varphi : I \vdash_{\mathcal{S},T} \varphi\} = Cn_{\mathcal{S}}(I \cup \{\neg L\varphi : T \not\vdash_{\mathcal{S}} \varphi\}).$$

The notion of a $T$-proof formalizes the way in which an agent reasons when deriving his beliefs. The intention is that *agent's beliefs are exactly those formulas that have a $T$-proof.* In order for the method of generating beliefs by constructing their $T$-proofs to be adequate, the context $T$ has to be chosen properly. Namely, $T$ must have the following properties:

**(A4)** If $I$ is consistent with $\mathcal{S}$, then $T$ is consistent with $\mathcal{S}$.

**(A5)** If a formula $\varphi$ has a $T$-proof from $I$ in $\mathcal{S}$, then $T \vdash_{\mathcal{S}} \varphi$.

**(A6)** If a formula $\varphi$ does not have a $T$-proof from $I$ in $\mathcal{S}$, then $T \not\vdash_{\mathcal{S}} \varphi$.

Let us now provide an intuitive motivation for conditions (A4) - (A6). Assume that $T$ is inconsistent with $\mathcal{S}$. Then it is easy to see that the notion of a $T$-proof from $I$ in $\mathcal{S}$ reduces to the standard notion of a proof from $I$ in $\mathcal{S}$. Hence, in such a case, the set of all beliefs (that is, formulas with $T$-proofs from $I$ in $\mathcal{S}$) coincides with $Cn_{\mathcal{S}}(I)$. Since, when $I$ is consistent with $\mathcal{S}$, the set $Cn_{\mathcal{S}}(I)$ may be not closed under negative introspection, condition (A1) may be violated. We require (A4) to guarantee that this problem will not occur.

Next, if (A5) is not satisfied, then consider a formula $\varphi$ possessing a $T$-proof from $I$ in $\mathcal{S}$ (that is, $\varphi$ is a belief of the agent). Assume that $T \not\vdash_{\mathcal{S}} \varphi$. Since $\varphi$ has a $T$-proof, $L\varphi$ has a $T$-proof, too. Hence, $L\varphi$ is a belief. Since $T \not\vdash_{\mathcal{S}} \varphi$, then $\neg L\varphi$ has a $T$-proof from $I$ in $\mathcal{S}$. Consequently, $\neg L\varphi$ is a belief, too. Thus, the set of all formulas with $T$-proofs from $I$ (that is, the set of all beliefs) is inconsistent, independently of the theory $I$. This is counterintuitive and prompts us to require (A5).

To see why (A6) must be accepted, consider a formula $\varphi$ that does not have a $T$-proof from $I$ in $\mathcal{S}$. Then $\varphi$ is not a belief. By (A1), the agent's belief set is closed under negative introspection. Thus, $\neg L\varphi$ and $L\neg L\varphi$ are beliefs. Suppose, contrary to (A6), that $T \vdash_{\mathcal{S}} \varphi$. Then $T \vdash_{\mathcal{S}} L\varphi$. Since $T$ is consistent with $\mathcal{S}$ (by (A4)), $T \not\vdash_{\mathcal{S}} \neg L\varphi$. Consequently, $\neg L\neg L\varphi$ has a $T$-proof from $I$ in $\mathcal{S}$. Thus, $\neg L\neg L\varphi$ is a belief, again yielding an inconsistent belief set independently of the theory $I$. Hence, we require (A6).

To summarize, conditions (A4) - (A6) imply that $T$ is a valid context if and only if $T$ is consistent with $\mathcal{S}$ and $Cn_{\mathcal{S}}(T)$ coincides with the set of all formulas having a $T$-proof from $I$ in $\mathcal{S}$. Formally, this requirement can be expressed as consistency of $T$ with $\mathcal{S}$ and

$$Cn_{\mathcal{S}}(T) = Cn_{\mathcal{S}}(I \cup \{\neg L\varphi : T \not\vdash_{\mathcal{S}} \varphi\}) \tag{9.1}$$

Hence, consequences in $\mathcal{S}$ of a valid context $T$ coincide with the belief set generated by means of $T$-proofs.

## 9.2 $\mathcal{S}$-expansions and their elementary properties

Our discussion in the previous section, in particular, equation (9.1), serves as a motivational basis for the definition of $\mathcal{S}$-expansions and nonmonotonic modal logic $\mathcal{S}$. We give these definitions in this section and study the corresponding notions throughout the remainder of this chapter.

**Definition 9.2** Given a modal logic $\mathcal{S}$ and a theory $I \subseteq \mathcal{L}_L$, a theory $T \subseteq \mathcal{L}_L$ is called an *$\mathcal{S}$-expansion* of $I$ if $T$ is consistent with $\mathcal{S}$ and satisfies the equation:

$$T = Cn_{\mathcal{S}}(I \cup \{\neg L\varphi : T \not\vdash_{\mathcal{S}} \varphi\}) \tag{9.2}$$

or, equivalently, if

$$T = \{\varphi \in \mathcal{L}_L : I \vdash_{\mathcal{S},T} \varphi\}. \qquad \blacktriangle$$

The *nonmonotonic modal logic* $\mathcal{S}$ is usually defined as the logic with the following (nonmonotonic) provability operator:

$$Cn^n_{\mathcal{S}}(I) = \bigcap\{T:\ T \text{ is an } \mathcal{S}\text{-expansion for } I\}.$$

In this book, we often use the term nonmonotonic logic $\mathcal{S}$ in a broader and less precise sense, as a denotation for a body of results concerned with properties of $\mathcal{S}$-expansions.

Requirement (A3) implies that belief sets are assigned only to theories consistent with $\mathcal{S}$. A solution $T$ to the equation (9.2) is inconsistent if and only if $I$ is inconsistent with $\mathcal{S}$. This is the reason why we restrict the class of $\mathcal{S}$-expansions only to theories consistent with $\mathcal{S}$.

In the informal terminology of the previous section, $\mathcal{S}$-expansions are exactly those theories that an ideally rational agent with full power of introspection might use as a context when reasoning. At the same time, those theories are exactly the belief sets such an agent may build using context-dependent proofs.

Since $\mathcal{S}$-expansions are closed under provability in $\mathcal{S}$, we get an equivalent and easier to use characterization of $\mathcal{S}$-expansions.

**Theorem 9.3** *A theory $T \subseteq \mathcal{L}_L$ is an $\mathcal{S}$-expansion for a theory $I \subseteq \mathcal{L}_L$ if and only if $T$ is consistent with $\mathcal{S}$ and*

$$T = Cn_{\mathcal{S}}(I \cup \{\neg L\varphi : \varphi \notin T\}) \tag{9.3}$$

Proof: Both equations (9.2) and (9.3) imply that $T$ is closed under provability in $\mathcal{S}$. Hence, for every $\varphi$

$$\varphi \in T \text{ if and only if } T \vdash_{\mathcal{S}} \varphi. \qquad \blacksquare$$

The definition of $\mathcal{S}$-expansions was tailored to make them suitable for representing the belief sets of an ideally rational agent with full introspection power. Thus, it is not surprising that $\mathcal{S}$-expansions are stable.

**Theorem 9.4** *Let $\mathcal{S}$ be a modal logic and let $I \subseteq \mathcal{L}_L$. Each $\mathcal{S}$-expansion of $I$ is a stable theory.*

Proof: In fact, the assertion follows only from the equation (9.3) and does not require the assumption of consistency in $\mathcal{S}$. Let $T$ be an $\mathcal{S}$-expansion for $I$. Then $T$ is closed under provability in $\mathcal{S}$. In particular, $T$ is closed under propositional consequence and under necessitation. To verify condition (St3) of the definition of stability, consider a formula $\psi \notin T$. Since $T = Cn_{\mathcal{S}}(I \cup \{\neg L\varphi : \varphi \notin T\})$, $\neg L\psi \in T$. ∎

Theorem 9.4 implies that for every theory $I \subseteq \mathcal{L}_L$ and for every modal logic $\mathcal{S}$, every $\mathcal{S}$-expansion $T$ of $I$ is closed under logic **S5**. What is, then, the role of the modal logic $\mathcal{S}$? Theorem 9.3 states that $\mathcal{S}$-expansions can be viewed as the results of the following process. First a context $T \subseteq \mathcal{L}_L$ is guessed. Then the result of negative introspection with respect to $T$, $\{\neg L\varphi : \varphi \notin T\}$, is added to the initial assumptions $I$. Next, the logic $\mathcal{S}$ is used to construct the theory $T' = Cn_{\mathcal{S}}(I \cup \{\neg L\varphi : \varphi \notin T\})$. Finally, $T$ and $T'$ are compared. If $T = T'$, $T$ is an $\mathcal{S}$-expansion of $I$. Thus, $\mathcal{S}$ can be regarded as a mechanism an agent uses to reconstruct $T$ out of $I$ and the result of negative introspection with respect to $T$.

Assume that instead of extending $I$ by formulas $\{\neg L\varphi : \varphi \notin T\}$ we add to $I$ formulas $\{M\varphi : \neg\varphi \notin T\}$ and define $T$ to be an $\mathcal{S}$-expansion for $I$ if $T$ is consistent with $\mathcal{S}$ and

$$T = Cn_{\mathcal{S}}(I \cup \{M\varphi : \neg\varphi \notin T\}). \tag{9.4}$$

Then, if $\mathcal{S}$ is a normal modal logic, both notions of $\mathcal{S}$-expansions are equivalent. Indeed, it follows by Theorem 7.17 that if $T$ is closed under propositional provability, then

$$Cn_{\mathcal{S}}(\{M\varphi : \neg\varphi \notin T\}) = Cn_{\mathcal{S}}(\{\neg L\varphi : \varphi \notin T\}). \tag{9.5}$$

But if $\mathcal{S}$ is subnormal, (9.5) is no longer true and solutions to the equation (9.4) may be different from the solutions to the equation (9.3). In particular, solutions to (9.4) may not be stable. For example, let $I = \emptyset$. Define

$$T_0 = Cn(\emptyset) \cap \mathcal{L}_{L,0},$$

and for $n \geq 0$,

$$T_{n+1} = Cn(T_n \cup \{L\varphi : \varphi \in [T_n]_{=n}\} \cup \{M\varphi : \neg\varphi \in [\mathcal{L}_L]_{=n} \setminus T_n\}) \cap \mathcal{L}_{L,n+1},$$

where for $U \subseteq \mathcal{L}_L$, $[U]_{=n}$ denotes the set of all formulas in $U$ with $L$-depth equal to $n$. Finally, define

$$T = \bigcup_{n=0}^{\infty} T_n.$$

The following properties are straightforward to prove:

**(a)** $T_m \subseteq T_n$ for $m \leq n$.

**(b)** $T$ is closed under propositional consequence.

**(c)** $T_{n+1} \cap \mathcal{L}_{L,n} = T_n$, and $T \cap \mathcal{L}_{L,n} = T_n$.

**(d)** $T \supseteq \{M\varphi : \neg\varphi \in \mathcal{L}_L \setminus T\}$.

Consider a formula $\varphi \in T$. Suppose that $d_L(\varphi) = n$. Then, by (c), $\varphi \in T_n$. Consequently, $L\varphi \in T_{n+1} \subseteq T$. Thus, we obtain the following property:

**(e)** $T$ is closed under necessitation.

From (b), (d) and (e) it follows that

$$T \supseteq Cn_{\mathbf{N}}(\{M\varphi : \neg\varphi \in \mathcal{L}_L \setminus T\}).$$

By an easy inductive reasoning and by (c), it follows that for every integer $n \geq 0$,

$$T_n \subseteq Cn_{\mathbf{N}}(\{M\varphi : \neg\varphi \in \mathcal{L}_L \setminus T\}).$$

Thus, $T$ satisfies (9.4) for $\mathcal{S} = \mathbf{N}$ and $I = \emptyset$. But $T$ is not stable. Let $p$ be any propositional variable. Clearly, $p \notin T_0$. By (c), $p \notin T$. Suppose $\neg Lp \in T$. Then, again by (c), $\neg Lp \in T_1$, a contradiction.

To summarize, it makes no difference which of the equations (9.3) or (9.4) is used to define an $\mathcal{S}$-expansion for $I$, if the logic $\mathcal{S}$ is normal. But, since some subnormal logics, especially logic **N**, play an important role in the theory of nonmonotonic reasoning, in order to have a notion of $\mathcal{S}$-expansion that has the property of being stable both for normal and subnormal logics $\mathcal{S}$, it is necessary to choose (9.3) as the definition of $\mathcal{S}$-expansions.

From now on, we will always assume that the modal logics under consideration are contained in **S5**. Under this restriction, the theory of $\mathcal{S}$-expansions is particularly interesting and well-understood.

Restriction to logics contained in **S5** immediately implies that in the definition of an $\mathcal{S}$-expansion the requirement that $T$ is consistent with $\mathcal{S}$ can be replaced by a weaker condition that it is propositionally consistent.

**Theorem 9.5** *Let $\mathcal{S}$ be a modal logic contained in* **S5** *and let $I \subseteq \mathcal{L}_L$. A theory $T$ is an $\mathcal{S}$-expansion of $I$ if and only if $T$ is propositionally consistent and satisfies the equation* (9.3).

Proof: If $T$ satisfies the equation (9.3) then $T$ is stable. By Theorem 8.4, $T$ is closed under provability in **S5**. Consequently, since $T$ is propositionally consistent, $T$ is consistent with **S5**. Since $\mathcal{S} \subseteq$ **S5**, $T$ is consistent with $\mathcal{S}$. Consequently, $T$ is an $\mathcal{S}$-expansion for $I$. ■

In the next result we establish a simple relation between classes of $\mathcal{S}$- and $\mathcal{T}$-expansions for $I$ if a logic $\mathcal{S}$ is contained in a logic $\mathcal{T}$ and both logics are contained in **S5**.

**Theorem 9.6** *Let $\mathcal{S}$ and $\mathcal{T}$ be two modal logics, $\mathcal{S} \subseteq \mathcal{T} \subseteq$ **S5**. Then each $\mathcal{S}$-expansion of $I$ is a $\mathcal{T}$-expansion for $I$.*

Proof: Let $T$ be an $\mathcal{S}$-expansion of $I$. By Theorem 9.5, $T$ is propositionally consistent and

$$T = Cn_{\mathcal{S}}(I \cup \{\neg L\varphi : \varphi \notin T\}).$$

Clearly, $I \cup \{\neg L\varphi : \varphi \notin T\} \subseteq T$ and, by Theorems 9.4 and 8.4, $T$ is closed under provability in **S5**. Since $\mathcal{T} \subseteq$ **S5**,

$$T \supseteq Cn_{\mathcal{T}}(I \cup \{\neg L\varphi : \varphi \notin T\}).$$

The converse inclusion follows immediately from the assumption that $\mathcal{S} \subseteq \mathcal{T}$. Thus, we have proved that $T$ is propositionally consistent and

$$T = Cn_{\mathcal{T}}(I \cup \{\neg L\varphi : \varphi \notin T\}).$$

By Theorem 9.5, $T$ is a $\mathcal{T}$-expansion for $I$. ■

Following Theorem 9.4 we observed that a modal logic $\mathcal{S}$ is used to reconstruct the theory $T$ out of $I \cup \{\neg L\varphi : \varphi \notin T\}$. If $T$ can be so reconstructed, $T$ is an $\mathcal{S}$-expansion of $I$. Let $\mathcal{S} \subseteq \mathcal{T}$ be two modal logics. Logic $\mathcal{T}$ provides more means for deriving formulas than $\mathcal{S}$ does. Theorem 9.6 states that providing an agent with additional reasoning tools for the reconstruction process potentially enlarges the class of belief sets.

Let $I \subseteq \mathcal{L}_L$ and consider a stable theory $T$ such that $I \subseteq T$. Then, for every modal logic $\mathcal{S}$ contained in **S5**, $T$ is closed under provability in $\mathcal{S}$. Consequently, we have

$$T \supseteq Cn_{\mathcal{S}}(I \cup \{\neg L\varphi : \varphi \notin T\}). \tag{9.6}$$

To prove that $T$ is an $\mathcal{S}$-expansion for $I$ one has to show that the converse inclusion holds, too. The following theorem gives a sufficient condition for this to be the case.

**Theorem 9.7** *Let $\mathcal{S}$ be a modal logic (not necessarily contained in* **S5***). Let $T = ST(S)$, for some set $S \subseteq \mathcal{L}$. If*

$$S \subseteq Cn_{\mathcal{S}}(I \cup \{\neg L\varphi : \varphi \notin T\})$$

*then*

$$T \subseteq Cn_{\mathcal{S}}(I \cup \{\neg L\varphi : \varphi \notin T\}).$$

Proof: Since $T = ST(S)$, $T$ is a stable theory. Therefore, by Theorem 8.23,

$$ST(S) = \bigcup_{n=0}^{\infty} ST_n(S).$$

where

$$ST_0(S) = Cn(S),$$

and for $n \geq 0$,

$$ST_{n+1}(S) = Cn(ST_n(S) \cup \{L\varphi : \varphi \in ST_n(S)\} \\ \cup \{\neg L\varphi : \varphi \in \mathcal{L}_{L,n} \setminus ST_n(S)\}) \cap \mathcal{L}_{L,n+1}.$$

We prove by induction on $n$ that $ST_n(S) \subseteq Cn_{\mathcal{S}}(I \cup \{\neg L\varphi : \varphi \notin T\})$.

For $n = 0$, $S \subseteq Cn_{\mathcal{S}}(I \cup \{\neg L\varphi : \varphi \notin T\})$, hence $Cn(S) \subseteq Cn_{\mathcal{S}}(I \cup \{\neg L\varphi : \varphi \notin T\})$, that is, $ST_0(S) \subseteq Cn_{\mathcal{S}}(I \cup \{\neg L\varphi : \varphi \notin T\})$.

Our inductive assumption is that $ST_n(S) \subseteq Cn_{\mathcal{S}}(I \cup \{\neg L\varphi : \varphi \notin T\})$.

Since $\mathcal{S}$ is a modal logic, it is closed under the necessitation rule. Therefore $\{L\varphi : \varphi \in ST_n(S)\} \subseteq Cn_{\mathcal{S}}(I \cup \{\neg L\varphi : \varphi \notin T\})$.

Finally, by Theorems 8.22 and 8.23, $\{\neg L\varphi : \varphi \in \mathcal{L}_{L,n} \setminus ST_n(S)\} \subseteq \{\neg L\varphi : \varphi \notin T\}$. Hence, it follows that $ST_{n+1}(S) \subseteq Cn_{\mathcal{S}}(I \cup \{\neg L\varphi : \varphi \notin T\})$.

Thus, by induction, $T \subseteq Cn_{\mathcal{S}}(I \cup \{\neg L\varphi : \varphi \notin T\})$. ■

Theorem 9.7 and inclusion (9.6) yield the following simple characterization of $\mathcal{S}$-expansions.

**Corollary 9.8** *Let $\mathcal{S}$ be a modal logic contained in* **S5** *and let $I \subseteq \mathcal{L}_L$. A theory $T \subseteq \mathcal{L}_L$ satisfies*

$$T = Cn_{\mathcal{S}}(I \cup \{\neg L\varphi : \varphi \notin T\})$$

*if and only if $T$ is stable, $I \subseteq T$ and $T \cap \mathcal{L} \subseteq Cn_{\mathcal{S}}(I \cup \{\neg L\varphi : \varphi \notin T\})$. In particular, a theory $T \subseteq \mathcal{L}_L$ is an $\mathcal{S}$-expansion for $I$ if and only if $T$ is stable consistent, $I \subseteq T$ and $T \cap \mathcal{L} \subseteq Cn_{\mathcal{S}}(I \cup \{\neg L\varphi : \varphi \notin T\})$.* ▲

Corollary 9.8, together with results from Section 9.5, will be used in Section 11.1 as the basis for an algorithm to compute *all* $\mathcal{S}$-expansions of a finite theory $I$. Below, we show how to use this corollary in the simpler task of checking if a given stable theory $ST(S)$, where $S$ is finite, is an $\mathcal{S}$-expansion for $I$.

**Example 9.9** Let $I = \{\neg LLq \supset Lp\}$. We will show that $ST(\{p\})$ is a **T**-expansion for $I$. First, observe that since $p \in ST(\{p\})$, $Lp \in ST(\{p\})$. Hence, $I \subseteq ST(\{p\})$. According to Corollary 9.8, to prove that $ST(\{p\})$ is a **T**-expansion for $I$ it suffices to show that

$$p \in Cn_{\mathbf{T}}(I \cup \{\neg L\varphi : \varphi \notin ST(\{p\})\}).$$

To this end, observe that $Lq \notin ST(\{p\})$. Hence, $\neg LLq \in \{\neg L\varphi : \varphi \notin ST(\{p\})\}$. By modus ponens, $Lp \in Cn_{\mathbf{T}}(I \cup \{\neg L\varphi : \varphi \notin ST(\{p\})\})$ and, by axiom schema t, $p \in Cn_{\mathbf{T}}(I \cup \{\neg L\varphi : \varphi \notin ST(\{p\})\})$. ▲

We will now characterize **S5**-expansions for a given theory $I$.

**Theorem 9.10** *Let $I \subseteq \mathcal{L}_L$. A theory $T \subseteq \mathcal{L}_L$ is an* **S5**-*expansion for $I$ if and only if $T$ is consistent, stable and $I \subseteq T$.*

Proof: Clearly, if $T$ is an **S5**-expansion for $I$ then $T$ is consistent and $I \subseteq T$. By Theorem 9.4, $T$ is stable. Conversely, if $T$ is stable then, by Theorem 8.27, $T = Cn_{\mathbf{S5}}(\{\neg L\varphi : \varphi \notin T\})$. Since $I \subseteq T$, $T = Cn_{\mathbf{S5}}(I \cup \{\neg L\varphi : \varphi \notin T\})$ and, since $T$ is consistent, $T$ is an **S5**-expansion for $I$. ■

Theorem 9.10 implies that the nonmonotonic modal logic **S5** coincides with the monotonic logic **S5**. Namely, we have the following corollary.

**Corollary 9.11** *Let $I \subseteq \mathcal{L}_L$. Then,*

$$Cn^n_{\mathbf{S5}}(I) = Cn_{\mathbf{S5}}(I).$$

Proof: Since each stable theory $T$ is closed under provability in **S5**, the inclusion $Cn_{\mathbf{S5}}(I) \subseteq Cn^n_{\mathbf{S5}}(I)$ follows. To prove the converse inclusion, consider a formula $\varphi \notin Cn_{\mathbf{S5}}(I)$. Then, there is an **S5**-model $\mathcal{M}$ such that $\mathcal{M} \models I$ and $\mathcal{M} \not\models \varphi$. Consequently, the stable theory $Th(\mathcal{M})$ contains $I$ and does not contain $\varphi$ (see Theorem 8.10). Thus, $\varphi \notin Cn^n_{\mathbf{S5}}(I)$. ■

Since we assume that the language $\mathcal{L}$ has infinitely many atoms, it is a consequence of Theorem 9.10 (and of Corollary 8.25(2)) that a consistent and finite theory $I$ has infinitely many (in fact, continuum many) **S5**-expansions. Thus, it is not possible to list them all. On the other hand, using the algorithms described in Chapter 8, given a finite theory $I$ and a *finite* set $S \subseteq \mathcal{L}$, we can decide whether $ST(S)$ is an **S5**-expansion of $I$.

## 9.3 Minimal model semantics

We will present now a semantic characterization of $\mathcal{S}$-expansions. In the most general case, this characterization is a simple consequence of the completeness of Kripke model semantics. Essentially, it is obtained by replacing in the equation (9.3) the syntactic operator $Cn_{\mathcal{S}}$ by a semantic operation of logical entailment $\models$. Consequently, this general semantic characterization of $\mathcal{S}$-expansions is so close to the original syntactic definition that it hardly provides any new insights into the nature of nonmonotonic logics. Furthermore, it is not easy to apply. There is, however, a good reason for developing this semantic approach. For a wide class of logics, including **K**, **T**, **S4**, **S5**, as well as **S4f**, **kd45** and **Sw5**, it can be reformulated in a way which is easier to use.

Before we continue, let us first discuss a general method of defining nonmonotonic logics semantically. An agent starts by selecting a class $\mathcal{K}$ of *models* that he will consult when reasoning. The models in $\mathcal{K}$ will be referred to as *intended* or *preferred*. There are two possibilities in which the class of models $\mathcal{K}$ can be used. In the first approach, the theory of each model can serve as a belief set. Such approach is often referred to as the *brave* reasoning. In the second one, the agent accepts as his belief set the intersection of the theories of all models in $\mathcal{K}$. This is usually called the *skeptical* mode of reasoning associated with $\mathcal{K}$. Assuming that the agent's knowledge is described by a theory $I$, each selected *model* has to satisfy all formulas in $I$, but may be also subject to additional requirements.

Classical reasoning falls into this scheme. For example, suppose that an agent whose initial assumptions are described by a theory $I$ is to reason by means of a modal logic $\mathcal{S}$. Let $\mathcal{C}$ be a class of (Kripke) models characterizing $\mathcal{S}$. To produce consequences of $I$ in logic $\mathcal{S}$, the agent should select for $\mathcal{K}$ all models from $\mathcal{C}$ which satisfy $I$. Then the formulas true in all models from $\mathcal{K}$ are exactly those that follow from $I$ in $\mathcal{S}$. In other words, classical modal logic is an example of skeptical reasoning.

In commonsense reasoning with logic $\mathcal{S}$, we would not consider all models in $\mathcal{C}$. Some of them may be incompatible with the agent's experience or be highly unlikely. These models should be eliminated. Let us denote the class of remaining models by $\mathcal{C}'$. If $\mathcal{S}$ is not complete with respect to $\mathcal{C}'$, the resulting reasoning system is no longer monotonic (neither in brave nor skeptical mode).

This technique of looking at *intended* models is well known in classical logic. It is used, for instance, when restricting attention to so-called $\omega$-models, or, more generally, models with a prescribed extension of a predicate.

We will apply these ideas to provide a semantic characterization of $\mathcal{S}$-expansions. Let us assume that logic $\mathcal{S}$ is characterized by a class of models $\mathcal{C}$. To define a nonmonotonic counterpart of $\mathcal{S}$ we need to select a subclass $\mathcal{K}$ of $\mathcal{C}$. To define $\mathcal{K}$ we need to take into account the properties (A1) - (A6), which were assumed at the beginning of this chapter. In addition, we will also assume that:

**(A7)** The agent's initial knowledge is described by a theory $I$.

The property (A1) requires that the theory consisting of all formulas true in a preferred model be stable, as the stability of a theory is precisely the property that formalizes full (positive and negative) introspection. Thus, **S5**-models, that is, models with a universal accessibility relation, are natural candidates for preferred models. To guarantee that all **S5**-models are available for the agent, we will require that all such models belong to $\mathcal{C}$. This requirement restricts the class of modal logics discussed in this book to those included in **S5**.

Since we assume (A7), the property (A3) is satisfied automatically, simply because theories inconsistent with $\mathcal{S}$ have no models in the class $\mathcal{C}$.

To understand the implications of the property (A2), let us consider an **S5**-model $\mathcal{M} = \langle M, V\rangle$ such that $\mathcal{M} \models I$. Suppose that there exists a model $\mathcal{N}$ in $\mathcal{C}$ such that

$$\mathcal{N} \models I \cup \{\neg L\varphi : \varphi \notin Th(\mathcal{M})\}; \tag{9.7}$$

Consider a formula $\varphi \notin Th(\mathcal{M})$. It follows from (9.7) that $\mathcal{N} \models \neg L\varphi$. Hence, $\mathcal{N} \not\models \varphi$ and $\varphi \notin Th(\mathcal{N})$. Consequently, $Th(\mathcal{N}) \subseteq Th(\mathcal{M})$. Suppose that $Th(\mathcal{N}) \neq Th(\mathcal{M})$. Then, not all formulas in $Th(\mathcal{M})$ can be proved in $\mathcal{S}$ from $I$ and from negative introspection with respect to $Th(\mathcal{M})$. Indeed, $\mathcal{N}$ is a model in $\mathcal{C}$ satisfying $I \cup \{\neg L\varphi : \varphi \notin Th(\mathcal{M})\}$ in which some of the formulas from $Th(\mathcal{M})$ are not true. Hence, $\mathcal{M}$ should not be selected as preferred.

This informal discussion leads us to the following definition.

**Definition 9.12** Let $\mathcal{S}$ be a modal logic contained in **S5**. Let $\mathcal{C}$ be a class of Kripke models characterizing $\mathcal{S}$ and containing all **S5**-models. An **S5**-model $\mathcal{M} \in \mathcal{C}$ is *$\mathcal{C}$-intended for $I$* if:

1. $\mathcal{M} \models I$.
2. For every model $\mathcal{N} \in \mathcal{C}$ such that $\mathcal{N} \models I \cup \{\neg L\varphi : \varphi \notin Th(\mathcal{M})\}$, $Th(\mathcal{N}) = Th(\mathcal{M})$. ▲

In the next theorem, we show that $\mathcal{S}$-expansions can be characterized as precisely the theories of intended models.

**Theorem 9.13** *Let $\mathcal{S}$ be a normal modal logic contained in* **S5**. *Let $\mathcal{C}$ be a class of Kripke models characterizing $\mathcal{S}$ and containing all* **S5**-*models. An* **S5**-*model $\mathcal{M}$ is a $\mathcal{C}$-intended model for $I$ if and only if $Th(\mathcal{M})$ is an $\mathcal{S}$-expansion for $I$.*

Proof: Assume that $\mathcal{M}$ is $\mathcal{C}$-intended for $I$. Define $T = Th(\mathcal{M})$. By Theorem 8.10, $T$ is stable. Hence, $T \supseteq \{\neg L\varphi : \varphi \notin T\}$. In addition, since $\mathcal{M}$ is $\mathcal{C}$-intended for $I$, $T \supseteq I$. Finally, since $\mathcal{S} \subseteq$ **S5**, $T$ is closed under provability in $\mathcal{S}$. Consequently,

$$T \supseteq Cn_{\mathcal{S}}(I \cup \{\neg L\varphi : \varphi \notin T\}).$$

Let now $\mathcal{N}$ be any model in $\mathcal{C}$ such that $\mathcal{N} \models I \cup \{\neg L\varphi : \varphi \notin T\}$. Then, since $\mathcal{M}$ is $\mathcal{C}$-intended for $I$, $Th(\mathcal{N}) = Th(\mathcal{M})$. Hence, every model of $I \cup \{\neg L\varphi : \varphi \notin T\}$ is a model of $T$. Since the class $\mathcal{C}$ characterizes $\mathcal{S}$,

$$T \subseteq Cn_{\mathcal{S}}(I \cup \{\neg L\varphi : \varphi \notin T\}).$$

Consequently, $T$ is an $\mathcal{S}$-expansion for $I$.

Conversely, let us assume that for some **S5**-model $\mathcal{M}$, $Th(\mathcal{M})$ is an $\mathcal{S}$-expansion for $I$. Denote $T = Th(\mathcal{M})$. Consider a model $\mathcal{N} \in \mathcal{C}$ such that $\mathcal{N} \models I \cup \{\neg L\varphi : \varphi \notin T\}$. Since $T = Cn_{\mathcal{S}}(I \cup \{\neg L\varphi : \varphi \notin T\})$, it follows that $T \subseteq Th(\mathcal{N})$. On the other hand, since $\mathcal{N} \models \{\neg L\varphi : \varphi \notin T\}$, if $\varphi \notin T$ then $\mathcal{N} \models \neg L\varphi$. Consequently, $\mathcal{N} \not\models \varphi$. Thus $\mathcal{L}_L \setminus T \subseteq \{\varphi : \mathcal{N} \not\models \varphi\}$. That is, $\{\varphi : \mathcal{N} \models \varphi\} \subseteq T$. Thus, $Th(\mathcal{N}) = T = Th(\mathcal{M})$ and $\mathcal{M}$ is $\mathcal{C}$-intended for $I$. ■

**Remark 9.14** The assumptions of Theorem 9.13 are not restrictive. Each normal modal logic is characterized by some class $\mathcal{C}$ of Kripke models (Corollary 7.47). Since we also require that $\mathcal{S}$ be contained in **S5**, without loss of generality we may assume that $\mathcal{C}$ contains all **S5**-models. Thus, Theorem 9.13 provides a characterization of $\mathcal{S}$-expansions for any normal modal logic $\mathcal{S}$ contained in **S5**. ▲

Our definition of a $\mathcal{C}$-intended model is not convenient to use. To determine that a model $\mathcal{M}$ is $\mathcal{C}$-intended, the theory $Th(\mathcal{M})$ of the model $\mathcal{M}$ has to be compared to the theories $Th(\mathcal{N})$ for other models $\mathcal{N}$ in $\mathcal{C}$. We will now show that for a wide class of logics, an equivalent definition of a $\mathcal{C}$-intended model can be given. This new description allows us to verify that a model is $\mathcal{C}$-intended without any explicit reference to the theory of the model.

**Definition 9.15** Let $\mathcal{M}_i = \langle M_i, R_i, V_i \rangle$, $i = 1, 2$, be two Kripke models. We say that $\mathcal{M}_2$ is *preferred over* $\mathcal{M}_1$ and write $\mathcal{M}_2 \sqsubset \mathcal{M}_1$ if:

1. $\mathcal{M}_2 = \mathcal{M} \odot \mathcal{M}_1$, for some Kripke model $\mathcal{M}$.

2. There is $a \in M_2 \setminus M_1$ and a finite set of propositional variables $P$ such that for each $b \in M_1$, $V_2(b)|_P \neq V_2(a)|_P$.

Let $\mathcal{C}$ be a class of Kripke models and let $I \subseteq \mathcal{L}_L$. We say that a model $\mathcal{M} \in \mathcal{C}$ is *$\mathcal{C}$-minimal for $I$* if $\mathcal{M} \models I$ and for every model $\mathcal{N} \in \mathcal{C}$ such that $\mathcal{N} \models I$, $\mathcal{N} \not\sqsubset \mathcal{M}$. In other words, $\mathcal{M}$ is $\mathcal{C}$-minimal for $I$ if it is a $\sqsubset$-minimal model of $I$ in $\mathcal{C}$. ▲

**Theorem 9.16** *Condition (2) of Definition 9.15 is equivalent to the condition:*

2'. *There is $a \in M_2 \setminus M_1$ and a formula $\varphi \in \mathcal{L}$ such that $\langle \mathcal{M}_2, a\rangle \not\models \varphi$ and $\langle \mathcal{M}_2, b\rangle \models \varphi$, for every $b \in M_1$.*

Proof: Assume that condition (2′) holds for some world $a \in M_2 \setminus M_1$ and for some formula $\varphi \in \mathcal{L}$. Let $P$ be the set of all propositional variables occurring in $\varphi$. Since $\varphi \in \mathcal{L}$, $\langle \mathcal{M}_2, a\rangle \not\models \varphi$, and $\langle \mathcal{M}_2, b\rangle \models \varphi$, for every $b \in M_1$, condition (2) follows.

Conversely, assume that condition (2) holds for some world $a \in M_2 \setminus M_1$ and for a set $P = \{p_1, \ldots, p_n\}$ of propositional variables. For every $i$, $1 \leq i \leq n$, define

$$\overline{p}_i = \begin{cases} p_i & \text{if } V_2(a)(p_i) = \mathbf{f}, \\ \neg p_i & \text{if } V_2(a)(p_i) = \mathbf{t}. \end{cases}$$

Next, define $\varphi = \overline{p}_1 \vee \ldots \vee \overline{p}_n$. Clearly, $\langle \mathcal{M}_2, a\rangle \not\models \varphi$ and $\langle \mathcal{M}_2, b\rangle \models \varphi$, for every $b \in M_1$. Thus, condition (2′) follows. ■

Under fairly weak conditions the notions of a $\mathcal{C}$-intended model for $I$ and a C-minimal model for $I$ coincide.

**Definition 9.17** Let $\mathcal{C}$ be a class of Kripke models. The class $\mathcal{C}$ is said to be *cluster-closed* if at least one of the following two conditions holds:

1. For every model $\mathcal{M} \in \mathcal{C}$ and every universal model (cluster) $\mathcal{N}$, the Kripke model $\mathcal{M} \odot \mathcal{N}$ belongs to $\mathcal{C}$.

2. Every model in $\mathcal{C}$ is of the form $\mathcal{M}_1 \odot \mathcal{M}_2$, where $\mathcal{M}_2$ is a universal Kripke model. Moreover, for every model from $\mathcal{C}$ of the form $\mathcal{M}_1 \odot \mathcal{M}_2$, where $\mathcal{M}_2$ is a universal Kripke model, and for every universal Kripke model $\mathcal{M}_2'$, with the universe of worlds disjoint from that of $\mathcal{M}_1$, the model $\mathcal{M}_1 \odot \mathcal{M}_2'$ is in $\mathcal{C}$. ▲

It is easy to see that the classes of all Kripke models, reflexive Kripke models, and transitive Kripke models satisfy condition (1) of Definition 9.17. The classes of **kd45**-models, **Sw5**-models, **S4f**-models and **S5**-models satisfy condition (2) of Definition 9.17.

For cluster-closed classes of Kripke models we have the following result.

**Theorem 9.18** *Let $I \subseteq \mathcal{L}_L$. Let $\mathcal{S}$ be a normal modal logic contained in* **S5**. *Assume that $\mathcal{S}$ is characterized by a cluster-closed class of models $\mathcal{C}$ containing all* **S5**-*models. Then, an* **S5**-*model $\mathcal{M} \in \mathcal{C}$ is $\mathcal{C}$-intended for $I$ if and only if $\mathcal{M}$ is $\mathcal{C}$-minimal for $I$.*

To prove Theorem 9.18 we need several lemmas.

**Lemma 9.19** *Let $\mathcal{N}$, $\mathcal{M}_1$ and $\mathcal{M}_2$ be Kripke models and let $Th(\mathcal{M}_1) = Th(\mathcal{M}_2)$. Then $Th(\mathcal{N} \odot \mathcal{M}_1) = Th(\mathcal{N} \odot \mathcal{M}_2)$.*

Proof: Let $\mathcal{N} = \langle N, R, V\rangle$ and let $\mathcal{M}_i = \langle M_i, R_i, V_i\rangle$, $i = 1, 2$. We first prove that for every $a \in N$ and for every formula $\varphi \in \mathcal{L}_L$,

$$\langle \mathcal{N} \odot \mathcal{M}_1, a\rangle \models \varphi \text{ if and only if } \langle \mathcal{N} \odot \mathcal{M}_2, a\rangle \models \varphi.$$

We proceed by induction on the complexity of a formula. If $\varphi$ is an atom, the claim follows because both models assign the same valuation to the world $a$, namely the valuation $V(a)$. In the induction step, if $\varphi = \neg\psi$, $\psi_1 \vee \psi_2$ or $\psi_1 \wedge \psi_2$, then the claim follows directly from the definition of the satisfiability relation in a world and from the induction hypothesis.

The only nontrivial case is when $\varphi = L\psi$. Assume that $\langle \mathcal{N} \odot \mathcal{M}_1, a\rangle \models L\psi$. Then, by the definition of concatenation of two Kripke models, $\mathcal{M}_1 \models \psi$. Consequently,

$$\mathcal{M}_2 \models \psi. \tag{9.8}$$

Moreover, for every $b \in N$ such that $(a, b) \in R$, $\langle \mathcal{N} \odot \mathcal{M}_1, b\rangle \models \psi$. By the induction hypothesis it follows that for every $b \in N$ such that $(a, b) \in R$, $\langle \mathcal{N} \odot \mathcal{M}_2, b\rangle \models \psi$. Consequently, by the definition of concatenation and (9.8), $\langle \mathcal{N} \odot \mathcal{M}_2, a\rangle \models L\psi$. The converse implication follows by a symmetry argument. Thus, the proof of the claim is complete.

Consider now $\varphi \in Th(\mathcal{N} \odot \mathcal{M}_1)$. Then $\varphi \in Th(\mathcal{M}_1)$ and for every $a \in N$, $\langle \mathcal{N} \odot \mathcal{M}_1, a\rangle \models \varphi$. Consequently, $\varphi \in Th(\mathcal{M}_2)$ and for every $a \in N$, $\langle \mathcal{N} \odot \mathcal{M}_2, a\rangle \models \varphi$. Hence, $\varphi \in Th(\mathcal{N} \odot \mathcal{M}_2)$ and $Th(\mathcal{N} \odot \mathcal{M}_1) \subseteq Th(\mathcal{N} \odot \mathcal{M}_2)$. The converse inclusion follows by a symmetry argument. ∎

**Lemma 9.20** *Let $\mathcal{M}$ and $\mathcal{N}$ be Kripke models such that $\mathcal{N} \models \{\neg L\varphi : \varphi \notin Th(\mathcal{M})\}$. Then, for every world $a$ of $\mathcal{N}$ and for every formula $\varphi \in \mathcal{L}_L$*

$$\langle \mathcal{N} \odot \mathcal{M}, a \rangle \models \varphi \text{ if and only if } \langle \mathcal{N}, a \rangle \models \varphi.$$

Proof: Let $\mathcal{N} = \langle N, R, V \rangle$ and $\mathcal{M} = \langle M, Q, W \rangle$. We proceed by induction on the complexity of $\varphi$. As in the previous proof, the only nontrivial part is in the induction step when $\varphi = L\psi$. We consider this case only and leave other details to the reader.

Let us assume first that $\langle \mathcal{N} \odot \mathcal{M}, a \rangle \models L\psi$. Then, in particular, for every $b \in N$ such that $(a, b) \in R$, $\langle \mathcal{N} \odot \mathcal{M}, b \rangle \models \psi$. By the induction hypothesis, for every $b \in N$ such that $(a, b) \in R$, $\langle \mathcal{N}, b \rangle \models \psi$, which means that $\langle \mathcal{N}, a \rangle \models L\psi$.

Conversely, assume that $\langle \mathcal{N}, a \rangle \models L\psi$. Then, for every $b \in N$ such that $(a, b) \in R$, $\langle \mathcal{N}, b \rangle \models \psi$. By the induction hypothesis, $\langle \mathcal{N} \odot \mathcal{M}, b \rangle \models \psi$, for every $b \in N$ such that $(a, b) \in R$. In addition, since $\mathcal{N} \not\models \neg L\psi$, $\psi \in Th(\mathcal{M})$. Hence, $\langle \mathcal{N} \odot \mathcal{M}, b \rangle \models \psi$, for every $b \in M$. Consequently, by the definition of concatenation, $\langle \mathcal{N} \odot \mathcal{M}, a \rangle \models L\psi$. ■

**Lemma 9.21** *Let $\mathcal{N}$ be a Kripke model and $\mathcal{M}$ be an* **S5**-*model. Assume that $\mathcal{N} \odot \mathcal{M} \not\subseteq \mathcal{M}$. Then $Th(\mathcal{M}) \subseteq Th(\mathcal{N} \odot \mathcal{M})$.*

Proof: We will prove by induction on $n$ that for all formulas $\varphi$ such that $d_L(\varphi) \leq n$, if $\varphi \in Th(\mathcal{M})$ then $\varphi \in Th(\mathcal{N} \odot \mathcal{M})$. Consider a non-negative integer $n$ and assume that the claim holds for all smaller non-negative integers. Since theories $Th(\mathcal{M})$ and $Th(\mathcal{N} \odot \mathcal{M})$ are closed under propositional provability, in the inductive step it suffices to prove the claim for each formula $\varphi$ of the form

$$\varphi_0 \vee L\varphi_1 \vee \ldots \vee L\varphi_k \vee \neg L\varphi_{k+1} \vee \ldots \vee \neg L\varphi_{k+m}, \tag{9.9}$$

where $\varphi_0 \in \mathcal{L}$ and $d_L(\varphi_i) \leq n - 1$, for every $i$, $1 \leq i \leq k + m$ (modal literals $L\varphi_i$, $\neg L\varphi_j$, or the modal-free part may be missing).

So, consider a formula $\varphi$ of the form (9.9) and assume that $\varphi \in Th(\mathcal{M})$. Since $\mathcal{M}$ is an **S5**-model, $Th(\mathcal{M})$ is stable and consistent. Hence, by Corollary 8.9, there are three possibilities:

**(a)** $\varphi_0 \in Th(\mathcal{M})$.

**(b)** $\varphi_i \in Th(\mathcal{M})$ for some $i$, $1 \leq i \leq k$.

**(c)** $\varphi_i \notin Th(\mathcal{M})$ for some $i$, $k + 1 \leq i \leq k + m$.

Case (a). Since $\varphi_0 \in Th(\mathcal{M})$, $\langle \mathcal{N} \odot \mathcal{M}, b\rangle \models \varphi_0$ for every $b \in M$. Since $\varphi_0 \in \mathcal{L}$ and $\mathcal{N} \odot \mathcal{M} \not\sqsubset \mathcal{M}$, it follows by Theorem 9.16 that $\langle \mathcal{N} \odot \mathcal{M}, a\rangle \models \varphi_0$ for every $a \in N$. Consequently, $\mathcal{N} \odot \mathcal{M} \models \varphi_0$ and $\varphi \in Th(\mathcal{N} \odot \mathcal{M})$. Observe that this argument also implies that the basis of the induction holds.

Case (b). Since $d_L(\varphi_i) < d_L(\varphi)$, it follows by the induction hypothesis that $\varphi_i \in Th(\mathcal{N} \odot \mathcal{M})$. Hence, $L\varphi_i \in Th(\mathcal{N} \odot \mathcal{M})$ and, consequently, $\varphi \in Th(\mathcal{N} \odot \mathcal{M})$.

Case (c). For some world $b \in M$, $\langle \mathcal{M}, b\rangle \not\models \varphi_i$. By Lemma 7.43(3), $\langle \mathcal{N} \odot \mathcal{M}, b\rangle \not\models \varphi_i$. By the definition of concatenation and since $\mathcal{M}$ is universal, $\mathcal{N} \odot \mathcal{M} \models \neg L\varphi_i$. Thus, $\neg L\varphi_i \in Th(\mathcal{N} \odot \mathcal{M})$ and, consequently, $\varphi \in Th(\mathcal{N} \odot \mathcal{M})$. ■

**Proof of Theorem 9.18**. Let $\mathcal{M} = \langle M, R, V\rangle$ be an **S5**-model. Assume first that $\mathcal{M}$ is $\mathcal{C}$-intended for $I$. We need to show that $\mathcal{M}$ is $\mathcal{C}$-minimal for $I$. Consider an arbitrary model $\mathcal{N} \in \mathcal{C}$ such that $\mathcal{N} \models I$. If there is no Kripke model $\mathcal{N}'$ such that $\mathcal{N} = \mathcal{N}' \odot \mathcal{M}$, then $\mathcal{N} \not\sqsubset \mathcal{M}$. Assume then that for some Kripke model $\mathcal{N}'$, $\mathcal{N} = \mathcal{N}' \odot \mathcal{M}$. By the definition of concatenation and since $\mathcal{M}$ is an **S5**-model, we have $\mathcal{N} \models \{\neg L\varphi : \varphi \notin Th(\mathcal{M})\}$. Since $\mathcal{M}$ is $\mathcal{C}$-intended for $I$, $Th(\mathcal{N}) = Th(\mathcal{M})$. Consider a formula $\varphi \in \mathcal{L}$ such that for every $b \in M$, $\langle \mathcal{M}, b\rangle \models \varphi$. Then $\varphi \in Th(\mathcal{M})$ and, consequently, $\varphi \in Th(\mathcal{N})$. Hence, for every world $a$ of $\mathcal{N}'$, $\langle \mathcal{N}, a\rangle \models \varphi$ and, by Theorem 9.16, $\mathcal{N} \not\sqsubset \mathcal{M}$. Consequently, $\mathcal{M}$ is $\mathcal{C}$-minimal for $I$.

Conversely, assume that $\mathcal{M}$ is $\mathcal{C}$-minimal for $I$. Consider a model $\mathcal{N} \in \mathcal{C}$ such that $\mathcal{N} \models I \cup \{\neg L\varphi : \varphi \notin Th(\mathcal{M})\}$. Let $\varphi \notin Th(\mathcal{M})$. Then $\mathcal{N} \models \neg L\varphi$. Consequently, $\varphi \notin Th(\mathcal{N})$ and $Th(\mathcal{N}) \subseteq Th(\mathcal{M})$ follows. Thus, to prove that $\mathcal{M}$ is $\mathcal{C}$-intended for $I$, it is enough to prove that $Th(\mathcal{M}) \subseteq Th(\mathcal{N})$.

Since $\mathcal{C}$ is cluster-closed, at least one of the following two cases holds:

**(a)** $\mathcal{N} \odot \mathcal{M} \in \mathcal{C}$.

**(b)** There is a Kripke model $\mathcal{N}'$ and an **S5**-model $\mathcal{M}'$ such that $\mathcal{N} = \mathcal{N}' \odot \mathcal{M}'$ and $\mathcal{N}' \odot \mathcal{M} \in \mathcal{C}$.

Case (a). Since $\mathcal{N} \models I$ and $\mathcal{M} \models I$, it follows by Lemma 9.20 that $\mathcal{N} \odot \mathcal{M} \models I$. Since $\mathcal{M}$ is $\mathcal{C}$-minimal for $I$ and $\mathcal{N} \odot \mathcal{M} \in \mathcal{C}$, it follows that $\mathcal{N} \odot \mathcal{M} \not\sqsubset \mathcal{M}$. By Lemma 9.21, $Th(\mathcal{M}) \subseteq Th(\mathcal{N} \odot \mathcal{M})$. Hence, by Lemma 9.20, $Th(\mathcal{M}) \subseteq Th(\mathcal{N})$.

Case (b). Since $\mathcal{N} \models \{\neg L\varphi : \varphi \notin Th(\mathcal{M})\}$, $\mathcal{M}' \models \{\neg L\varphi : \varphi \notin Th(\mathcal{M})\}$. Then, $Th(\mathcal{M}') \subseteq Th(\mathcal{M})$. Both $\mathcal{M}'$ and $\mathcal{M}$ are **S5**-models. By Theorem 8.10, $Th(\mathcal{M}')$ and $Th(\mathcal{M})$ are stable. Then, by Theorem 8.6, $Th(\mathcal{M}') = Th(\mathcal{M})$. By Lemma 9.19, $Th(\mathcal{N}' \odot \mathcal{M}) = Th(\mathcal{N}' \odot \mathcal{M}') = Th(\mathcal{N})$. In particular, $\mathcal{N}' \odot \mathcal{M} \models I$. By the $\mathcal{C}$-minimality of $\mathcal{M}$ for $I$, $\mathcal{N}' \odot \mathcal{M} \not\sqsubset \mathcal{M}$. Hence, by Lemma 9.21, $Th(\mathcal{M}) \subseteq Th(\mathcal{N}' \odot \mathcal{M}) = Th(\mathcal{N})$. ■

Theorems 9.13 and 9.18 have several corollaries which provide powerful tools for investigating nonmonotonicity. We will frequently use them in the remainder of the book. The first corollary is a straightforward consequence of these theorems and its proof is omitted.

**Corollary 9.22** *Let $\mathcal{S}$ be a normal modal logic contained in* **S5**. *Assume that $\mathcal{S}$ is characterized by a cluster-closed class $\mathcal{C}$ of models such that $\mathcal{C}$ contains all* **S5**-*models. For every* **S5**-*model $\mathcal{M}$, $Th(\mathcal{M})$ is an $\mathcal{S}$-expansion for $I$ if and only if $\mathcal{M}$ is $\mathcal{C}$-minimal for $I$.* ▲

Let us recall (see Section 8.2) that given a consistent stable theory $T$, by $\mathcal{M}_T$ we denote the **S5**-model $\langle M, V\rangle$, where $M$ consists of all propositional valuations in which all modal-free formulas from $T$ are true and $V(w) = w$ for every $w \in M$. This model is referred to as *canonical* for $T$. The key property of the canonical model $\mathcal{M}_T$ is that $Th(\mathcal{M}_T) = T$. Our next corollary is a simple restatement of Corollary 9.22 tailored to the case of canonical models.

**Corollary 9.23** *Let $\mathcal{S}$ be a normal modal logic contained in* **S5**. *Assume that $\mathcal{S}$ is characterized by a cluster-closed class $\mathcal{C}$ of models such that $\mathcal{C}$ contains all* **S5**-*models. Let $T$ be a stable theory. Theory $T$ is an $\mathcal{S}$-expansion for $I$ if and only if $\mathcal{M}_T$ is $\mathcal{C}$-minimal for $I$.* ▲

In order to apply Corollary 9.23, one has to verify that $\mathcal{M}_T$ is $\mathcal{C}$-minimal. The problem whether a canonical model $\mathcal{M}_T$ is $\mathcal{C}$-minimal can be decided by considering an easier condition to deal with than condition (2) of Definition 9.15. To see this, we introduce yet another notion of preference of one Kripke model over another.

**Definition 9.24** Let $\mathcal{M}_i = \langle M_i, R_i, V_i\rangle$, $i = 1, 2$, be two Kripke models. We say that $\mathcal{M}_2$ is *weakly preferred over* $\mathcal{M}_1$ and write $\mathcal{M}_2 \sqsubset_w \mathcal{M}_1$ if:

**(1)** $\mathcal{M}_2 = \mathcal{M} \odot \mathcal{M}_1$, for some Kripke model $\mathcal{M}$.

**(2)** There is $a \in M_2 \setminus M_1$ such that for each $b \in M_1$, $V_2(b) \neq V_2(a)$.

Let $\mathcal{C}$ be a class of Kripke models and let $I \subseteq \mathcal{L}_L$. We say that a model $\mathcal{M} \in \mathcal{C}$ is *strongly $\mathcal{C}$-minimal for $I$* if $\mathcal{M} \models I$ and for every model $\mathcal{N} \in \mathcal{C}$ such that $\mathcal{N} \models I$, $\mathcal{N} \not\sqsubset_w \mathcal{M}$. In other words, $\mathcal{M}$ is strongly $\mathcal{C}$-minimal for $I$ if it is a $\sqsubset_w$-minimal model of $I$ in $\mathcal{C}$. ▲

Condition (2) of Definition 9.15 implies condition (2) of Definition 9.24. Hence, for every two Kripke models $\mathcal{M}_1$ and $\mathcal{M}_2$, if $\mathcal{M}_1 \sqsubset \mathcal{M}_2$ then

$\mathcal{M}_1 \sqsubset_w \mathcal{M}_2$ and, consequently, if a model $\mathcal{M} \in \mathcal{C}$ is strongly $\mathcal{C}$-minimal for $I$, then $\mathcal{M}$ is $\mathcal{C}$-minimal for $I$. In general, this implication cannot be reversed. Yet, it turns out that for canonical models, both notions of minimality are equivalent, that is, a canonical **S5**-model $\mathcal{M}$ is $\mathcal{C}$-minimal for $I$ in $\mathcal{C}$ if and only if $\mathcal{M}$ is strongly $\mathcal{C}$-minimal for $I$ in $\mathcal{C}$.

**Corollary 9.25** *Let $\mathcal{S}$ be a normal modal logic contained in* **S5***. Assume that $\mathcal{S}$ is characterized by a cluster-closed class $\mathcal{C}$ of models such that $\mathcal{C}$ contains all* **S5***-models. Let $T$ be a stable theory. Theory $T$ is an $\mathcal{S}$-expansion for $I$ if and only if the canonical model $\mathcal{M}_T$ is strongly $\mathcal{C}$-minimal for $I$ in $\mathcal{C}$.*

Proof: It is enough to prove that for every Kripke model $\mathcal{N}$,

$$\mathcal{N} \sqsubset \mathcal{M}_T \text{ if and only if } \mathcal{N} \sqsubset_w \mathcal{M}_T.$$

The implication from left to right is obvious and follows immediately from the definitions of the relations $\sqsubset$ and $\sqsubset_w$. So, assume that $\mathcal{N} \sqsubset_w \mathcal{M}_T$. Let $\mathcal{N} = \langle N, R, V\rangle$. Then, there is $a \in N \setminus M_T$ ($M_T$ denotes the set of worlds of the canonical model $\mathcal{M}_T$) such that for every $b \in M_T$, $V(a) \neq V(b)$. Since $M_T$ consists of all valuations of $\mathcal{L}$ which satisfy $T \cap \mathcal{L}$, it follows that for some formula $\varphi \in T \cap \mathcal{L}$, $V(a)(\varphi) = \mathbf{f} \neq \mathbf{t} = V(b)(\varphi)$, for every $b \in M_T$. Consequently, by Theorem 9.16, $\mathcal{N} \sqsubset \mathcal{M}_T$. ■

**Remark 9.26** Our results providing semantic characterization of expansions hold for many "standard" modal logics including **K**, **T**, **k4**, **S4**, **S5**, **kd45** and also for the logics **S4f** and **Sw5**, as each of these logics is characterized by a cluster-closed class of models containing all **S5**-models. ▲

In order to facilitate applications of Corollary 9.25 for logics **S4f**, **kd45**, **Sw5** we will describe now what it means for an **S5**-model to be strongly $\mathcal{C}$-minimal, for the classes of models $\mathcal{C}$ characterizing these logics.

**Theorem 9.27** *Let $I \subseteq \mathcal{L}_L$ and let $\mathcal{M} = \langle M, V\rangle$ be an* **S5***-model for $I$.*

1. *$\mathcal{M}$ is strongly $\mathcal{C}_{\mathbf{S4f}}$-minimal for $I$ if and only if for every* **S4f***-model $\langle M', M, V'\rangle$ for $I$ such that $V'|_M = V$ we have $\{V'(m): m \in M'\} \subseteq \{V(m): m \in M\}$.*

2. *$\mathcal{M}$ is strongly $\mathcal{C}_{\mathbf{Sw5}}$-minimal for $I$ if and only if for every* **Sw5***-model $\langle a, M, V'\rangle$ for $I$ such that $V'|_M = V$ we have $V'(a) = V(m)$, for some $m \in M$.*

3. *$\mathcal{M}$ is strongly $C_{\mathbf{kd45}}$-minimal for $I$ if and only if for every* **kd45**-*model $\langle a, M, V' \rangle$ for $I$ such that $V'|_M = V$ we have $V'(a) = V(m)$, for some $m \in M$.* ▲

By combining Corollary 9.25 and Theorem 9.27 we obtain the following result containing characterizations of $\mathcal{S}$-expansions for $\mathcal{S} =$ **S4f**, **kd45** and **Sw5**.

**Corollary 9.28** *Let $I \subseteq \mathcal{L}_L$. Let $T$ be a stable theory containing $I$ and let $\mathcal{M} = \langle M, V \rangle$ be the canonical model for $T$.*

1. *The theory $T$ is an* **S4f**-*expansion for $I$ if and only if for every* **S4f**-*model $\langle M', M, V' \rangle$ for $I$ such that $V'|_M = V$ we have $\{V'(m): m \in M'\} \subseteq \{V(m): m \in M\}$.*
2. *The theory $T$ is an* **Sw5**-*expansion for $I$ if and only if for every* **Sw5**-*model $\langle a, M, V' \rangle$ for $I$ such that $V'|_M = V$ we have $V'(a) = V(m)$, for some $m \in M$.*
3. *The theory $T$ is a* **kd45**-*expansion for $I$ if and only if for every* **kd45**-*model $\langle a, M, V' \rangle$ for $I$ such that $V'|_M = V$ we have $V'(a) = V(m)$, for some $m \in M$.* ▲

**Example 9.29** Consider a theory $I = \{\neg Lp \supset q, \neg Lq \supset p\}$. We will show that $ST(\{p\})$ is an **S4f**-expansion of $I$. We will use Corollary 9.28. Let $\mathcal{M} = \langle M, V \rangle$ be the canonical model for $ST(\{p\})$. It is easy to see that $I \subseteq ST(\{p\})$, that is, $\mathcal{M}$ is a model for $I$. Consider an **S4f**-model $\mathcal{N} = \langle M', M, V' \rangle$, where $M' \cap M = \emptyset$ and $V'|_M = V$, such that $\mathcal{N} \models I$. Suppose that for some $a \in M'$, $V'(a)(p) = \mathbf{f}$. Since there is a world $b \in M$ in which $q$ is not true ($q \notin ST(\{p\})$), $\langle \mathcal{N}, a \rangle \models \neg Lq$. Consequently, $\langle \mathcal{N}, a \rangle \models \neg(\neg Lq \supset p)$, a contradiction. Hence, for every world $a \in M'$, $V'(a)(p) = \mathbf{t}$. Since $M$ consists of all valuations satisfying $p$, $\{V'(a): a \in M'\} \subseteq \{V(b): b \in M\}$. By Corollary 9.28, $ST(\{p\})$ is an **S4f**-expansion for $I$. ▲

We conclude this section with two simple applications of the minimal model characterization of $\mathcal{S}$-expansions.

**Definition 9.30**

1. A formula of the form

$$L\alpha \wedge LM\beta_1 \wedge \ldots \wedge LM\beta_n \supset \gamma, \tag{9.10}$$

where $\alpha$, $\beta_i$, $1 \leq i \leq n$ and $\gamma$ are from $\mathcal{L}$, is called a *default clause.*

2. A formula of the form

$$L(a_1 \wedge \ldots \wedge a_m) \wedge LMb_1 \wedge \ldots \wedge LMb_n \supset c, \tag{9.11}$$

where all $a_i$, $1 \leq i \leq m$, $b_i$, $1 \leq i \leq n$ and $c$ are literals from $\mathcal{L}$, is called an *lp-clause* (where lp stands for logic programming). ▲

It turns out that formulas of the form (9.10) appear naturally as modal encodings of defaults. They allow us to embed default logic into modal nonmonotonic logics. This property motivated our choice of the name for formulas of the form (9.10). Similarly, lp-clauses are natural encodings of *CN*-clauses, that is, clauses of logic programs with classical negation. The details of embedding default logic and logic programming into modal nonmonotonic logics are discussed in Chapter 12.

**Theorem 9.31** *Let $\mathcal{S}$ be a modal logic such that $\mathcal{S} \subseteq$ **S4f**. For every theory $I$ consisting of default clauses and for every $\mathcal{S}$-expansion $T$ for $I$, if $S' \subseteq T \cap \mathcal{L}$ is closed under propositional consequence and $I \subseteq ST(S')$ then $ST(S') = T$.*

Proof: Let $\mathcal{M} = \langle M, V \rangle$ be the canonical model for the stable theory $T$ and let $\mathcal{M}' = \langle M', V' \rangle$ be the canonical model for the stable theory $ST(S')$. Since $S' \subseteq T \cap \mathcal{L}$, it follows from the definition of canonical models for stable theories that $M' \supseteq M$.

Let us consider a Kripke model $\mathcal{M}'' = \langle M'', V'' \rangle$ isomorphic to $\mathcal{M}'$ and such that $M \cap M'' = \emptyset$. Clearly, we have $\mathcal{M} \models I$ and $\mathcal{M}'' \models I$. Define $\mathcal{N} = \mathcal{M}'' \odot \mathcal{M}$. We will show that $\mathcal{N} \models I$. To this end, consider a formula $\varphi \in I$, where $\varphi$ is of the form (9.10). Let $a \in M \cup M''$. Clearly, if $a \in M$, then $\langle \mathcal{N}, a \rangle \models \varphi$. So, assume that $a \in M''$.

First, assume that $\langle \mathcal{N}, a \rangle \models \neg L\alpha$. Then, clearly, $\langle \mathcal{N}, a \rangle \models \varphi$.

Next, consider the possibility that for some $i$, $1 \leq i \leq n$, $\neg\beta_i \in S'$. Then $\mathcal{M}'' \models \neg\beta_i$ and, since $S' \subseteq T \cap \mathcal{L}$, $\mathcal{M} \models \neg\beta_i$. Since $\beta_i \in \mathcal{L}$, it follows that $\mathcal{N} \models \neg\beta_i$. Consequently, $\mathcal{N} \models \neg LM\beta_i$ and $\mathcal{N} \models \varphi$. In particular, $\langle \mathcal{N}, a \rangle \models \varphi$.

Finally, assume that $\langle \mathcal{N}, a \rangle \models L\alpha$ and that for every $i$, $1 \leq i \leq n$, $\neg\beta_i \notin S'$. The first assumption implies that $\langle \mathcal{N}, b \rangle \models \alpha$ for every $b \in M''$. That is, $\alpha \in S'$. Consequently, $L\alpha \in ST(S')$. The second assumption implies that $M\beta_i \in ST(S')$, for every $i$, $1 \leq i \leq n$. Consequently, $LM\beta_i \in ST(S')$, for every $i$, $1 \leq i \leq n$. Since $\varphi \in ST(S')$, $\gamma \in ST(S')$ and so $\gamma \in S'$. Thus, $\langle \mathcal{N}, a \rangle \models \gamma$ and $\langle \mathcal{N}, a \rangle \models \varphi$.

Since $T$ is an $\mathcal{S}$-expansion for $I$, $T$ is an **S4f**-expansion for $I$ (Theorem 9.6). By Corollary 9.28, every valuation of $\mathcal{M}''$ is assigned to some world in $M$. Since $\mathcal{M}'$ and $\mathcal{M}''$ are isomorphic, $M' \subseteq M$. Consequently, $M = M'$, $S' = T \cap \mathcal{L}$ and, finally, $ST(S') = T$. ■

**Theorem 9.32** *Let $\mathcal{S}$ be a modal logic such that $\mathcal{S} \subseteq$ **Sw5** and let $I$ be a theory consisting of lp-clauses. Let $S$ be a consistent set of literals such that $ST(S)$ is an **Sw5**-expansion for $I$. For every set of literals $S'$, if $S' \subseteq S$ and $I \subseteq ST(S')$ then $S' = S$.*

Proof: We proceed similarly as in the proof of Theorem 9.31. Let $\mathcal{M} = \langle M, V\rangle$ be the canonical **S5** model for the stable theory $T = ST(S)$ and let $u$ be a valuation such that for every literal $s$, $u(s) = \mathbf{t}$ if and only if $s \in S'$. Since $S' \subseteq S$ and $S$ is consistent, $S'$ is consistent and such valuation exists.

Let $r$ be any element not in $M$. Define an **Sw5**-model $\mathcal{M}' = \langle r, M, V'\rangle$, where $V'(r) = u$ and $V'(m) = V(m)$, for $m \in M$.

We will show that $\mathcal{M}' \models I$. To this end, consider a formula $\varphi \in I$, where $\varphi$ is of the form (9.11). Consider a world $m \in \{r\} \cup M$. Clearly, if $m \in M$, then $\langle \mathcal{M}', m\rangle \models \varphi$ (because $\langle \mathcal{M}, m\rangle \models \varphi$ and it is easy to see that for the formulas of the form (9.11) and worlds from $M$ the satisfaction is preserved from $\mathcal{M}$ to $\mathcal{M}'$). So, assume that $m = r$.

First, assume that for some $i$, $1 \leq i \leq m$, $\langle \mathcal{M}', r\rangle \models \neg La_i$. Then, clearly, $\langle \mathcal{M}', r\rangle \models \neg L(a_1 \wedge \ldots \wedge a_m)$. Hence, $\langle \mathcal{M}', r\rangle \models \varphi$.

Next, consider the possibility that for some $i$, $1 \leq i \leq n$, $\neg b_i \in S'$. Then, $\neg b_i \in S$ and, for every $m \in M$, $\langle \mathcal{M}, m\rangle \models \neg b_i$. Consequently, $\langle \mathcal{M}', r\rangle \models \neg LMb_i$. Hence, $\langle \mathcal{M}', r\rangle \models \varphi$.

Finally, assume that for every $i$, $1 \leq i \leq m$, $\langle \mathcal{M}', r\rangle \models La_i$, and that for every $i$, $1 \leq i \leq n$, $\neg b_i \notin S'$. It follows that for every $i$, $1 \leq i \leq m$, $u(a_i) = \mathbf{t}$ (by definition of $u$ and the fact that the accessibility relation of $\mathcal{M}'$ is reflexive) and, consequently, $a_i \in S'$. By stability of $ST(S')$, $L(a_1 \wedge \ldots \wedge a_m) \in ST(S')$. Furthermore, for every $i$, $1 \leq i \leq n$, $Mb_i \in ST(S')$ and $LMb_i \in ST(S')$. Since $\varphi \in ST(S')$, $c \in ST(S')$ and so $c \in S'$. Thus, $u(c) = \mathbf{t}$, $\langle \mathcal{M}', r\rangle \models c$ and $\langle \mathcal{M}', r\rangle \models \varphi$.

Since $ST(S)$ is an $\mathcal{S}$-expansion for $I$, $ST(S)$ is an **Sw5**-expansion for $I$. By Corollary 9.28, $u$ is assigned to some world in $M$. Hence, $S \subseteq S'$ and, consequently, $S' = S$. ■

**Remark 9.33** The notion of a cluster-closed class of Kripke models can be extended to the case of Kripke structures. Furthermore, Corollaries 9.22,

9.23 and 9.25 can be extended to the case of modal logics characterized by cluster-closed classes of Kripke structures. In particular, logics **N** and **t** are characterized by cluster-closed class of Kripke structures and have minimal-model semantics. In the case of **N** we will provide more details in the next chapter. ▲

## 9.4 Consistency with respect to introspection

The results of the previous section provide a semantic characterization of $\mathcal{S}$-expansions of a given theory for a wide class of logics. In this section and in the next one we describe a syntactic characterization of $\mathcal{S}$-expansions.

One of the results of Section 9.2 asserts that $\mathcal{S}$-expansions are stable. Therefore, a possible way to characterize $\mathcal{S}$-expansions is to establish necessary and sufficient conditions for a stable theory to be an $\mathcal{S}$-expansion. This is the approach we take up now.

**Definition 9.34** Let $I \subseteq \mathcal{L}_L$. By $MA(I)$ we denote the set of formulas:

$$\{\varphi\colon L\varphi \text{ is a subformula of a formula from } I\}.$$

A formula built of propositional variables from $\mathcal{L}$ and of modal atoms of the form $L\varphi$, where $\varphi \in MA(I)$ is called an *I-formula*. ▲

Formulas from $MA(I)$ play a key role in the effective characterization of $\mathcal{S}$-expansions which will be described in the next section. Namely, $\mathcal{S}$-expansions will be shown to correspond to certain partitions of $MA(I)$.

We will first describe a subclass of the class of all stable theories containing $I$ which, for many logics $\mathcal{S}$, contains all $\mathcal{S}$-expansions for $I$.

**Definition 9.35** Let $I \subseteq \mathcal{L}_L$. A pair $(\Phi, \Psi)$ of subsets of $MA(I)$ is called *consistent with I under introspection* or *introspection-consistent with I* if the following conditions are met:

**(IC1)** $\Phi \cup \Psi = MA(I)$ and $\Phi \cap \Psi = \emptyset$. Any such pair will be referred to as a *partition of* $MA(I)$.

**(IC2)** Theory $I \cup \{\neg L\varphi\colon \varphi \in \Phi\} \cup \{L\psi\colon \psi \in \Psi\} \cup \Psi$ is propositionally consistent.

**(IC3)** For each $\varphi \in \Phi$, $I \cup \{\neg L\varphi\colon \varphi \in \Phi\} \cup \{L\psi\colon \psi \in \Psi\} \cup \Psi \not\vdash \varphi$. ▲

Each stable and consistent theory containing $I$ determines a partition of $MA(I)$ which is introspection-consistent with $I$. This is the content of our next theorem.

**Theorem 9.36** *Let $I \subseteq \mathcal{L}_L$ and let $T \subseteq \mathcal{L}_L$ be a stable and consistent theory such that $I \subseteq T$. Then the partition $(MA(I) \setminus T, MA(I) \cap T)$ is introspection-consistent with $I$.*

Proof: Let $\Phi = MA(I) \setminus T$ and $\Psi = MA(I) \cap T$. Since $T$ is stable and contains $I$,

$$I \cup \{\neg L\varphi : \varphi \in \Phi\} \cup \{L\psi : \psi \in \Psi\} \cup \Psi \subseteq T.$$

Theory $T$ is propositionally consistent. Thus, (IC2) follows. Furthermore, if $\varphi \in \Phi$, then $\varphi \notin T$. Since $T$ is closed under propositional consequence, (IC3) holds, too. Condition (IC1) follows immediately from the definition of $\Phi$ and $\Psi$. ■

**Remark 9.37** Observe that two (or even infinitely many) different stable theories may determine the same partition of $MA(I)$. For example, consider $I = \emptyset$. Clearly, $MA(I) = \emptyset$ and $(\emptyset, \emptyset)$ is the only partition of $MA(I)$. On the other hand, there are continuum many stable theories containing $I = \emptyset$. ▲

We will now see that with each partition $(\Phi, \Psi)$ of $MA(I)$ which is introspection-consistent with $I$, one can associate a stable and consistent theory $T$ containing $I$ and such that $\Phi = MA(I) \setminus T$ and $\Psi = MA(I) \cap T$.

**Definition 9.38** Let $I \subseteq \mathcal{L}_L$ and let a partition $(\Phi, \Psi)$ of $MA(I)$ be introspection-consistent with $I$. Define $M_I(\Phi, \Psi)$ to consist of all propositional valuations $v$ of $\mathcal{L}_L$ such that

$$v(I \cup \{\neg L\varphi : \varphi \in \Phi\} \cup \{L\psi : \psi \in \Psi\} \cup \Psi) = \mathbf{t}.$$

For a valuation $v \in M_I(\Phi, \Psi)$, let $U(\Phi, \Psi)(v)$ be a propositional valuation of $\mathcal{L}$ such that

$$U(\Phi, \Psi)(v) = v|_{\mathcal{L}}.$$

An **S5**-model

$$\mathcal{M}_I(\Phi, \Psi) = \langle M_I(\Phi, \Psi), U(\Phi, \Psi) \rangle$$

is called *canonical* for the partition $(\Phi, \Psi)$. The theory $Th(\mathcal{M}_I(\Phi, \Psi))$ is called *canonical* for the partition $(\Phi, \Psi)$ and is denoted by $T_I(\Phi, \Psi)$. ▲

Note that by (IC2), $M_I(\Phi,\Psi) \neq \emptyset$ and $\mathcal{M}_I(\Phi,\Psi)$ and $T_I(\Phi,\Psi)$ are well-defined.

The following theorem collects the most important properties of canonical models and canonical theories associated with partitions.

**Theorem 9.39** *Let $I \subseteq \mathcal{L}_L$. If a partition $(\Phi,\Psi)$ of $MA(I)$ is introspection-consistent with $I$, then:*

1. *The theory $T_I(\Phi,\Psi)$ is stable.*
2. *For every $I$-formula $\varphi$ and for every world (valuation) $v \in M_I(\Phi,\Psi)$, $\langle \mathcal{M}_I(\Phi,\Psi), v\rangle \models \varphi$ if and only if $v(\varphi) = \mathbf{t}$.*
3. *$I \cup \{\neg L\varphi: \varphi \in \Phi\} \cup \{L\psi: \psi \in \Psi\} \cup \Psi \subseteq T_I(\Phi,\Psi)$.*
4. *$\Phi = MA(I) \setminus T_I(\Phi,\Psi)$ and $\Psi = MA(I) \cap T_I(\Phi,\Psi)$.*
5. *$T_I(\Phi,\Psi) = ST(Cn(I \cup \{\neg L\varphi: \varphi \in \Phi\} \cup \{L\psi: \psi \in \Psi\} \cup \Psi) \cap \mathcal{L})$.*

Proof: (1) The assertion follows directly from Theorem 8.10.

(2) We proceed by induction on the complexity of a formula. If $\varphi \in \mathcal{L}$, then the claim is a direct consequence of the definition of $U(\Phi,\Psi)$.

Consider a formula $\varphi$ and assume that the assertion holds for all formulas of $\mathcal{L}_L$ shorter than $\varphi$. If $\varphi = \neg\psi$, $\psi_1 \vee \psi_2$ or $\psi_1 \wedge \psi_2$, then the assertion follows directly from the induction hypothesis. So, assume that $\varphi = L\psi$. We have the following sequence of equivalent statements:
**(a)** $\langle \mathcal{M}_I(\Phi,\Psi), v\rangle \not\models \varphi$.
**(b)** there exists $w \in M_I(\Phi,\Psi)$ such that $\langle \mathcal{M}_I(\Phi,\Psi), w\rangle \not\models \psi$.
**(c)** $w(\psi) = \mathbf{f}$.
**(d)** $I \cup \{\neg L\varphi: \varphi \in \Phi\} \cup \{L\psi: \psi \in \Psi\} \cup \Psi \not\vdash \psi$.
**(e)** $\psi \in \Phi$.
**(f)** $v(L\psi) = \mathbf{f}$.
The equivalences between (a) and (b), and (c) and (d) are obvious. The equivalence of (b) and (c) follows from the induction hypothesis. Statements (d) and (e) are equivalent because $\psi \in MA(I)$. Finally, (e) and (f) are equivalent because $v \in M_I(\Phi,\Psi)$.

In particular, it follows that (a) and (f) are equivalent, which completes the induction step.

(3) Let $\eta \in I \cup \{\neg L\varphi: \varphi \in \Phi\} \cup \{L\psi: \psi \in \Psi\} \cup \Psi$. By the definition of $M_I(\Phi,\Psi)$, for every valuation $v \in M_I(\Phi,\Psi)$, $v(\eta) = \mathbf{t}$. Thus, by (2), for every $v \in M_I(\Phi,\Psi)$, $\langle \mathcal{M}_I(\Phi,\Psi), v\rangle \models \eta$. Consequently, $\eta \in T_I(\Phi,\Psi)$.

(4) Consider $\varphi \in MA(I) \setminus T_I(\Phi,\Psi)$. By the stability of $T_I(\Phi,\Psi)$ (part (1)), $\neg L\varphi \in T_I(\Phi,\Psi)$. Since $T_I(\Phi,\Psi)$ is consistent, $L\varphi \notin T_I(\Phi,\Psi)$. Thus, $\varphi \notin \Psi$. Since $\varphi \in MA(I)$, it follows that $\varphi \in \Phi$. Conversely, if $\varphi \in \Phi$ then, by (3), $\neg L\varphi \in T_I(\Phi,\Psi)$. Since $T_I(\Phi,\Psi)$ is stable and consistent, $\varphi \notin T_I(\Phi,\Psi)$. Thus, $\Phi = MA(I) \setminus T_I(\Phi,\Psi)$. The second equality follows from the fact that $(\Phi,\Psi)$ is a partition of $MA(I)$.

(5) By (2), it follows that if $\varphi \in \mathcal{L}$, then $\varphi \in T_I(\Phi,\Psi)$ if and only if

$$\varphi \in Cn(I \cup \{\neg L\varphi : \varphi \in \Phi\} \cup \{L\psi : \psi \in \Psi\} \cup \Psi).$$

Thus, the assertion follows from the definition of the operator $ST$ (Definition 8.20). ■

The last part of Theorem 9.39 implies that there is an easy test which allows us to determine whether a modal-free formula $\xi$ belongs to $T_I(\Phi,\Psi)$. It is enough to determine whether

$$I \cup \{\neg L\varphi : \varphi \in \Phi\} \cup \{L\psi : \psi \in \Psi\} \cup \Psi \vdash \xi.$$

This test can be generalized to arbitrary formulas. Namely, we have the following result.

**Theorem 9.40** *Let $I \subseteq \mathcal{L}_L$ and $\xi \in \mathcal{L}_L$. If a partition $(\Phi,\Psi)$ of $MA(I)$ is introspection-consistent with $I$, then $\xi \in T_I(\Phi,\Psi)$ if and only if*

$$I \cup \{\neg L\varphi : \varphi \in \Phi\} \cup \{L\psi : \psi \in \Psi\} \cup \Psi \cup \\ \{L\alpha : \alpha \in MA(\{\xi\}) \cap T_I(\Phi,\Psi)\} \cup \{\neg L\alpha : \alpha \in MA(\{\xi\}) \setminus T_I(\Phi,\Psi)\} \vdash \xi.$$

Proof: First, let us denote the theory

$$I \cup \{\neg L\varphi : \varphi \in \Phi\} \cup \{L\psi : \psi \in \Psi\} \cup \Psi \cup \\ \{L\alpha : \alpha \in MA(\{\xi\}) \cap T_I(\Phi,\Psi)\} \cup \{\neg L\alpha : \alpha \in MA(\{\xi\}) \setminus T_I(\Phi,\Psi)\}$$

by $\Sigma$. To prove the assertion we will proceed by induction on the $L$-depth of $\xi$. The basis of induction follows directly from Theorem 9.39(5) and from the observation that $MA(\{\xi\}) = \emptyset$, for every modal-free formula $\xi$.

Let us then consider a formula $\xi \in \mathcal{L}_L$ with positive $L$-depth, and assume that the assertion holds for every formula of smaller $L$-depth. Since $\xi$ and all its propositionally equivalent normal forms have the same set of modal atoms and the same $L$-depth, it is clear that without loss of generality we may assume that $\xi$ is of the form

$$L\alpha_1 \vee \ldots \vee L\alpha_k \vee \neg L\beta_1 \vee \ldots \vee \neg L\beta_m \vee \gamma,$$

where $\alpha_i$ and $\beta_i$ have $L$-depths smaller than the $L$-depth of $\xi$ and $\gamma$ is modal-free.

By Theorem 9.39, $\Sigma \subseteq T_I(\Phi,\Psi)$. Hence, if $\Sigma \vdash \xi$, then $\xi \in T_I(\Phi,\Psi)$ (recall that $T_I(\Phi,\Psi)$ is closed under propositional provability). This proves the sufficiency part of the assertion.

Conversely, assume that $\xi \in T_I(\Phi,\Psi)$. Since $T_I(\Phi,\Psi)$ is stable and consistent, there are three possibilities (Corollary 8.9):

1. $\alpha_i \in T_I(\Phi,\Psi)$ for some $i$, $1 \leq i \leq k$.
2. $\beta_i \notin T_I(\Phi,\Psi)$ for some $i$, $1 \leq i \leq m$.
3. $\gamma \in T_I(\Phi,\Psi)$.

(1) Since $\alpha_i \in MA(\{\xi\})$, $L\alpha_i \in \Sigma$. Consequently, $\Sigma \vdash \xi$.

(2) Similarly, since $\beta_i \in MA(\{\xi\})$, $\neg L\beta_i \in \Sigma$. Hence, $\Sigma \vdash \xi$.

(3) In this case we show that $\Sigma \vdash \gamma$ in the same way in which we dealt with the basis step. Hence, also in this case $\Sigma \vdash \xi$. ∎

The role of partitions of $MA(I)$ which are introspection-consistent with $I$ is explained in the next theorem. It turns out that for many modal logics $\mathcal{S}$, every $\mathcal{S}$-expansion for $I$ is generated by such a partition.

**Theorem 9.41** *Let $\mathcal{S}$ be any (not necessarily normal) modal logic contained in* **kd45** *or* **Sw5**. *Let $I \subseteq \mathcal{L}_L$. If a theory $T \subseteq \mathcal{L}_L$ is an $\mathcal{S}$-expansion for $I$, then $(\Phi,\Psi)$, where $\Phi = MA(I) \setminus T$ and $\Psi = MA(I) \cap T$, is introspection-consistent with $I$ and $T = T_I(\Phi,\Psi)$.*

Proof: It follows from Theorem 9.6 that it suffices to prove the assertion for $\mathcal{S} =$ **kd45** and $\mathcal{S} =$ **Sw5**. Both cases can be dealt with in essentially the same way. Therefore, we present the argument only for the case of $\mathcal{S} =$ **kd45**, leaving the other case to the reader.

Let $T$ be a **kd45**-expansion for $I$. Define $\Phi = MA(I) \setminus T$ and $\Psi = MA(I) \cap T$. Since $T$ is stable and consistent, it follows from Theorem 9.36 that $(\Phi,\Psi)$ is introspection-consistent with $I$. We will show that $T = T_I(\Phi,\Psi)$. Since $T_I(\Phi,\Psi)$ is stable, it suffices to show that $T$ and $T_I(\Phi,\Psi)$ have the same modal-free parts (Corollary 8.19).

First, consider $\eta \in T_I(\Phi,\Psi) \cap \mathcal{L}$. By Theorem 9.39(5),

$$\eta \in Cn(I \cup \{\neg L\varphi : \varphi \in \Phi\} \cup \{L\psi : \psi \in \Psi\} \cup \Psi).$$

Since $T$ is stable, consistent, and contains $I$,

$$I \cup \{\neg L\varphi : \varphi \in \Phi\} \cup \{L\psi : \psi \in \Psi\} \cup \Psi \subseteq T.$$

Consequently, $\eta \in T$.

Next, suppose that $\eta \in T \cap \mathcal{L}$. Consequently,

$$I \cup \{\neg L\varphi : \varphi \notin T\} \vdash_{\mathbf{kd45}} \eta. \tag{9.12}$$

Consider an arbitrary valuation $v$ of $\mathcal{L}_L$ such that

$$v(I \cup \{\neg L\varphi : \varphi \in \Phi\} \cup \{L\psi : \psi \in \Psi\} \cup \Psi) = \mathbf{t}.$$

Let $\mathcal{M} = \langle M, U\rangle$ be an **S5**-model such that $T = Th(\mathcal{M})$ (Theorem 8.12). Let $a$ be an element not in $M$. Define a **kd45**-model $\mathcal{M}' = \langle a, M, V\rangle$, where $V(a) = v|_{\mathcal{L}}$ and $V(b) = U(b)$, for $b \in M$.

For every $b \in M$,

$$\langle \mathcal{M}', b\rangle \models I \cup \{\neg L\varphi : \varphi \notin T\}.$$

It is also easy to see that

$$\langle \mathcal{M}', a\rangle \models \{\neg L\varphi : \varphi \notin T\}.$$

By induction on the complexity of a formula one can show that for every $I$-formula $\varphi$,

$$\langle \mathcal{M}', a\rangle \models \varphi \text{ if and only if } v(\varphi) = \mathbf{t}. \tag{9.13}$$

Consequently, it follows that $\langle \mathcal{M}', a\rangle \models I$. Since $\mathcal{M} \models I \cup \{\neg L\varphi : \varphi \notin T\}$,

$$\mathcal{M}' \models I \cup \{\neg L\varphi : \varphi \notin T\}$$

and, by (9.12), $\mathcal{M}' \models \eta$. In particular, $\langle \mathcal{M}', a\rangle \models \eta$. Since $\eta \in \mathcal{L}$, $\eta$ is an $I$-formula and, by (9.13), $v(\eta) = \mathbf{t}$. Since $v$ was an arbitrary valuation of $\mathcal{L}_L$, it follows that $I \cup \{\neg L\varphi : \varphi \in \Phi\} \cup \{L\psi : \psi \in \Psi\} \cup \Psi \models \eta$, that is,

$$\eta \in Cn(I \cup \{\neg L\varphi : \varphi \in \Phi\} \cup \{L\psi : \psi \in \Psi\} \cup \Psi).$$

Hence, by Theorem 9.39, $\eta \in T_I(\Phi, \Psi)$. ∎

Let $\mathcal{S}$ be contained in **kd45** or **Sw5**. Consider a consistent theory $I \subseteq \mathcal{L}$. Then, it follows from Corollary 9.8 that $ST(I)$ is an $\mathcal{S}$-expansion for $I$. Since $MA(I) = \emptyset$, it follows from Theorem 9.41 that $I$ has no other $\mathcal{S}$-expansions. Hence, we obtain the following result.

**Corollary 9.42** *Let $\mathcal{S}$ be a modal logic contained in* **kd45** *or* **Sw5** *and let $S \subseteq \mathcal{L}$ be consistent. Then $ST(S)$ is the unique $\mathcal{S}$-expansion for $S$.* ▲

We have already seen that the nonmonotonic logic **S5** coincides with the monotonic **S5**. It follows from Corollary 9.42 that for every modal logic $\mathcal{S}$

contained in **kd45** or **Sw5**, the nonmonotonic modal logic $\mathcal{S}$, when applied to a modal-free theory $S$, generates the same modal-free consequences of $S$ as the monotonic modal logic $\mathcal{S}$ and, in fact, as the propositional logic.

**Corollary 9.43** *Let $\mathcal{S}$ be a modal logic contained in* **kd45** *or* **Sw5**. *Then for every $S \subseteq \mathcal{L}$,*

$$Cn^n_{\mathcal{S}}(S) \cap \mathcal{L} = Cn_{\mathcal{S}}(S) \cap \mathcal{L} = Cn(S).$$

▲

Theorem 9.41 has another important consequence. Consider a *finite* theory $I \subseteq \mathcal{L}_L$. Then, $MA(I)$ is finite. Hence, the set of all possible partitions of $MA(I)$ is finite, as is the set of all partitions of $MA(I)$ which are introspection-consistent with $I$. Consequently, we have the following result.

**Corollary 9.44** *Let $\mathcal{S}$ be a modal logic contained in* **kd45** *or* **Sw5**. *Then, every finite theory $I \subseteq \mathcal{L}_L$ has finitely many $\mathcal{S}$-expansions.* ▲

The assertions of Theorem 9.41 and Corollary 9.44 fail for the logic **S5**. Indeed, consider the theory $I = \emptyset$. Then, $I$ has continuum many **S5**-expansions — each consistent stable theory is an **S5**-expansion for $I$ (Theorem 9.10). On the other hand, $MA(I)$ has exactly one partition. In fact, our next result shows that the logics **kd45** and **Sw5** are maximal logics for which the assertions of Theorem 9.41 and Corollary 9.44 hold.

**Theorem 9.45** *Let $\mathcal{S}$ be any modal logic contained in* **S5** *and properly containing* **kd45** *or* **Sw5**. *Then there is continuum many $\mathcal{S}$-expansions for $I = \emptyset$.*

Proof: We have already seen that the assertion holds for $\mathcal{S} = \mathbf{S5}$. Therefore, in the remainder of the proof we assume that $\mathcal{S}$ is properly contained in **S5**.

First, we consider the case when $\mathcal{S}$ properly contains **kd45**. By Theorem 7.59, there are the following three possibilities:

1. $\mathcal{S} = S_{\langle\omega\rangle} \cap S_{\langle 0\rangle}$.
2. $\mathcal{S} = S_{\langle\omega\rangle} \cap S_{\langle 0\rangle} \cap S_{\langle 0,n\rangle}$, for some integer $n \geq 1$.
3. $\mathcal{S} = S_{\langle\omega\rangle} \cap S_{\langle 0,n\rangle}$, for some integer $n \geq 1$.

Consider an axiom schema $\eta_n$, $n \geq 1$:

$$(\neg L\psi_1 \wedge \neg L(\psi_1 \supset \psi_2) \wedge \ldots \wedge \neg L(\psi_1 \wedge \ldots \wedge \psi_n \supset \psi_{n+1}) \wedge L\varphi) \supset \varphi. \quad (9.14)$$

Clearly, each instance of the schema $\eta_n$ is a propositional consequence of a suitable instance of schema t and, consequently, holds in $S_{\langle\omega\rangle} = \mathbf{S5}$. It is also easy to see that $\eta_n$ holds in the logic $S_{\langle 0\rangle}$ (since every formula of the form $L\psi$ is a theorem of this logic).

Consider now a **kd45**-model $\mathcal{M} = \langle a, M, U\rangle$, where $|M| = n$. Clearly, $\mathcal{M}$ has index $\langle 0, n\rangle$.

Suppose now that for each $i$, $0 \leq i \leq n$, there is a world $b_i \in M$ such that $\langle \mathcal{M}, b_i\rangle \not\models \psi_1 \wedge \ldots \wedge \psi_i \supset \psi_{i+1}$. Notice that $\psi_1 \wedge \ldots \wedge \psi_i \supset \psi_{i+1}$ does not hold in a world $b \in M$ precisely when $\langle \mathcal{M}, b\rangle \models \psi_j$, $1 \leq j \leq i$, and $\langle \mathcal{M}, b\rangle \not\models \psi_{i+1}$. Hence, all worlds $b_i$ are distinct, which implies that $|M| > n$, a contradiction.

Thus, there is $i$, $0 \leq i \leq n$, such that for every world $b \in M$, $\langle \mathcal{M}, b\rangle \models \psi_1 \wedge \ldots \wedge \psi_i \supset \psi_{i+1}$. Consequently, for every $b \in M \cup \{a\}$, $\langle \mathcal{M}, b\rangle \models L(\psi_1 \wedge \ldots \wedge \psi_i \supset \psi_{i+1})$. Thus, $\mathcal{M} \models \eta_n$.

Since each instance of $\eta_n$ is a theorem of $S_{\langle\omega\rangle}$, $S_{\langle 0\rangle}$ and $S_{\langle 0,n\rangle}$, it follows that there exists $n \geq 1$ such that each instance of $\eta_n$ is a theorem of $\mathcal{S}$. Let us fix one such $n$ for the remainder of the proof.

Let $P = \{p_1, \ldots, p_{n+1}\}$ be a set of $n+1$ distinct propositional variables of $\mathcal{L}$. Consider a set $S$ of propositional variables of $\mathcal{L}$ disjoint from $P$. We will show that $T = ST(S)$ is an $\mathcal{S}$-expansion for $\emptyset$. Let $q \in S$. The formula

$$(\neg Lp_1 \wedge \neg L(p_1 \supset p_2) \wedge \ldots \wedge \neg L(p_1 \wedge \ldots \wedge p_n \supset p_{n+1}) \wedge Lq) \supset q$$

is a theorem of $\mathcal{S}$. In addition,

$$\neg Lp_1, \neg L(\neg p_1 \vee p_2), \ldots, \neg L(\neg p_1 \vee \ldots \vee \neg p_n \vee p_{n+1}) \in \{\neg L\varphi : \varphi \notin ST(S)\}.$$

Also,

$$\neg L \neg Lq \in \{\neg L\varphi : \varphi \notin ST(S)\}.$$

By schema 5,

$$Lq \in Cn_{\mathcal{S}}(\{\neg L\varphi : \varphi \notin ST(S)\}).$$

Putting all these facts together, and using normality of $\mathcal{S}$, we obtain

$$q \in Cn_{\mathcal{S}}(\{\neg L\varphi : \varphi \notin ST(S)\}).$$

By Corollary 9.8, $ST(S)$ is an $\mathcal{S}$-expansion for $\emptyset$. Since there are infinitely many propositional variables in $\mathcal{L}$, there is a continuum many different sets $S$ of propositional variables which are disjoint from $P$.

The case of the logic $\mathcal{S}$ properly containing **Sw5** can be dealt with similarly. By Theorem 7.59, $\mathcal{S} = S_{\langle\omega\rangle} \cap S_{\langle 1,n\rangle}$, for some integer $n \geq 1$. We also make use of the fact that the following axiom schema holds in $\mathcal{S}$:

$$(L\neg L\psi_1 \wedge L\neg L(\psi_1 \supset \psi_2) \wedge \ldots \wedge L\neg L(\psi_1 \wedge \ldots \wedge \psi_n \supset \psi_{n+1}) \wedge \neg L\neg L\varphi) \supset \varphi.$$

Finally, proceeding similarly as before, we show that for every set $S$ of propositional variables disjoint from some fixed set $P$ of propositional variables of cardinality $n+1$, the theory $ST(S)$ is an $\mathcal{S}$-expansion for $\emptyset$. ■

The implication in Theorem 9.41 cannot be reversed. In other words, the condition of Theorem 9.41 is necessary but, in general, not sufficient for a theory to be an $\mathcal{S}$-expansion for $I$. This is not particularly surprising; one would expect that properties of $\mathcal{S}$-expansions for $I$ should depend on the underlying logic $\mathcal{S}$, while the class of canonical theories for partitions which are introspection-consistent with $I$ is independent of $\mathcal{S}$. Below we exhibit two theories which show that for every logic $\mathcal{S}$ contained in **kd45** or **Sw5** the converse to Theorem 9.41 fails.

**Example 9.46 (a)** Let $I = \{Lp\}$. The partition $(\emptyset, \{p\})$ is introspection-consistent with $I$. We leave an easy verification of that fact to the reader. On the other hand, it is easy to see that the corresponding canonical stable theory $T_I(\emptyset, \{p\}) = ST(\{p\})$ is not a **kd45**-expansion for $I$. Indeed, let $a$ be any element not in $M_I(\emptyset, \{p\})$. Define a **kd45**-model $\mathcal{M} = \langle a, M_I(\emptyset, \{p\}), V\rangle$, where $V(a)$ is a propositional valuation such that $V(a)(p) = \mathbf{f}$, and for $v \in M_I(\emptyset, \{p\})$, $V(v) = U(\emptyset, \{p\})(v)$. Clearly, $\mathcal{M}$ is a **kd45**-model such that $\mathcal{M} \models \{Lp\} \cup \{\neg L\varphi : \varphi \notin ST(\{p\})\}$ and $\mathcal{M} \not\models p$. Thus,

$$p \notin Cn_{\mathbf{kd45}}(\{Lp\} \cup \{\neg L\varphi : \varphi \notin ST(\{p\})\})$$

and $ST(\{p\})$ is not a **kd45**-expansion for $I$. By Theorem 9.6, for every logic $\mathcal{S} \subseteq \mathbf{kd45}$, $ST(\{p\})$ is not an $\mathcal{S}$-expansion for $I$. Thus, the converse to Theorem 9.41 fails for all modal logics contained in **kd45**.

**(b)** The case of logics contained in **Sw5** can be dealt with in the same way. Consider the theory $I = \{Lp \supset p\}$. Reasoning as above, one can show that $I$ has only one **Sw5**-expansion, $ST(\emptyset)$, but two partitions of $MA(I)$ introspection-consistent with $I$. Thus, the converse to Theorem 9.41 fails for all modal logics contained in **Sw5**. ▲

## 9.5 Characterization of $\mathcal{S}$-expansions

It follows from Theorem 9.41 and Example 9.46 that if a logic $\mathcal{S}$ is contained in **kd45** or **Sw5** then, in order to characterize $\mathcal{S}$-expansions, we need to narrow down the class of all canonical theories for partitions which are introspection-consistent with $I$.

**Definition 9.47** Let $\mathcal{S}$ be a modal logic contained in **S5**. Let $I \subseteq \mathcal{L}_L$. A partition $(\Phi, \Psi)$ of $MA(I)$ is called *$\mathcal{S}$-perfect for I* if it is introspection-consistent with $I$ and, for each $\psi \in \Psi$,

$$I \cup \{\neg L\varphi : \varphi \in \Phi\} \vdash_{\mathcal{S}} \psi.$$

▲

What does it mean that the partition $(\Phi, \Psi)$ of $MA(I)$ is $\mathcal{S}$-perfect? Roughly, it means that the negative introspection $\{\neg L\varphi : \varphi \in \Phi\}$ allows us (in the presence of $I$) to reconstruct the positive side ($\Psi$) of the partition. In the process of reconstruction we use the modal logic $\mathcal{S}$.

The class of $\mathcal{S}$-perfect partitions for $I$ plays the key role in our considerations in this section. It turns out that if $\mathcal{S} \subseteq$ **S5** then each $\mathcal{S}$-perfect partition of $MA(I)$ generates an $\mathcal{S}$-expansion for $I$. Moreover, under some additional assumptions, each $\mathcal{S}$-expansion is generated by an $\mathcal{S}$-perfect partition for $I$.

**Theorem 9.48** *Let $\mathcal{S}$ be a modal logic contained in* **S5**. *Let $I \subseteq \mathcal{L}_L$. If a partition $(\Phi, \Psi)$ of $MA(I)$ is $\mathcal{S}$-perfect for $I$ then $T_I(\Phi, \Psi)$ is an $\mathcal{S}$-expansion of $I$.*

Proof: Since $(\Phi, \Psi)$ is $\mathcal{S}$-perfect for $I$, it follows that

$$I \cup \{\neg L\varphi : \varphi \in \Phi\} \vdash_{\mathcal{S}} \{L\psi : \psi \in \Psi\} \cup \Psi.$$

By Theorem 9.39(4),

$$I \cup \{\neg L\varphi : \varphi \in \Phi\} \subseteq I \cup \{\neg L\varphi : \varphi \notin T_I(\Phi, \Psi)\}.$$

Thus,

$$I \cup \{\neg L\varphi : \varphi \in \Phi\} \cup \{L\psi : \psi \in \Psi\} \cup \Psi \subseteq Cn_{\mathcal{S}}(I \cup \{\neg L\varphi : \varphi \notin T_I(\Phi, \Psi)\}),$$

and the assertion follows from Corollary 9.8 and Theorem 9.39(5). ■

The implication in Theorem 9.48 cannot be reversed.

**Example 9.49** **(a)** Let $\mathcal{S} =$ **S5**. The empty theory has only one **S5**-perfect partition, but continuum many **S5**-expansions.

**(b)** Next assume that $\mathcal{S} =$ **kd45**. Let $I = \{Lp \supset p\}$, for some propositional letter $p \in \mathcal{L}$. Observe that the stable theory $ST(\{p\})$ contains $I$. In addition,

$$p \in Cn_{\mathbf{kd45}}(I \cup \{\neg L\varphi : \varphi \notin ST(\{p\})\}).$$

Indeed, $\neg Lp \notin ST(\{p\})$. Thus, $\neg L\neg Lp \in \{\neg L\varphi : \varphi \notin ST(\{p\})\}$. By the axiom schema 5, $Lp \in Cn_{\mathbf{kd45}}(I \cup \{\neg L\varphi : \varphi \notin ST(\{p\})\})$. Consequently,

$p \in Cn_{\mathbf{kd45}}(I \cup \{\neg L\varphi : \varphi \notin ST(\{p\})\})$. By Corollary 9.8, it follows that $ST(\{p\})$ is a **kd45**-expansion of $I$.

By Theorem 9.41, $ST(\{p\}) = T_I(\Phi, \Psi)$, where $\Phi = MA(I) \setminus ST(\{p\}) = \emptyset$ and $\Psi = MA(I) \cap ST(\{p\}) = \{p\}$. Partition $(\emptyset, \{p\})$ is not **kd45**-perfect. Indeed, we have that

$$I \nvdash_{\mathbf{kd45}} p,$$

which can be easily verified using the semantic characterization of logic **kd45**. Clearly, $(\{p\}, \emptyset)$ and $(\emptyset, \{p\})$ are the only partitions of $MA(I)$. Partition $(\emptyset, \{p\})$ is not **kd45**-perfect and partition $(\{p\}, \emptyset)$ generates the stable theory $T_I(\{p\}, \emptyset) = ST(\emptyset)$. Hence, there is no partition $(\Phi', \Psi')$ of $MA(I)$ which is **kd45**-perfect for $I$ and such that $ST(\{p\}) = T_I(\Phi', \Psi')$.

**(c)** For the case of the logic **Sw5**, consider the theory $I = \{L(q \supset p) \supset q, Lq \supset p\}$. This theory has two **Sw5**-expansions: $ST(\emptyset)$ and $ST(\{p, q\})$. To see that the second theory is an **Sw5**-expansion for $I$ notice first that $I \subseteq ST(\{p, q\})$. Observe also that

$$\neg L \neg Lq \in \{\neg L\varphi : \varphi \notin ST(\{p, q\})\}.$$

Thus, by schema w5,

$$q \supset Lq \in Cn_{\mathbf{Sw5}}(I \cup \{\neg L\varphi : \varphi \notin ST(\{p, q\})\}).$$

Consequently,

$$q \supset p \in Cn_{\mathbf{Sw5}}(I \cup \{\neg L\varphi : \varphi \notin ST(\{p, q\})\}).$$

Since $L(q \supset p) \supset q \in I$,

$$q \in Cn_{\mathbf{Sw5}}(I \cup \{\neg L\varphi : \varphi \notin ST(\{p, q\})\}).$$

Finally, since $Lq \supset p \in I$,

$$p \in Cn_{\mathbf{Sw5}}(I \cup \{\neg L\varphi : \varphi \notin ST(\{p, q\})\}).$$

Since $p, q \in Cn_{\mathbf{Sw5}}(I \cup \{\neg L\varphi : \varphi \notin ST(\{p, q\})\})$ and since $I \subseteq ST(\{p, q\})$, $ST(\{p, q\})$ is an **Sw5**-expansion of $I$ (Corollary 9.8).

There are four possible partitions of $MA(I) = \{q \supset p, q\}$. One of these partitions, $(\{q \supset p, q\}, \emptyset)$ is **Sw5**-perfect and it yields the **Sw5**-expansion $T_I(\{q \supset p, q\}, \emptyset) = ST(\emptyset)$.

Partitions $(\{q \supset p\}, \{q\})$ and $(\{q\}, \{q \supset p\})$ do not satisfy condition (IC3), so they are not even introspection-consistent with the theory $I$. Partition $(\emptyset, \{q \supset p, q\})$ is introspection-consistent with $I$, but it is not **Sw5**-perfect, as neither $I \vdash_{\mathbf{Sw5}} q \supset p$ nor $I \vdash_{\mathbf{Sw5}} q$ holds. ▲

Even though for several logics including **kd45**, **Sw5** and **S5**, the converse to Theorem 9.48 fails, there is a wide class of logics for which it holds.

**Theorem 9.50** *Let $\mathcal{S}$ be a normal modal logic contained in* **S4f** *and characterized by a cluster-closed class of models. Then:*

1. *If $T$ is an $\mathcal{S}$-expansion for $I$ then $(\Phi, \Psi)$, where $\Phi = MA(I) \setminus T$ and $\Psi = MA(I) \cap T$, is $\mathcal{S}$-perfect and $T = T_I(\Phi, \Psi)$.*
2. *If a partition $(\Phi, \Psi)$ of $MA(I)$ is $\mathcal{S}$-perfect for $I$, then $T_I(\Phi, \Psi)$ is an $\mathcal{S}$-expansion for $I$.*

In order to prove Theorem 9.50, we need several auxiliary results.

**Lemma 9.51** *Let $(\Phi, \Psi)$ be a partition of $MA(I)$. Let $\mathcal{M}_1 = \langle M_1, R_1, V_1 \rangle$ and $\mathcal{M}_2 = \langle M_2, R_2, V_2 \rangle$ be Kripke models such that $M_1 \cap M_2 = \emptyset$, $\mathcal{M}_1 \models \{\neg L\varphi : \varphi \in \Phi\}$, and $\mathcal{M}_2 \models \Psi$. Then, for each $I$-formula $\varphi$ and for each $a \in M_1$,*

$$\langle \mathcal{M}_1, a \rangle \models \varphi \text{ if and only if } \langle \mathcal{M}_1 \odot \mathcal{M}_2, a \rangle \models \varphi.$$

Proof: We proceed by induction on the complexity of an $I$-formula. If $\varphi \in \mathcal{L}$, then the claim follows directly from the definition of concatenation of two Kripke models. Consider now an $I$-formula $\varphi$ and assume that the claim holds for all shorter $I$-formulas. The case when $\varphi$ is constructed from simpler formulas by means of a boolean connective is straightforward. So, assume that $\varphi = L\eta$. Since $\varphi$ is an $I$-formula, it follows that $\eta \in MA(I)$. Moreover, $\eta$ is an $I$-formula, too.

Suppose first that $\langle \mathcal{M}_1, a \rangle \models L\eta$. Then, for every $b \in M_1$ such that $(a, b) \in R_1$, $\langle \mathcal{M}_1, b \rangle \models \eta$. Let $R$ denote the accessibility relation of $\mathcal{M}_1 \odot \mathcal{M}_2$. By the induction hypothesis and by the definition of $R$, it follows that for every $b \in M_1$ such that $(a, b) \in R$,

$$\langle \mathcal{M}_1 \odot \mathcal{M}_2, b \rangle \models \eta.$$

Next, observe that since $\langle \mathcal{M}_1, a \rangle \models L\eta$ and $\mathcal{M}_1 \models \{\neg L\varphi : \varphi \in \Phi\}$, $\eta \in \Psi$. Since $\mathcal{M}_2 \models \Psi$, by the definition of concatenation it follows that for every $b \in M_2$,

$$\langle \mathcal{M}_1 \odot \mathcal{M}_2, b \rangle \models \eta.$$

Thus, $\eta$ holds in every world $b \in M_1 \cup M_2$ such that $(a, b) \in R$. Consequently, $\langle \mathcal{M}_1 \odot \mathcal{M}_2, a \rangle \models L\eta$.

Next, assume that $\langle \mathcal{M}_1, a\rangle \not\models L\eta$. Then, there is a world $b \in M_1$ such that $(a,b) \in R_1$ and $\langle \mathcal{M}_1, b\rangle \not\models \eta$. By the induction hypothesis,

$$\langle \mathcal{M}_1 \odot \mathcal{M}_2, b\rangle \not\models \eta.$$

By the definition of $R$, it follows that $(a,b) \in R$. Thus, $\langle \mathcal{M}_1 \odot \mathcal{M}_2, a\rangle \not\models L\eta$. ■

**Lemma 9.52** *Let $(\Phi, \Psi)$ be an* **S5***-perfect for $I$ partition of $MA(I)$. Let $\mathcal{M}_1 = \langle M_1, R_1, V_1\rangle$ be a Kripke model and let $\mathcal{M}_i = \langle M_i, V_i\rangle$, $i = 2, 3$, be* **S5***-models. Assume that:*

1. $M_1 \cap (M_2 \cup M_3) = \emptyset$.
2. $\mathcal{M}_1 \odot \mathcal{M}_2 \models I \cup \{\neg L\varphi : \varphi \in \Phi\}$.
3. $\mathcal{M}_3 \models \{\neg L\varphi : \varphi \in \Phi\} \cup \{L\psi : \psi \in \Psi\}$.

*Then, for each $I$-formula $\psi$ and for each $a \in M_1$,*

$$\langle \mathcal{M}_1 \odot \mathcal{M}_2, a\rangle \models \psi \quad \textit{if and only if} \quad \langle \mathcal{M}_1 \odot \mathcal{M}_3, a\rangle \models \psi.$$

Proof: We proceed by induction on the complexity of $\psi$. The only non-trivial part of the proof is in the induction step when $\psi$ is of the form $L\eta$. We describe the argument for this case only. Other details are omitted.

By $R_2$ and $R_3$ let us denote the accessibility relations of $\mathcal{M}_1 \odot \mathcal{M}_2$ and $\mathcal{M}_1 \odot \mathcal{M}_3$, respectively.

Since $\psi$ is an $I$-formula, it follows that $\eta \in MA(I)$. In addition, $\eta$ is an $I$-formula.

Assume first that $\langle \mathcal{M}_1 \odot \mathcal{M}_2, a\rangle \models L\eta$. Then, for each $c \in M_1 \cup M_2$ such that $(a,c) \in R_2$, we have $\langle \mathcal{M}_1 \odot \mathcal{M}_2, c\rangle \models \eta$. If $c \in M_1$ then, by the induction hypothesis, we have $\langle \mathcal{M}_1 \odot \mathcal{M}_3, c\rangle \models \eta$. Consider now $c \in M_3$. Since $\mathcal{M}_1 \odot \mathcal{M}_2 \models \{\neg L\varphi : \varphi \in \Phi\}$, we have $\eta \in \Psi$. Hence $\mathcal{M}_3 \models L\eta$. Consequently, by the definition of concatenation of models, $\langle \mathcal{M}_1 \odot \mathcal{M}_3, c\rangle \models \eta$. Thus, $\langle \mathcal{M}_1 \odot \mathcal{M}_3, a\rangle \models L\eta$.

Conversely, assume $\langle \mathcal{M}_1 \odot \mathcal{M}_3, a\rangle \models L\eta$. Then $\mathcal{M}_3 \models L\eta$, which implies that $\eta \in \Psi$. Since $\mathcal{M}_1 \odot \mathcal{M}_2 \models I \cup \{\neg L\varphi : \varphi \in \Phi\}$, $\mathcal{M}_2 \models I \cup \{\neg L\varphi : \varphi \in \Phi\}$. Since $(\Phi, \Psi)$ is **S5**-perfect for $I$, $\mathcal{M}_2 \models \eta$. Thus, for each $c \in M_2$, $\langle \mathcal{M}_1 \odot \mathcal{M}_2, c\rangle \models \eta$. Next, assume that $c \in M_1$ and $(a,c) \in R_2$. Then $(a,c) \in R_3$. Hence, $\langle \mathcal{M}_1 \odot \mathcal{M}_3, c\rangle \models \eta$. By the induction hypothesis, $\langle \mathcal{M}_1 \odot \mathcal{M}_2, c\rangle \models \eta$. Thus, $\langle \mathcal{M}_1 \odot \mathcal{M}_2, a\rangle \models L\eta$. ■

**Lemma 9.53** *Let $\mathcal{S} \subseteq$ **S4f** and let $T$ be an $\mathcal{S}$-expansion for $I$. Then the partition $(MA(I) \setminus T, MA(I) \cap T)$ is **S5**-perfect for $I$.*

Proof: Denote $\Phi = MA(I) \setminus T$ and $\Psi = MA(I) \cap T$. For each modal logic $\mathcal{S} \subseteq$ **S4f** an $\mathcal{S}$-expansion for $I$ is an **S4f**-expansion for $I$. Thus, it suffices to prove the lemma for $\mathcal{S} =$ **S4f**. By Theorem 9.41, it follows that $T = T_I(\Phi', \Psi')$ for some introspection-consistent with $I$ partition $(\Phi', \Psi')$ of $MA(I)$. By Theorem 9.39(4), it follows that $\Phi' = \Phi$ and $\Psi' = \Psi$. Hence, $(\Phi, \Psi)$ is introspection-consistent with $I$. To prove the lemma, it is then enough to show that for each $\psi \in \Psi$, $I \cup \{\neg L\varphi : \varphi \in \Phi\} \vdash_{\mathbf{S5}} \psi$.

To this end, let us consider an **S5**-model $\mathcal{M}_1 = \langle M_1, V_1 \rangle$ such that $\mathcal{M}_1 \models I \cup \{\neg L\varphi : \varphi \in \Phi\}$. Since $T$ is stable and consistent, there is an **S5**-model $\mathcal{M}_2 = \langle M_2, V_2 \rangle$ such that $Th(\mathcal{M}_2) = T$ (Theorem 8.12). By Lemma 9.51 and since $\mathcal{M}_2 \models I$, $\mathcal{M}_1 \odot \mathcal{M}_2 \models I$. Clearly $\mathcal{M}_2 \models \{\neg L\varphi : \varphi \notin T\}$, and therefore, since every point of $M_2$ is accessible from every point of $M_1$, $\mathcal{M}_1 \odot \mathcal{M}_2 \models \{\neg L\varphi : \varphi \notin T\}$. Hence, we have that $\mathcal{M}_1 \odot \mathcal{M}_2$ is an **S4f**-model $\mathcal{M}_1 \odot \mathcal{M}_2 \models I \cup \{\neg L\varphi : \varphi \notin T\}$. Since $T$ is an **S4f**-expansion for $I$, $\mathcal{M}_1 \odot \mathcal{M}_2 \models T$. Since $\psi \in \Psi$, $\psi \in T$ and $\mathcal{M}_1 \odot \mathcal{M}_2 \models \psi$. Since $\psi$ is an $I$-formula, it follows from Lemma 9.51 that $\mathcal{M}_1 \models \psi$. Since $\mathcal{M}_1$ is an arbitrary **S5**-model such that $\mathcal{M}_1 \models I \cup \{\neg L\varphi : \varphi \in \Phi\}$, $I \cup \{\neg L\varphi : \varphi \in \Phi\} \vdash_{\mathbf{S5}} \psi$ follows. ■

**Lemma 9.54** *Let $\mathcal{S} \subseteq$ **S4f** be a normal modal logic characterized by a cluster-closed class of models. Let $T$ be an $\mathcal{S}$-expansion for $I$. Then $(MA(I) \setminus T, MA(I) \cap T)$ is $\mathcal{S}$-perfect for $I$.*

Proof: Let $\mathcal{C}$ be a cluster-closed class of models that characterizes $\mathcal{S}$. Define $\Phi = MA(I) \setminus T$ and $\Psi = MA(I) \cap T$. By Lemma 9.53, $(\Phi, \Psi)$ is **S5**-perfect for $I$. In particular, $(\Phi, \Psi)$ is introspection-consistent with $I$. Thus, to prove the lemma, it suffices to show that for each model $\mathcal{M} = \langle M, R, V \rangle \in \mathcal{C}$, if $\mathcal{M} \models I \cup \{\neg L\varphi : \varphi \in \Phi\}$ then $\mathcal{M} \models \Psi$.

Consider then a model $\mathcal{M} = \langle M, R, V \rangle \in \mathcal{C}$ such that $\mathcal{M} \models I \cup \{\neg L\varphi : \varphi \in \Phi\}$. Since theory $T$ is stable, there is an **S5**-model $\mathcal{K} = \langle K, U \rangle$ such that $Th(\mathcal{K}) = T$. Without loss of generality we can assume $M \cap K = \emptyset$. By the definition of a cluster-closed class of models, there are two possibilities:

**(a)** $\mathcal{M} \odot \mathcal{K} \in \mathcal{C}$, or

**(b)** $\mathcal{M} = \mathcal{M}_1 \odot \mathcal{M}_2$, where $\mathcal{M}_2$ is an **S5**-model, and $\mathcal{M}_1 \odot \mathcal{K} \in \mathcal{C}$.

Case (a). Since $\mathcal{K} \models I$, it follows that $\langle \mathcal{M} \odot \mathcal{K}, a \rangle \models I$, for every $a \in K$. From Lemma 9.51 it follows that $\langle \mathcal{M} \odot \mathcal{K}, a \rangle \models I$, for every $a \in M$. Thus, $\mathcal{M} \odot \mathcal{K} \models I$. In addition, if $\varphi \notin T$ then $\mathcal{K} \models \neg L\varphi$ and, consequently, $\mathcal{M} \odot \mathcal{K} \models \neg L\varphi$.

Hence, $\mathcal{M} \odot \mathcal{K} \models I \cup \{\neg L\varphi : \varphi \notin T\}$. Since $T$ is an $\mathcal{S}$-expansion for $I$ and $\mathcal{M} \odot \mathcal{K} \in \mathcal{C}$, $\mathcal{M} \odot \mathcal{K} \models T$. Since $\Psi \subseteq T$, $\mathcal{M} \odot \mathcal{K} \models \Psi$. By Lemma 9.51, $\mathcal{M} \models \Psi$.

Case (b). Since $Th(\mathcal{K}) = T$, $\mathcal{K} \models \{\neg L\varphi : \varphi \in \Phi\} \cup \{L\psi : \psi \in \Psi\}$. By Lemma 9.52, $\langle \mathcal{M}_1 \odot \mathcal{K}, a\rangle \models I$, for every $a \in M_1$. Similarly as in (a) we argue that $\mathcal{M}_1 \odot \mathcal{K} \models \{\neg L\varphi : \varphi \notin T\}$ and obtain that $\mathcal{M}_1 \odot \mathcal{K} \models \Psi$. Since $\Psi$ consists of $I$-formulas, applying Lemma 9.52 again, we obtain $\langle \mathcal{M}, a\rangle \models \Psi$, for every $a \in M_1$.

We also have that $\mathcal{M} \models I \cup \{\neg L\varphi : \varphi \in \Phi\}$. Consequently, $\mathcal{M}_2 \models I \cup \{\neg L\varphi : \varphi \in \Phi\}$. By Lemma 9.53, $(\Phi, \Psi)$ is **S5**-perfect for $I$. Since $\mathcal{M}_2$ is an **S5**-model, it follows that $\mathcal{M}_2 \models \Psi$. Consequently, $\langle \mathcal{M}, a\rangle \models \Psi$, for every $a \in M_2$. Thus, $\mathcal{M} \models \Psi$.

Summarizing, in both cases $\mathcal{M} \models \Psi$ and the assertion follows. ■

**Proof of Theorem 9.50**: Part (2) of the assertion follows from Theorem 9.48. To prove part (1) of the theorem, consider an $\mathcal{S}$-expansion $T$ for $I$. Put $\Phi = MA(I) \setminus T$ and $\Psi = MA(I) \cap T$. By Theorem 9.41, $T = T_I(\Phi, \Psi)$ and, by Lemma 9.54, $(\Phi, \Psi)$ is $\mathcal{S}$-perfect for $I$. ■

**Corollary 9.55** *Let $\mathcal{S}$ be a modal logic contained in* **S4f** *and characterized by a cluster-closed class of models. Let $I$ be consistent with $\mathcal{S}$. Let $S \subseteq \mathcal{L}$. Then, $ST(S)$ is an $\mathcal{S}$-expansion for $I$ if and only if $I \subseteq ST(S)$ and $S \subseteq Cn_{\mathcal{S}}(I \cup \{\neg L\varphi : \varphi \in \Phi\})$, where $\Phi = MA(I) \setminus ST(S)$.*

Proof: If $S \subseteq Cn_{\mathcal{S}}(I \cup \{\neg L\varphi : \varphi \in \Phi\})$, then $S \subseteq Cn_{\mathcal{S}}(I \cup \{\neg L\varphi : \varphi \notin ST(S)\})$. Since $I$ is consistent with $\mathcal{S}$, $S$ is consistent. By Corollary 9.8, $ST(S)$ is an $\mathcal{S}$-expansion for $I$.

Conversely, assume that $ST(S)$ is an $\mathcal{S}$-expansion for $I$. Then, $I \subseteq ST(S)$. In addition, by Lemma 9.54, $(\Phi, \Psi)$, where $\Psi = MA(I) \setminus \Phi$, is $\mathcal{S}$-perfect for $I$. Consequently,

$$I \cup \{\neg L\varphi : \varphi \in \Phi\} \vdash_{\mathcal{S}} \{L\psi : \psi \in \Psi\} \cup \Psi.$$

Since $(\Phi, \Psi)$ is introspection-consistent with $I$, Theorem 9.39 implies that

$$Cn(S) = Cn(I \cup \{\neg L\varphi : \varphi \in \Phi\} \cup \{L\psi : \psi \in \Psi\} \cup \Psi) \cap \mathcal{L}.$$

Thus, the inclusion $S \subseteq Cn_{\mathcal{S}}(I \cup \{\neg L\varphi : \varphi \in \Phi\})$ follows. ■

**Remark 9.56** The assumptions of Theorem 9.50 and Corollary 9.55 hold for each of the modal logics **K**, **T**, **S4**, and **S4f**. As we mentioned in Remark 9.33, the notion of a cluster-closed class of Kripke models can be extended to the case of Kripke structures. It is then possible to extend Theorem 9.50 and Corollary 9.55 to cover the case of modal logics (including subnormal modal logics) that are characterized by a a cluster-closed class of Kripke structures. The case of logic **N** will be dealt with in the next chapter. ▲

The assertion of Theorem 9.50 fails for any modal logic properly containing **S4f**.

**Lemma 9.57** *Let $\mathcal{S} = S_{\langle n,\omega\rangle} \cap S_{\langle \omega,m\rangle}$, for some positive integers m and n. Let I be a theory consisting of the following $n+1$ formulas:*

$$\begin{array}{ll} \varphi_1 = & Lp_1 \supset p_2 \\ \varphi_2 = & L(p_1 \supset p_2) \supset p_3 \\ \dots & \dots \\ \varphi_{n+1} = & L(p_1 \wedge \ldots \wedge p_n \supset p_{n+1}) \supset p_1, \end{array}$$

*where $p_1, \ldots, p_{n+1}$ are pairwise different propositional variables. Then the theory $ST(\{p_1, \ldots, p_{n+1}\})$ is an $\mathcal{S}$-expansion of $I$.*

Proof: Denote $ST(\{p_1, \ldots, p_{n+1}\})$ by $T$. Clearly, $I \subseteq T$. Thus, by Corollary 9.8, it is sufficient to prove that

$$\{p_1, \ldots, p_{n+1}\} \subseteq Cn_{\mathcal{S}}(I \cup \{\neg L\psi : \psi \notin T\}).$$

Consider an **S4f**-model $\mathcal{M} = \langle M_1, M_2, V\rangle$ with index $\langle n,\omega\rangle$ or $\langle \omega, m\rangle$, such that

$$\mathcal{M} \models I \cup \{\neg L\psi : \psi \notin T\}.$$

We will show that for every $i$, $1 \leq i \leq n+1$,

$$\mathcal{M} \models p_i.$$

Assume that $\mathcal{M}$ has index $\langle \omega, m\rangle$. Consider $m+1$ distinct propositional variables $q_1, \ldots, q_{m+1}$, such that for all $i, j$, $p_i \neq q_j$. Then, for each $j$, $1 \leq j \leq m+1$,

$$q_1 \wedge \ldots \wedge q_{j-1} \supset q_j \notin T.$$

Consequently, for each $j$, $1 \leq j \leq m+1$,

$$\mathcal{M} \models \neg L(q_1 \wedge \ldots \wedge q_{j-1} \supset q_j).$$

In particular, there is a world $a_j \in M_2$ such that

$$\langle \mathcal{M}, a_j\rangle \not\models q_1 \wedge \ldots \wedge q_{j-1} \supset q_j.$$

Thus, in this world $a_j$, all $q_1, \ldots, q_{j-1}$ are true and $q_j$ is false. Consequently, all worlds $a_j$, $1 \leq j \leq m+1$, are distinct, a contradiction.

So, $\mathcal{M}$ has index $\langle n, \omega \rangle$. First consider a world $a \in M_2$ and assume that for some $i$, $1 \leq i \leq n+1$, $\langle \mathcal{M}, a \rangle \not\models p_i$. Then, $\mathcal{M} \models \neg L p_i$ and $\mathcal{M} \models L \neg L p_i$. On the other hand, $\neg L p_i \notin T$. Thus, $\mathcal{M} \models \neg L \neg L p_i$, a contradiction. Thus, for every $i$, $1 \leq i \leq n+1$, $p_i$ is true in every world $a \in M_2$.

Suppose now that for some world $a \in M_1$, and for some $i$, $1 \leq i \leq n+1$, $\langle \mathcal{M}, a \rangle \not\models p_i$. If $i > 1$, then, since $\langle \mathcal{M}, a \rangle \models \varphi_{i-1}$, it follows that there is a world $b \in M_1 \cup M_2$ such that $\langle \mathcal{M}, b \rangle \not\models p_1 \wedge \ldots \wedge p_{i-2} \supset p_{i-1}$. We proved above that if $b \in M_2$ then $\langle \mathcal{M}, b \rangle \models p_{i-1}$. Hence, $b \in M_1$ and, $\langle \mathcal{M}, b \rangle \not\models p_{i-1}$. Thus, it follows that there is a world $b_1 \in M_1$ such that $\langle \mathcal{M}, b_1 \rangle \not\models p_1$. Now, reasoning in the same way and using the fact that $\mathcal{M} \models \varphi_{n+1}$, we show that there is a world $b_{n+1} \in M_1$ such that $\langle \mathcal{M}, b_{n+1} \rangle \models p_i$, $1 \leq i \leq n$, and $\langle \mathcal{M}, b_{n+1} \rangle \not\models p_{n+1}$. Applying this method repeatedly, we show that for each $j$, $1 \leq j \leq n+1$, there is a world $b_j \in M_1$ such that $\langle \mathcal{M}, b_j \rangle \models p_i$, $1 \leq i \leq j-1$ and $\langle \mathcal{M}, b_j \rangle \not\models p_j$. All these worlds $b_j$ are distinct, a contradiction. Thus, in every world of $\mathcal{M}$, all variables $p_i$ are true. ■

**Theorem 9.58** *Let $\mathcal{S}$ be a modal logic contained in* **S5** *and properly containing* **S4f**. *Then there is a theory $I$ having an $\mathcal{S}$-expansion $T$ such that for every partition $(\Phi, \Psi)$ of $MA(I)$, which is $\mathcal{S}$-perfect for $I$, $T \neq T_I(\Phi, \Psi)$.*

Proof: Let $\mathcal{S}$ be a logic satisfying the assumptions of the theorem. By Theorem 7.60, for some indices $\langle n, \omega \rangle$ and $\langle \omega, m \rangle$, the logic $S_{\langle n, \omega \rangle} \cap S_{\langle \omega, m \rangle}$ is contained in $\mathcal{S}$. Consider the theory $I$ defined in Lemma 9.57. By Lemma 9.57 and Theorem 9.6, $T = ST(\{p_1, \ldots, p_{n+1}\})$ is an $\mathcal{S}$-expansion of $I$. Assume that $T = T_I(\Phi, \Psi)$, where $(\Phi, \Psi)$ is a partition of $MA(I)$ $\mathcal{S}$-perfect for $I$. Then, by Theorem 9.39(4), $\Phi = \emptyset$ and $\Psi = MA(I)$. It is straightforward to show that $(\Phi, \Psi)$ is not **S5**-perfect for $I$. Consequently, $(\Phi, \Psi)$ is not $\mathcal{S}$-perfect for $I$, a contradiction. ■

## 9.6 Comments and remarks

The first nonmonotonic modal system based on the fixpoint equation (9.3) was introduced by McDermott and Doyle [MD80]. In this system, the role of $\mathcal{S}$ was played by propositional logic. McDermott and Doyle discovered some counterintuitive properties of their logic. Subsequently, McDermott [McD82], in an effort to eliminate deficiencies of the logic from [MD80], introduced the notion of an $\mathcal{S}$-expansion for an arbitrary modal logic $\mathcal{S}$. He investigated in detail the case when $\mathcal{S}$ = **S5**. In particular, he proved that nonmonotonic modal logic **S5** coincides with the monotonic **S5** (Corollary 9.11). He also mentioned logics **T** and **S4** as possible bases of nonmonotonic modal systems. McDermott also obtained several general results. For example, he proved

Theorem 9.6 as well as completeness and soundness results for modal systems in which provability from an arbitrary theory is considered, and necessitation is applied to every formula that can be derived. The nonmonotonic logics based on the notion of an $\mathcal{S}$-expansion are called in this book *McDermott-Doyle family of nonmonotonic logics.*

The section on minimal model semantics is based on the work of Schwarz [Sch92]. An equivalent semantics for the special case of the nonmonotonic logic **kd45** (autoepistemic logic) was found earlier by Moore [Moo84] (Corollary 9.28(3)). Levesque [Lev90] obtained the same characterization of **kd45**-expansions in the terminology of his "only-knowing" approach.

The notion of consistency with respect to introspection (under a different name) is due to Schwarz [Shv90], as are the results on $\mathcal{S}$-expansions presented in Sections 9.4 and 9.5. The notion of consistency with respect to introspection was found independently by Niemelä [Nie88]. He also found Theorem 9.40, which we will later use to establish correctness of algorithms for verifying membership of formulas in $\mathcal{S}$-expansions.

# 10. Nonmonotonic logic of pure necessitation and autoepistemic logics

In this chapter we study several important nonmonotonic modal logics from the family of logics introduced by McDermott and Doyle, and discussed in Chapter 9.

First, we study the nonmonotonic modal logic **N**. Logic **N** plays a special role among modal logics. It is the weakest of them. Consequently, it imposes the strongest discipline in the process of reconstructing a putative theory $T$ from the initial assumptions and the result of the negative introspection with respect to $T$. We extend the results from Chapter 9, given there for normal modal logics, to the case of logic **N**.

Next, we consider nonmonotonic modal logics **kd45** and **Sw5**. As we will see, these logics can be given interpretations as the nonmonotonic logic of belief, and the nonmonotonic logic of reflexive knowledge, respectively.

Results of Section 9.5, do not hold for logics **kd45** and **Sw5**. We find counterparts to Theorem 9.50 for logics **kd45** and **Sw5** in Section 10.2.

Nonmonotonic logics **kd45** and **Sw5** are maximal with respect to several natural properties; in a sense they are on the opposite side from **N** in the spectrum of nonmonotonic modal logics of McDermott and Doyle.

In Section 10.3 we investigate autoepistemic logic introduced by Moore [Moo85]. This logic turns out to be closely related to nonmonotonic **kd45**, but can be defined without reference to any particular modal logic. Similar results are proved for reflexive autoepistemic logic of Schwarz [Sch91a] and nonmonotonic modal logic **Sw5**.

## 10.1 Nonmonotonic logic N

A natural question to ask is what is the relationship between different modal nonmonotonic logics $\mathcal{S}$. We have already obtained some results providing insights into this problem. For example, we know that if two modal logics $\mathcal{S}$ and $\mathcal{T}$ are contained in **S5** and $\mathcal{S} \subseteq \mathcal{T}$, then for every theory $I \subseteq \mathcal{L}$, if a theory $T \subseteq \mathcal{L}_L$ is an $\mathcal{S}$-expansion for $I$ then $T$ is a $\mathcal{T}$-expansion for $I$ (Theorem 9.6). The following related question seems to be of fundamental importance:

> Given two logics $\mathcal{T}$ and $\mathcal{S}$, can nonmonotonic logic $\mathcal{S}$ be faithfully expressed within nonmonotonic logic $\mathcal{T}$?

We need to make this question more precise. We say that a nonmonotonic logic $\mathcal{S}$ is faithfully expressed within nonmonotonic logic $\mathcal{T}$ if there is an operator $emb: \mathcal{P}(\mathcal{L}_L) \rightarrow \mathcal{P}(\mathcal{L}_L)$ such that for every theory $I \subseteq \mathcal{L}_L$, a theory $T$ is an $\mathcal{S}$-expansion for $I$ if and only if $T$ is a $\mathcal{T}$-expansion for $emb(I)$.

An important property of the logic **N** is that every nonmonotonic modal logic $\mathcal{S}$ can be embedded into the nonmonotonic logic **N**. For a modal logic $\mathcal{S}$, define an operator $emb_{\mathcal{S}}: \mathcal{P}(\mathcal{L}_L) \rightarrow \mathcal{P}(\mathcal{L}_L)$ by

$$emb_{\mathcal{S}}(I) = I \cup Cn_{\mathcal{S}}(\emptyset). \tag{10.1}$$

In other words, $emb_{\mathcal{S}}(I)$ consists of $I$ and all theorems of $\mathcal{S}$.

**Theorem 10.1** *For every modal logic $\mathcal{S}$ and for every theory $I \subseteq \mathcal{L}_L$, a theory $T \subseteq \mathcal{L}_L$ is an $\mathcal{S}$-expansion for $I$ if and only if $T$ is an* **N***-expansion for $emb_{\mathcal{S}}(I)$.*

Proof: The result is a straightforward consequence of the following property of the provability operators $Cn_{\mathcal{S}}$ and $Cn_{\mathbf{N}}$: for every theory $J \subseteq \mathcal{L}_L$

$$Cn_{\mathcal{S}}(J) = Cn_{\mathbf{N}}(J \cup Cn_{\mathcal{S}}(\emptyset)) = Cn_{\mathbf{N}}(emb_{\mathcal{S}}(J)).$$

Since for every theory $T \subseteq \mathcal{L}_L$, $emb_{\mathcal{S}}(I \cup \{\neg L\varphi : \varphi \notin T\}) = emb_{\mathcal{S}}(I) \cup \{\neg L\varphi : \varphi \notin T\}$, it follows that

$$Cn_{\mathcal{S}}(I \cup \{\neg L\varphi : \varphi \notin T\}) = Cn_{\mathbf{N}}(emb_{\mathcal{S}}(I) \cup \{\neg L\varphi : \varphi \notin T\}). \qquad \blacksquare$$

**Remark 10.2** Clearly, the assertion of Theorem 10.1 remains true if in the definition of $emb_{\mathcal{S}}$ the set $Cn_{\mathcal{S}}(\emptyset)$ is replaced by any set of formulas $F$ such that $Cn_{\mathbf{N}}(F) = Cn_{\mathcal{S}}(\emptyset)$. For example, if $\mathcal{S}$ is axiomatizable by a scheme $C$, one might take for $F$ the set of all instances of $C$. ▲

Theorem 10.1 implies that every nonmonotonic modal logic can be faithfully expressed in the nonmonotonic modal logic **N**. Let us note, however, that the theory $emb_{\mathcal{S}}(I)$, produced by the embedding, is infinite regardless of the size of $I$.

Similarly, the same argument shows that if $\mathcal{T} \subseteq \mathcal{S}$ then nonmonotonic logic $\mathcal{S}$ can be faithfully expressed in nonmonotonic modal logic $\mathcal{T}$ by the embedding $emb_{\mathcal{S}}$.

We move on to the main topic of this section, that is, the problem of characterizing **N**-expansions. Results proved earlier for normal logics and providing a minimal model characterization of $\mathcal{S}$-expansions (Section 9.3), as well as the characterization of $\mathcal{S}$-expansions as theories generated by $\mathcal{S}$-perfect partitions (Section 9.5) can be extended to the case of the (subnormal) logic **N**. In fact, due to similarities between the Kripke-style semantics of **N**-structures and regular Kripke semantics for normal logics, only small modifications in the proofs given for normal modal logics are needed.

**Definition 10.3** An **S5**-model $\mathcal{M}$ is **N**-*intended for* $I$ if:

1. $\mathcal{M} \models I$.
2. For every **N**-structure $\mathcal{N}$ such that $\mathcal{N} \models I \cup \{\neg L\varphi : \varphi \notin Th(\mathcal{M})\}$, $Th(\mathcal{N}) = Th(\mathcal{M})$. ▲

In the next theorem, we show that $\mathcal{N}$-expansions can be characterized as precisely the theories of **N**-intended models.

**Theorem 10.4** *Let* $\mathcal{M}$ *be an* **S5**-*model. Theory* $Th(\mathcal{M})$ *is an* **N**-*expansion for a theory* $I$ *if and only if* $\mathcal{M}$ *is* **N**-*intended for* $I$.

Proof: The proof is obtained by a straightforward modification of the proof of Theorem 9.13.

Define $T = Th(\mathcal{M})$. Assume that $\mathcal{M}$ is **N**-intended for $I$. By Theorem 8.10, $T$ is stable. Hence, $T \supseteq \{\neg L\varphi : \varphi \notin T\}$. In addition, since $\mathcal{M}$ is **N**-intended for $I$, $T \supseteq I$. Finally, since $\mathbf{N} \subseteq \mathbf{S5}$, $T$ is closed under provability in **N**. Consequently,

$$T \supseteq Cn_{\mathbf{N}}(I \cup \{\neg L\varphi : \varphi \notin T\}).$$

Let now $\mathcal{N}$ be any **N**-structure such that $\mathcal{N} \models I \cup \{\neg L\varphi : \varphi \notin T\}$. Then, since $\mathcal{M}$ is **N**-intended for $I$, $\mathcal{N} \models T$. By the completeness theorem for the logic **N** (Corollary 7.39(1)),

$$T \subseteq Cn_{\mathbf{N}}(I \cup \{\neg L\varphi : \varphi \notin T\}).$$

Consequently, $T$ is an **N**-expansion for $I$.

Conversely, let us assume that $T = Th(\mathcal{M})$ is an **N**-expansion for $I$. Clearly, $\mathcal{M} \models I$. Consider an **N**-structure $\mathcal{N}$ such that $\mathcal{N} \models I \cup \{\neg L\varphi : \varphi \notin T\}$. Since

$$T = Cn_{\mathbf{N}}(I \cup \{\neg L\varphi : \varphi \notin T\}),$$

it follows that $T \subseteq Th(\mathcal{N})$. On the other hand, since $\mathcal{N} \models \{\neg L\varphi : \varphi \notin T\}$, $Th(\mathcal{N}) \subseteq T$. Thus, $Th(\mathcal{N}) = T = Th(\mathcal{M})$ and $\mathcal{M}$ is **N**-intended for $I$. ■

We will next introduce the notion of an **N**-*minimal* structure for $I$. To this end, we modify Definition 9.15. But first, we need more terminology.

**Definition 10.5** Let $\mathcal{M}_1 = \langle M_1, \{Q_\varphi\}_{\varphi \in \mathcal{L}_L}, V_1 \rangle$ and $\mathcal{M}_2 = \langle M_2, R_2, V_2 \rangle$ be an **N**-structure and a Kripke model, respectively, with disjoint sets of worlds. The *concatenation* $\mathcal{M}_1 \odot \mathcal{M}_2$ of $\mathcal{M}_1$ and $\mathcal{M}_2$ is the **N**-structure $\langle M, \{R_\varphi\}_{\varphi \in \mathcal{L}_L}, V \rangle$, where $M = M_1 \cup M_2$, $R_\varphi = Q_\varphi \cup (M_1 \times M_2) \cup R_2$ and

$$V(m) = \begin{cases} V_1(m) & \text{if } m \in M_1, \\ V_2(m) & \text{if } m \in M_2. \end{cases}$$

▲

**Definition 10.6** Let $\mathcal{M} = \langle M, V_1 \rangle$ be an **S5**-model and let $\mathcal{N}$ be an **N**-structure with the set of worlds $N$ and the valuation function $V_2$. We say that $\mathcal{N} \odot \mathcal{M}$ is *preferred over* $\mathcal{M}$ ($\mathcal{N} \odot \mathcal{M} \sqsubset \mathcal{M}$, in symbols) if:

> There is a world $a \in N$ and there is a finite set $P$ of propositional variables such that for every $b \in M$, $V_2(a)|_P \neq V_1(b)|_P$. ▲

We have the following equivalent characterization of the relation $\sqsubset$. It can be obtained in a similar fashion as Theorem 9.16. We omit the details.

**Theorem 10.7** *Let $\mathcal{M} = \langle M, V_1 \rangle$ be an* **S5***-model and let $\mathcal{N}$ be an* **N***-structure with the set of worlds $N$ and the valuation function $V_2$. Then $\mathcal{N} \odot \mathcal{M} \sqsubset \mathcal{M}$ if and only if the following condition is met:*

> *There is $a \in N$ and a formula $\varphi \in \mathcal{L}$ such that $\langle \mathcal{N} \odot \mathcal{M}, a \rangle \not\models \varphi$ and $\langle \mathcal{M}, b \rangle \models \varphi$, for every $b \in M$.* ▲

Finally, we are ready to introduce the notion of an **N**-minimal model.

**Definition 10.8** An **S5**-model $\mathcal{M}$ is **N**-*minimal* for a theory $I \subseteq \mathcal{L}_L$ if $\mathcal{M} \models I$ and for every **N**-structure $\mathcal{N}$ such that $\mathcal{N} \odot \mathcal{M} \models I$, $\mathcal{N} \odot \mathcal{M} \not\sqsubseteq \mathcal{M}$. ▲

As in the case of normal modal logics considered in the previous chapter, it turns out that **N**-expansions correspond to **N**-minimal **S5**-models.

**Theorem 10.9** *Let $\mathcal{M}$ be an* **S5**-*model. The theory $Th(\mathcal{M})$ is an* **N**-*expansion for a theory $I$ if and only if $\mathcal{M}$ is* **N**-*minimal for $I$.*

In the proof, we will need two auxiliary results analogous to Lemmas 9.20 and 9.21.

**Lemma 10.10** *Let $\mathcal{N}$ and $\mathcal{M}$ be an* **N**-*structure and a Kripke model, respectively, such that $\mathcal{N} \models \{\neg L\varphi : \varphi \notin Th(\mathcal{M})\}$. Then, for every world $a$ of $\mathcal{N}$ and for every formula $\varphi \in \mathcal{L}_L$,*

$$\langle \mathcal{N} \odot \mathcal{M}, a\rangle \models \varphi \quad \textit{if and only if} \quad \langle \mathcal{N}, a\rangle \models \varphi.$$

Proof: The proof is essentially the same as that of Lemma 9.20. Let $\mathcal{N} = \langle N, \{Q_\psi\}_{\psi \in \mathcal{L}_L}, V\rangle$ and $\mathcal{M} = \langle M, R, W\rangle$. We proceed by induction on the complexity of $\varphi$. The only nontrivial part is in the induction step when $\varphi = L\psi$. We consider this case only and leave other details to the reader.

Let us assume first that $\langle \mathcal{N} \odot \mathcal{M}, a\rangle \models L\psi$. Then, in particular, for every $b \in N$ such that $(a,b) \in Q_\psi$, $\langle \mathcal{N} \odot \mathcal{M}, b\rangle \models \psi$. By the induction hypothesis, for every $b \in N$ such that $(a,b) \in Q_\psi$, $\langle \mathcal{N}, b\rangle \models \psi$ which means that $\langle \mathcal{N}, a\rangle \models L\psi$.

Conversely, assume that $\langle \mathcal{N}, a\rangle \models L\psi$. Thus, for every $b \in N$ such that $(a,b) \in Q_\psi$, $\langle \mathcal{N}, b\rangle \models \psi$. By the induction hypothesis, $\langle \mathcal{N} \odot \mathcal{M}, b\rangle \models \psi$, for every $b \in N$ such that $(a,b) \in Q_\psi$. In addition, since for each $\eta \notin Th(\mathcal{M})$, $\mathcal{N} \models \neg L\eta$, it follows that $\psi \in Th(\mathcal{M})$. Hence, $\langle \mathcal{N} \odot \mathcal{M}, b\rangle \models \psi$, for every $b \in M$. Consequently, by the definition of concatenation, $\langle \mathcal{N} \odot \mathcal{M}, a\rangle \models L\psi$ follows. ■

To prove the next lemma, one has to literally repeat the proof of Lemma 9.21 and take into account that the term concatenation refers now to the concatenation of an **N**-structure and a Kripke model. We leave verification of that to the reader.

**Lemma 10.11** *Let $\mathcal{N}$ and $\mathcal{M}$ be an* **N**-*structure and an* **S5**-*model, respectively, such that $\mathcal{N} \odot \mathcal{M} \not\sqsubset \mathcal{M}$. Then $Th(\mathcal{M}) \subseteq Th(\mathcal{N} \odot \mathcal{M})$.* ▲

**Proof of Theorem 10.9**: Let $\mathcal{M} = \langle M, V\rangle$ be an **S5**-model. Assume first that $Th(\mathcal{M})$ is an **N**-expansion for $I$. Then, by Theorem 10.4, $\mathcal{M}$ is **N**-intended for $I$. We need to show that $\mathcal{M}$ is **N**-minimal for $I$. Suppose to the contrary that for some **N**-structure $\mathcal{N}$, say $\mathcal{N} = \langle N, \{Q_\psi\}_{\psi \in \mathcal{L}_L}, W\rangle$, such that $\mathcal{N} \odot \mathcal{M} \models I$, $\mathcal{N} \odot \mathcal{M} \sqsubset \mathcal{M}$. Since $\mathcal{M} \models \{\neg L\varphi : \varphi \notin Th(\mathcal{M})\}$, it follows that $\mathcal{N} \odot \mathcal{M} \models \{\neg L\varphi : \varphi \notin Th(\mathcal{M})\}$. Since $\mathcal{M}$ is **N**-intended for $I$, $Th(\mathcal{N} \odot \mathcal{M}) = Th(\mathcal{M})$. In particular, for every formula $\varphi \in \mathcal{L}$, if for every $b \in M$, $\langle \mathcal{M}, b\rangle \models \varphi$ then for every $a \in N$, $\langle \mathcal{N} \odot \mathcal{M}, a\rangle \models \varphi$. By Theorem 10.7, $\mathcal{N} \odot \mathcal{M} \not\sqsubset \mathcal{M}$, a contradiction. Hence, $\mathcal{M}$ is **N**-minimal for $I$.

Conversely, assume that $\mathcal{M}$ is **N**-minimal for $I$. Consider an **N**-structure $\mathcal{N}$ such that $\mathcal{N} \models I \cup \{\neg L\varphi : \varphi \notin Th(\mathcal{M})\}$. Let $\varphi \notin Th(\mathcal{M})$. Then $\mathcal{N} \models \neg L\varphi$. Consequently, $\varphi \notin Th(\mathcal{N})$ and $Th(\mathcal{N}) \subseteq Th(\mathcal{M})$ follows. Thus, to prove that $\mathcal{M}$ is **N**-intended for $I$, it is enough to prove that $Th(\mathcal{M}) \subseteq Th(\mathcal{N})$.

Since $\mathcal{N} \models I$, it follows by Lemma 10.10 that $\langle \mathcal{N} \odot \mathcal{M}, a\rangle \models I$ for every world $a \in N$. In addition, $\mathcal{M} \models I$. Hence, $\mathcal{N} \odot \mathcal{M} \models I$. Since $\mathcal{M}$ is **N**-minimal for $I$, it follows that $\mathcal{N} \odot \mathcal{M} \not\sqsubset \mathcal{M}$. By Lemma 10.11, $Th(\mathcal{M}) \subseteq Th(\mathcal{N} \odot \mathcal{M})$. Hence, by Lemma 10.10, we have $Th(\mathcal{M}) \subseteq Th(\mathcal{N})$. Therefore, $\mathcal{M}$ is an **N**-intended model for $I$, and by Theorem 10.4 $Th(\mathcal{M})$ is an **N**-expansion of $I$. ■

Theorem 10.9 implies that there is a *one-to-one* correspondence between **N**-minimal models for $I$ and **N**-expansions for $I$, thus yielding an elegant semantic characterization of **N**-expansions parallel to that given in Corollary 9.23.

**Corollary 10.12** *Let $T$ be a stable theory. The theory $T$ is an* **N**-*expansion for $I$ if and only if $\mathcal{M}_T$ is* **N**-*minimal for $I$.* ▲

This result can be further strengthened.

**Definition 10.13** Let $\mathcal{M} = \langle M, V_1\rangle$ be an **S5**-model and let $\mathcal{N}$ be an **N**-structure with the set of worlds $N$ and the valuation function $V_2$. We say that $\mathcal{N} \odot \mathcal{M}$ is *weakly preferred over* $\mathcal{M}$ ($\mathcal{N} \odot \mathcal{M} \sqsubset_w \mathcal{M}$, in symbols) if there is $a \in N$ such that for every $b \in M$, $V_2(a) \neq V_1(b)$.

We say that an **S5**-model $\mathcal{M}$ is *strongly* **N**-*minimal* for a theory $I \subseteq \mathcal{L}_L$ if $\mathcal{M} \models I$ and for every **N**-structure $\mathcal{N}$ such that $\mathcal{N} \odot \mathcal{M} \models I$, $\mathcal{N} \odot \mathcal{M} \not\sqsubset_w \mathcal{M}$. ▲

It is evident that if an **S5**-model is strongly **N**-minimal for $I$ then it is **N**-minimal for $I$, but not conversely. Yet the assertion of Corollary 10.12 remains true if **N**-minimality is replaced in the statement by strong **N**-minimality.

**Theorem 10.14** *Let $T$ be a stable theory. The theory $T$ is an* **N**-*expansion for $I$ if and only if $\mathcal{M}_T$ is strongly* **N**-*minimal for $I$.* ▲

Similarly, only rather minor modifications in the proofs are needed to extend Theorem 9.50 to the case of logic **N**. We have the following result.

**Theorem 10.15** *A theory $T$ is an* **N**-*expansion for $I$ if and only if $T = T_I(\Phi, \Psi)$ for some partition $(\Phi, \Psi)$ of $MA(I)$ which is* **N**-*perfect for $I$.*

In the proof we use Theorem 9.41 and Lemma 9.53, which apply to logic **N**. In addition, we use two other lemmas which are counterparts of Lemmas 9.51 and 9.54.

**Lemma 10.16** *Assume that $(\Phi, \Psi)$ is a partition of $MA(I)$. Let $\mathcal{M}_1 = \langle M_1, \{Q_\psi\}_{\psi \in \mathcal{L}_L}, V_1\rangle$ be an* **N**-*structure and let $\mathcal{M}_2 = \langle M_2, R_2, V_2\rangle$ be a Kripke model such that $M_1 \cap M_2 = \emptyset$, $\mathcal{M}_1 \models \{\neg L\varphi : \varphi \in \Phi\}$, and $\mathcal{M}_2 \models \Psi$. Then, for each $I$-formula $\varphi$ and for each $a \in M_1$,*

$$\langle \mathcal{M}_1, a\rangle \models \varphi \quad \textit{if and only if} \quad \langle \mathcal{M}_1 \odot \mathcal{M}_2, a\rangle \models \varphi.$$

▲

The proof of this lemma is almost a literal copy of the proof of Lemma 9.51. We leave verification of the details to the reader.

**Lemma 10.17** *Let a theory $T \subseteq \mathcal{L}_L$ be an* **N**-*expansion for a theory $I \subseteq \mathcal{L}_L$. Then $(MA(I) \setminus T, MA(I) \cap T)$ is* **N**-*perfect for $I$.*

Proof: Define $\Phi = MA(I) \setminus T$ and $\Psi = MA(I) \cap T$. Since $T$ is stable, it follows by Lemma 9.53 that $(\Phi, \Psi)$ is **S5**-perfect for $I$. In particular, $(\Phi, \Psi)$ is introspection-consistent with $I$. Thus, to prove the lemma, it suffices to show, that for each **N**-structure $\mathcal{M} = \langle M, \{R_\psi\}_{\psi \in \mathcal{L}_L}, V\rangle$, if $\mathcal{M} \models I \cup \{\neg L\varphi : \varphi \in \Phi\}$, then $\mathcal{M} \models \Psi$.

Theory $T$ is stable, so there is an **S5**-model $\mathcal{K} = \langle K, U\rangle$ such that $Th(\mathcal{K}) = T$. Since $I \subseteq T$, it follows that $\langle \mathcal{K}, a\rangle \models I$, for every $a \in K$, and so $\langle \mathcal{M} \odot \mathcal{K}, a\rangle \models I$, for every $a \in K$. Since $\Psi \subseteq T$, $\mathcal{K} \models \Psi$. In addition, we have that $\mathcal{M} \models I$. Hence, it follows from Lemma 10.16 that $\langle \mathcal{M} \odot \mathcal{K}, a\rangle \models I$, for every $a \in M$. Thus, $\mathcal{M} \odot \mathcal{K} \models I$. In addition, if $\varphi \notin T$ then $\mathcal{K} \models \neg L\varphi$ and, consequently, $\mathcal{M} \odot \mathcal{K} \models \neg L\varphi$. Hence, $\mathcal{M} \odot \mathcal{K} \models I \cup \{\neg L\varphi : \varphi \notin T\}$. Since $T$ is an **N**-expansion for $I$, $\mathcal{M} \odot \mathcal{K} \models T$. Since $\Psi \subseteq T$, $\mathcal{M} \odot \mathcal{K} \models \Psi$. By Lemma 10.16, $\mathcal{M} \models \Psi$. ■

**Proof of Theorem 10.15**: The "if" part of the assertion follows from Theorem 9.48. To prove the "only if" part of the theorem, consider an **N**-expansion $T$ for $I$. Put $\Phi = MA(I) \setminus T$ and $\Psi = MA(I) \cap T$. By Theorem 9.41, $T = T_I(\Phi, \Psi)$ and, by Lemma 10.17, $(\Phi, \Psi)$ is **N**-perfect for $I$. ■

**Remark 10.18** Theorem 10.9, Corollary 10.12 and Theorem 10.14 can be extended to the case of subnormal modal logics contained in **S5** and characterized by cluster-closed class of Kripke structures containing all **S5**-models. Theorem 10.15 can be extended to the case of subnormal modal logics contained in **S4f** and characterized by cluster-closed class of Kripke structures. ▲

## 10.2 Nonmonotonic logics kd45 and Sw5

Nonmonotonic logics **kd45** and **Sw5** have several identical or similar properties. We showed in Section 9.4 that both are maximal logics for which the assertions of Theorem 9.41 and Corollaries 9.42 and 9.44 are valid. In this section, we will show that both logics can be given a characterization in terms of the propositional consequence operator and that a similar modification of the concept of an introspection consistent partition yields characterizations of **kd45**- and **Sw5**-expansions. Finally, we will formulate and prove a theorem which provides a precise explanation of the nature of the relationship between these two logics. We start with the nonmonotonic logic **kd45**.

**Theorem 10.19** *A consistent theory $T$ is a* **kd45**-*expansion for $I$ if and only if*

$$T = Cn(I \cup \{L\varphi : \varphi \in T\} \cup \{\neg L\varphi : \varphi \notin T\}). \tag{10.2}$$

Proof: It is easy to see that if $T \subseteq \mathcal{L}_L$ is stable and consistent, then the theory

$$Cn(\{L\varphi : \varphi \in T\} \cup \{\neg L\varphi : \varphi \notin T\})$$

contains all instances of axiom schemata k, d, 4 and 5. Consider for example a formula $\psi = \neg L\eta \vee LL\eta$, equivalent to an instance of 4. If $\eta \notin T$, then $\neg L\eta \in \{\neg L\varphi : \varphi \notin T\}$ and so $\psi \in Cn(\{\neg L\varphi : \varphi \notin T\})$. If $\eta \in T$, then $L\eta \in T$. Consequently, $LL\eta \in \{L\varphi : \varphi \in T\}$ and $\psi \in Cn(\{L\varphi : \varphi \in T\})$. The remaining axiom schemata can be dealt with similarly.

In addition, $\{\neg L\varphi : \varphi \notin T\} \vdash_5 \{L\varphi : \varphi \in T\}$. Indeed, let $\varphi \in T$. Then, since $T$ is stable and consistent, $\neg L\varphi \notin T$. Since $\neg L \neg L\varphi \in \{\neg L\varphi : \varphi \notin T\}$ and $\neg L \neg L\varphi \vdash_5 L\varphi$, our claim follows.

Next, let us notice that if $T$ is stable and $I \subseteq T$, then $Cn(I \cup \{L\varphi : \varphi \in T\} \cup \{\neg L\varphi : \varphi \notin T\})$ is closed under necessitation.

Our observations imply that for a stable and consistent theory $T$ such that $I \subseteq T$,

$$Cn(I \cup \{L\varphi : \varphi \in T\} \cup \{\neg L\varphi : \varphi \notin T\}) = Cn_{\mathbf{kd45}}(I \cup \{\neg L\varphi : \varphi \notin T\}).$$

Thus, $T = Cn(I \cup \{L\varphi : \varphi \in T\} \cup \{\neg L\varphi : \varphi \notin T\})$ if and only if $T = Cn_{\mathbf{kd45}}(I \cup \{\neg L\varphi : \varphi \notin T\})$. ■

We have seen that Theorem 9.50 fails for the logics **kd45** and **Sw5**. More precisely, there are theories $I$ with expansions not corresponding to any **kd45**-perfect partition of $MA(I)$. The same is true for the logic **Sw5**. A different way to narrow down the class of partitions that are introspection-consistent with $I$ is needed to describe **kd45**-expansions and **Sw5**-expansions.

**Definition 10.20** Let $I \subseteq \mathcal{L}_L$. A partition $(\Phi, \Psi)$ of $MA(I)$ is called *propositionally perfect for I with respect to a theory* $U \subseteq \mathcal{L}_L$, if:

1. $(\Phi, \Psi)$ is introspection-consistent with $I$.
2. For each $\psi \in \Psi$,
$$I \cup \{\neg L\varphi : \varphi \in \Phi\} \cup U \vdash \psi.$$
▲

Our next theorem characterizes **kd45**-expansions in terms of those partitions of $MA(I)$ which are propositionally perfect for $I$ with respect to the theory $\{L\psi : \psi \in \Psi\}$.

**Theorem 10.21** *Let* $I \subseteq \mathcal{L}_L$.

1. *If a theory $T$ is a* **kd45**-*expansion for $I$ then $(\Phi, \Psi)$, where $\Phi = MA(I) \setminus T$ and $\Psi = MA(I) \cap T$, is propositionally perfect for $I$ with respect to $\{L\psi : \psi \in \Psi\}$, and $T = T_I(\Phi, \Psi)$.*
2. *If a partition $(\Phi, \Psi)$ of $MA(I)$ is propositionally perfect for $I$ with respect to $\{L\psi : \psi \in \Psi\}$ then $T_I(\Phi, \Psi)$ is a* **kd45**-*expansion for $I$.*

Proof: We first prove part (2) of the theorem. Consider a partition $(\Phi, \Psi)$ of $MA(I)$, propositionally perfect for $I$ with respect to $\{L\psi : \psi \in \Psi\}$. Define

$I' = I \cup \{L\psi : \psi \in \Psi\}$. Then, for each $\psi \in \Psi$, $I' \cup \{\neg L\varphi : \varphi \in \Phi\} \vdash \psi$. This implies that $I' \cup \{\neg L\varphi : \varphi \in \Phi\} \vdash_{\mathbf{kd45}} \psi$. Consequently, $(\Phi, \Psi)$ is **kd45**-perfect for $I'$. Thus, by Theorem 9.48, $T_{I'}(\Phi, \Psi)$ is a **kd45**-expansion of $I'$. To complete this part of the reasoning, we will show that $T_{I'}(\Phi, \Psi)$ is a **kd45**-expansion for $I$ and that $T_{I'}(\Phi, \Psi) = T_I(\Phi, \Psi)$. Put $T = T_{I'}(\Phi, \Psi)$. By Theorem 10.19,

$$T = Cn(I \cup \{L\psi : \psi \in \Psi\} \cup \{L\psi : \psi \in T\} \cup \{\neg L\varphi : \varphi \notin T\}).$$

Since $\Psi \subseteq T_{I'}(\Phi, \Psi) = T$ (Theorem 9.39), we have $\{L\psi : \psi \in \Psi\} \subseteq \{L\psi : \psi \in T\}$. Consequently,

$$T = Cn(I \cup \{L\psi : \psi \in T\} \cup \{\neg L\varphi : \varphi \notin T\}).$$

that is, $T$ is a **kd45**-expansion for $I$ (Theorem 10.19). It remains to be shown that $T_{I'}(\Phi, \Psi) = T_I(\Phi, \Psi)$, but this follows immediately from the definition of canonical theories.

We pass on now to the proof of part (1) of the assertion. Let $T$ be a **kd45**-expansion for $I$. Let $\Psi = T \cap MA(I)$ and $\Phi = MA(I) \setminus T$. By Theorem 9.36, $(\Phi, \Psi)$ is introspection-consistent with $I$.

Next, we prove that for every $I$-formula $\eta$, if

$$I \cup \{\neg L\varphi : \varphi \notin T\} \cup \{L\psi : \psi \in T\} \vdash \eta, \tag{10.3}$$

then

$$I \cup \{\neg L\varphi : \varphi \in \Phi\} \cup \{L\psi : \psi \in \Psi\} \vdash \eta. \tag{10.4}$$

To this end we consider a propositional valuation $V$ evaluating the premises of (10.4) as **t** and define a valuation $W$ as follows:

1. $W(p) = V(p)$, for all propositional atoms.
2. $W(L\varphi) = \mathbf{f}$, for $L\varphi \notin T$.
3. $W(L\varphi) = \mathbf{t}$, for $L\varphi \in T$.

The valuations $V$ and $W$ have the following property:

$$\text{for each } I\text{-formula } \varphi,\ W(\varphi) = V(\varphi). \tag{10.5}$$

This property can be easily proved by the induction on the complexity of $\varphi$. The basis for the induction is a direct consequence of part 1 of the definition of $W$ In the inductive step, the case when $\varphi$ is built of one or two simpler formulas by means of a boolean connective follows easily from the definition of a propositional valuation and from the induction hypothesis. The only remaining case is when $\varphi = L\zeta$.

Assume first that $W(\varphi) = \mathbf{t}$. Then, by the definition of $W$, $L\zeta \in T$. Now, $\varphi$ is an $I$-formula. Therefore, $\zeta \in MA(I)$. Since $T$ is stable, $\zeta \in \Psi$. Thus, $V(L\zeta) = \mathbf{t}$.

Next, assume that $W(\varphi) = \mathbf{f}$. Then $L\zeta \notin T$. Since $\varphi$ is an $I$-formula and since $T$ is stable, $\zeta \in \Phi$. Consequently, $V(L\zeta) = \mathbf{f}$.

It follows from (10.5) that each formula in $I$ is evaluated by $W$ as $\mathbf{t}$. Hence, all the premises in (10.3) are evaluated by $W$ as $\mathbf{t}$ and, therefore, $W(\eta) = \mathbf{t}$. Since $\eta$ is an $I$-formula, $V(\eta) = \mathbf{t}$.

To sum up, we have shown that for each valuation $V$ satisfying the premises of (10.4), $V(\eta) = \mathbf{t}$. Consequently, (10.4) holds.

Now, we continue with the proof of the theorem. Observe that $T = T_I(\Phi, \Psi)$ and that $(\Phi, \Psi)$ is introspection-consistent with $I$ (Theorem 9.41). Consider $\eta \in \Psi$. Then $\eta \in T$ and, since $T$ is a **kd45**-expansion for $I$, it follows by Theorem 10.19 that

$$I \cup \{\neg L\varphi : \varphi \notin T\} \cup \{L\psi : \psi \in T\} \vdash \eta.$$

Since $\eta \in \Psi$, $\eta$ is an $I$-formula. Hence,

$$I \cup \{\neg L\varphi : \varphi \in \Phi\} \cup \{L\psi : \psi \in \Psi\} \vdash \eta,$$

which implies that $(\Phi, \Psi)$ is propositionally perfect for $I$ with respect to $\{L\psi : \psi \in \Psi\}$. ■

From Theorems 10.19 and 10.21 one can derive characterizations of **kd45**-expansions analogous to characterizations of $\mathcal{S}$-expansions given by Corollaries 9.8 and 9.55.

**Corollary 10.22** *A consistent theory $T$ is a* **kd45***-expansion for $I$ if and only if $T$ is stable, $I \subseteq T$ and $T \cap \mathcal{L} \subseteq Cn(I \cup \{L\varphi: \varphi \in T\} \cup \{\neg L\psi: \psi \notin T\})$.*

Proof: The "only if" part of the assertion follows immediately from Theorem 10.19. To prove the "if" part, observe that by Theorem 8.26,

$$T = Cn((T \cap \mathcal{L}) \cup \{L\varphi: \varphi \in T\} \cup \{\neg L\psi: \psi \notin T\}).$$

Since $T \cap \mathcal{L} \subseteq Cn(I \cup \{L\varphi: \varphi \in T\} \cup \{\neg L\psi: \psi \notin T\})$,

$$T \subseteq Cn(I \cup \{L\varphi: \varphi \in T\} \cup \{\neg L\psi: \psi \notin T\}).$$

Next, observe that by stability of $T$ we have

$$T \supseteq \{L\varphi: \varphi \in T\} \cup \{\neg L\psi: \psi \notin T\}.$$

Since $I \subseteq T$,

$$T \supseteq Cn(I \cup \{L\varphi : \varphi \in T\} \cup \{\neg L\psi : \psi \notin T\}).$$

Thus,

$$T = Cn(I \cup \{L\varphi : \varphi \in T\} \cup \{\neg L\psi : \psi \notin T\}),$$

and, by Theorem 10.19, $T$ is a **kd45**-expansion for $I$. ■

**Corollary 10.23** *Let $S \subseteq \mathcal{L}$ be a consistent theory. Then, $ST(S)$ is a* **kd45**-*expansion for $I$ if and only if $I \subseteq ST(S)$ and $S \subseteq Cn(I \cup \{\neg L\varphi : \varphi \in \Phi\} \cup \{L\psi : \psi \in \Psi\})$, where $\Phi = MA(I) \setminus ST(S)$ and $\Psi = MA(I) \setminus \Phi$.*

Proof: Assume that $S \subseteq Cn(I \cup \{\neg L\varphi : \varphi \in \Phi\} \cup \{L\psi : \psi \in \Psi\})$. Then $S \subseteq Cn(I \cup \{\neg L\varphi : \varphi \notin ST(S)\} \cup \{L\psi : \psi \in ST(S)\})$. Since $I \subseteq ST(S)$, it follows by Corollary 10.22 that $ST(S)$ is a **kd45**-expansion for $I$.

Conversely, assume that $ST(S)$ is a **kd45**-expansion for $I$. Then, by Theorem 10.21, $ST(S) = T_I(\Phi, \Psi)$. In addition, $(\Phi, \Psi)$ is introspection-consistent with $I$. Hence, by Theorem 9.39,

$$Cn(S) = Cn(I \cup \{\neg L\varphi : \varphi \in \Phi\} \cup \{L\psi : \psi \in \Psi\} \cup \Psi) \cap \mathcal{L}.$$

Consequently, the inclusion $S \subseteq Cn(I \cup \{\neg L\varphi : \varphi \in \Phi\} \cup \{L\psi : \psi \in \Psi\})$ follows. ■

We pass on to the nonmonotonic logic **Sw5**. All results we prove here are counterparts of the results proved above for the nonmonotonic logic **kd45**.

**Theorem 10.24** *A consistent theory $T$ is an* **Sw5**-*expansion for $I$ if and only if*

$$T = Cn(I \cup \{\varphi \equiv L\varphi : \varphi \in T\} \cup \{\neg L\varphi : \varphi \notin T\}). \tag{10.6}$$

Proof: The proof is very similar to that of Theorem 10.19. It is easy to see that if $T \subseteq \mathcal{L}_L$ is stable and consistent, then the theory

$$Cn(\{\varphi \equiv L\varphi : \varphi \in T\} \cup \{\neg L\varphi : \varphi \notin T\})$$

contains all instances of axiom schemata k, t, 4 and w5. Here we will only show that an instance $\alpha = L(\beta \supset \gamma) \supset (L\beta \supset L\gamma)$ of the schema k is contained in $Cn(\{\psi \equiv L\psi : \psi \in T\} \cup \{\neg L\varphi : \varphi \notin T\})$. If $\beta \notin T$ or $\beta \supset \gamma \notin T$ then $\alpha \in Cn(\{\neg L\varphi : \varphi \notin T\})$. Therefore, assume that $\beta \in T$ and $\beta \supset \gamma \in T$. Then $\gamma \in T$, as well. Hence, we have

$$\beta \equiv L\beta,\ \gamma \equiv L\gamma,\ ((\beta \supset \gamma) \equiv L(\beta \supset \gamma)) \in \{\varphi \equiv L\varphi : \varphi \in T\}.$$

Since

$$\beta \equiv L\beta,\ \gamma \equiv L\gamma,\ ((\beta \supset \gamma) \equiv L(\beta \supset \gamma)) \vdash \alpha,$$

the claim follows. The remaining axiom schemata can be dealt with similarly.

Next, observe that $\{\neg L\varphi : \varphi \notin T\} \vdash_{\mathbf{Sw5}} \{\varphi \equiv L\varphi : \varphi \in T\}$. Indeed, let $\psi \in T$. Since $T$ is stable and consistent, we have $\neg L\psi \notin T$. Therefore, $\neg L \neg L\psi \in \{\neg L\varphi : \varphi \notin T\}$. We also have $\vdash_{\mathbf{Sw5}} \neg L \neg L\psi \supset (\psi \supset L\psi)$ and $\vdash_{\mathbf{Sw5}} L\psi \supset \psi$. Thus, $\{\neg L\varphi : \varphi \notin T\} \vdash_{\mathbf{Sw5}} \psi \equiv L\psi$ follows.

Next, let us notice that if $T$ is stable and $I \subseteq T$, then $Cn(I \cup \{\varphi \equiv L\varphi : \varphi \in T\} \cup \{\neg L\varphi : \varphi \notin T\})$ is closed under necessitation. Indeed, let $\psi \in Cn(I \cup \{\varphi \equiv L\varphi : \varphi \in T\} \cup \{\neg L\varphi : \varphi \notin T\})$. Since $T$ is stable and contains $I$, $Cn(I \cup \{\varphi \equiv L\varphi : \varphi \in T\} \cup \{\neg L\varphi : \varphi \notin T\}) \subseteq T$. Clearly, $\psi \in T$. Consequently, $\psi \equiv L\psi \in Cn(I \cup \{\varphi \equiv L\varphi : \varphi \in T\} \cup \{\neg L\varphi : \varphi \notin T\})$. Hence, $L\psi \in Cn(I \cup \{\varphi \equiv L\varphi : \varphi \in T\} \cup \{\neg L\varphi : \varphi \notin T\})$.

Our observations imply that for a stable and consistent theory $T$ such that $I \subseteq T$,

$$Cn(I \cup \{\varphi \equiv L\varphi : \varphi \in T\} \cup \{\neg L\varphi : \varphi \notin T\}) = Cn_{\mathbf{Sw5}}(I \cup \{\neg L\varphi : \varphi \notin T\}),$$

which implies the assertion. ■

**Theorem 10.25** *Let $I \subseteq \mathcal{L}_L$.*

1. *If a theory $T$ is an* **Sw5***-expansion for $I$, then $(\Phi, \Psi)$, where $\Phi = MA(I) \setminus T$ and $\Psi = MA \cap T$, is propositionally perfect for $I$ with respect to $\{\psi \equiv L\psi : \psi \in \Psi\}$, and $T = T_I(\Phi, \Psi)$.*
2. *If a partition $(\Phi, \Psi)$ of $MA(I)$ is propositionally perfect for $I$ with respect to $\{\psi \equiv L\psi : \psi \in \Psi\}$, then $T_I(\Phi, \Psi)$ is an* **Sw5***-expansion for $I$.*

Proof: Consider a partition $(\Phi, \Psi)$ of $MA(I)$, propositionally perfect for $I$ with respect to $\{\psi \equiv L\psi : \psi \in \Psi\}$. Define $I' = I \cup \{\psi \equiv L\psi : \psi \in \Psi\}$. Then, for each $\psi \in \Psi$, $I' \cup \{\neg L\varphi : \varphi \in \Phi\} \vdash \psi$. This implies that $I' \cup \{\neg L\varphi : \varphi \in \Phi\} \vdash_{\mathbf{Sw5}} \psi$. Since $(\Phi, \Psi)$ is introspection consistent for $I$, it is introspection-consistent for $I'$. Hence, $(\Phi, \Psi)$ is **Sw5**-perfect for $I'$ and, by Theorem 9.48, $T_{I'}(\Phi, \Psi)$ is an **Sw5**-expansion for $I'$. Denote $T = T_{I'}(\Phi, \Psi)$. By Theorem 10.24,

$$T = Cn(I \cup \{\psi \equiv L\psi : \psi \in \Psi\} \cup \{\psi \equiv L\psi : \psi \in T\} \cup \{\neg L\varphi : \varphi \notin T\}).$$

But $\Psi \subseteq T_{I'}(\Phi, \Psi) = T$. Thus,

$$\{\psi \equiv L\psi : \psi \in \Psi\} \subseteq \{\psi \equiv L\psi : \psi \in T\}.$$

Consequently,

$$T = Cn(I \cup \{\psi \equiv L\psi : \psi \in T\} \cup \{\neg L\varphi : \varphi \notin T\}),$$

that is, $T$ is an **Sw5**-expansion for $I$. It remains to show that $T_{I'}(\Phi, \Psi) = T_I(\Phi, \Psi)$, but this follows immediately from the definition of canonical theories and the equality:

$$Cn(I \cup \{\neg L\varphi : \varphi \in \Phi\} \cup \{L\psi : \psi \in \Psi\} \cup \Psi) = \\ Cn(I \cup \{\psi \equiv L\psi : \psi \in \Psi\} \cup \{\neg L\varphi : \varphi \in \Phi\} \cup \{L\psi : \psi \in \Psi\} \cup \Psi).$$

Hence, the proof of part (2) of the assertion is complete.

Next, we will prove part (1) of the assertion. Let $T$ be an **Sw5**-expansion for $I$. Let $\Psi = T \cap MA(I)$ and $\Phi = MA(I) \setminus T$. First we prove that for every $I$-formula $\eta$, if

$$I \cup \{\neg L\varphi : \varphi \notin T\} \cup \{\psi \equiv L\psi : \psi \in T\} \vdash \eta, \tag{10.7}$$

then

$$I \cup \{\neg L\varphi : \varphi \in \Phi\} \cup \{\psi \equiv L\psi : \psi \in \Psi\} \vdash \eta. \tag{10.8}$$

Consider any propositional valuation $V$ evaluating the premises of (10.8) as **t**. Define a valuation $W$ recursively with respect to the complexity of a formula as follows:

1. $W(p) = V(p)$, for all propositional atoms.
2. $W(L\varphi) = \mathbf{f}$, for $L\varphi \notin T$.
3. $W(L\varphi) = W(\varphi)$, for $L\varphi \in T$.

As in the proof of Theorem 10.21, it follows easily by induction on the complexity of a formula that the following property holds:

$$\text{for each } I\text{-formula } \varphi,\ W(\varphi) = V(\varphi). \tag{10.9}$$

Property (10.9) implies in particular that each formula in $I$ is evaluated by $W$ as **t**. Thus, all the premises in (10.7) are evaluated by $W$ as **t** and, consequently, $W(\eta) = \mathbf{t}$. Since $\eta$ is an $I$-formula, again by the property (10.9), $V(\eta) = \mathbf{t}$. Thus, for each valuation $V$ satisfying the premises of (10.8), $V(\eta) = \mathbf{t}$ and (10.8) follows.

Consider an arbitrary formula $\eta \in \Psi$. Then, $\eta \in T$ and, since $T$ is an **Sw5**-expansion, we have

$$I \cup \{\neg L\varphi : \varphi \notin T\} \cup \{\psi \equiv L\psi : \psi \in T\} \vdash \eta.$$

The formula $\eta$ is an $I$-formula and so, by the claim we have just proved, it follows that

$$I \cup \{\neg L\varphi : \varphi \in \Phi\} \cup \{\psi \equiv L\psi : \psi \in \Psi\} \vdash \eta. \tag{10.10}$$

Since $T$ is an **Sw5**-expansion for $I$, it follows that $T = T_I(\Phi, \Psi)$ and that $(\Phi, \Psi)$ is introspection-consistent with $I$ (Theorem 9.41). Hence, by (10.10), $(\Phi, \Psi)$ is propositionally perfect for $I$ with respect to $\{\psi \equiv L\psi : \psi \in \Psi\}$. ■

**Corollary 10.26** *A consistent theory $T$ is an* **Sw5**-*expansion for $I$ if and only if $T$ is stable, $I \subseteq T$ and $T \cap \mathcal{L} \subseteq Cn(I \cup \{\neg L\varphi : \varphi \notin T\} \cup \{\psi \equiv L\psi : \psi \in T\})$.*

Proof: The "only if" part of the assertion follows immediately from Theorem 10.24. To prove the "if" part, observe that by Corollary 9.42 and Theorem 10.24

$$T = Cn((T \cap \mathcal{L}) \cup \{\neg L\varphi : \varphi \notin T\} \cup \{\psi \equiv L\psi : \psi \in T\}).$$

From now on, the proof follows that of Corollary 10.22 with the modification that the reference to Theorem 10.19 is replaced by the reference to Theorem 10.24, and $\{L\psi : \psi \in T\}$ is replaced by $\{\psi \equiv L\psi : \psi \in T\}$. ■

The next result is analogous to Corollary 10.23 and can be proved in a similar way. We leave the proof for the reader.

**Corollary 10.27** *Let $S \subseteq \mathcal{L}$ be consistent. Then, $ST(S)$ is an* **Sw5**-*expansion for $I$ if and only if $I \subseteq ST(S)$ and $S \subseteq Cn(I \cup \{\neg L\varphi : \varphi \in \Phi\} \cup \{\psi \equiv L\psi : \psi \in \Psi\})$, where $\Phi = MA(I) \setminus ST(I)$ and $\Psi = MA(I) \setminus \Phi$.* ▲

We will now introduce embeddings of (monotonic) **kd45** into **Sw5**, and conversely. The corresponding translations $(\cdot)^B$ and $(\cdot)^K$ are defined below.

For a formula $\varphi \in \mathcal{L}_L$, we define the formula $\varphi^B$ recursively as follows:

1. For a propositional variable $p$, $p^B = p$.
2. For $\varphi = \neg\psi$, $\varphi^B = \neg\psi^B$; for $\varphi = \psi_1 \vee \psi_2$, $\varphi^B = \psi_1^B \vee \psi_2^B$; for $\varphi = \psi_1 \wedge \psi_2$, $\varphi^B = \psi_1^B \wedge \psi_2^B$; and for $\varphi = \psi_1 \supset \psi_2$, $\varphi^B = \psi_1^B \supset \psi_2^B$.
3. For $\varphi = L\psi$, $\varphi^B = L(\psi^B) \wedge \psi^B$.

For a theory $I \subseteq \mathcal{L}_L$, we define

$$I^B = \{\varphi^B : \varphi \in I\}.$$

Next, for a formula $\varphi \in \mathcal{L}_L$, we define the formula $\varphi^K$ recursively as follows:

1. For a propositional variable $p$, $p^K = p$.
2. For $\varphi = \neg\psi$, $\varphi^K = \neg\psi^K$; for $\varphi = \psi_1 \vee \psi_2$, $\varphi^K = \psi_1^K \vee \psi_2^K$; for $\varphi = \psi_1 \wedge \psi_2$, $\varphi^K = \psi_1^K \wedge \psi_2^K$; and for $\varphi = \psi_1 \supset \psi_2$, $\varphi^K = \psi_1^K \supset \psi_2^K$.
3. For $\varphi = L\psi$, $\varphi^K = ML(\psi^K)$.

In other words, $\varphi^K$ is the formula obtained from $\varphi$ by simultaneously replacing each occurrence of $L$ with $ML$. Similarly as before, for a theory $I \subseteq \mathcal{L}_L$, we put

$$I^K = \{\varphi^K : \varphi \in I\}.$$

The intuition underlying these definitions should be clear by now. We will use the operator $\varphi \mapsto \varphi^B$ to embed the nonmonotonic logic **Sw5** into the nonmonotonic logic **kd45**. Logic **Sw5** satisfies axiom t. Consequently, logic **Sw5** treats the modal operator $L$ as the operator of *knowledge* (in **Sw5**, $Lp$ allows us to derive $p$). On the other hand, logic **kd45** does not satisfy schema t, but satisfies schema 5 which allows us to derive $Lp$ from $\neg L \neg Lp$. Such properties can be attributed to the modality of *believing* rather than *knowing*. If nonmonotonic logic **Sw5** is to be interpreted within the nonmonotonic logic **kd45** we need to express the modality of *knowing* in terms of the modality of *believing*. One possibility is to translate a formula $Lp$ (where $p$ does not contain $L$), interpreted as "I know p" in the logic **Sw5**, into $Lp \wedge p$, which is interpreted as "I believe in $p$ and $p$ is true" in the logic **kd45**. Such an approach gives rise to our definition of the mapping $\varphi \mapsto \varphi^B$.

Similarly, in order to embed the nonmonotonic logic **kd45** into the nonmonotonic logic **Sw5**, we need to express the modality of *believing* by means of the modality of knowing. If $Lp$ means "I know $p$", then $\neg L \neg Lp$ means "I do not know that I do not know $p$". It can equivalently be expressed as $MLp$ — "it is possible that I know $p$" which is one way of expressing a belief in $p$.

**Lemma 10.28** *Let $\mathcal{M} = \langle M, V \rangle$ be an* **S5**-*model. Then, for every formula $\varphi \in \mathcal{L}_L$ and for every $m \in M$ the following three conditions are equivalent:*

**(i)** $\langle \mathcal{M}, m \rangle \models \varphi$.

**(ii)** $\langle \mathcal{M}, m \rangle \models \varphi^B$.

**(iii)** $\langle \mathcal{M}, m \rangle \models \varphi^K$.

Proof: We proceed by induction on the complexity of $\varphi$. The basis of the induction follows directly from the definitions of $\varphi^B$ and $\varphi^K$. If $\varphi = \neg\psi$, $\psi_1 \vee \psi_2$, $\psi_1 \wedge \psi_2$ or $\varphi = \psi_1 \supset \psi_2$, then the induction step follows easily from the induction hypothesis and from the definition of the satisfaction relation. So, assume that $\varphi = L\psi$ and that $\langle \mathcal{M}, m\rangle \models \varphi$. Then, for every world $m' \in M$ $\langle \mathcal{M}, m'\rangle \models \psi$. By the induction hypothesis, for every world $m' \in M$,

$$\langle \mathcal{M}, m'\rangle \models \psi^B.$$

Thus, $\langle \mathcal{M}, m\rangle \models L\psi^B \wedge \psi^B$ follows. Similarly, it follows by the induction hypothesis, that for every world $m' \in M$,

$$\langle \mathcal{M}, m'\rangle \models \psi^K.$$

Consequently, for every world $m' \in M$,

$$\langle \mathcal{M}, m'\rangle \models L\psi^K,$$

and

$$\langle \mathcal{M}, m'\rangle \models \neg L \neg L\psi^K = ML\psi^K = \varphi^K.$$

The converse implications are equally easy to prove. We leave the details for the reader. ■

For a **kd45**-model $\mathcal{M} = \langle a, M, V\rangle$ we define $\mathcal{M}^*$ to be the **Sw5**-model $\langle a, M, V\rangle$. (That is, $\mathcal{M}^*$ is obtained from $\mathcal{M}$ by making $a$ accessible from itself.) This operation establishes a *one-to-one* correspondence between the classes of **kd45**- and **Sw5**-models. We have the following lemma.

**Lemma 10.29** *Let $\mathcal{M} = \langle m_0, M, V\rangle$ be a* **kd45***-model. For every formula $\varphi \in \mathcal{L}_L$ and for every $m \in M \cup \{m_0\}$, we have:*

1. *$\langle \mathcal{M}^*, m\rangle \models \varphi$ if and only if $\langle \mathcal{M}, m\rangle \models \varphi^B$.*
2. *$\langle \mathcal{M}^*, m\rangle \models \varphi^K$ if and only if $\langle \mathcal{M}, m\rangle \models \varphi$.*

Proof: (1) We proceed by induction on the complexity of a formula. The basis of the induction follows directly from the definition of $\varphi^B$. In the inductive step, the only more difficult case is when $\varphi = L\psi$. We discuss this case here; the details concerning other cases are omitted.

Let $m \in M \cup \{m_0\}$. Assume first that $\langle \mathcal{M}^*, m\rangle \models \varphi$. Then, for every $m' \in M$, $\langle \mathcal{M}^*, m'\rangle \models \psi$. By the induction hypothesis, $\langle \mathcal{M}, m'\rangle \models \psi^B$. Consequently, $\langle \mathcal{M}, m\rangle \models L\psi^B$. If $m \in M$ then, as we have seen, $\langle \mathcal{M}, m\rangle \models \psi^B$. If $m = m_0$, then we have $\langle \mathcal{M}^*, m\rangle \models \psi$ (since $\langle \mathcal{M}^*, m\rangle \models L\psi$). Again by the induction hypothesis, $\langle \mathcal{M}, m\rangle \models \psi^B$. Hence, $\langle \mathcal{M}, m\rangle \models L\psi^B \wedge \psi^B (= \varphi^B)$.

Conversely, assume that $\langle \mathcal{M}, m\rangle \models L\psi^B \wedge \psi^B (= \varphi^B)$. In particular, $\langle \mathcal{M}, m'\rangle \models \psi^B$, for every $m' \in M$. Thus, by the induction hypothesis,

$\langle \mathcal{M}^*, m' \rangle \models \psi$, for every $m' \in M$. If $m \in M$, then $\langle \mathcal{M}^*, m \rangle \models L\psi$ follows. So, assume that $m = m_0$. Then, $\langle \mathcal{M}, m \rangle \models \psi^B$ and, by the induction hypothesis, it follows that $\langle \mathcal{M}^*, m \rangle \models \psi$. Consequently, $\langle \mathcal{M}^*, m \rangle \models L\psi$.

The proof of (2) is similar and is left for the reader as an exercise. ■

**Theorem 10.30** *A theory $T \subseteq \mathcal{L}_L$ is an* **Sw5**-*expansion of a theory $I$ if and only if $T$ is a* **kd45**-*expansion of $I^B$.*

Proof: Let us denote the classes of **kd45**- and **Sw5**-models by $\mathcal{C}_{\mathbf{kd45}}$ and $\mathcal{C}_{\mathbf{Sw5}}$, respectively: Clearly, both classes are cluster-closed. Consequently, by Corollary 9.23, it suffices to show that the canonical model for $T$, $\mathcal{M}_T$, is $\mathcal{C}_{\mathbf{Sw5}}$-minimal for $I$ if and only if $\mathcal{M}_T$ is $\mathcal{C}_{\mathbf{kd45}}$-minimal for $I^B$. To this end, first observe that the equivalence

$$\mathcal{M}_T \models I \text{ if and only if } \mathcal{M}_T \models I^B$$

follows from Lemma 10.28.

Next, observe that by Lemma 10.29(1), for every **kd45**-model $\mathcal{M}$ and for every world $m$ of $\mathcal{M}$,

$$\langle \mathcal{M}^*, m \rangle \models I \text{ if and only if } \langle \mathcal{M}, m \rangle \models I^B.$$

Since the mapping $\mathcal{M} \mapsto \mathcal{M}^*$ is a *one-to-one* correspondence between classes $\mathcal{C}_{\mathbf{kd45}}$ and $\mathcal{C}_{\mathbf{Sw5}}$, the theorem follows. ■

Reasoning in the same way and using Lemma 10.29(2) in place of Lemma 10.29(1), we prove a result on interpretability of the nonmonotonic logic **kd45** in the nonmonotonic logic **Sw5**.

**Theorem 10.31** *A theory $T \subseteq \mathcal{L}_L$ is a* **kd45**-*expansion of a theory $I$ if and only if $T$ is an* **Sw5**-*expansion of $I^K$.* ▲

In fact a stronger result can easily be obtained now — interpretability of the nonmonotonic logic **kd45** in *any* nonmonotonic modal logic $\mathcal{S} \subseteq$ **Sw5**. Before we prove it, we need a lemma.

**Lemma 10.32** *Assume that $\mathcal{M} = \langle M, V \rangle$ is an* **S5**-*model and that $\mathcal{N} = \langle N, \{R_\varphi\}_{\varphi \in \mathcal{L}_L}, U \rangle$ is an* **N**-*structure. If $M \cap N = \emptyset$, then for every world $m \in N$ and for every formula $\varphi \in \mathcal{L}_L$,*

$$\langle \mathcal{N} \odot \mathcal{M}, m \rangle \models \varphi^K \textit{ if and only if } \langle \mathcal{M}^m, m \rangle \models \varphi,$$

*where $\mathcal{M}^m = \langle m, M, W \rangle$ is a* **kd45**-*model such that $W|_M = V$ and $W(m) = U(m)$.*

Proof: We will proceed, as usual in such cases, by induction on the complexity of a formula. The basis of the induction is obvious and follows immediately from the definition of $W$. In the inductive step, the cases when $\varphi$ is built of one or two simpler formulas by means of boolean connectives is also straightforward. Here, we will provide details only for the last remaining case when $\varphi = L\psi$. Let us assume that $\langle \mathcal{N} \odot \mathcal{M}, m\rangle \models \varphi^K$. Then $\langle \mathcal{N} \odot \mathcal{M}, m\rangle \models \neg L \neg L\psi^K$. By the definitions of concatenation and the satisfaction relation, it follows that $\langle \mathcal{N} \odot \mathcal{M}, m'\rangle \models \psi^K$, for every $m' \in M$. Hence, for every $m' \in M$ we have $\langle \mathcal{M}, m'\rangle \models \psi^K$. By Lemma 10.28, for every $m' \in M$ we have $\langle \mathcal{M}, m'\rangle \models \psi$. Thus, by the definition of $\mathcal{M}^m$, $\langle \mathcal{M}^m, m\rangle \models L\psi$.

Conversely, if $\langle \mathcal{M}^m, m\rangle \models L\psi$, then $\langle \mathcal{M}, m'\rangle \models \psi$, for every $m' \in M$. By Lemma 10.28, it follows that $\langle \mathcal{M}, m'\rangle \models \psi^K$, for every $m' \in M$. By the definitions of concatenation and the satisfaction relation, $\langle \mathcal{N} \odot \mathcal{M}, m\rangle \models \neg L \neg L\psi^K$, which completes the proof. ■

**Theorem 10.33** *Let $\mathcal{S}$ be a modal logic contained in* **Sw5**. *A theory $T \subseteq \mathcal{L}_L$ is a* **kd45**-*expansion of a theory $I$ if and only if $T$ is an $\mathcal{S}$-expansion of $I^K$.*

Proof: If $T$ is an $\mathcal{S}$-expansion for $I^K$, then $T$ is an **Sw5**-expansion for $I^K$. By Theorem 10.31, $T$ is a **kd45**-expansion for $I$.

Conversely, assume that $T$ is a **kd45**-expansion for $I$. To complete the proof, it suffices to show that $T$ is an **N**-expansion for $I^K$. To this end, we will once more employ the method of minimal model semantics.

Consider the canonical model for $T$, $\mathcal{M}_T$. Clearly, since $T$ contains $I$, it follows from Lemma 10.28 that $\mathcal{M}_T \models I^K$. Let now $\mathcal{N} = \langle N, \{R_\varphi\}_{\varphi \in \mathcal{L}_L}, U\rangle$ be an **N**-structure such that $\mathcal{N} \odot \mathcal{M}_T \models I^K$.

Consider a world $m \in N$ and a formula $\varphi \in \mathcal{L}$ such that $\langle \mathcal{N} \odot \mathcal{M}_T, m\rangle \models \varphi$. Since $\varphi = \varphi^K$, it follows by Lemma 10.32 that $\langle (\mathcal{M}_T)^m, m\rangle \models \varphi$. Since $T$ is a **kd45**-expansion for $I$, $\mathcal{M}_T$ is $\mathcal{C}_{\mathbf{kd45}}$-minimal for $I$ (Corollary 9.23). By Lemma 10.32, $\langle (\mathcal{M}_T)^m, m\rangle \models I$. Since $I \subseteq T$, $\mathcal{M}_T \models I$. Hence, $(\mathcal{M}_T)^m \models I$. By $\mathcal{C}_{\mathbf{kd45}}$-minimality of $\mathcal{M}_T$. there exists a world $m'$ of $\mathcal{M}_T$ such that $\langle \mathcal{M}_T, m'\rangle \models \varphi$. Thus, $\mathcal{M}_T$ is **N**-minimal for $I$ and, by Corollary 10.12, $T$ is an **N**-expansion for $I^K$. ■

**Corollary 10.34** *A theory $T \subseteq \mathcal{L}_L$ is a* **kd45**-*expansion of a theory $I$ if and only if $T$ is an* **N**-*expansion of $I^K$.* ▲

## 10.3 Autoepistemic logic; context semantics

The proposal of McDermott and Doyle for nonmonotonic modal logics for reasoning about beliefs of an agent capable of *introspection*, which we have been studying in this chapter, has been subject to criticism. The main reason was the presence of some counterintuitive properties of the two nonmonotonic logics in the McDermott and Doyle family that historically were investigated first, namely, the logic obtained by taking for $\mathcal{S}$ the logic of propositional calculus in $\mathcal{L}_L$ (in this book this system is not considered at all) and the nonmonotonic modal logic **S5**. To mention briefly the main deficiencies of these logics, the former admits consistent expansions for a theory $\{\neg Lp, p\}$ and the latter associates infinitely many expansions with the empty theory.

Autoepistemic logic is a modal nonmonotonic formalism proposed by Moore [Moo84, Moo85] as an alternative to the proposal of McDermott and Doyle and a solution to the deficiencies mentioned above. But, as we will prove shortly, autoepistemic logic somewhat ironically turns out to be just one of many modal nonmonotonic logics that can be defined through the identity (9.3). However, the approach used by Moore is different from that of McDermott and Doyle and warrants careful study. An attractive feature of this approach is that it is based on a modification of the semantics of the propositional calculus in $\mathcal{L}_L$ and does not make any direct references to any modal logic.

We start with an intuitive motivation for the approach. As always in such cases, we will describe how an agent capable of introspecting on her beliefs might go around building her total belief set ("expansion") grounded in a collection of her original beliefs.

First, let us note that if the agent considers as models all truth assignments to atoms of $\mathcal{L}$ and to modal atoms $L\varphi$, $\varphi \in \mathcal{L}_L$ , then the resulting logic is simply the logic of propositional calculus in the language $\mathcal{L}_L$. However, our agent is capable of introspection, and some such models are counterintuitive and should be abandoned.

To be specific, consider a situation where a theory $T \subseteq \mathcal{L}_L$ is the collection of the agent's beliefs. Then, a valuation $v$ such that $v(L\varphi) = \mathbf{t}$, where $\varphi$ is a statement the agent does not believe in ($\varphi \notin T$), disagrees with her negative introspection. She can introspect on her belief set $T$ and verify that $\varphi$ is not her belief. A model making the statement "I believe in $\varphi$" true is thus counterintuitive from the agent's point of view and should be disregarded. Similarly, such valuations $v$ in which $v(L\varphi) = \mathbf{f}$ for some $\varphi \in T$ are also counterintuitive, as they are contradictory with the results of the agent's positive introspection.

These informal observations suggest that when studying belief sets of an agent capable of introspection one should restrict oneself to models that are consistent with the results of both positive and negative introspection. Formally, we have the following definition.

**Definition 10.35** Let $v$ be a valuation of atoms of $\mathcal{L}$ and of modal atoms of $\mathcal{L}_L$, and let $T$ be a theory in $\mathcal{L}_L$. Valuation $v$ is called an *autoepistemic interpretation with respect to* $T$ or simply an *autoepistemic* $T$*-interpretation* if for every formula $\varphi \in \mathcal{L}_L$,

$$v(L\varphi) = \mathbf{t} \quad \text{if and only if} \quad \varphi \in T.$$

The theory $T$ is called the *context* of the interpretation. ▲

The concept of autoepistemic interpretation in the standard way yields the corresponding notion of entailment. Given a context $T$, a theory $I \subseteq \mathcal{L}_L$ and a formula $\varphi \in \mathcal{L}_L$, we say that $I$ $T$*-entails* $\varphi$ if for every autoepistemic $T$-interpretation $v$ such that $v(I) = \mathbf{t}$, $v(\varphi) = \mathbf{t}$. We will write

$$I \models_T \varphi$$

to denote that a theory $I$ $T$-entails a formula $\varphi$.

Several simple properties of $T$-entailment are gathered in the following theorem. The first of them states that the relation of $T$-entailment can be expressed entirely in terms of the entailment relation $\models$ of propositional calculus in $\mathcal{L}_L$. It follows from the observation that a valuation $v$ is an autoepistemic $T$-interpretation if and only if $v(\{L\varphi : \varphi \in T\} \cup \{\neg L\varphi : \varphi \notin T\}) = \mathbf{t}$. Consequently, the set of all formulas $T$-entailed by $I$ is closed under propositional consequence. Finally, by the completeness theorem for propositional calculus, it follows that the set of all formulas $T$-entailed by $I$ can equivalently be described by means of the propositional provability operator.

**Theorem 10.36** *Let* $I, T \subseteq \mathcal{L}_L$.

1. *For every* $\varphi \in \mathcal{L}_L$,

$$I \models_T \varphi \quad \textit{if and only if} \quad I \cup \{L\psi : \psi \in T\} \cup \{\neg L\psi : \psi \notin T\} \models \varphi.$$

2. *The theory* $\{\varphi : I \models_T \varphi\}$ *is closed under propositional consequence;*
3. $\{\varphi : I \models_T \varphi\} = Cn(I \cup \{L\psi : \psi \in T\} \cup \{\neg L\psi : \psi \notin T\})$. ▲

Another important property of autoepistemic $T$-interpretations is that they are uniquely determined by the context $T$ and by a valuation of atoms of $\mathcal{L}$.

**Theorem 10.37** *Let $v$ and $v'$ be two autoepistemic $T$-interpretations. Then $v = v'$ if and only if $v|_{\mathcal{L}} = v'|_{\mathcal{L}}$.*

Proof: Clearly, if $v = v'$ then $v|_{\mathcal{L}} = v'|_{\mathcal{L}}$. To prove the converse implication, we prove that if $v|_{\mathcal{L}} = v'|_{\mathcal{L}}$, then for every formula $\varphi \in \mathcal{L}_L$, $v(\varphi) = v'(\varphi)$. To this end, we proceed by induction on the length of the formula. If $\varphi \in \mathcal{L}$, then $v(\varphi) = v'(\varphi)$ follows from the assumption. Consider now a formula $\varphi \notin \mathcal{L}$ and assume that the claim holds for all formulas of $\mathcal{L}_L$ which are shorter than $\varphi$. There are several cases to consider. First assume that $\varphi = \varphi_1 \vee \varphi_2$. By the induction hypothesis, $v(\varphi_i) = v'(\varphi_i)$, $i = 1, 2$. Consequently, $v(\varphi) = v'(\varphi)$. The cases when $\varphi = \varphi_1 \wedge \varphi_2$, $\varphi = \psi_1 \supset \psi_2$, and $\varphi = \neg\psi$ are equally straightforward. The last case to consider is when $\varphi = L\psi$. In this case, since both $v$ and $v'$ are autoepistemic $T$-interpretations, if $\psi \in T$, then both $v(\varphi) = \mathbf{t}$ and $v'(\varphi) = \mathbf{t}$, and if $\psi \notin T$, then both $v(\varphi) = \mathbf{f}$ and $v'(\varphi) = \mathbf{f}$. Thus, $v(\varphi) = v'(\varphi)$. ■

The proof of the previous result suggests a way of reconstructing an autoepistemic $T$-interpretation $v$ out of $T$ and $v|_{\mathcal{L}}$ and leads to a different but equivalent notion of entailment.

**Definition 10.38** A pair $(u, T)$, where $u$ is a valuation of atoms of $\mathcal{L}$ and $T \subseteq \mathcal{L}_L$, is called a *context interpretation with the context* $T$. The corresponding entailment relation $\models_{(u,T)}$ is defined recursively as follows:

1. For an atom $p$ of $\mathcal{L}$, $\models_{(u,T)} p$ if and only if $u(p) = \mathbf{t}$.
2. $\models_{(u,T)} \neg\varphi$ if and only if $\not\models_{(u,T)} \varphi$.
3. $\models_{(u,T)} \varphi \wedge \psi$ if and only if $\models_{(u,T)} \varphi$ and $\models_{(u,T)} \psi$.
4. $\models_{(u,T)} \varphi \vee \psi$ if and only if $\models_{(u,T)} \varphi$ or $\models_{(u,T)} \psi$.
5. $\models_{(u,T)} \varphi \supset \psi$ if and only if $\not\models_{(u,T)} \varphi$ or $\models_{(u,T)} \psi$.
6. $\models_{(u,T)} L\varphi$ if and only if $\varphi \in T$. ▲

Given a context interpretation $(w, T)$, we now define a valuation $v_{w,T}$ of $\mathcal{L}_L$ by setting

$$v_{w,T}(\varphi) = \mathbf{t} \quad \text{if and only if} \quad \models_{(w,T)} \varphi,$$

where $\varphi$ is an atom of $\mathcal{L}$ or a modal atom of $\mathcal{L}_L$. We have the following result establishing the correspondence between context and autoepistemic interpretations.

**Theorem 10.39** *Let $T \subseteq \mathcal{L}_L$.*

1. *For every valuation $w$ of $\mathcal{L}$, $v_{w,T}$ is an autoepistemic $T$-interpretation.*
2. *For every autoepistemic $T$-interpretation $u$,*

$$u = v_{w,T},$$

*where $w = u|_{\mathcal{L}}$.*

Proof: Assertion (1) follows directly from the definition of $v_{w,T}$ and from part (5) of the definition of context interpretations. Assertion (2) follows easily by induction on the length of a formula. ■

Theorem 10.39 has the following obvious corollary.

**Corollary 10.40** *Let $T \subseteq \mathcal{L}_L$, $I \subseteq \mathcal{L}_L$ and let $\varphi \in \mathcal{L}_L$. Then $I \models_T \varphi$ if and only if $\models_{(v,T)} \varphi$ for every propositional valuation $v$ such that $\models_{(v,T)} I$.* ▲

Thus, the semantics of autoepistemic and context interpretations are equivalent. In the rest of this chapter we will use them interchangingly.

The theory $\{\varphi: I \models_T \varphi\}$ plays a crucial role in our further discussion. It can be regarded as the collection of the formulas an agent capable of introspecting can derive from $I$ (justify) based on what she believes in and does not believe in. The key question that has to be asked is: Which theories $T$ should be considered by the agent as candidates for the process of building her belief set $\{\varphi: I \models_T \varphi\}$ on the basis of some initial assumptions $I$? Before giving a general answer, let us first discuss some examples.

**Example 10.41** Consider the theory $I = \{\neg Lq \supset p, \neg Lp \supset q\}$. Can a consistent belief set $T$ contain both $p$ and $q$? To answer this question, it is more convenient to use the concept of context interpretations. Let us assume that $p, q \in T$. Consider a valuation $v$ of $\mathcal{L}$ such that $v(p) = v(q) = \mathbf{f}$. Since $p, q \in T$, $\models_{(v,T)} Lp$ and $\models_{(v,T)} Lq$. Consequently,

$$\models_{(v,T)} I.$$

By Corollary 10.40, $p \notin \{\varphi: I \models_T \varphi\}$. Similarly, $q \notin \{\varphi: I \models_T \varphi\}$. Intuitively, this means that an agent capable of introspecting on her beliefs would not be able to regard $p$ and $q$ as true, contrary to the fact that both $p$ and $q$ are her beliefs. Thus, a theory $T$ containing $p$ and $q$ cannot serve as the agent's belief set, as not all of the beliefs in $T$ can be justified.

On the other hand, can a context $T$, which contains neither $p$ nor $q$, be a belief set of the agent? Consider any autoepistemic $T$-interpretation $v$. Clearly, $v(Lp) = v(Lq) = \mathbf{f}$. If $v(I) = \mathbf{t}$, then $v(p) = v(q) = \mathbf{t}$. Thus, $p, q \in \{\varphi: I \models_T \varphi\}$. In this case, the agent must regard both $p$ and $q$ as true even though both are missing from her belief set. Hence, a theory $T$ containing neither $p$ nor $q$ also is not a suitable candidate for the agent's belief set, as it allows the justification of facts in which she does not believe. ▲

This example illustrates that a theory $T$, to be a candidate for the belief set of an agent with capabilities to introspect on her beliefs, must coincide with exactly those beliefs that she can justify on the basis of what she believes in and what she does not believe in, that is, with $\{\varphi: I \models_T \varphi\}$. This observation leads directly to the following definition.

**Definition 10.42** Let $I \subseteq \mathcal{L}_L$. A theory $T$ is a *stable expansion* or, simply, an *expansion* for $I$ if $T$ is consistent and

$$T = \{\varphi: I \models_T \varphi\}.$$

▲

The nonmonotonic logic based on the notion of a stable expansion is called the *autoepistemic logic.*

It follows directly from Theorem 10.36(3) that the semantic definition of the concept of an expansion has a syntactic counterpart.

**Theorem 10.43** *A consistent theory $T \subseteq \mathcal{L}_L$ is a stable expansion of $I \subseteq \mathcal{L}_L$ if and only if*

$$T = Cn(I \cup \{L\varphi: \varphi \in T\} \cup \{\neg L\varphi: \varphi \notin T\}). \tag{10.11}$$

▲

Equation (10.11) may possess inconsistent solutions. For instance, a theory $I = \{Lp \supset q, Lr \supset \neg q\}$ possesses just one stable expansion. It is a *consistent* solution of equation (10.11) and equals to $ST(\emptyset)$. However, it is easy to see that the inconsistent theory, $\mathcal{L}_L$, is also a solution to equation (10.11). On the other hand theory $I$ is consistent in propositional logic (that is, $I$ does not entail $\perp$ in propositional logic).

Theorems 10.43 and 10.19 imply the following corollary.

**Corollary 10.44**

1. *A consistent theory $T$ is a stable expansion of $I$ if and only if it is a* **kd45**-*expansion of $I$.*

2. *Autoepistemic logic coincides with the nonmonotonic logic* **kd45**. ▲

It is clear now, by the results characterizing **kd45**-expansions, that the theory from Example 10.41 has two stable expansions: $ST(\{p\})$ and $ST(\{q\})$.

The key idea of the approach presented above is to restrict the class of all propositional interpretations (in the language $\mathcal{L}_L$) to the class of autoepistemic interpretations. But there are at least two other natural classes of interpretations that lead to interesting nonmonotonic logics. We describe them briefly below.

The first class can be obtained by extending the class of autoepistemic interpretations. One might argue that the requirement that $v(L\varphi) = \mathbf{t}$ if and only if $\varphi \in T$ is too strong. It excludes, in particular, interpretations $v$ such that $v(L\varphi) = \mathbf{f}$, for some $\varphi \in T$ with $v(\varphi) = \mathbf{f}$. However, let us consider an agent who uses introspection to construct interpretations of $\mathcal{L}_L$ from interpretations of $\mathcal{L}$ in the following recursive manner. If $\varphi \notin T$ then the formula $\neg L\varphi$ is evaluated as **f**. This is exactly what happens in the case of autoepistemic interpretations. When $\varphi \in T$, the situation changes. When $\varphi \in T$, in order to assign a value to $L\varphi$, the agent considers the value of $\varphi$. If $\varphi$ has been evaluated as **t**, $L\varphi$ is also evaluated as **t**. If $\varphi$ has been evaluated as **f**, the agent is free to assigns to $L\varphi$ an arbitrary truth value. Such an agent may, in particular, construct an interpretation $v$ of $\mathcal{L}_L$ such that $v(L\varphi) = \mathbf{f}$, for some $\varphi \in T$ with $v(\varphi) = \mathbf{f}$ as well as an interpretation $w$ of $\mathcal{L}_L$ such that $w(L\psi) = \mathbf{t}$, for some $\psi \in T$ with $w(\psi) = \mathbf{f}$.

**Definition 10.45** Let $v$ be a valuation of atoms of $\mathcal{L}$ and of modal atoms of $\mathcal{L}_L$, and let $T$ be a theory in $\mathcal{L}_L$. Valuation $v$ is called a *weak autoepistemic interpretation with respect to $T$* or a *weak autoepistemic $T$-interpretation* if for every formula $\varphi \in \mathcal{L}_L$:

1. If $\varphi \notin T$, then $v(L\varphi) = \mathbf{f}$.

2. If $\varphi \in T$, then $v(L\varphi) = \mathbf{t}$ providing $v(\varphi) = \mathbf{t}$. ▲

The corresponding notion of entailment depends on the theory $T$ and will be denoted $\models^w_T$. Formally, for a theory $I \subseteq \mathcal{L}_L$ and for a formula $\varphi \in \mathcal{L}_L$, $I \models^w_T \varphi$ if for every weak autoepistemic $T$-interpretation $v$ such that $v(I) = \mathbf{t}$, $v(\varphi) = \mathbf{t}$. A theory $T$ is a *strong expansion* for $I$ if

$$T = \{\varphi : I \models^{w}_{T} \varphi\}.$$

The next theorem gives two characterizations of strong expansions.

**Theorem 10.46**

1. *A consistent theory $T \subseteq \mathcal{L}_L$ is a strong expansion for $I \subseteq \mathcal{L}_L$ if and only if*
$$T = Cn(I \cup \{\varphi \supset L\varphi : \varphi \in T\} \cup \{\neg L\varphi : \varphi \notin T\}).$$
2. *A consistent theory $T \subseteq \mathcal{L}_L$ is a strong expansion for $I \subseteq \mathcal{L}_L$ if and only if $T$ is a* **w5***-expansion for $I$, where* **w5** *denotes the modal logic specified by the schema* w5.

Proof: Part (1) follows from an easy observation that an interpretation $v$ is a weak autoepistemic $T$-interpretation if and only if

$$v(\{\varphi \supset L\varphi : \varphi \in T\} \cup \{\neg L\varphi : \varphi \notin T\}) = \mathbf{t}.$$

Part (2) can be proved in the same way as Theorems 10.19 and 10.24. ■

The second class of interpretations of $\mathcal{L}_L$ can be obtained by restricting the class of weak autoepistemic $T$-interpretations. Here, we require that an agent, when constructing an interpretation, proceeds as follows: if $\varphi \notin T$ then evaluate $L\varphi$ to $\mathbf{f}$; otherwise evaluate $L\varphi$ to $\mathbf{f}$ if $v(\varphi) = \mathbf{f}$, and to $\mathbf{t}$ it $v(\varphi) = \mathbf{t}$.

**Definition 10.47** Let $v$ be a valuation of atoms of $\mathcal{L}$ and of modal atoms of $\mathcal{L}_L$, and let $T$ be a theory in $\mathcal{L}_L$. Valuation $v$ is called a *reflexive autoepistemic interpretation with respect to $T$* or a *reflexive autoepistemic $T$-interpretation* if for every formula $\varphi \in \mathcal{L}_L$:

1. If $\varphi \notin T$, then $v(L\varphi) = \mathbf{f}$.
2. If $\varphi \in T$, then $v(L\varphi) = v(\varphi)$. ▲

The corresponding notion of entailment will be denoted $\models^{r}_{T}$ and, for a theory $I \subseteq \mathcal{L}_L$ and for a formula $\varphi \in \mathcal{L}_L$, we write $I \models^{r}_{T} \varphi$ if for every reflexive autoepistemic $T$-interpretation $v$ such that $v(I) = \mathbf{t}$, $v(\varphi) = \mathbf{t}$. A theory $T$ is a *reflexive expansion* for $I$ if

$$T = \{\varphi : I \models^{r}_{T} \varphi\}.$$

The following result, analogous to Theorem 10.46, holds.

**Theorem 10.48**

1. *A consistent theory* $T \subseteq \mathcal{L}_L$ *is a reflexive expansion for* $I \subseteq \mathcal{L}_L$ *if and only if*
$$T = Cn(I \cup \{\varphi \equiv L\varphi : \varphi \in T\} \cup \{\neg L\varphi : \varphi \notin T\}).$$

2. *A consistent theory* $T \subseteq \mathcal{L}_L$ *is a reflexive expansion for* $I \subseteq \mathcal{L}_L$ *if and only if* $T$ *is an* **Sw5**-*expansion for* $I$. ▲

The nonmonotonic logic based on the notion of a reflexive expansion is called the *reflexive autoepistemic logic* or the *logic of reflexive knowledge.*

To sum up the results of this section, the method of context-sensitive semantics is a powerful tool for defining nonmonotonic modal formalisms. Three arguably most natural classes of autoepistemic interpretations lead to three logics in the McDermott-Doyle family. Two of them — nonmonotonic **kd45** and nonmonotonic **Sw5** — are quite well known and were investigated earlier in the book. The third is related to the subnormal logic **w5**. The fact that these nonmonotonic logics can be defined without any reference to any modal system by a simple modification of the semantics of propositional calculus is an important advantage. It seems to be a challenging open problem, with potentially far-reaching consequences, to decide whether any other nonmonotonic logic in the McDermott-Doyle family admits similar characterization.

## 10.4 Comments and remarks

The nonmonotonic logic **N** was introduced by the authors [MT90]. It was further investigated by Fitting and the authors in [FMT92]. The minimal knowledge characterization of the nonmonotonic modal logic **N** was obtained by Schwarz and Truszczyński [ST93]. The characterization of **N**-expansions for a theory $I$ in terms of **N**-perfect partitions for $I$ is due to Schwarz and the authors and can be found in [MST91].

The concept of context semantics presented in Section 10.3 is due to Moore [Moo85].

Autoepistemic logic was introduced by Moore [Moo84, Moo85] and, in a slightly different guise, by Levesque [Lev90]. Let us note that the definition of stable expansions given in this book is slightly different from the original definition of Moore. We require that stable expansions are *consistent* solutions to (10.11) while Moore does not make this stipulation. The nonmonotonic logic **kd45** was shown to be equivalent to autoepistemic logic by Schwarz

[Shv90]. The characterization of stable expansions given in this chapter is taken from the same paper. A similar characterization was found independently by Niemelä [Nie88].

Weak autoepistemic interpretations and the notion of a strong expansion were introduced by the authors in [MT90].

The reflexive autoepistemic logic was introduced by Schwarz [Sch91a]. He proved that the reflexive autoepistemic logic coincides with the nonmonotonic logic **Sw5**. He also found the connections between reflexive autoepistemic and autoepistemic logics, and characterized reflexive expansions.

# 11. Topics in modal nonmonotonic logic

In this chapter we study a variety of loosely connected topics in nonmonotonic modal logic. We start with the discussion of algorithmic issues related to the problem of computing $\mathcal{S}$-expansions. Next, we study the relations between modal nonmonotonic logics. It turns out that often, despite being based on different modal *monotonic* logics, the corresponding modal *nonmonotonic* logics coincide. In the following section we investigate the effects of adding new definitions to a theory $I \subseteq \mathcal{L}_L$ on the collection of $\mathcal{S}$-expansions. In the last three sections, we consider the issues of bounding negative introspection. In particular, we study bounding negative introspection to formulas from $\mathcal{L}$, and to formulas of $L$-depths at most 1.

## 11.1 Algorithms to compute $\mathcal{S}$-expansions

In this section, we utilize results obtained earlier to investigate the following general problem:

> Given a modal logic $\mathcal{S}$, how to compute all $\mathcal{S}$-expansions of a finite theory $I$?

Clearly, in order to compute *all* $\mathcal{S}$-expansions of $I$, the set of all $\mathcal{S}$-expansions of $I$ must be finite. In addition, a method is needed to uniquely identify an $\mathcal{S}$-expansion by means of some finite representation.

Both requirements are satisfied for logics $\mathcal{S}$ contained in **kd45** or **Sw5**. Indeed, by Corollary 9.44, a finite theory has finitely many $\mathcal{S}$-expansions and, by Theorem 9.41, each $\mathcal{S}$-expansion of a finite theory $I$ can uniquely be represented by a partition $(\Phi, \Psi)$ of $MA(I)$, introspection-consistent with $I$. Moreover, if a logic $\mathcal{S}$ properly contains logic **kd45** or **Sw5** and is contained in **S5**. then a finite theory $I$ may have infinitely many $\mathcal{S}$-expansions (Theorem 9.45). Therefore, in the remainder of this section we focus on logics contained in **kd45** or **Sw5**.

We start with an auxiliary algorithm that decides whether a given partition $(\Phi, \Psi)$ of $MA(I)$ is introspection-consistent with $I$. The algorithm has to check conditions (IC2) and (IC3) of Definition 9.35. To this end, in both cases, one has to determine whether a certain theory is propositionally consistent in $\mathcal{L}_L$. We first show how to reduce such task to checking consistency of theories in the language $\mathcal{L}$.

Consider an $I$-formula $\varphi$. Replace in $\varphi$ each occurrence of $L\psi$ by $\bot$ if $\psi \in \Phi$ and by $\top$ if $\psi \in \Psi$ and simplify the resulting formula by applying standard rewriting rules:

$$\begin{aligned}
\neg\top &\mapsto \bot \\
\neg\bot &\mapsto \top \\
\top \wedge \alpha &\mapsto \alpha \\
\bot \wedge \alpha &\mapsto \bot \\
\top \vee \alpha &\mapsto \top \\
\bot \vee \alpha &\mapsto \alpha \\
\top \supset \alpha &\mapsto \alpha \\
\bot \supset \alpha &\mapsto \top
\end{aligned}$$

Denote the result by $\varphi_{(\Phi,\Psi)}$. Observe that $\varphi_{(\Phi,\Psi)} \in \mathcal{L}$. For a theory $A$ consisting of $I$-formulas, define $A_{(\Phi,\Psi)} = \{\varphi_{(\Phi,\Psi)} : \varphi \in A\}$.

**Theorem 11.1** *Let $I \subseteq \mathcal{L}_L$ and let $(\Phi, \Psi)$ be a partition of $MA(I)$. Then, for every theory $A$ consisting of $I$-formulas, $A \cup \{\neg L\varphi : \varphi \in \Phi\} \cup \{L\psi : \psi \in \Psi\}$ is propositionally consistent (in $\mathcal{L}_L$) if and only if $A_{(\Phi,\Psi)}$ is propositionally consistent (in $\mathcal{L}$).*

Proof: First, let us observe that the definition of $\varphi_{(\Phi,\Psi)}$ immediately implies that for every valuation $v$ of $\mathcal{L}_L$:

**(a)** If $\varphi = \neg\psi$, then $v(\varphi_{(\Phi,\Psi)}) = \mathbf{t}$ if and only if $v(\psi_{(\Phi,\Psi)}) = \mathbf{f}$.

**(b)** If $\varphi = \psi^1 \vee \psi^2$, then $v(\varphi_{(\Phi,\Psi)}) = \mathbf{t}$ if and only if $v(\psi^1_{(\Phi,\Psi)}) = \mathbf{t}$ or $v(\psi^2_{(\Phi,\Psi)}) = \mathbf{t}$.

**(c)** If $\varphi = \psi^1 \wedge \psi^2$, then $v(\varphi_{(\Phi,\Psi)}) = \mathbf{t}$ if and only if $v(\psi^1_{(\Phi,\Psi)}) = \mathbf{t}$ and $v(\psi^2_{(\Phi,\Psi)}) = \mathbf{t}$.

**(d)** If $\varphi = \psi^1 \supset \psi^2$, then $v(\varphi_{(\Phi,\Psi)}) = \mathbf{t}$ if and only if $v(\psi^1_{(\Phi,\Psi)}) = \mathbf{f}$ or $v(\psi^2_{(\Phi,\Psi)}) = \mathbf{t}$.

Using (a) - (d), one can prove by the induction on the complexity of a formula that if $v$ is a valuation of $\mathcal{L}_L$ such that $v(\{\neg L\varphi : \varphi \in \Phi\} \cup \{L\psi : \psi \in \Psi\}) = \mathbf{t}$ then for every $I$-formula $\varphi$, $v(\varphi) = v(\varphi_{(\Phi,\Psi)})$. Consequently, if $A \cup \{\neg L\varphi : \varphi \in \Phi\} \cup \{L\psi : \psi \in \Psi\}$ is consistent, $A_{(\Phi,\Psi)}$ is consistent, as well.

---

**IC**$(I,\Phi,\Psi)$
*Input:* A finite theory $I \subseteq \mathcal{L}_L$ and a partition $(\Phi,\Psi)$ of $MA(I)$;
*Output:* The algorithm returns **true** if $(\Phi,\Psi)$ is introspection-consistent with $I$ and **false**, otherwise.

compute $(I \cup \Psi)_{(\Phi,\Psi)}$ and $(I \cup \Psi \cup \{\neg\varphi\})_{(\Phi,\Psi)}$, for all $\varphi \in \Phi$;
**if** at least one of these theories is propositionally inconsistent
    **then return{false}**
    **else return{true}**
**fi**

---

**Fig. 11.1.** Algorithm **IC**

To prove the converse implication, consider a valuation $w$ of $\mathcal{L}$ such that $w(A_{(\Phi,\Psi)}) = \mathbf{t}$. Let $v$ be any valuation of $\mathcal{L}_L$ such that $v|_{\mathcal{L}} = w$ and $v(\{\neg L\varphi: \varphi \in \Phi\} \cup \{L\psi: \psi \in \Psi\}) = \mathbf{t}$ (clearly, such valuations exist). Since $v|_{\mathcal{L}} = w$ and $\varphi_{(\Phi,\Psi)} \in \mathcal{L}$, $v(\varphi_{(\Phi,\Psi)}) = w(\varphi_{(\Phi,\Psi)})$. We have already shown earlier that if $\varphi$ is an $I$-formula then $v(\varphi) = v(\varphi_{(\Phi,\Psi)})$. Thus, for every $I$-formula $\varphi$, $v(\varphi) = w(\varphi_{(\Phi,\Psi)})$. Consequently, $A \cup \{\neg L\varphi: \varphi \in \Phi\} \cup \{L\psi: \psi \in \Psi\}$ is consistent. ∎

Theorem 11.1 implies the correctness of algorithm **IC**. This algorithm decides whether a given partition of $MA(I)$ is introspection-consistent with $I$. It is presented in Figure 11.1.

Our first algorithm to compute all $\mathcal{S}$-expansions for a finite theory $I$ is based on Theorem 9.50. The assumptions of this theorem restrict the applicability of the algorithm to those logics $\mathcal{S}$ that meet the following conditions:

**(a)** $\mathcal{S} \subseteq$ **S4f**.

**(b)** $\mathcal{S}$ is characterized by a cluster-closed class of models.

In addition, we need to assume that:

**(c)** There is an automated theorem prover for logic $\mathcal{S}$, that is, a procedure that, given $\varphi \in \mathcal{L}_L$ and a finite theory $A \subseteq \mathcal{L}_L$, decides whether $A \vdash_{\mathcal{S}} \varphi$.

The idea behind the algorithm can roughly be described as follows. One has to consider all partitions $(\Phi,\Psi)$ of $MA(I)$. For each such partition, one has to check whether it is $\mathcal{S}$-perfect for $I$. If so, the theory $T_I(\Phi,\Psi)$ or, to be

**All_ExpsA**$(I)$

*Input:* A finite theory $I \subseteq \mathcal{L}_L$;

*Output:* The list of all $\mathcal{S}$-expansions for $I$; each $\mathcal{S}$-expansion on the list is represented by the corresponding $\mathcal{S}$-perfect partition for $I$.

compute $MA(I)$;<br>
**for** each $\Phi \subseteq MA(I)$ **do**<br>
    $\Psi := MA(I) \setminus \Phi$;<br>
    **if** IC$(I, \Phi, \Psi)$ and $I \cup \{\neg\varphi : \varphi \in \P i\} \vdash_{\mathcal{S}} \Psi$ **then**<br>
        output $(\Phi, \Psi)$ and report that $T_I(\Phi, \Psi)$ is an $\mathcal{S}$-expansion for $I$<br>
    **fi**<br>
**rof**

**Fig. 11.2.** Algorithm **All_ExpsA**

precise, the partition $(\Phi, \Psi)$, is reported as an $\mathcal{S}$-expansion. A more detailed description of the algorithm is given in Figure 11.2.

Algorithm **All_ExpsA** is computationally very complex. Let $|MA(I)| = k$. In such case the algorithm considers $2^k$ partitions $(\Phi, \Psi)$ of $MA(I)$. For each such partition there are up to $|\Psi|$ calls to the theorem prover for logic $\mathcal{S}$ and up to $|\Phi| + 1$ calls to a theorem prover for the propositional calculus (during the call to the algorithm **IC**). Since $|\Phi| = O(k)$ and $|\Psi| = O(k)$, altogether there are $O(k2^k)$ calls to the theorem prover for $\mathcal{S}$ and $O(k2^k)$ calls to the theorem prover for the propositional calculus.

Next, we describe an alternative method to compute all $\mathcal{S}$-expansions for a finite theory $I$. Theorems 9.39(5) and 9.41 imply that each $\mathcal{S}$-expansion for $I$ (assuming that $\mathcal{S}$ is contained in **kd45** or **Sw5**) is of the form $ST(S)$, for some set $S$ consisting of propositional formulas built of atoms occurring in formulas from $I$. To see whether a consistent stable theory $ST(S)$ is an $\mathcal{S}$-expansion for $I$ ($\mathcal{S} \subseteq$ **S5**), it is enough to decide whether $I \subseteq ST(S)$ and $S \subseteq Cn_{\mathcal{S}}(I \cup \{\neg L\varphi : \varphi \notin ST(S)\})$ (Corollary 9.8).

We will first consider the case when $\mathcal{S} \subseteq$ **S4f**. If $\mathcal{S}$ is characterized by a cluster-closed class of models, checking that $S \subseteq Cn_{\mathcal{S}}(I \cup \{\neg L\varphi : \varphi \notin ST(S)\})$ can be replaced by checking whether $S \subseteq Cn_{\mathcal{S}}(I \cup \{\neg L\varphi : \varphi \in \Phi\})$, where $\Phi = MA(I) \setminus ST(S)$ (Corollary 9.55).

It follows from our discussion that if a logic $\mathcal{S}$ satisfies conditions (a) - (c) given above, then the algorithm **All_ExpsB** described in Figure 11.3 correctly outputs all $\mathcal{S}$-expansions for a finite theory $I$.

---

**All_ExpsB**$(I)$
*Input:* A finite theory $I \subseteq \mathcal{L}_L$;
*Output:* The list of all $\mathcal{S}$-expansions for $I$.

compute $MA(I)$;
A:= the set of all formulas in conjunctive normal form built of propositional letters occurring in $I$ such that no clause contains the same literal occurring twice, no clause contains two contradictory literals and no clause in a formula subsumes another clause in the formula;
($\star$ clearly, $A$ is finite $\star$)
**for** each $S \subseteq A$ **do**
  $\Phi := MA(I) \setminus ST(S)$;
  **if** $I \subseteq ST(S)$ and $I \cup \{\neg L\varphi : \varphi \in \Phi\} \vdash_{\mathcal{S}} S$ **then**
    output $S$ and report that $ST(S)$ is an $\mathcal{S}$-expansion for $I$
  **fi**
**rof**

---

**Fig. 11.3.** Algorithm **All_ExpsB**

To produce $\Phi$ and to check if $I \subseteq ST(S)$, one can use any of the algorithms developed in Chapter 8 which, given a formula $\varphi \in \mathcal{L}_L$ and a finite set $S \subseteq \mathcal{L}$, decide whether $\varphi \in ST(S)$.

Similarly as the algorithm **All_ExpsA**, the algorithm **All_ExpsB** is computationally complex. The size of $S$ may be exponential in the size of $I$ and, to make matters worse, all subsets of $A$ have to be considered. For each such subset $S$ there are up to $|S|$ calls to a theorem prover for logic $\mathcal{S}$ and $|I|+|MA(I)|$ calls to one of the algorithms **InStabA**, **InStabB**, **InStabC** or **InStabD**, which are quite complex themselves. However, this method might prove more efficient if the size of the set $A$ can be substantially reduced. We will discuss one such case at the end of this section.

The algorithms **All_ExpsA** and **All_ExpsB** can be modified to yield algorithms computing all **kd45**-expansions and **Sw5**-expansions for a finite theory $I$. To this end, the condition in the **if** instruction in the algorithm **All_ExpsA** has to be changed to:

- IC$(I,\Phi,\Psi)$ and $I \cup \{\neg L\varphi : \varphi \in \Phi\} \cup \{L\psi : \psi \in \Psi\} \vdash \Psi$, in order to compute all **kd45**-expansions.

- IC$(I,\Phi,\Psi)$ and $I \cup \{\neg L\varphi : \varphi \in \Phi\} \cup \{\psi \equiv L\psi : \psi \in \Psi\} \vdash \Psi$, in order to compute all **Sw5**-expansions.

The resulting algorithms are referred to as **kd45_expsA** and **Sw5_expsA**, respectively.

To modify the algorithm **All_ExpsB**, the condition in the **if** instruction has to be changed to:

- $I \subseteq ST(S)$ and $I \cup \{\neg L\varphi : \varphi \in \Phi\} \cup \{L\psi : \psi \in \Psi\} \vdash S$, in order to compute all **kd45**-expansions.
- $I \subseteq ST(S)$ and $I \cup \{\neg L\varphi : \varphi \in \Phi\} \cup \{\psi \equiv L\psi : \psi \in \Psi\} \vdash S$, in order to compute all **Sw5**-expansions.

The resulting algorithms are referred to as **kd45_expsB** and **Sw5_expsB**, respectively.

The correctness of the algorithms **kd45_expsA** and **Sw5_expsA** follows from Theorems 10.21 and 10.25. Corollaries 10.23 and 10.27 imply the correctness of the algorithms **kd45_expsB** and **Sw5_expsB**.

Now we will discuss a method which may improve the efficiency of the algorithms of type **B** for computing $\mathcal{S}$-expansions, where $\mathcal{S} \subseteq$ **kd45**.

**Definition 11.2** An *epistemic clause* is any formula of the form

$$L\alpha_1 \wedge \ldots \wedge L\alpha_k \wedge \neg L\beta_1 \wedge \ldots \wedge \neg L\beta_m \supset \gamma, \tag{11.1}$$

where $\gamma \in \mathcal{L}$. Formula $\gamma$ is referred to as the *head* of an epistemic clause (11.1). For a theory $I$ consisting of epistemic clauses, we define $H(I)$ to be the collection of the heads of all clauses from $I$. ▲

**Definition 11.3** Let $I$ be a theory consisting of epistemic clauses and let $T$ be stable.

1. We say that a formula $\varphi \in H(I)$ has an *I-support* in $T$ if for some clause
$$L\alpha_1 \wedge \ldots \wedge L\alpha_k \wedge \neg L\beta_1 \wedge \ldots \wedge \neg L\beta_m \supset \varphi$$
from $I$, $\alpha_i \in T$, $1 \leq i \leq k$, and $\beta_i \notin T$, $1 \leq i \leq m$.
2. We define $S_T(I) = \{\varphi \in H(I) : \varphi$ has an $I$-support in $T\}$. ▲

We have the following useful lemma

**Lemma 11.4** *Let $T$ be a consistent stable theory. Let $I$ consist of epistemic clauses. Then*

$$Cn(I \cup \{L\varphi : \varphi \in T\} \cup \{\neg L\varphi : \varphi \notin T\}) \cap \mathcal{L} = Cn(S_T(I)).$$

Proof: Inclusion $\supseteq$ is obvious. To prove the converse inclusion, we proceed as follows. Assume that a formula $\psi$ belongs to the left-hand side but does not belong to the right-hand side. Then there must be a valuation $V$ of $\mathcal{L}$ such that $V$ evaluates all the formulas in $S_T(I)$ as $\mathbf{t}$, but $V(\psi) = \mathbf{f}$. Extend this valuation $V$ to a valuation $W$ of the entire $\mathcal{L}_L$ by setting $W(p) = V(p)$ for atoms $p$ of $\mathcal{L}$, and setting $W(L\varphi) = \mathbf{t}$ precisely when $\varphi \in T$. It is easy to see that $W$ evaluates $\{L\varphi : \varphi \in T\} \cup \{\neg L\varphi : \varphi \notin T\}$ as $\mathbf{t}$.

Let $C = (L\alpha_1 \wedge \ldots \wedge L\alpha_k \wedge \neg L\beta_1 \wedge \ldots \wedge \neg L\beta_m \supset \varphi)$ be a formula from $I$. If $\varphi \in S_T(I)$ then $W(\varphi) = V(\varphi) = \mathbf{t}$. Consequently, $W(C) = \mathbf{t}$. If $\varphi \notin S_T(I)$, then for some $i$, $1 \leq i \leq k$, $W(L\alpha_i) = \mathbf{f}$, or for some $j$, $1 \leq j \leq m$, $W(\neg L\beta_j) = \mathbf{f}$. In both cases $W(C) = \mathbf{t}$. Hence, $W(I) = \mathbf{t}$.

It follows that $W$ evaluates $Cn(I \cup \{L\varphi : \varphi \in T\} \cup \{\neg L\varphi : \varphi \notin T\})$ as $\mathbf{t}$. Therefore $W(\psi) = \mathbf{t}$. On the other hand, $W$ coincides with $V$ on all the formulas of $\mathcal{L}$. Hence $W(\psi) = \mathbf{f}$, a contradiction. ■

We have the following result characterizing **kd45**-expansions for theories consisting of epistemic clauses.

**Theorem 11.5** *Let $I$ be a theory consisting of epistemic clauses.*

1. *A consistent theory $T$ is a* **kd45***-expansion for $I$ if and only if $I \subseteq T$ and $T = ST(S)$ for some set $S \subseteq H(I)$ such that each $\gamma \in S$ has an $I$-support in $T$.*
2. *A consistent theory $T$ is a* **kd45***-expansion for $I$ if and only if $T = ST(S_T(I))$.*

Proof: (1) Let us assume first that $I \subseteq T$ and $T = ST(S)$ for some set $S \subseteq H(I)$ such that each $\gamma \in S$ has an $I$-support in $T$. It follows easily from the definition of $S$ that

$$S \subseteq Cn(I \cup \{\neg L\varphi : \varphi \notin T\} \cup \{L\psi : \psi \in T\}).$$

Thus, by Corollary 10.22, $T$ is a **kd45**-expansion for $I$.

Conversely, assume that $T$ is a **kd45**-expansion for $I$. Then $T$ is consistent. Let $S = S_T(I)$. It is an easy exercise to show that

$$\begin{aligned} T &= Cn(I \cup \{\neg L\varphi : \varphi \notin T\} \cup \{L\psi : \psi \in T\}) \\ &= Cn(S \cup \{\neg L\varphi : \varphi \notin T\} \cup \{L\psi : \psi \in T\}) \end{aligned}$$

and that

$$Cn(S \cup \{\neg L\varphi : \varphi \notin T\} \cup \{L\psi : \psi \in T\}) \cap \mathcal{L} = Cn(S).$$

Hence, $T \cap \mathcal{L} = Cn(S)$ and, consequently, $T = ST(S)$.

(2) The last paragraph of the proof of (1) proves also the necessity part of (2). To prove sufficiency, assume that $T = ST(S_T(I))$. Then, in order to prove that $T$ is a **kd45**-expansion for $I$ it is enough, by part (1), to prove that $I \subseteq ST(S_T(I)) = T$.

So assume that $C = L\alpha_1 \wedge \ldots \wedge L\alpha_k \wedge \neg L\beta_1 \wedge \ldots \wedge \neg L\beta_m \supset \gamma$ belongs to $I$. If for some $1 \leq i \leq k$, $\alpha_i \notin T$ then $\neg L\alpha_i \in T$ and so $C \in T$. Next, if for some $1 \leq j \leq m$, $\beta_j \in T$ then $L\beta_j \in T$, and again $C \in T$. Finally, if for all $1 \leq i \leq k$, $\alpha_i \in T$, and for all $1 \leq j \leq m$, $\beta_j \notin T$ then, by the definition, $\gamma \in S_T(I)$. Hence, again $C \in ST(S_T(I)) = T$. ■

Theorem 11.5 allows us to restrict significantly the set of all possible sets $S$ for which $ST(S)$ has to be tested for expansion in the algorithm **kd45_expsB**, if $I$ consists of epistemic clauses. Namely, it is sufficient to take for $A$ the set of all the heads of clauses in $I$ instead of a much larger set of all logically nonequivalent formulas in the conjunctive normal form built of propositional letters occurring in $I$. The same modification can also be used in the algorithm **All_ExpsB** computing $\mathcal{S}$-expansions of a theory $I$ consisting of epistemic clauses, if $\mathcal{S}$ is contained in **kd45**. The correctness of this approach follows from the observation that for each such logic $\mathcal{S}$, an $\mathcal{S}$-expansion for $I$ is a **kd45**-expansion for $I$ (Theorem 9.6) and, consequently, it is of the form $ST(S)$, for some set $S \subseteq H(I)$ such that each $\gamma \in S$ has an $I$-support in $T$.

Knowledge bases arising in applications often consist of epistemic clauses. Therefore, the modification of the algorithms of type **B** that we have just described seems to be especially useful and should be the method of choice if the underlying modal logic $\mathcal{S}$ is contained in the logic **kd45**.

In this section we described algorithms for computing $\mathcal{S}$-expansions for a theory $I$ in the following three cases:

1. $\mathcal{S} \subseteq$ **S4f**, $\mathcal{S}$ is characterized by a cluster-closed class of models (or structures) and there is an algorithm for testing whether $A \vdash_{\mathcal{S}} \varphi$, given a finite set $A$ of formulas and a formula $\varphi$.
2. $\mathcal{S} =$ **kd45**.
3. $\mathcal{S} =$ **Sw5**.

In all other cases, problem of computing $\mathcal{S}$-expansions remains open. In the case when $\mathcal{S}$ is contained in **kd45** or **Sw5** one could use algorithm

**All_ExpsB**, if only a computable operator $M$ can be found such that for a finite set $S \subseteq \mathcal{L}$ and a finite theory $I \subseteq \mathcal{L}$, $M(S,I)$ is a *finite* theory with the property: $S \subseteq Cn_S(I \cup \{\neg L\varphi: \varphi \notin T\})$ if and only if $S \subseteq Cn_S(I \cup \{\neg L\varphi: \varphi \in M(I,S)\})$.

## 11.2 Monotonic modal logics with the same nonmonotonic counterpart

At first glance it may appear that the realm of nonmonotonic modal logics is as rich as that of monotonic modal logics. After all, for each monotonic logic we have its nonmonotonic counterpart. Somewhat surprisingly, it turns out not to be the case. The highlight of this section is Theorem 11.8 which states that the nonmonotonic logic **kd45** coincides with the nonmonotonic logic $\mathcal{S}$, for every modal logic $\mathcal{S}$ such that $\mathbf{5} \subseteq \mathcal{S} \subseteq \mathbf{kd45}$, where by **5** we denote the modal logic containing schema 5 only. In particular, nonmonotonic **5** is identical with nonmonotonic **k45** and with nonmonotonic **kd45**.

The meaning of this result — the collapse into each other of nonmonotonic modal logics **5** and **kd45** — is deeply intriguing; it indicates that different monotonic logics may generate the same notion of expansion, and consequently the same nonmonotonic consequence operation! Therefore, the hierarchy of nonmonotonic modal logics is less diversified than that of monotonic modal logics.

On a closer inspection these observations seem to be less puzzling. After all, nonmonotonic $\mathcal{S}$-consequences of a theory $I$ are often strictly larger than monotonic $\mathcal{S}$-consequences of $I$ due to the powerful principle of allowing as premises in proofs formulas expressing *possibility* that some other formulas might be true, that is, formulas from $\{\neg L\varphi: \varphi \notin T\}$. Our first result provides simple sufficient conditions for nonmonotonic logics $\mathcal{S}_1$ and $\mathcal{S}_2$ to coincide.

**Theorem 11.6** *Let $\mathcal{S}_1 \subseteq \mathcal{S}_2 \subseteq$ **S5** be two modal logics. If for every stable and consistent theory $T \subseteq \mathcal{L}_L$, and for every instance $\psi$ of every axiom schema of $\mathcal{S}_2$, $\{\neg L\varphi: \varphi \notin T\} \vdash_{\mathcal{S}_1} \psi$, then nonmonotonic logics $\mathcal{S}_1$ and $\mathcal{S}_2$ are identical. That is, for every theory $I \subseteq \mathcal{L}_L$, a theory $T$ is an $\mathcal{S}_1$-expansion for $I$ if and only if $T$ is an $\mathcal{S}_2$-expansion for $I$.*

Proof: By Theorem 9.6, if $T$ is an $\mathcal{S}_1$-expansion for $I$ then $T$ is an $\mathcal{S}_2$-expansion for $I$. Conversely, assume that $T$ is an $\mathcal{S}_2$-expansion for $I$. Then $T$ is consistent and

$$T = Cn_{\mathcal{S}_2}(I \cup \{\neg L\varphi: \varphi \notin T\}).$$

Clearly, $T$ is stable, it contains $I$ and is closed under provability in $\mathcal{S}_1$. Hence, $T \supseteq Cn_{\mathcal{S}_1}(I \cup \{\neg L\varphi: \varphi \notin T\})$. Since for every instance $\psi$ of every axiom schema of $\mathcal{S}_2$, $\{\neg L\varphi: \varphi \notin T\} \vdash_{\mathcal{S}_1} \psi$,

$$Cn_{\mathcal{S}_2}(I \cup \{\neg L\varphi: \varphi \notin T\}) \subseteq Cn_{\mathcal{S}_1}(I \cup \{\neg L\varphi: \varphi \notin T\}).$$

Hence, $T = Cn_{\mathcal{S}_1}(I \cup \{\neg L\varphi: \varphi \notin T\})$ and, consequently, $T$ is an $\mathcal{S}_1$-expansion for $I$. ■

**Definition 11.7** Two modal logics $\mathcal{S}$ and $\mathcal{T}$ are *nonmonotonically equivalent* if for every theory $I \subseteq \mathcal{L}_L$, a theory $T$ is an $\mathcal{S}$-expansion for $I$ if and only if $T$ is a $\mathcal{T}$-expansion for $I$. Every maximal collection of nonmonotonically equivalent modal logics is called a *range.* ▲

Our first two results provide some information on the ranges containing logics **kd45** and **Sw5** and, at the same time, show that the notion of a range is nontrivial, that is, that there are ranges containing more than one modal logic.

**Theorem 11.8** *Let $\mathcal{S}$ be a modal logic such that* **5** $\subseteq \mathcal{S} \subseteq$ **kd45**. *Let $I \subseteq \mathcal{L}_L$. A consistent theory $T \subseteq \mathcal{L}_L$ is a* **kd45***-expansion for $I$ if and only if $T$ is an $\mathcal{S}$-expansion for $I$.*

Proof: A careful analysis of the proof of Theorem 10.19 shows that the assumptions of Theorem 11.6 are satisfied for $\mathcal{S}_1 =$ **5** and $\mathcal{S}_2 =$ **kd45**, which immediately implies the assertion. ■

Our next result exhibits a large fragment of the range containing logic **Sw5**. In order to state it, we denote by **tw5** the modal logic based only on schemata t and w5. Let us observe that the logic **tw5** is not contained in **S4**.

**Theorem 11.9** *Let $\mathcal{S}$ be a modal logic such that* **tw5** $\subseteq \mathcal{S} \subseteq$ **Sw5**. *Let $I \subseteq \mathcal{L}_L$. Then a consistent theory $T \subseteq \mathcal{L}_L$ is an* **Sw5***-expansion for $I$ if and only if $T$ is an $\mathcal{S}$-expansion for $I$.*

Proof: It is sufficient to prove that if $T$ is stable and consistent, then $Cn(\{\varphi \equiv L\varphi : \varphi \in T\} \cup \{\neg L\varphi: \varphi \notin T\})$ contains all instances of axiom schemata k, t, 4 and w5, and that for any $\varphi \in T$, $\{\neg L\varphi: \varphi \notin T\} \vdash_{\mathbf{tw5}} \varphi \equiv L\varphi$. All this is quite straightforward and the proof is similar to the proof of Theorems 10.24. We omit the details. ■

Let us remark that the fact that nontrivial ranges exist is, potentially, quite beneficial. For instance a range may contain very different logics. Some may have nice automated theorem provers or lower complexity for the provability problem, whereas others may have elegant semantics. In such a case, the *nonmonotonic* modal logic associated with a range, may have several different mechanisms for semantical and for syntactical manipulations. The freedom to select a monotonic modal logic from a range may be advantageous in practical implementations. One example where it may be so (assuming the complexity class PSPACE does not collapse to the first level of the polynomial hierarchy) is described in the next theorem.

**Theorem 11.10** *Let $I \subseteq \mathcal{L}_L$ be finite. A theory $T \subseteq \mathcal{L}_L$ is an* **S4***-expansion of $I$ if and only if $T$ is an* **S4f***-expansion of $I$.*

Proof: By Theorem 9.6, every **S4**-expansion for $I$ is an **S4f**-expansion for $I$ (even without an assumption that $I$ is finite). Theorem 9.50 implies also that to prove the converse implication it is enough to show that for any partition $(\Phi, \Psi)$, introspection-consistent with $I$, if

$$I \cup \{\neg L\varphi : \varphi \in \Phi\} \vdash_{\mathbf{S4f}} \Psi \tag{11.2}$$

then

$$I \cup \{\neg L\varphi : \varphi \in \Phi\} \vdash_{\mathbf{S4}} \Psi. \tag{11.3}$$

Hence, let $(\Phi, \Psi)$ be a partition of $MA(I)$ introspection-consistent with $I$. Assume also that (11.2) holds. To prove (11.3) it is enough to show that for every *finite* **S4**-model $\mathcal{M}$, if $\mathcal{M} \models I \cup \{\neg L\varphi : \varphi \in \Phi\}$, then $\mathcal{M} \models \Psi$ (since $I$ is finite, the set of **S4**-consequences of $I$ is characterized by the class of *finite* **S4**-models of $I$, see Theorem 7.53).

So, let $\mathcal{M} = \langle M, R, V \rangle$ be a finite **S4**-model such that $\mathcal{M} \models I \cup \{\neg L\varphi : \varphi \in \Phi\}$. By a *rank* of a world $a \in M$, $r(a)$, we mean the maximal integer $n$ such that there exists a sequence of worlds $a_0, \ldots, a_n$, where $a_0 = a$ and for each $i$, $0 \leq i < n$, $(a_i, a_{i+1}) \in R$, but not $(a_{i+1}, a_i) \in R$. Since $M$ is finite, each world has a rank. We prove by the induction on $r(a)$, that for each $a \in M$ $(\mathcal{M}, a) \models \Psi$.

Let $r(a) = 0$. Put $M_a = \{b : (b, a) \in R \text{ and } (a, b) \in R\}$. Consider the Kripke model $\mathcal{M}_a = \langle M_a, R_a, V_a \rangle$, where $R_a$ and $V_a$ are the restrictions of $R$ and $V$ to $M_a$. Clearly, $\mathcal{M}_a$ is an **S4f**-model (in fact, even an **S5**-model). Since $r(a) = 0$, there is no world $b \in M$ such that $(a, b) \in R$ but $(b, a) \notin R$. It follows that for every formula $\vartheta$ and for every world $b \in \mathcal{M}_a$, $(\mathcal{M}, b) \models \vartheta$ if and only if $(\mathcal{M}_a, b) \models \vartheta$. In particular, since $\mathcal{M} \models I \cup \{\neg L\varphi : \varphi \in \Phi\}$, $\mathcal{M}_a \models I \cup \{\neg L\varphi : \varphi \in \Phi\}$. Since (11.2) holds and $\mathcal{M}_a$ is an **S4f**-model, $\mathcal{M}_a \models \Psi$. Hence, $(\mathcal{M}_a, a) \models \Psi$. Consequently, $(\mathcal{M}, a) \models \Psi$.

Consider now a world $a \in M$ with $r(a) = k + 1$, and assume that for each world $b \in M$ such that $r(b) \leq k$ $(\mathcal{M}, b) \models \Psi$.

Define $M_0 = \{c \in M : (a,c) \in R \text{ and } (c,a) \in R\}$ and $M_1 = \{c \in M : (a,c) \in R \text{ and } (c,a) \notin R\}$. Clearly, $M_0$ and $M_1$ are disjoint and for each $c$, $(a,c) \in R$ if and only if $c \in M_0 \cup M_1$.

Define an **S4f**-model $\mathcal{N} = \langle M_0, M_1, U\rangle$, where $U$ is the restriction of $V$ to $M_0 \cup M_1$. Let us denote $N = M_0 \cup M_1$ and $Q = (M_0 \times N) \cup (M_1 \times M_1)$ (that is, $Q$ is the accessibility relation of $\mathcal{N}$). Observe that for $b \in N$ and for every world $c \in M$, if $(b,c) \in R$ then $(b,c) \in Q$. Moreover, if $b \in M_0$ then $(b,c) \in R$ if and only if $(b,c) \in Q$.

We will prove now that for each $b \in N$ and for each $I$-formula $\eta$, $(\mathcal{M}, b) \models \eta$ if and only if $(\mathcal{N}, b) \models \eta$. The proof will be by induction on the complexity of $\eta$. (So, we have a double induction here. The external induction on $r(a)$ and, within its induction step, an internal induction on the complexity of a formula.) The basis of the (internal) induction is obvious. The only nontrivial case in the induction step is when $\eta = L\zeta$.

Assume $(\mathcal{N}, b) \models L\zeta$. Consider an arbitrary world $c \in M$ such that $(b,c) \in R$. Then, $c \in N$ and $(b,c) \in Q$. Hence, $(\mathcal{N}, c) \models \zeta$. By the induction hypothesis, $(\mathcal{M}, c) \models \zeta$. Consequently, $(\mathcal{M}, b) \models L\zeta$.

Conversely, assume that $(\mathcal{M}, b) \models L\zeta$. If $b \in M_0$, then, since for each $c$, $(b,c) \in R$ if and only if $(b,c) \in Q$, it follows directly from the (internal) induction hypothesis that $(\mathcal{N}, b) \models L\zeta$.

Assume then that $b \in M_1$. Since $\mathcal{M} \models \{\neg L\varphi : \varphi \in \Phi\}$, we have that $\zeta \in \Psi$ (since $\zeta$ is an $I$-formula, it belongs to $MA(I)$). By the definition of the model $\mathcal{N}$, we have that for each $c \in M_1$, $r(c) \leq k$. Thus, applying the (external) induction hypothesis, we obtain that for each $c \in M_1$, $(\mathcal{M}, c) \models \zeta$. By the (internal) induction hypothesis, for each $c \in M_1$ (that is, for each $c$ such that $(b,c) \in Q$), $(\mathcal{N}, c) \models \zeta$. Hence $(\mathcal{N}, b) \models L\zeta$.

Thus, we have proved that for each $I$-formula $\eta$ and for each $b \in N$, $(\mathcal{M}, b) \models \eta$ if and only if $(\mathcal{N}, b) \models \eta$. Since $I \cup \{\neg L\varphi : \varphi \in \Phi\}$ consists of $I$-formulas, we obtain that $\mathcal{N} \models I \cup \{\neg L\varphi : \varphi \in \Phi\}$. But $\mathcal{N}$ is an **S4f**-model. Since (11.2) holds, $\mathcal{N} \models \Psi$. In particular, $(\mathcal{N}, a) \models \Psi$ and, since $\Psi$ consists of $I$-formulas, $(\mathcal{M}, a) \models \Psi$. This completes the proof of the induction step of the main induction. ■

Theorem 11.10 implies that when considering computational properties of the *nonmonotonic* modal logic **S4** one can select the modal logic **S4f** as the *monotonic* basis for this logic. It may be beneficial since the provability problem for **S4f** is on the first level of the polynomial hierarchy, whereas it is PSPACE-complete for **S4**. The reader is referred to Chapter 13 for a more detailed treatment of complexity issues in nonmonotonic logics.

Let us mention that Theorem 11.10 is not true without the assumption about $I$ being finite. The reference is given in Section "Comments and remarks" at the end of this chapter.

It turns out that if we restrict the class of theories $I$ by imposing *syntactic conditions*, ranges may further coagulate. Four important classes of theories we study below are:

- The class of theories consisting of epistemic clauses with only positive literals of $L$-depth equal to 1 in the antecedents (such theories will be referred to as *epistemic Horn clauses*).
- The class of theories consisting of epistemic clauses with only negative literals in the antecedents.
- The class of modal-free theories.
- The class of *strongly stratified* theories.

Our first result deals with the case of theories consisting of *epistemic Horn clauses*, that is, clauses of the form

$$L\varphi_1 \wedge \ldots \wedge L\varphi_k \supset \gamma,$$

where all $\varphi_i$ and $\gamma$ are from $\mathcal{L}$. We have the following result.

**Theorem 11.11** *Let $I$ consist of epistemic Horn clauses. Then for every theory $I \subseteq \mathcal{L}_L$ and every modal logic $\mathcal{S} \subseteq$ **S4f**, a theory $T \subseteq \mathcal{L}_L$ is an $\mathcal{S}$-expansion for $I$ if and only if $T$ is an **S4f**-expansion for $I$.*

Proof: It follows from Theorem 9.6 that to prove the theorem it is sufficient to show that if a theory $T$ is an **S4f**-expansion for $I$, then $T$ is an **N**-expansion for $I$. To this end, we will use characterizations of **N**- and **S4f**-expansions given in Theorem 10.14 and Corollary 9.28.

Let us assume that a stable theory $T$ such that $I \subseteq T$ is not an **N**-expansion for $I$. Let $\mathcal{M}_T = \langle M_T, V_T\rangle$ be the canonical **S5**-model for $T$ (see Section 8.2). Then, by Theorem 10.14, there is an **N**-structure $\mathcal{N} = \langle N, \{R_\varphi\}_{\varphi\in\mathcal{L}_L}, V\rangle$, where $N \cap M_T = \emptyset$, such that:

**(a)** $\mathcal{N} \odot \mathcal{M}_T \models I$.

**(b)** For some $a \in N$ and for every $b \in M_T$, $V(a) \neq V_T(b)$.

Consider an **S4f**-model $\mathcal{N}' \odot \mathcal{M}_T$, where $\mathcal{N}'$ is the **S5**-model $\langle N, V\rangle$. Clearly, condition (b) holds for $\mathcal{N}' \odot \mathcal{M}_T$. Let

$$C = (L\varphi_1 \wedge \ldots \wedge L\varphi_k \supset \gamma)$$

be a clause from $I$. Clearly, $\mathcal{M}_T \models C$ and, consequently, $\langle \mathcal{N}' \odot \mathcal{M}_T, m\rangle \models C$ for every world $m \in M_T$. Consider a world $m \in N$. Since $\mathcal{N} \odot \mathcal{M}_T \models C$, at least one of the following two conditions is satisfied:

- $V(m)(\gamma) = \mathbf{t}$.
- There exists a world $n \in N \cup M_T$ in which at least one of $\varphi_i$ is not true.

Clearly, in both cases, $\langle \mathcal{N}' \odot \mathcal{M}_T, m\rangle \models C$. Thus, by Corollary 9.28, $T$ is not an **S4f**-expansion for $I$. ■

Next, we move on to the case of theories consisting of epistemic clauses without positive modal literals in the antecedents, that is, formulas of the form

$$\neg L\varphi_1 \wedge \ldots \wedge \neg L\varphi_k \supset \gamma,$$

where $\gamma \in \mathcal{L}$.

**Theorem 11.12** *Let $I$ consist of epistemic clauses without positive modal literals in the antecedents. Then for every modal logic $\mathcal{S}$ contained in* **kd45**, *a theory $T$ is an $\mathcal{S}$-expansion for $I$ if and only if $T$ is a* **kd45**-*expansion for $I$.*

Proof: It is sufficient to prove that if a theory $T$ is a **kd45**-expansion for $I$, then $T$ is an **N**-expansion for $I$. So, consider a **kd45**-expansion $T$ for $I$. Then

$$T = Cn(I \cup \{\neg L\varphi : \varphi \notin T\} \cup \{L\psi : \psi \in T\}).$$

Since $I$ consists only of clauses with negative literals in the antecedents, it is easy to show that

$$Cn(I \cup \{\neg L\varphi : \varphi \notin T\}) \cap \mathcal{L} = Cn(I \cup \{\neg L\varphi : \varphi \notin T\} \cup \{L\psi : \psi \in T\}) \cap \mathcal{L}.$$

Hence, $T \cap \mathcal{L} \subseteq Cn_{\mathbf{N}}(I \cup \{\neg L\varphi : \varphi \notin T\})$. Since $I \subseteq T$, Corollary 9.8 implies that $T$ is an **N**-expansion for $I$. (Let us mention here that this theorem could also be proved by a semantical argument similar to that used to prove the previous result.) ■

This result has two simple corollaries.

**Corollary 11.13** *Let $I$ consist of epistemic clauses without positive modal literals in the antecedents. Then a theory $T$ is an* **N**-*expansion for $I$ if and only if $T$ is an* **kd45**-*expansion for $I$.* ▲

**Corollary 11.14** *Let $I$ consist of epistemic clauses without positive modal literals in the antecedents. If a theory $T$ is a* **kd45**-*expansion of $I$ then $T$ is an $\mathcal{S}$-expansion of $I$ for every modal logic $\mathcal{S}$ contained in* **S5**.

Proof: The result follows from Corollary 11.13 and Theorem 9.6. ■

Another range can be found if additional restrictions are imposed on epistemic clauses without positive modal literals in the antecedents. Let us define the notions of *positive* and *negative* occurrence of a subformula $\eta$ in a formula $\psi$ recursively as follows:

**(a)** If $\psi = \eta$ then we say that $\eta$ occurs positively in $\psi$.

**(b)** If $\psi = \neg\psi'$ then each positive (respectively, negative) occurrence of $\eta$ in $\psi'$ is a negative (respectively, positive) occurrence of $\eta$ in $\psi$.

**(c)** If $\psi = \psi' \wedge \psi''$ or $\psi = \psi' \vee \psi''$, then each positive (respectively, negative) occurrence of $\eta$ in $\psi'$ or $\psi''$ is a positive (respectively, negative) occurrence of $\eta$ in $\psi$.

**(d)** If $\psi = \psi' \supset \psi''$ then each positive (respectively, negative) occurrence of $\eta$ in $\psi'$ is a negative (respectively, positive) occurrence of $\eta$ in $\psi$, and each positive (respectively, negative) occurrence of $\eta$ in $\psi''$ is a positive (respectively, negative) occurrence of $\eta$ in $\psi$.

**(e)** If $\psi = L\psi'$ then each positive (respectively, negative) occurrence of $\eta$ in $\psi'$ is a positive (respectively, negative) occurrence of $\eta$ in $\psi$.

Recall that an *epistemic clause* with only negative modal literals in the antecedent is of the form

$$\neg L\psi_1 \wedge \ldots \wedge \neg L\psi_k \supset \gamma. \tag{11.4}$$

But not every formula of the form (11.4) is free of negative occurrences of modal atoms. For example, a formula $\neg L \neg L\varphi \supset \gamma$ contains a negative occurrence of the modal atom $L\varphi$. Thus, the class of theories consisting of epistemic clauses without negative occurrences of modal atoms is a proper subclass of the class of theories consisting of epistemic clauses of the form (11.4).

**Theorem 11.15** *Let $I$ consist only of epistemic clauses without negative occurrences of modal atoms. Then for every modal logic $\mathcal{S}$ such that* $\mathbf{t} \subseteq \mathcal{S} \subseteq$ **Sw5**, *a theory $T$ is an $\mathcal{S}$-expansion for $I$ if and only if $T$ is a* **t**-*expansion for $I$.*

Proof: Clearly, it is enough to show that every **Sw5**-expansion for $I$ is a **t**-expansion for $I$. We proceed similarly as in the proof of Theorem 11.11. Consider a stable theory $T$ and assume that $T$ is not a **t**-expansion for $I$. Let $\mathcal{M}_T = \langle M_T, V_T \rangle$ be the canonical **S5**-model for $T$. Then, by Corollary 9.25 (see Remark 9.33), there is a **t**-structure $\mathcal{N} = \langle N, \{R_\varphi\}_{\varphi \in \mathcal{L}_L}, V \rangle$, where $N \cap M_T = \emptyset$, such that:

**(a)** $\mathcal{N} \odot \mathcal{M}_T \models I$.

**(b)** For some $a \in N$ and for every $b \in M_T$, $V(a) \neq V_T(b)$.

Consider an **Sw5**-model $\mathcal{M} = \langle a, M_T, U \rangle$, where $U(a) = V(a)$ and $U|_{M_T} = V_T$. Clearly, condition (b) holds for $\mathcal{M}$. To prove that $T$ is not an **Sw5**-expansion for $I$ it suffices to show that $\mathcal{M} \models I$. To this end, one can prove by induction on the complexity of a formula that for every formula $\psi$ without negative occurrences of modal atoms

$$\langle \mathcal{M}, a \rangle \models \psi \text{ if and only if } \langle \mathcal{N} \odot \mathcal{M}_T \rangle \models \psi.$$

We leave the details for the reader. ∎

The next result deals with the case of modal-free theories and is simply a reformulation of Corollary 9.42.

**Theorem 11.16** *Let $I \subseteq \mathcal{L}$ be consistent. For each logic $\mathcal{S}$ (possibly subnormal) such that $\mathcal{S} \subseteq$ **kd45** or $\mathcal{S} \subseteq$ **Sw5**, the theory $ST(I)$ is the only $\mathcal{S}$-expansion for $I$.* ▲

Theorem 11.16 can be generalized to a much wider class of theories. A theory $I$ is *strongly stratified* if:

**STRAT1** Each formula $\varphi$ in $I$ is of the form $a(\varphi) \supset c(\varphi)$, where $c(\varphi) \in \mathcal{L}$, and $\{c(\varphi) : \varphi \in I\}$ is consistent.

**STRAT2** For some ordinal $\alpha$, $I$ has a partition $\{I_\beta\}_{\beta<\alpha}$ (that is, sets $I_\beta$ are nonempty and pairwise disjoint, and their union is $I$), called *stratification* such that for every $\beta < \alpha$ and for each $\varphi \in I_\beta$, propositional letters occurring in $c(\varphi)$ do not occur in formulas of $\bigcup_{\gamma<\beta} I_\gamma$, and do not occur in formulas of $I_\beta$ within the scope of the operator $L$.

This notion of stratification is closely related to stratification of logic programs discussed in detail in Chapter 6.

We will now show that if a theory $I$ is strongly stratified then, for every modal nonmonotonic logic $\mathcal{S}$ contained in **kd45** or **Sw5**, $I$ has exactly one $\mathcal{S}$-expansion. It will follow then from Theorem 9.6 that all these $\mathcal{S}$-expansions, where $\mathcal{S}$ ranges over the set of all modal logics contained in **kd45** or **Sw5**, coincide.

We will first show that if $I$ is strongly stratified than $I$ has exactly one **kd45**-expansion. We do this in two steps. First, we will prove that a strongly stratified theory possesses at least one **kd45** expansion, and then that in fact it possesses at most one. Each argument is based on a technical lemma.

Consider a family of sets $\{A_{\beta,m}\}_{\beta<\alpha, m=0,1,\ldots}$, where $\alpha$ is an ordinal. Following the conventions introduced in Section 6.7 we define

$$A_{\leq\beta,<n} = \bigcup\{A_{\gamma,m}: \gamma \leq \beta,\ m < n\},$$

$$A_\beta = \bigcup_{m=0}^{\infty} A_{\beta,m},$$

and

$$A_{<\beta} = \bigcup_{\gamma<\beta} A_\gamma.$$

Sets $A_{\leq\beta}$ and $A_{\leq\beta,<n}$ are defined analogously.

We will first show that if $I$ is strongly stratified then it has at least one **kd45**-expansion. To this end, we will construct a partition $(\Phi, \Psi)$ of $MA(I)$ such that $(\Phi, \Psi)$ is propositionally perfect for $I$ with respect to $\{L\psi: \psi \in \Psi\}$.

Let $\{I_\beta\}_{\beta<\alpha}$ be a stratification of $I$. For every ordinal $\beta < \alpha$ and for every non-negative integer $m$, we define sets $\Psi_{\beta,m}$ and $\Phi_{\beta,m}$ recursively as follows. Let $\eta \in MA(I_\beta) \setminus MA(I_{<\beta})$ be such that $d_L(\eta) = m$. Then $\eta$ is included in $\Psi_{\beta,m}$ if $\eta$ is a propositional consequence of the set

$$I_{<\beta} \cup \{\neg L\varphi: \varphi \in \Phi_{<\beta} \cup \Phi_{\leq\beta,<m}\} \cup \{L\psi: \psi \in \Psi_{<\beta} \cup \Psi_{\leq\beta,<m}\},$$

and $\eta$ is included into $\Phi_{\beta,m}$ otherwise. In addition, we require that no other formulas be included in $\Psi_{\beta,m}$ and $\Phi_{\beta,m}$.

Define

$$\Psi = \bigcup_{\beta<\alpha} \Psi_\beta, \quad \text{and} \quad \Phi = \bigcup_{\beta<\alpha} \Phi_\beta.$$

It follows directly from the construction that $(\Phi, \Psi)$ is a partition of $MA(I)$.

Next, for every propositional variable $p$ we define the *rank* of $p$, $r(p)$, as follows. If $p$ does not occur in $c(\varphi)$ for any $\varphi \in I$, we set $r(p) = 0$. If $p$ occurs in $c(\varphi)$ for some formula $\varphi \in I_\beta$, then we put $r(p) = \beta$. By the definition of strongly stratified theories, such an index $\beta$ is unique. The definition of rank can be extended to all formulas of $\mathcal{L}_L$. Namely, for $\varphi \in \mathcal{L}_L$, we define $r(\varphi) = \max\{r(p) : p \text{ occurs in } \varphi\}$.

**Lemma 11.17** *The partition $(\Phi, \Psi)$ of $MA(I)$ is propositionally perfect for $I$ with respect to $\{L\psi : \psi \in \Psi\}$.*

Proof: Consider a formula $\eta \in \Psi$. Then, for some $\beta < \alpha$ and for some non-negative integer $m$, $\eta \in \Psi_{\beta,m}$. Since $I_{<\beta} \subseteq I$, $\Phi_{<\beta} \cup \Phi_{\leq\beta,<m} \subseteq \Phi$ and $\Psi_{<\beta} \cup \Psi_{\leq\beta,<m} \subseteq \Psi$, it follows that

$$I \cup \{\neg L\varphi : \varphi \in \Phi\} \cup \{L\psi : \psi \in \Psi\} \vdash \eta. \tag{11.5}$$

To complete the proof, it is now enough to show that $(\Phi, \Psi)$ is introspection-consistent with $I$. As we have already mentioned, $(\Phi, \Psi)$ is a partition of $MA(I)$. Thus we need to show that conditions (IC2) and (IC3) of Definition 9.35 are satisfied by $(\Phi, \Psi)$.

Since $\{c(\varphi) : \varphi \in I\}$ is consistent, there is a valuation $W$ of $\mathcal{L}$ such that $W(\{c(\varphi) : \varphi \in I\}) = \mathbf{t}$. Extend this valuation to $\mathcal{L}_L$ by defining $W(L\vartheta) = \mathbf{f}$ if $\vartheta \in \Phi$, and $W(L\vartheta) = \mathbf{t}$ otherwise. It is easy to see that $W(I \cup \{\neg L\varphi : \varphi \in \Phi\} \cup \{L\psi : \psi \in \Psi\}) = \mathbf{t}$. By (11.5), $W(\eta) = \mathbf{t}$ for every $\eta \in \Psi$. Hence, the set

$$I \cup \{\neg L\varphi : \varphi \in \Phi\} \cup \{L\psi : \psi \in \Psi\} \cup \Psi$$

is propositionally consistent.

To prove condition (IC3), consider a formula $\eta \in \Phi$. Then, for some $\beta < \alpha$ and for some non-negative integer $m$, $\eta \in \Phi_{\beta,m}$. It follows that $L\eta$ is a subformula of a formula from $I_\beta$. Consequently, $r(\eta) < \beta$.

By the definition of $\Phi_{\beta,m}$, there is a valuation $V$ of $\mathcal{L}_L$ such that

$$V(I_{<\beta} \cup \{\neg L\varphi : \varphi \in \Phi_{<\beta} \cup \Phi_{\leq\beta,<m}\} \cup \{L\psi : \psi \in \Psi_{<\beta} \cup \Psi_{\leq\beta,<m}\}) = \mathbf{t}$$

and $V(\eta) = \mathbf{f}$. Let $W$, as before, denote a valuation of $\mathcal{L}$ such that $W(\{c(\varphi) : \varphi \in I\}) = \mathbf{t}$.

Define a valuation $U$ as follows. For a propositional variable $p$, if $r(p) < \beta$ then put $U(p) = V(p)$. Otherwise, put $U(p) = W(p)$. For a modal atom $L\varphi$, put $U(L\varphi) = \mathbf{f}$, if $\varphi \in \Phi$, and $U(L\varphi) = \mathbf{t}$ otherwise.

Observe that if $\varphi \in I_\gamma$, $\gamma \geq \beta$, then $U(c(\varphi)) = W(c(\varphi)) = \mathbf{t}$. Thus $U(\varphi) = \mathbf{t}$. Consider $\varphi \in I_\gamma$ with $\gamma < \beta$. For each propositional variable $p$ occurring in $\varphi$, $r(p) < \gamma$. Hence $U(p) = V(p)$. Let $L\vartheta$ be a modal atom occurring in $\varphi$. Then $\vartheta \in MA(I_{<\beta})$. Consequently, $\vartheta \in \Phi_{<\beta} \cup \Psi_{<\beta}$. If $\vartheta \in \Phi_{<\beta}$, then $V(L\vartheta) = \mathbf{f} = U(L\vartheta)$ and, if $\vartheta \in \Psi_{<\beta}$, $V(L\vartheta) = \mathbf{t} = U(L\vartheta)$. Hence $U$ and $V$ coincide on all propositional and modal atoms occurring in $\varphi$. Since $V(\varphi) = \mathbf{t}$, $U(\varphi) = \mathbf{t}$. Thus $U(I) = \mathbf{t}$. In addition, by the definition of $U$, $U(\{\neg L\varphi : \varphi \in \Phi\} \cup \{L\psi : \psi \in \Psi\}) = \mathbf{t}$. By (11.5), $U(\Psi) = \mathbf{t}$.

Now let $p$ be a propositional atom occurring in $\eta$. Then $r(p) \leq r(\eta) < \beta$ and $U(p) = V(p)$. Next, consider a modal atom $L\vartheta$ occurring in $\eta$. Then

$\vartheta \in MA(I_{\leq\beta})$ and $d_L(\vartheta) \leq m-1$. Consequently, $\vartheta \in (\Phi_{<\beta} \cup \Phi_{\beta,<m}) \cup (\Psi_{<\beta} \cup \Psi_{\beta,<m})$. As before, it follows that $U(L\vartheta) = V(L\vartheta)$. Hence $U$ and $V$ agree on all propositional and modal atoms occurring in $\eta$. Consequently, $V(\eta) = U(\eta) = \mathbf{f}$. Since, as we have seen, $U(I \cup \{\neg L\varphi : \varphi \in \Phi\} \cup \{L\psi : \psi \in \Psi\} \cup \Psi) = \mathbf{t}$, condition (IC3) follows. ■

**Theorem 11.18** *Let $I$ be a strongly stratified theory. Then $I$ has at least one* **kd45**-*expansion.*

Proof: It follows from Lemma 11.17 that there is at least one partition $(\Phi, \Psi)$ of $MA(I)$ such that $(\Phi, \Psi)$ is propositionally perfect for $I$ with respect to $\{L\psi : \psi \in \Psi\}$. Consequently, the assertion follows from Theorem 10.21. ■

Next, we will prove that each strongly stratified theory has exactly one **kd45**-expansion. To this end, we need the following lemma.

**Lemma 11.19** *Let $\{I_\beta\}_{\beta<\alpha}$ be a stratification of a strongly stratified theory $I \subseteq \mathcal{L}_L$. Let $(\Phi, \Psi)$ be a partition of $MA(I)$ that is propositionally perfect for $I$ with respect to $\{L\psi : \psi \in \Psi\}$. Define $\Phi_{\beta,m}$ to consist of all formulas $\eta \in MA(I_\beta) \setminus MA(I_{<\beta})$ such that $d_L(\eta) = m$ and $\eta \in \Phi$, and define $\Psi_{\beta,m}$ to consist of all formulas $\eta \in MA(I_\beta) \setminus MA(I_{<\beta})$ such that $d_L(\eta) = m$ and $\eta \in \Psi$.*

1. *If $\eta \in \Psi_{\beta,m}$ then*

$$I_{<\beta} \cup \{\neg L\varphi : \varphi \in \Phi_{<\beta} \cup \Phi_{\leq\beta,<m}\} \cup \{L\psi : \psi \in \Psi_{<\beta} \cup \Psi_{\leq\beta,<m}\} \vdash \eta.$$

2. *If $\eta \in \Phi_{\beta,m}$, then*

$$I_{<\beta} \cup \{\neg L\varphi : \varphi \in \Phi_{<\beta} \cup \Phi_{\leq\beta,<m}\} \cup \{L\psi : \psi \in \Psi_{<\beta} \cup \Psi_{\leq\beta,<m}\} \not\vdash \eta.$$

Proof: (1) Observe that $\eta \in MA(I_{\leq\beta})$. Hence, $L\eta$ is a subformula of a formula from $I_{\leq\beta}$. Hence, $r(\eta) < \beta$.

Let $V$ be an arbitrary propositional valuation such that $V(I_{<\beta}) = \mathbf{t}$, $V(L\varphi) = \mathbf{f}$, for every $\varphi \in \Phi_{<\beta} \cup \Phi_{\leq\beta,<m}$, and $V(L\psi) = \mathbf{t}$ for each $\psi \in \Psi_{<\beta} \cup \Psi_{\leq\beta,<m}$.

Since $\{c(\varphi) : \varphi \in I\}$ is consistent, there is a valuation $W$ such that $W(c(\varphi)) = \mathbf{t}$ for each $\varphi \in I$. Define a valuation $U$ as follows. For a propositional variable $p$, if $r(p) < \beta$ then put $U(p) = V(p)$. Otherwise, put $U(p) = W(p)$. For a modal atom $L\varphi$, put $U(L\varphi) = \mathbf{f}$, if $\varphi \in \Phi$, and $U(L\varphi) = \mathbf{t}$ otherwise.

Observe that if $\varphi \in I_\gamma$, $\gamma \geq \beta$, then $U(c(\varphi)) = W(c(\varphi)) = \mathbf{t}$. Thus, $U(\varphi) = \mathbf{t}$. Consider $\varphi \in I_\gamma$ with $\gamma < \beta$. For each propositional variable $p$ occurring in $\varphi$, $r(p) \leq \gamma < \beta$. Hence, $U(p) = V(p)$. Let $L\vartheta$ be a modal atom occurring in $\varphi$. Then, $\vartheta \in MA(I_{<\beta})$ and, consequently, $\vartheta \in \Phi_{<\beta} \cup \Psi_{<\beta}$. If $\vartheta \in \Phi_{<\beta}$, then $V(L\vartheta) = \mathbf{f} = U(L\vartheta)$ and, if $\vartheta \in \Psi_{<\beta}$, $V(L\vartheta) = \mathbf{t} = U(L\vartheta)$. Hence, $U$ and $V$ coincide on all propositional and modal atoms occurring in $\varphi$. Since $V(\varphi) = \mathbf{t}$, $U(\varphi) = \mathbf{t}$. Thus, $U(I) = \mathbf{t}$. In addition, by the definition of $U$, $U(\{\neg L\varphi : \varphi \in \Phi\} \cup \{L\psi : \psi \in \Psi\}) = \mathbf{t}$. Since $(\Phi, \Psi)$ is propositionally perfect for $I$ with respect to $\{L\psi : \psi \in \Psi\}$, $U(\eta) = \mathbf{t}$.

Now, let $p$ be a propositional atom occurring in $\eta$. Then, $r(p) \leq r(\eta) < \beta$ and $U(p) = V(p)$. Next, consider a modal atom $L\vartheta$ occurring in $\eta$. Then, $\vartheta \in MA(I_{\leq\beta})$ and $d_L(\vartheta) = m - 1$. Consequently, $\vartheta \in (\Phi_{<\beta} \cup \Phi_{\leq\beta,<m}) \cup (\Psi_{<\beta} \cup \Psi_{\leq\beta,<m})$. As before, it follows that $U(L\vartheta) = V(L\vartheta)$. Hence, $U$ and $V$ agree on all propositional and modal atoms occurring in $\eta$. Consequently, $V(\eta) = U(\eta) = \mathbf{t}$. This completes the proof of (1).

To prove (2) observe that if $\eta \in \Phi_{\beta,m}$, then $\eta \in \Phi$. Since $(\Phi, \Psi)$ is introspection-consistent with $I$,

$$I \cup \{\neg L\varphi : \varphi \in \Phi\} \cup \{L\psi : \psi \in \Psi\} \cup \Psi \not\vdash \eta.$$

In particular,

$$I_{<\beta} \cup \{\neg L\varphi : \varphi \in \Phi_{<\beta} \cup \Phi_{\leq\beta,<m}\} \cup \{L\psi : \psi \in \Psi_{<\beta} \cup \Psi_{\leq\beta,<m}\} \not\vdash \eta.$$

■

**Theorem 11.20** *A strongly stratified theory I has exactly one* **kd45**-*expansion.*

Proof: By Theorem 11.18, $I$ has a **kd45**-expansion. Thus, by Theorem 10.21, to prove the assertion it is sufficient to show that if partitions $(\Phi^i, \Psi^i)$, $i = 1, 2$, of $MA(I)$ are propositionally perfect for $I$ with respect to $\{L\psi : \psi \in \Psi^i\}$, $i = 1, 2$, then $\Psi^1 = \Psi^2$. But, by an easy induction, it follows from Lemma 11.19, that for all $\beta < \alpha$ and for all non-negative integers $m$

$$\Psi^1_{\beta,m} = \Psi^2_{\beta,m} \text{ and } \Phi^1_{\beta,m} = \Phi^2_{\beta,m}.$$

Consequently, $\Psi^1 = \Psi^2$. ■

Finally, we will prove that a strongly stratified theory has the same unique $\mathcal{S}$-expansion for every logic $\mathcal{S}$ contained in **kd45** or **Sw5**.

Since the transformation $(\cdot)^B$ (see Section 10.2) preserves stratification, given a theory $I \subseteq \mathcal{L}_L$, if $I$ is strongly stratified so is $I^B$. Then $I^B$ has a unique **kd45**-expansion, and so, by Theorem 10.30, $I$ has a unique **Sw5**-expansion. In this fashion we get the following result.

**Theorem 11.21** *A strongly stratified theory $I$ has exactly one* **Sw5**-*expansion.* ▲

**Theorem 11.22** *Let $I \subseteq \mathcal{L}$ be strongly stratified. For each logic $\mathcal{S}$ such that $\mathcal{S} \subseteq$* **kd45** *or $\mathcal{S} \subseteq$* **Sw5**, *the theory $I$ has exactly one $\mathcal{S}$-expansion and for all such logics $\mathcal{S}$, these unique $\mathcal{S}$-expansions of $I$ coincide.*

Proof: Assume that $I$ is strongly stratified and $(\Phi, \Psi)$ is propositionally perfect for $I$ with respect to $\{L\psi : \psi \in \Psi\}$. Then, for each $\psi \in \Psi$, we have $I \cup \{\neg L\varphi : \varphi \in \Phi\} \vdash_{\mathbf{N}} \psi$. Indeed, this claim follows easily by induction on $(r, m)$, where $r = r(\psi)$ and $m = d_L(\psi)$ — both the basis of the induction and the induction step follow from Lemma 11.19. In other words, for a strongly stratified theory $I$, if $(\Phi, \Psi)$ is propositionally perfect for $I$ with respect to $\{L\psi : \psi \in \Psi\}$ then $(\Phi, \Psi)$ is **N**-perfect for $I$. Consequently, by Theorems 10.21 and 10.15, each **kd45**-expansion of $I$ is an **N**-expansion for $I$. Since also conversely, each **N**-expansion for $I$ is a **kd45**-expansion for $I$, the classes of **N**-expansions and **kd45**-expansions coincide for strongly stratified theories and consist of exactly one theory.

Since this unique **N**-expansion is also an **Sw5**-expansion, the assertion follows by Theorem 11.21. ■

## 11.3 Adding new definitions to a theory

An important property of every monotonic modal logic $\mathcal{S}$ is that when a new symbol $q$ is added to the language and a new definition $q \equiv \eta$, where $\eta$ is a formula in the original language, is added to $I$, the resulting theory $I^{q,\eta}$ is consistent if and only if $I$ is consistent. More generally, the set of consequences of $I^{q,\eta}$ is a *conservative* extension of the set of consequences of the original theory $I$.

To be precise, let $q$ be a propositional variable distinct from all variables in a propositional language $\mathcal{L}$. By $\mathcal{L}^q$ we denote the extension of $\mathcal{L}$ by $q$ and by $\mathcal{L}^q_L$ the modal language corresponding to $\mathcal{L}^q$. We say that a theory $U \subseteq \mathcal{L}^q_L$ is a *conservative* extension of a theory $T \subseteq \mathcal{L}_L$ if $T = U \cap \mathcal{L}_L$.

Consider a modal logic $\mathcal{S}$ (recall that by the definition of a modal logic, the set of theorems of $\mathcal{S}$ is closed under the rule of uniform substitution). Let $Cn_{\mathcal{S}}$ denote, as usual, the consequence operator in $\mathcal{S}$ in the language $\mathcal{L}_L$, and let $Cn^q_{\mathcal{S}}$ be the consequence operator in $\mathcal{S}$ in the language $\mathcal{L}^q_L$. Consider a formula $\eta \in \mathcal{L}_L$. For a theory $I \subseteq \mathcal{L}_L$, define

$$I^{q,\eta} = I \cup \{q \equiv \eta\}.$$

Since the set of instances of the axioms is closed under the rule of uniform substitution, it is easy to show by induction on the length of a derivation that

$$Cn_{\mathcal{S}}(I) = Cn^q_{\mathcal{S}}(I^{q,\eta}) \cap \mathcal{L}_L.$$

In other words, $Cn^q_{\mathcal{S}}(I^{q,\eta})$ is a conservative extension of $Cn_{\mathcal{S}}(I)$.

In this section we will show that nonmonotonic modal logics possess a similar property: there is a natural *one-to-one* and *onto* mapping between $\mathcal{S}$-expansions for $I$ and $I^{q,\eta}$. Moreover, $\mathcal{S}$-expansions for $I^{q,\eta}$ are conservative extensions of corresponding $\mathcal{S}$-expansions for $I$.

Let $p$ be a propositional variable and let $\psi$ and $\varphi$ be formulas. We write $\psi(p/\varphi)$ to denote the result of substituting $\varphi$ for $p$ uniformly in $\psi$. Clearly, if $p$ does not occur in $\psi$ then $\psi(p/\varphi) = \psi$. For a theory $T \subseteq \mathcal{L}_L$ define

$$T^{q/\eta} = \{\psi \in \mathcal{L}^q_L : \psi(q/\eta) \in T\}.$$

The following result collects some properties of the relationship between theories $T$ and $T^{q/\eta}$.

**Lemma 11.23**

1. *Theory $T^{q/\eta}$ is a conservative extension of $T$, that is, $T^{q/\eta} \cap \mathcal{L}_L = T$.*
2. *If $T$ contains all tautologies, $q \equiv \eta \in T^{q/\eta}$.*
3. *Theory $T$ is stable if and only if $T^{q/\eta}$ is stable.*
4. *Let $I \subseteq \mathcal{L}_L$. The mapping $T \mapsto T^{q/\eta}$ is a one-to-one and onto mapping between stable theories in $\mathcal{L}_L$ containing $I$ and stable theories in $\mathcal{L}^q_L$ containing $I^{q,\eta}$.*

Proof: (1) Let $\psi \in \mathcal{L}_L$. By the definition of $T^{q/\eta}$, $\psi \in T^{q/\eta}$ if and only if $\psi(q/\eta) \in T$. Since $\psi \in \mathcal{L}_L$, $\psi(q/\eta) = \psi$ and the equality follows.

(2) This claim is a simple consequence of the definition of $T^{q/\eta}$.

(3) If $T^{q/\eta}$ is stable, stability of $T$ follows directly from (1). Conversely, assume that $T$ is stable. We will prove that $T^{q/\eta}$ is stable, too. First, we will show that $T^{q/\eta}$ is closed under propositional consequence. Consider a formula $\psi \in \mathcal{L}^q_L$ such that $T^{q/\eta} \vdash \psi$. By induction on the length of a derivation, we will show that $\psi \in T^{q/\eta}$. If $\psi$ has a proof of length 1, then $\psi$ is a tautology in $\mathcal{L}^q_L$ or belongs to $T^{q/\eta}$. In the first case $\psi(q/\eta)$ is also a tautology and belongs to $\mathcal{L}_L$. Since $T$ is stable, $T$ is closed under propositional consequence and so contains all tautologies. In particular, $\psi(q/\eta) \in T$. Hence $\psi \in T^{q/\eta}$. In the second case, when $\psi \in T^{q/\eta}$, there is nothing to prove.

Assume next that $\psi$ has a proof of length $n > 1$ from $T^{q/\eta}$ in propositional calculus, say $\psi_1, \ldots, \psi_n (= \psi)$. Assume also that every formula from $\mathcal{L}_L^q$ with a propositional proof from $T^{q/\eta}$ of length less than $n$ is in $T^{q/\eta}$. If $\psi_n$ is a tautology or if $\psi_n \in T^{q/\eta}$, we reason as before. Otherwise, $\psi_n$ is the result of applying modus ponens to formulas $\psi_i$ and $\psi_j$, for some $i, j < n$. By the induction hypothesis, $\psi_i, \psi_j \in T^{q/\eta}$. Thus $\psi_i(q/\eta), \psi_j(q/\eta) \in T$. Applying modus ponens to $\psi_i(q/\eta)$ and $\psi_j(q/\eta)$, we get $\psi_n(q/\eta)$. Since $T$ is closed under propositional consequence, $\psi_n(q/\eta) \in T$. Hence, $\psi_n \in T^{q/\eta}$. Summing up, $T^{q/\eta}$ is closed under propositional consequence.

Consider now a formula $\psi \in T^{q/\eta}$. Then $\psi(q/\eta) \in T$ and, consequently, $L(\psi(q/\eta)) \in T$. Since $L(\psi(q/\eta)) = (L\psi)(q/\eta)$, it follows that $L\psi \in T^{q/\eta}$.

Finally, assume that $\psi \notin T^{q/\eta}$. Then $\psi(q/\eta) \notin T$. Consequently, we have that $\neg L(\psi(q/\eta)) \in T$. Similarly as before, $\neg L(\psi(q/\eta)) = (\neg L\psi)(q/\eta)$. Hence $\neg L\psi \in T^{q/\eta}$.

Summing up, $T^{q/\eta}$ is stable.

(4) If $I \subseteq \mathcal{L}_L$ and a stable theory $T \subseteq \mathcal{L}_L$ contains $I$, then $T^{q/\eta}$ is stable (by (3)) and contains $I^{q,\eta}$ (by (1) and (2)). By (1), if $T_1 \neq T_2$, then $T_1^{q/\eta} \neq T_2^{q/\eta}$. Finally, consider a stable theory $T' \subseteq \mathcal{L}_L^q$ such that $I^{q,\eta} \subseteq T'$. Define $T = T' \cap \mathcal{L}_L$. Clearly $T$ is stable and $I \subseteq T$. We will show that $T^{q/\eta} = T'$. Assume that $\psi \in T'$. Since $q \equiv \eta \in T'$, $\psi(q/\eta) \in T'$. But $\psi(q/\eta) \in \mathcal{L}_L$. Hence, $\psi(q/\eta) \in T$ and, consequently, $\psi \in T^{q/\eta}$. Since both $T'$ and $T^{q/\eta}$ are stable, the equality $T^{q/\eta} = T'$ follows. ■

**Lemma 11.24** *Let $I \subseteq \mathcal{L}_L$. For a theory $T \subseteq \mathcal{L}_L$ we have*

$$T = Cn_{\mathbf{N}}(I \cup A \cup \{\neg L\varphi : \varphi \in \mathcal{L}_L \setminus T\}) \tag{11.6}$$

*if and only if*

$$T^{q/\eta} = Cn^q_{\mathbf{N}}(I^{q,\eta} \cup A^q \cup \{\neg L\varphi : \varphi \in \mathcal{L}^q_L \setminus T^{q/\eta}\}), \tag{11.7}$$

*where $A = Cn_{\mathbf{N}}(\emptyset)$ and $A^q = Cn^q_{\mathbf{N}}(\emptyset)$.*

Proof: First, let us observe that

$$A = A^q \cap \mathcal{L}_L. \tag{11.8}$$

Now suppose that (11.6) holds. Then $T$ is stable and contains $I \cup A$. By Lemma 11.23(4), $T^{q/\eta}$ is stable and $I^{q,\eta} \subseteq T^{q/\eta}$. Consider a formula $\psi \in A^q$. Since $\mathcal{S}$ is closed under uniform substitution, $\psi(q/\eta) \in A^q$. Formula $\psi(q/\eta)$ belongs to $\mathcal{L}_L$. By (11.8), $\psi(q/\eta) \in A$. Hence, $\psi(q/\eta) \in T$ and, consequently, $\psi \in T^{q/\eta}$. Summarizing, $T^{q/\eta}$ is stable and $I^{q,\eta} \cup A^q \subseteq T^{q/\eta}$.

Consider a formula $\psi \in T^{q/\eta} \cap \mathcal{L}^q$. Then $\psi(q/\eta) \in T$. Hence,

$$\psi(q/\eta) \in Cn_{\mathbf{N}}(I \cup A \cup \{\neg L\varphi : \varphi \in \mathcal{L}_L \setminus T\}).$$

Clearly,

$$Cn_{\mathbf{N}}(I \cup A \cup \{\neg L\varphi : \varphi \in \mathcal{L}_L \setminus T\}) \subseteq Cn^q_{\mathbf{N}}(I^{q,\eta} \cup A^q \cup \{\neg L\varphi : \varphi \in \mathcal{L}^q_L \setminus T^{q/\eta}\}).$$

Thus,

$$\psi(q/\eta) \in Cn^q_{\mathbf{N}}(I^{q,\eta} \cup A^q \cup \{\neg L\varphi : \varphi \in \mathcal{L}^q_L \setminus T^{q/\eta}\}).$$

Since $q \equiv \eta \in I^{q,\eta}$, it follows that

$$\psi \in Cn^q_{\mathbf{N}}(I^{q,\eta} \cup A^q \cup \{\neg L\varphi : \varphi \in \mathcal{L}^q_L \setminus T^{q/\eta}\}).$$

Thus, by Corollary 9.8, $T^{q/\eta}$ satisfies (11.7).

Next, assume that $T^{q/\eta}$ satisfies (11.7). Since $I^{q,\eta} \cup A^q \subseteq T^{q/\eta}$, it follows from Lemma 11.23(1) and from (11.8) that $I \cup A \subseteq T$. In addition, $T^{q/\eta}$ is stable. Thus $T$ is stable, as well.

Consider an $\mathbf{N}$-structure $\mathcal{N} = \langle N, \{R_\varphi\}_{\varphi \in \mathcal{L}_L}, V\rangle$ satisfying $I \cup A \cup \{\neg L\varphi : \varphi \in \mathcal{L}_L \setminus T\}$. Define $\mathcal{N}^{q,\eta} = \langle N, \{Q_\varphi\}_{\varphi \in \mathcal{L}^q_L}, U\rangle$, where $Q_\varphi = R_{\varphi(q/\eta)}$, for $\varphi \in \mathcal{L}^q_L$, and, for every world $m \in N$, $U(m)$ is an extension of $V(m)$ to $\mathcal{L}^q$ defined by letting $U(m)(q) = \mathbf{t}$ if and only if $\langle \mathcal{N}, m\rangle \models \eta$.

Clearly, $\mathcal{N}^{q,\eta} \models q \equiv \eta$. Moreover, it is easy to prove by induction on the complexity of a formula that for every world $m \in N$ and for every formula $\psi \in \mathcal{L}^q_L$,

$$\langle \mathcal{N}^{q,\eta}, m\rangle \models \psi \text{ if and only if } \langle \mathcal{N}, m\rangle \models \psi(q/\eta). \tag{11.9}$$

Hence $\mathcal{N}^{q,\eta} \models I^{q,\eta} \cup A^q$. Let $\varphi \notin T^{q/\eta}$. Then $\varphi(q/\eta) \notin T$ and, consequently, $\mathcal{N} \models \neg L\varphi(q/\eta)$. Let $m \in N$. Then there exists $m' \in N$ such that $(m, m') \in R_{\varphi(q/\eta)}$ and $\langle \mathcal{N}, m'\rangle \not\models \varphi(q/\eta)$. Since $R_{\varphi(q/\eta)} = Q_\varphi$, by (11.9) it follows that $\langle \mathcal{N}^{q,\eta}, m\rangle \models \neg L\varphi$. Hence $\mathcal{N}^{q,\eta} \models I^{q,\eta} \cup A^q \cup \{\neg L\varphi : \varphi \in \mathcal{L}^q_L \setminus T^{q/\eta}\}$. From (11.7) it follows that $\mathcal{N}^{q,\eta} \models T^{q/\eta}$. Since $T \cap \mathcal{L} \subseteq T^{q/\eta}$, $\mathcal{N}^{q,\eta} \models T \cap \mathcal{L}$. By (11.9), $\mathcal{N} \models T \cap \mathcal{L}$. Thus $T \cap \mathcal{L} \subseteq Cn_{\mathbf{N}}(I \cup A \cup \{\neg L\varphi : \varphi \in \mathcal{L}_L \setminus T\})$. Since $T$ is stable and contains $I \cup A$, by Corollary 9.8, $T$ satisfies equality (11.6). ■

**Theorem 11.25** *Let $I \subseteq \mathcal{L}_L$. A theory $T$ is an $\mathcal{S}$-expansion for $I$ (in $\mathcal{L}_L$) if and only if $T^{q/\eta}$ is an $\mathcal{S}$-expansion for $I^{q,\eta}$ (in $\mathcal{L}^q_L$).*

Proof: Let $A$ be the set of instances of axiom schemata of $\mathcal{S}$. The following four identities are equivalent ($A^q$ was defined in Lemma 11.24):

$$T = Cn_{\mathcal{S}}(I \cup \{\neg L\varphi : \varphi \in \mathcal{L}_L \setminus T\})$$

$$T = Cn_{\mathbf{N}}(I \cup A \cup \{\neg L\varphi : \varphi \in \mathcal{L}_L \setminus T\})$$

$$T^{q/\eta} = Cn^q_{\mathbf{N}}(I^{q,\eta} \cup A^q \cup \{\neg L\varphi : \varphi \in \mathcal{L}^q_L \setminus T^{q/\eta}\})$$

$$T^{q/\eta} = Cn^q_{\mathcal{S}}(I^{q,\eta} \cup \{\neg L\varphi : \varphi \in \mathcal{L}^q_L \setminus T^{q/\eta}\}).$$

The equivalences of the first and the second, and the third and the fourth equalities follow from Theorem 10.1. The equivalence of the second and the third equalities follows from Lemma 11.24. Since $T$ is consistent if and only if $T^{q/\eta}$ is consistent, the assertion follows. ■

## 11.4 Bounding the depth of introspection

When defining $\mathcal{S}$-expansions we allow an agent to make use of all formulas expressing what is possible with respect to a context $T$, that is, of all formulas of the form $\neg L\varphi$, where $\varphi \notin T$. As a result, we get defeasibility of an agent's reasoning. But let us observe that if a formula $\psi$ has a derivation that relies entirely on agent's initial assumptions $I$ and does not refer to any formula $\neg L\varphi$, where $\varphi \notin T$, then $\psi \in Cn^n_{\mathcal{S}}(I')$, for every theory $I'$ such that $I \subseteq I'$. On the other hand, if a derivation of $\psi$ relies on a formula $\neg L\varphi$ and $\varphi$ is added to $I$, it may happen that $\psi \notin Cn^n_{\mathcal{S}}(I \cup \{\varphi\})$. Thus, an obvious way to make nonmonotonic derivations more robust with respect to additions of new information to the set of initial assumptions might be to restrict the class of formulas of the form $\neg L\varphi$ that the agent is allowed to use in reasoning.

**Definition 11.26** Let $\mathcal{S}$ be a modal logic and let $X$ be a collection of formulas from $\mathcal{L}_L$. A theory $T \subseteq \mathcal{L}_L$ is an $(\mathcal{S}, X)$-*expansion* for a theory $I \subseteq \mathcal{L}_L$ if $T$ is consistent and

$$T = Cn_{\mathcal{S}}(I \cup \{\neg L\varphi : \varphi \in X \setminus T\}). \tag{11.10}$$

▲

We now discuss properties of $(\mathcal{S}, X)$-expansions. It turns out that if $\mathcal{L} \subseteq X$ then many properties of $\mathcal{S}$-expansions carry over to the more general concept of $(\mathcal{S}, X)$-expansions. The results of this section rely on Theorem 8.17, which assumes normality of the underlying modal logic $\mathcal{S}$. Without this assumption, solutions to (11.10) need not be stable. This is the reason why we require that $\mathcal{S}$ be normal throughout this section.

**Theorem 11.27** *Let logic $\mathcal{S}$ be a normal modal logic. Let $X$ be a set of formulas such that $\mathcal{L} \subseteq X$. For every theory $I \subseteq \mathcal{L}_L$, and $T \subseteq \mathcal{L}_L$, if $T$ satisfies equation (11.10), then $T$ is stable.*

Proof: Assume that $T$ satisfies the equation (11.10). If $T$ is inconsistent with $\mathcal{S}$, then $T$ is stable. Hence, assume that $T$ is consistent with $\mathcal{S}$. Consider now the canonical model $\mathcal{M} = \mathcal{M}_{\mathcal{S}}^{can}(I \cup \{\neg L\varphi : \varphi \in X \setminus T\})$. Then $T = Th(\mathcal{M})$.

Since $\mathcal{L} \subseteq X$, $\mathcal{M} \models \{\neg L\varphi : \varphi \in \mathcal{L} \setminus T\}$. In addition, $T = Th(\mathcal{M})$. Hence, by Theorem 8.17, $T = Th(\mathcal{M})$ is stable. ■

**Corollary 11.28** *Let $\mathcal{S}$ be a normal modal logic. Let $X$ be a set of formulas such that $\mathcal{L} \subseteq X$. For a theory $I \subseteq \mathcal{L}_L$, if $T \subseteq \mathcal{L}_L$ is an $(\mathcal{S}, X)$-expansion of $I$, then $T$ is stable.* ▲

The next result gives a characterization of $(\mathcal{S}, X)$-expansions similar to that of Corollary 9.8.

**Theorem 11.29** *Let $\mathcal{S}$ be a normal modal logic contained in* **S5**. *Let $X$ be a set of formulas such that $\mathcal{L} \subseteq X$. A theory $T \subseteq \mathcal{L}_L$ is an $(\mathcal{S}, X)$-expansion of a theory $I \subseteq \mathcal{L}_L$ if and only if $T$ is stable, consistent, $I \subseteq T$ and $T \cap \mathcal{L} \subseteq Cn_{\mathcal{S}}(I \cup \{\neg L\varphi : \varphi \in X \setminus T\})$.*

Proof: The necessity part of the assertion follows directly from Theorem 11.27 and from the definition of an $(\mathcal{S}, X)$-expansion. Conversely, assume that $T$ is consistent and stable, $I \subseteq T$ and $T \cap \mathcal{L} \subseteq Cn_{\mathcal{S}}(I \cup \{\neg L\varphi : \varphi \in X \setminus T\})$. Put $U = Cn_{\mathcal{S}}(I \cup \{\neg L\varphi : \varphi \in X \setminus T\})$ and $S = T \cap \mathcal{L}$. Theory $T$ is stable. Therefore, $T$ is closed under provability in **S5**. Since $T$ is stable, $I \subseteq T$ and $\mathcal{S}$ is contained in **S5**, we have $U \subseteq T$. Since $T$ is consistent and stable, $T$ is consistent with **S5**. Logic $\mathcal{S}$ is contained in **S5**. Hence, both $U$ and $I \cup \{\neg L\varphi : \varphi \in X \setminus T\}$ are consistent with $\mathcal{S}$. Consider the canonical model $\mathcal{M} = \mathcal{M}_{\mathcal{S}}^{can}(I \cup \{\neg L\varphi : \varphi \in X \setminus T\})$. Then $Th(\mathcal{M}) = U$. Since $S \subseteq U$ and $\mathcal{L} \subseteq X$, $\mathcal{M} \models S \cup \{\neg L\varphi : \varphi \in \mathcal{L} \setminus T\}$. By Theorem 8.17, it follows that $U = Th(\mathcal{M})$ is stable and that $U \cap \mathcal{L} = S$. Since both $U$ and $T$ are stable, $U = T$. Since $T$ is consistent, $T$ is an $(\mathcal{S}, X)$-expansion for $I$. ■

Stability of $(\mathcal{S}, X)$-expansions implies a natural relationship between $(\mathcal{S}, X)$-expansions for different sets $X$. The next result states that the smaller the set of formulas to which the negation-as-failure-to-prove rule is applied, the smaller the class of $(\mathcal{S}, X)$-expansions.

**Theorem 11.30** *Let $\mathcal{S}$ be a normal modal logic contained in* **S5**. *Let $X$ and $Y$ be sets of formulas such that $\mathcal{L} \subseteq X \subseteq Y$. If a theory $T \subseteq \mathcal{L}_L$ is an $(\mathcal{S}, X)$-expansion for a theory $I \subseteq \mathcal{L}_L$, then $T$ is an $(\mathcal{S}, Y)$-expansion for $I$.*

Proof: Assume that a theory $T \subseteq \mathcal{L}_L$ is an $(\mathcal{S}, X)$-expansion for a theory $I$. Then $T$ is stable and consistent, $I \subseteq T$ and $T \cap \mathcal{L} \subseteq Cn_{\mathcal{S}}(I \cup \{\neg L\varphi : \varphi \in X \setminus T\})$. Since $X \subseteq Y$, $T \cap \mathcal{L} \subseteq Cn_{\mathcal{S}}(I \cup \{\neg L\varphi : \varphi \in Y \setminus T\})$. Hence, by Theorem 11.29, $T$ is an $(\mathcal{S}, Y)$-expansion for $I$. ■

**Corollary 11.31** *Let $\mathcal{S}$ be a normal modal logic contained in* **S5**. *Let $X$ be a set of formulas such that $\mathcal{L} \subseteq X$. If a theory $T \subseteq \mathcal{L}_L$ is an $(\mathcal{S}, X)$-expansion for a theory $I \subseteq \mathcal{L}_L$, then $T$ is an $\mathcal{S}$-expansion for $I$.* ▲

The final result of this section is a generalization of Theorem 9.6. The proof is similar and relies on stability of expansions. We leave it to the reader.

**Theorem 11.32** *Let $\mathcal{S}$ and $\mathcal{T}$ be normal modal logics such that $\mathcal{S} \subseteq \mathcal{T} \subseteq$* **S5**. *Let $\mathcal{L} \subseteq X$. Then, for every $I \subseteq \mathcal{L}_L$, if a theory $T \subseteq \mathcal{L}_L$ is an $(\mathcal{S}, X)$-expansion for $I$, then $T$ is a $(\mathcal{T}, X)$-expansion of $I$.* ▲

## 11.5 Ground $\mathcal{S}$-expansions

In this section, we focus our attention on the case when negative introspection is bounded to formulas from $\mathcal{L}$. An $(\mathcal{S}, \mathcal{L})$-expansion for a theory $I$ is called a *ground* $\mathcal{S}$-expansion for $I$. Clearly, all of the results presented in the previous section hold for ground expansions. In addition, for several modal logics $\mathcal{S}$, ground $\mathcal{S}$-expansions can be characterized as $\mathcal{S}$-expansions satisfying some minimality criterion.

**Definition 11.33** Let $T$ and $T'$ be two stable theories. We say that $T \trianglelefteq T'$ if $T \cap \mathcal{L} \subseteq T' \cap \mathcal{L}$; ▲

It is immediate that the relation $\trianglelefteq$ is a partial ordering of the set of all stable theories. Reflexivity and transitivity are entirely obvious, and the antisymmetry of $\trianglelefteq$ follows from Corollary 8.19.

**Definition 11.34**

1. Let $I \subseteq \mathcal{L}_L$. A stable theory $T$ is *minimal* for $I$ if $T$ is minimal with respect to the partial ordering $\trianglelefteq$ among all stable theories containing $I$.
2. An $\mathcal{S}$-expansion $T$ for a theory $I$ is *minimal* if for every stable theory $T'$ such that $I \subseteq T'$, $T \trianglelefteq T'$. ▲

Note that in order for an $\mathcal{S}$-expansion for $I$ to be minimal it is not enough that there is no $\mathcal{S}$-expansion for $I$ with a "smaller" modal-free part. It is necessary that there is no stable theory containing $I$ with a "smaller" modal-free part. It is easy to see that ground $\mathcal{S}$-expansions are minimal.

**Theorem 11.35** *Let $\mathcal{S}$ be a normal modal logic contained in* **S5**. *If a theory $T \subseteq \mathcal{L}_L$ is a ground $\mathcal{S}$-expansion for a theory $I \subseteq \mathcal{L}_L$, then $T$ is minimal.*

Proof: Consider a ground $\mathcal{S}$-expansion $T$ of $I$. Then the theory $T$ satisfies

$$T = Cn_{\mathcal{S}}(I \cup \{\neg L\varphi : \varphi \in \mathcal{L} \setminus T\}).$$

Let $T'$ be a stable theory such that $I \subseteq T'$ and $T' \cap \mathcal{L} \subseteq T \cap \mathcal{L}$. Then we have

$$T \cap \mathcal{L} \subseteq T = Cn_{\mathcal{S}}(I \cup \{\neg L\varphi : \varphi \in \mathcal{L} \setminus T\}) \subseteq T'.$$

The last inclusion follows from the inclusions $I \subseteq T'$ and $T' \cap \mathcal{L} \subseteq T \cap \mathcal{L}$, and stability of $T'$. Hence, $T' \cap \mathcal{L} = T \cap \mathcal{L}$, and $T$ is minimal. ■

Next, we study the converse statement. First, we deal with the case of the logic **S5**.

**Theorem 11.36** *A theory $T \subseteq \mathcal{L}_L$ is a ground* **S5**-*expansion for a theory $I \subseteq \mathcal{L}_L$ if and only if $T$ is a minimal* **S5**-*expansion for $I$ (or, equivalently, a minimal stable theory for $I$).*

Proof: Necessity follows from Theorem 11.35. So, assume that $T$ is a minimal **S5**-expansion for $I$. Consider an **S5**-model $\mathcal{M}$ of $I \cup \{\neg L\varphi : \varphi \in \mathcal{L} \setminus T\}$. Clearly, $Th(\mathcal{M})$ is stable and contains $I$. In addition, $Th(\mathcal{M}) \cap \mathcal{L} \subseteq T \cap \mathcal{L}$. By minimality of $T$, $Th(\mathcal{M}) \cap \mathcal{L} = T \cap \mathcal{L}$. Hence, $\mathcal{M} \models T \cap \mathcal{L}$ and, consequently, $T \cap \mathcal{L} \subseteq Cn_{\mathbf{S5}}(I \cup \{\neg L\varphi : \varphi \in \mathcal{L} \setminus T\})$. Since $T$ is stable, consistent and contains $I$, Theorem 11.29 implies that $T$ is a ground **S5**-expansion of $I$. ■

Now we are in a position to prove a more general result.

**Definition 11.37** Let $R \subseteq M \times M$ be a binary relation. We say that $R$ has the *terminal cluster property* if for every $m$ in the domain of $R$ there is a cluster $Y \subseteq M$ ($Y \subseteq M$ is a cluster of $R$ if $Y$ is a maximal subset of $M$ such that $Y \times Y \subseteq R$) such that:

1. For every $m' \in Y$, $(m, m') \in R$.

2. For every $m' \in Y$ and every $m'' \in M \setminus Y$, $(m', m'') \notin R$.

Such a cluster $Y$ is called a *terminal cluster for m.* ▲

**Theorem 11.38** *Let $\mathcal{S}$ be a normal modal logic contained in* **S5** *and characterized by a class $\mathcal{C}$ of models with an accessibility relation satisfying the terminal cluster property. Then a theory $T \subseteq \mathcal{L}_L$ is a ground $\mathcal{S}$-expansion for a theory $I \subseteq \mathcal{L}_L$ if and only if $T$ is a minimal $\mathcal{S}$-expansion of $I$.*

Proof: The necessity part is an immediate consequence of Theorem 11.35. To prove the sufficiency part of the theorem, assume that $T$ is a minimal $\mathcal{S}$-expansion of $I$. Then $T$ is stable and $I \subseteq T$. Consequently,

$$Cn_{\mathcal{S}}(I \cup \{\neg L\varphi: \varphi \in \mathcal{L} \setminus T\}) \subseteq T.$$

Since $T = Cn_{\mathcal{S}}(I \cup \{\neg L\varphi: \varphi \in \mathcal{L}_L \setminus T\})$, to prove the converse inclusion it suffices to show that the set of formulas $A = \{\neg L\varphi: \varphi \in \mathcal{L}_L \setminus T\}$ is contained in $Cn_{\mathcal{S}}(I \cup \{\neg L\varphi: \varphi \in \mathcal{L} \setminus T\})$.

First, observe that $T$ is a minimal **S5**-expansion for $I$, that is, $T$ is a $\trianglelefteq$-minimal stable theory containing $I$. By Theorem 11.36, $T$ is a ground **S5**-expansion for $I$. Thus $A \subseteq Cn_{\mathbf{S5}}(I \cup \{\neg L\varphi: \varphi \in \mathcal{L} \setminus T\})$. Let $\mathcal{M} = (M, R, V) \in \mathcal{C}$ be any model of $I \cup \{\neg L\varphi: \varphi \in \mathcal{L} \setminus T\}$. Consider any terminal cluster $Y \subseteq M$. Let $\mathcal{M}_Y$ be the restriction of the model $\mathcal{M}$ to the set of worlds $Y$. Clearly, $\mathcal{M}_Y$ is an **S5**-model and $\mathcal{M}_Y \models I \cup \{\neg L\varphi: \varphi \in \mathcal{L} \setminus T\}$. Hence $\mathcal{M}_Y \models A$.

Consider an arbitrary world $m \in M$. Since $T$ is consistent, there is $\psi \in \mathcal{L}_L \setminus T$ such that $\langle \mathcal{M}, m \rangle \models \neg L\psi$. Hence $m$ is in the domain of $R$. Since $R$ has the terminal cluster property, there is at least one terminal cluster for $m$. Let $Y$ be a terminal cluster for $m$. Let $\varphi \in \mathcal{L}_L \setminus T$. We have just proved that $\mathcal{M}_Y \models A$. Therefore, $\mathcal{M}_Y \models \neg L\varphi$ and, consequently, there is a world $m' \in Y$ such that $\langle \mathcal{M}_Y, m' \rangle \not\models \varphi$. Since $Y$ is a terminal cluster for $m$, $(m, m') \in R$. Thus it follows that $\langle \mathcal{M}, m \rangle \models \neg L\varphi$ and $\mathcal{M} \models A$. Since $\mathcal{S}$ is characterized by the class of models $\mathcal{C}$,

$$A \subseteq Cn_{\mathcal{S}}(I \cup \{\neg L\varphi: \varphi \in \mathcal{L} \setminus T\}).$$

■

Notice that the characterization of ground $\mathcal{S}$-expansions given in Theorem 11.38 applies, for example, to logics **S4f**, **Sw5** and **kd45**.

The assertion of Theorem 11.38 is not true for an arbitrary normal modal logic $\mathcal{S}$. A counterexample, in the case of logic **T**, is given below.

**Example 11.39** Consider the theory

$$I = \{\neg L\neg Lp \supset q, \neg L\neg Lq \supset r, \neg L\neg Lr \supset p, \neg Lp \wedge \neg Lq \supset r\},$$

where $p$, $q$ and $r$ are three distinct atoms. It is easy to show that $ST(\{p,q,r\})$ is a minimal **T**-expansion of $I$. On the other hand, none of $p$, $q$ and $r$ is contained in $Cn_{\mathbf{T}}(I \cup \{\neg L\varphi : \varphi \in \mathcal{L} \setminus ST(\{p,q,r\})\})$. Hence, $ST(\{p,q,r\})$ is not a ground **T**-expansion of $I$. ▲

Elegant properties of ground $\mathcal{S}$-expansions notwithstanding, they have a serious drawback. Unlike $\mathcal{S}$-expansions, ground $\mathcal{S}$-expansions need not be preserved after a theory is extended by a new definition.

**Example 11.40** Consider the theory $I = \{\neg L \neg Lp \supset p\}$. It is easy to see that this theory has exactly two $\mathcal{S}$-expansions, $ST(\emptyset)$ and $ST(\{p\})$, for every modal logic $\mathcal{S}$ contained in **Sw5** or **kd45**. If $\mathcal{S}$ is normal then, in addition, $ST(\emptyset)$ is a ground $\mathcal{S}$-expansion for $I$ while $ST(\{p\})$ is not. Let us introduce a new propositional variable $r$ and let us add to $I$ the formula $r \equiv Lp$. We will denote the resulting theory by $I^*$. We will show that $I^*$ has two ground $\mathcal{S}$-expansions for every normal modal logic contained in **Sw5** or **kd45**. To this end, it is enough to show that $I^*$ has two ground **K**-expansions.

First observe that $I^* \subseteq ST(\{\neg r\})$. Moreover, $\neg r \in Cn_{\mathbf{K}}(I^* \cup \{\neg L\varphi : \varphi \in \mathcal{L} \setminus ST(\{\neg r\})\})$. By Theorem 11.29, $ST(\{\neg r\})$ is a ground **K**-expansion for $I^*$.

Similarly, it is easy to see that $I^* \subseteq ST(\{p,r\})$. We need to show that $\{p,r\} \subseteq Cn_{\mathbf{K}}(I^* \cup \{\neg L\varphi : \varphi \in \mathcal{L} \setminus ST(\{p,r\})\})$. First observe that using the schema $k$ and necessitation we can get, out of the formula $\neg Lp \supset \neg r$ (which is a "half" of $r \equiv Lp$), the formula $L\neg Lp \supset L\neg r$. Thus the formula $L\neg Lp \supset L\neg r$ belongs to $Cn_{\mathbf{K}}(I^* \cup \{\neg L\varphi : \varphi \in \mathcal{L} \setminus ST(\{p,r\})\})$. Consequently, $\neg L\neg r \supset \neg L\neg Lp$ belongs to $Cn_{\mathbf{K}}(I^* \cup \{\neg L\varphi : \varphi \in \mathcal{L} \setminus ST(\{p,r\})\})$. Since $\neg r \notin ST(\{p,r\})$, $\neg L\neg r$ belongs to $Cn_{\mathbf{K}}(I^* \cup \{\neg L\varphi : \varphi \in \mathcal{L} \setminus ST(\{p,r\})\})$. Therefore $\neg L\neg Lp$ belongs to $Cn_{\mathbf{K}}(I^* \cup \{\neg L\varphi : \varphi \in \mathcal{L} \setminus ST(\{p,r\})\})$. It follows easily that both $p$ and $r$ belong to $Cn_{\mathbf{K}}(I^* \cup \{\neg L\varphi : \varphi \in \mathcal{L} \setminus ST(\{p,r\})\})$. As before, it follows that $ST(\{p,r\})$ is a ground **K**-expansion for $I^*$. Thus adding a definition resulted in increasing the number of ground expansions. ▲

## 11.6 Bounding introspection to $\mathcal{L} \cup \{\neg L\varphi : \varphi \in \mathcal{L}\}$

We will now prove a result which states, perhaps somewhat surprisingly, that for many logics, including some with especially interesting nonmonotonic counterparts like **S4f**, **Sw5** and **kd45**, even after a significant restriction of the set of additional assumptions $\neg L\varphi$ used in the process of reasoning, the resulting notion of expansion is the same as that given by Definition 9.2.

Specifically, let us define

$$\mathcal{C}_1 = \mathcal{L} \cup \{\neg L\varphi : \varphi \in \mathcal{L}\}. \tag{11.11}$$

In this section we show that for a wide class of normal modal logics the notions of $\mathcal{S}$-expansion and $(\mathcal{S}, \mathcal{C}_1)$-expansion coincide.

The significance of this result becomes apparent when we recall that in Section 11.4 we proved that bounding introspection to formulas of $\mathcal{L}$ alone changes considerably the properties of the resulting notion of expansion.

Before we formulate the result, let us prove the following lemma.

**Lemma 11.41**

1. $M\varphi \wedge ML\psi \vdash_{\mathbf{K4}} M(\varphi \wedge L\psi)$.
2. $ML\varphi \wedge ML\psi \vdash_{\mathbf{K4}} ML(\varphi \wedge \psi)$.

Proof: (1) By Theorem 7.17, $\vdash_{\mathbf{K}} ML\varphi \wedge LL\psi \supset M(L\varphi \wedge L\psi)$. Hence, by Theorem 7.17,

$$\vdash_{\mathbf{K}} M(ML\varphi \wedge LL\psi) \supset MM(L\varphi \wedge L\psi). \tag{11.12}$$

Applying Theorem 7.17 to the axiom schema 4 we get

$$\vdash_{\mathbf{K4}} ML\psi \supset MLL\psi. \tag{11.13}$$

Applying necessitation we get $M\varphi \wedge ML\psi \vdash_{\mathbf{K}} LM\varphi$. Hence, by Theorem 7.17 and (11.13),

$$M\varphi \wedge ML\psi \vdash_{\mathbf{K4}} M(M\varphi \wedge LL\psi). \tag{11.14}$$

Again by Theorem 7.17, $\vdash_{\mathbf{K}} M\varphi \wedge LL\psi \supset M(\varphi \wedge L\psi)$. Applying Theorem 7.17 we get

$$\vdash_{\mathbf{K}} M(M\varphi \wedge LL\psi) \supset MM(\varphi \wedge L\psi). \tag{11.15}$$

Combining (11.14) and (11.15), and applying schema, 4 we get

$$M\varphi \wedge ML\psi \vdash_{\mathbf{K4}} M(\varphi \wedge L\psi).$$

(2) follows immediately from (1) using Theorem 7.17. ∎

**Theorem 11.42** *Let $\mathcal{S}$ be any modal logic such that* $\mathbf{K4} \subseteq \mathcal{S} \subseteq \mathbf{S5}$. *Then a theory $T$ is an $\mathcal{S}$-expansion for a theory $I$ if and only if $T$ is propositionally consistent and*

$$T = Cn_{\mathcal{S}}(I \cup \{\neg L\varphi : \varphi \in \mathcal{C}_1 \setminus T\}), \tag{11.16}$$

*that is, if and only if $T$ is an $(\mathcal{S}, \mathcal{C}_1)$-expansion for $I$.*

Proof: Let us assume that $T$ is an $\mathcal{S}$-expansion for $I$. We will first prove that for each formula $\eta \in \mathcal{L}_L$, if $\eta \notin T$ then

$$\{\neg L\varphi : \varphi \in \mathcal{L} \setminus T\} \cup \{ML\varphi : \varphi \in T \cap \mathcal{L}\} \vdash_{\mathbf{K4}} \neg L\eta \tag{11.17}$$

and if $\eta \in T$, then

$$\{\neg L\varphi : \varphi \in \mathcal{L} \setminus T\} \cup \{ML\varphi : \varphi \in T \cap \mathcal{L}\} \vdash_{\mathbf{K4}} ML\eta. \tag{11.18}$$

Let us denote

$$A = \{\neg L\varphi : \varphi \in \mathcal{L} \setminus T\} \cup \{ML\varphi : \varphi \in T \cap \mathcal{L}\}.$$

Hence, our first goal is to prove that for each formula $\eta \in \mathcal{L}_L \setminus T$, $A \vdash_{\mathbf{K4}} \neg L\eta$ and for each $\eta \in T$, $A \vdash_{\mathbf{K4}} ML\eta$.

We proceed by induction on the depth of $\eta$. If $d_L(\eta) = 0$, then the assertion is straightforward since if $\eta \notin T$ then $\neg L\eta \in A$, and if $\eta \in T$ then $ML\eta \in A$. So, consider $\eta$ such that $d_L(\eta) = n > 0$ and assume the assertion holds for all formulas of $L$-depth smaller than $n$. Without loss of generality, we may assume that $\eta$ is in a conjunctive normal form, that is

$$\eta = \eta_1 \wedge \ldots \wedge \eta_p,$$

where each $\eta_i$ is of the form

$$\psi_0 \vee L\psi_1 \vee \ldots \vee L\psi_k \vee \neg L\psi_{k+1} \vee \ldots \vee \neg L\psi_{k+m}, \tag{11.19}$$

$\psi_0 \in \mathcal{L}$ and $d_L(\psi_i) \leq n - 1$, $1 \leq i \leq k + m$.

First, assume that $\eta \notin T$. Then there is $i$, $1 \leq i \leq p$, such that $\eta_i \notin T$. Since $T$ is stable and consistent, it follows by Corollary 8.9, that $\psi_j \notin T$, for every $j$, $0 \leq j \leq k$, and $\psi_j \in T$, for every $j$, $k < j \leq k + m$.

By the definition of $A$, $A \vdash_{\mathbf{K4}} \neg L\psi_0$. Since $\vdash_{\mathbf{K}} \neg L\psi_0$ if and only if $\vdash_{\mathbf{K}} M\neg\psi_0$,

$$A \vdash_{\mathbf{K4}} M\neg\psi_0.$$

By the induction hypothesis, for every $j$ such that $k < j \leq k + m$,

$$A \vdash_{\mathbf{K4}} ML\psi_j.$$

Hence, by Lemma 11.41(1),

$$A \vdash_{\mathbf{K4}} M(\neg\psi_0 \wedge L\psi_{k+1} \wedge \ldots \wedge L\psi_{k+m}). \tag{11.20}$$

By the induction hypothesis and by necessitation, for every $j$ such that $1 \leq j \leq k$,

$$A \vdash_{\mathbf{K4}} L\neg L\psi_j.$$

By (11.20) and by repeated application of Theorem 7.17,

$$A \vdash_{\mathbf{K4}} M(\neg\psi_0 \wedge \neg L\psi_1 \wedge \ldots \wedge \neg L\psi_k \wedge L\psi_{k+1} \wedge \ldots \wedge L\psi_{k+m}).$$

Consequently,

$$A \vdash_{\mathbf{K4}} \neg L(\psi_0 \vee L\psi_1 \vee \ldots \vee L\psi_k \vee \neg L\psi_{k+1} \vee \ldots \vee \neg L\psi_{k+m}) = \neg L\eta_i.$$

Since $\eta \supset \eta_i$ is a tautology, by Theorem 7.17 $A \vdash_{\mathbf{K4}} \neg L\eta$.

Next, assume that $\eta \in T$. Then $\eta_i \in T$, for every $i$, $1 \leq i \leq p$. By Lemma 11.41(2), to complete the induction step, it is sufficient to prove that $A \vdash_{\mathbf{K4}} ML\eta_i$, for every $i$, $1 \leq i \leq p$. Since $\eta_i \in T$, it follows by Corollary 8.9 that there are three possibilities:

Case 1. $\psi_0 \in T$. Then $ML\psi_0 \in A$. Since $\psi_0 \supset \eta_i$ is a tautology, Theorem 7.17 implies $A \vdash_{\mathbf{K4}} ML\eta_i$.

Case 2. $\psi_j \in T$, for some $j$, $1 \leq j \leq k$. By the induction hypothesis, $A \vdash_{\mathbf{K4}} ML\psi_j$. Since $L\psi_j \supset \eta_i$ is a tautology, we get $\vdash_{\mathbf{K}} LL\psi_j \supset L\eta_i$ (Theorem 7.17). By 4, $\vdash_{\mathbf{K4}} L\psi_j \supset L\eta_i$. Thus $\vdash_{\mathbf{K4}} ML\psi_j \supset ML\eta_i$. Consequently, $A \vdash_{\mathbf{K4}} ML\eta_i$.

Case 3. $\psi_j \notin T$, for some $j$, $k < j \leq k+m$. By the induction hypothesis, $A \vdash_{\mathbf{K4}} \neg L\psi_j$. Consequently, $A \vdash_{\mathbf{K4}} M\neg\psi_j$ and $A \vdash_{\mathbf{K4}} LL\neg L\psi_j$. By Theorem 7.17,

$$A \vdash_{\mathbf{K4}} M(L\neg L\psi_j \wedge \neg\psi_j).$$

Consequently, $A \vdash_{\mathbf{K4}} ML\neg L\psi_j$. Since $\neg L\psi_j \supset \eta_i$ is a tautology, $A \vdash_{\mathbf{K4}} ML\eta_i$.

Now, the assertion of the theorem follows easily. Recall that we assumed that $T$ is an $\mathcal{S}$-expansion for $I$. Since

$$\{\neg L\varphi: \varphi \in \mathcal{L} \setminus T\} \cup \{ML\varphi: \varphi \in T \cap \mathcal{L}\} \subseteq \{\neg L\varphi: \varphi \notin T\}, \tag{11.21}$$

the inclusion

$$T \supseteq Cn_{\mathcal{S}}(I \cup \{\neg L\varphi: \varphi \in \mathcal{L} \setminus T\} \cup \{ML\varphi: \varphi \in T \cap \mathcal{L}\})$$

follows. The converse inclusion follows by (11.17). Hence,

$$T = Cn_{\mathcal{S}}(I \cup \{\neg L\varphi: \varphi \in \mathcal{L} \setminus T\} \cup \{ML\varphi: \varphi \in T \cap \mathcal{L}\}).$$

Since $T$ is an $\mathcal{S}$-expansion, $T$ is consistent and stable. Thus,

$$\{\neg L\varphi: \varphi \in \mathcal{L} \setminus T\} \cup \{ML\varphi: \varphi \in T \cap \mathcal{L}\}) = \{\neg L\varphi: \varphi \in \mathcal{C}_1 \setminus T\}.$$

Consequently, $T$ is consistent and satisfies the equation (11.16).

The converse implication follows from Corollary 11.31. ■

## 11.7 Comments and remarks

Algorithms **All_ExpsA** and **All_ExpsB** for computing $\mathcal{S}$-expansions given in this chapter were found by Schwarz and the authors [MST91]. Earlier, an algorithm for computing autoepistemic expansions was proposed by Niemelä [Nie88]. The characterization of autoepistemic expansions contained in Theorem 11.5 and the algorithm it implies were given by the authors [MT91].

The concept of a *range* was proposed in [MST91] and several ranges of modal logics with the same nonmonotonic counterpart were exhibited there. Theorem 11.15 is due to Schwarz [Sch92].

The idea of stratification in the context of modal nonmonotonic logics is due to Gelfond [Gel87]. The results on stratified theories given in Section 11.2 were proved in [MT91] and [MST91].

The problem of adding new definitions to a theory was investigated by Schwarz and Truszczyński [ST93].

The material in Section 11.4 on $(\mathcal{S}, X)$-expansions were obtained in [Tru91c]. Ground **kd45**-expansions (under the name of *moderately grounded* expansions) were introduced by Konolige [Kon88]. The concept of a ground $\mathcal{S}$-expansion was introduced by Tiomkin and Kaminski [TK90, Kam91]. It was also investigated in [Tru91c]. The results on bounding the depth of introspection to formulas in the class $\mathcal{C}_1$ are due to Schwarz [Sch91b].

# 12. Relations among nonmonotonic formalisms

Our book has two main themes: nonmonotonic formalisms based on the language of propositional logic (default logic and logic programming and its extensions fall into this category) and modal nonmonotonic logics. Until now we have not discussed the question of interconnections between them. We will do so in this chapter. First, we will show that several modal nonmonotonic logics studied in Chapter 7 can be regarded as generalizations of default logic. An important consequence of this result is yet another semantical characterization of extensions. Even stronger results can be obtained for logic programming and, in particular, new characterizations of stable models for logic programs and stable answer sets for logic programs with classical negation are presented. Next, we show that default logic with weak extensions naturally corresponds to autoepistemic logic (that is, the nonmonotonic logic **kd45**). Finally, we introduce the notion of *truth maintenance* and briefly discuss its relation to logic programming and modal nonmonotonic logics.

## 12.1 Expressing default reasoning in modal nonmonotonic logics

Let us recall that default logic is concerned with properties of *default theories*, that is, pairs $(D, W)$, where $W$ is a set of formulas from $\mathcal{L}$ and $D$ is a collection of nonstandard inference rules of the form

$$\frac{\alpha : M\beta_1, \ldots, M\beta_n}{\gamma}. \tag{12.1}$$

In order to apply a default of the form (12.1), all its premises must be established. That is, $\alpha$ has to be established (proved) and for every $i$, $1 \leq i \leq n$, it has to be established that $\beta_i$ is *possible* or *consistent*. To establish the possibility of a formula, we consider a theory $S \subseteq \mathcal{L}$, called a *context*, and define a formula $\beta$ to be possible (consistent) with respect to $S$ if $S \not\vdash \neg\beta$.

Nonmonotonic modal logics are also defined in terms of derivations *with respect to a context*, a theory in the language $\mathcal{L}_L$. However, in modal nonmonotonic logics, the context is used in a slightly different way than in default logic. In default logic, a context is used to establish nonstandard premises (justifications) of defaults, but its elements are not directly subject to the rules of propositional consequence. In the case of modal nonmonotonic logics, a context provides additional assumptions of the form $\neg L\varphi$, which might be interpreted as *$\varphi$ is not known, $\neg\varphi$ is possible* or *$\neg\varphi$ is consistent.* These additional assumptions are treated in exactly the same way as the initial assumptions of an agent and are used as premises in derivations.

This short discussion points to a similarity between default logic and modal nonmonotonic logics. Both are based on the notion of a context-dependent proof. But the way in which contexts are used is not the same. A natural question arises:

> What is the precise relationship between default logic and nonmonotonic modal logics?

In the remainder of this section, we will provide an answer to this question. We will exhibit an encoding of defaults and default theories by modal formulas and modal theories, respectively, under which there is a one-to-one correspondence between extensions and $\mathcal{S}$-expansions, for suitably chosen modal logics $\mathcal{S}$.

As we have already said, a default $d$ of the form (12.1) can intuitively be interpreted as follows:

> if it is known that $\alpha$ and if it is known that each of $\beta_1, \ldots, \beta_n$ is possible, then conclude $\gamma$.

This interpretation can be expressed by the modal formula

$$emb(d) = L\alpha \wedge LM\beta_1 \wedge \ldots \wedge LM\beta_n \supset \gamma. \tag{12.2}$$

Let us recall that we studied clauses of the form (12.2) in Chapter 9 (Theorem 9.31), where we referred to them as *default* clauses.

For a default theory $(D, W)$, we define now its *modal image* $emb(D, W)$ by

$$emb(D, W) = W \cup \{emb(d) \colon d \in D\}. \tag{12.3}$$

The next result shows that the encoding of default theories provided by (12.3) faithfully embeds default logic into modal nonmonotonic logics $\mathcal{S}$, for every modal logic $\mathcal{S}$ such that $\mathbf{N} \subseteq \mathcal{S} \subseteq \mathbf{S4f}$.

**Theorem 12.1** *Let $\mathcal{S}$ be a modal logic such that* $\mathbf{N} \subseteq \mathcal{S} \subseteq \mathbf{S4f}$ *and let* $S \subseteq \mathcal{L}$ *be consistent and closed under propositional consequence. Then* $S$ *is an extension for a default theory* $(D, W)$ *if and only if* $ST(S)$ *is an* $\mathcal{S}$-*expansion for* $emb(D, W)$.

Proof: Let us recall that in Chapter 3 (Definition 3.24) we defined

$$D_S = \left\{ \frac{\alpha}{\gamma} : \frac{\alpha : M\beta_1, \ldots, M\beta_n}{\gamma} \in D \text{ and } \neg\beta_i \notin S,\ i = 1, \ldots, n \right\}.$$

Assume first that $S$ is an extension for a $(D, W)$. Then, by Corollary 3.26, $S = Cn^{D_S}(W)$. Hence, for every formula $\varphi \in S$ there exists a sequence $\varphi_1, \ldots, \varphi_n$ of formulas such that (see Definition 3.1):

**(i)** $\varphi = \varphi_n$.

**(ii)** For every $i$, $1 \leq i \leq n$, at least one of the following conditions holds:

1. $\varphi_i \in W$.
2. $\varphi_i$ is a tautology.
3. $\varphi_i$ is the result of applying modus ponens to $\varphi_j$, $\varphi_k$, where $j, k < i$.
4. for some $j < i$, the rule $\frac{\varphi_j}{\varphi_i}$ belongs to $D_S$.

In Chapter 3, such a sequence is called a proof of $\varphi$ from $W$ in the system $PC + D_S$.

We will show, by induction on the length of a proof in $PC + D_S$, that if $\varphi \in S$ then $\varphi \in Cn_{\mathbf{N}}(emb(D, W) \cup \{\neg L\varphi : \varphi \notin ST(S)\})$. To this end, assume that the claim holds for all formulas in $S$ with proofs of length $< n$.

If $\varphi_n$ (that is $\varphi$) is a member of $W$ or is a tautology, then, evidently, $\varphi \in Cn_{\mathbf{N}}(emb(D, W) \cup \{\neg L\varphi : \varphi \notin ST(S)\})$. This also establishes the basis of the induction.

If $\varphi_n$ is obtained from $\varphi_j$, $\varphi_k$, where $j, k < n$, by means of modus ponens, then the induction hypothesis applies and $\varphi_i, \varphi_j \in Cn_{\mathbf{N}}(emb(D, W) \cup \{\neg L\varphi : \varphi \notin ST(S)\})$. Consequently, $\varphi \in Cn_{\mathbf{N}}(emb(D, W) \cup \{\neg L\varphi : \varphi \notin ST(S)\})$.

The last possibility is that $\varphi_n$ is obtained from $\varphi_j$, $j < n$, by means of the rule $\frac{\varphi_j}{\varphi_n} \in D_S$. In such case, there is a default $d = \frac{\varphi_j : M\beta_1, \ldots, M\beta_m}{\varphi_n} \in D$ such that for all $i$, $1 \leq i \leq m$, $\neg\beta_i \notin S$. Since $S$ is closed under propositional consequence, it follows from the definition of $ST(S)$ that $ST(S) \cap \mathcal{L} = S$. Hence, for every $i$, $1 \leq i \leq m$, $\neg\beta_i \notin ST(S)$ and, consequently,

$$\neg L\neg\beta_i \in Cn_{\mathbf{N}}(emb(D, W) \cup \{\neg L\varphi : \varphi \notin ST(S)\}).$$

Since $M\beta_i = \neg L \neg \beta_i$ and $Cn_{\mathbf{N}}(emb(D,W) \cup \{\neg L\varphi : \varphi \notin ST(S)\})$ is closed under necessitation, for every $i$, $1 \leq i \leq m$,

$$LM\beta_i \in Cn_{\mathbf{N}}(emb(D,W) \cup \{\neg L\varphi : \varphi \notin ST(S)\}).$$

By the induction hypothesis, $\varphi_j \in Cn_{\mathbf{N}}(emb(D,W) \cup \{\neg L\varphi : \varphi \notin ST(S)\})$. Thus,

$$L\varphi_j \in Cn_{\mathbf{N}}(emb(D,W) \cup \{\neg L\varphi : \varphi \notin ST(S)\}).$$

Moreover, we have that $emb(d) = L\varphi_j \wedge LM\beta_1 \wedge \ldots \wedge LM\beta_m \supset \varphi_n$ belongs to $Cn_{\mathbf{N}}(emb(D,W) \cup \{\neg L\varphi : \varphi \notin ST(S)\})$. Thus,

$$\varphi_n \in Cn_{\mathbf{N}}(emb(D,W) \cup \{\neg L\varphi : \varphi \notin ST(S)\}),$$

which completes our inductive reasoning and proves that

$$S \subseteq Cn_{\mathbf{N}}(emb(D,W) \cup \{\neg L\varphi : \varphi \notin ST(S)\}).$$

Next, observe that $emb(D,W) \subseteq ST(S)$. Indeed, since $S$ is an extension for $(D,W)$, $W \subseteq S \subseteq ST(S)$. Consider a default

$$d = \frac{\alpha : M\beta_1, \ldots, M\beta_n}{\gamma},$$

from $D$. To prove that $emb(d) \in ST(S)$, we proceed as follows.

If $\alpha \notin S$, then $\neg L\alpha \in ST(S)$. Consequently, in this case, $emb(d) \in ST(S)$. If for some $i$, $1 \leq i \leq n$, $\neg\beta_i \in S$ then $L\neg\beta_i \in ST(S)$. Since $S$ is consistent, it follows that $\neg L \neg \beta_i \notin ST(S)$. Consequently, $M\beta_i \notin ST(S)$, $\neg LM\beta_i \in ST(S)$ and $emb(d) \in ST(S)$. Finally, assume that $\alpha \in S$ and for every $i$, $1 \leq i \leq n$, $\neg\beta_i \notin S$. Since $S$ is an extension for $(D,W)$, it follows by Theorem 3.49 that $S$ is closed under defaults from $D$. Consequently, $S$ is closed under $d$ and hence $\gamma \in S$. But then $\gamma \in ST(S)$ and $emb(d) \in ST(S)$.

Corollary 9.8 implies that $ST(S)$ is an **N**-expansion for $emb(D,W)$. Since $\mathbf{N} \subseteq \mathcal{S}$, it follows by Theorem 9.6 that $ST(S)$ is an $\mathcal{S}$-expansion for $emb(D,W)$.

Next, we will prove the converse statement. Let us assume that $ST(S)$ is an $\mathcal{S}$-expansion for $emb(D,W)$. By Theorem 9.6, $ST(S)$ is an **S4f**-expansion for $emb(D,W)$.

Define

$$I = W \cup \left\{ L\alpha \supset \gamma : \frac{\alpha}{\gamma} \in D_S \right\}.$$

Consider a formula

$$L\alpha \wedge LM\beta_1 \wedge \ldots \wedge LM\beta_n \supset \gamma$$

from $emb(D,W)$. If for all $i$, $1 \le i \le n$, $\neg\beta_i \notin S$, then $L\alpha \supset \gamma \in I$ and, consequently,

$$L\alpha \wedge LM\beta_1 \wedge \ldots \wedge LM\beta_n \supset \gamma \in Cn_{\mathbf{S4f}}(I \cup \{\neg L\varphi: \varphi \notin ST(S)\}). \quad (12.4)$$

If for some $i$, $1 \le i \le n$, $\neg\beta_i \in S$, then $L\neg\beta_i \in ST(S)$. By the consistency of $S$, $M\beta_i \notin ST(S)$ and $\neg LM\beta_i \in \{\neg L\varphi: \varphi \notin ST(S)\}$. Hence, (12.4) holds in this case, too. Consequently,

$$emb(D,W) \subseteq Cn_{\mathbf{S4f}}(I \cup \{\neg L\varphi: \varphi \notin ST(S)\})$$

and

$$Cn_{\mathbf{S4f}}(emb(D,W) \cup \{\neg L\varphi: \varphi \notin ST(S)\}) \subseteq Cn_{\mathbf{S4f}}(I \cup \{\neg L\varphi: \varphi \notin ST(S)\}).$$

We will now prove the converse inclusion. Clearly,

$$W \subseteq Cn_{\mathbf{S4f}}(emb(D,W) \cup \{\neg L\varphi: \varphi \notin ST(S)\}).$$

Consider a formula $L\alpha \supset \gamma \in I$. Then, for some $\beta_1, \ldots, \beta_n$ such that $\neg\beta_i \notin S$, $1 \le i \le n$,

$$L\alpha \wedge LM\beta_1 \wedge \ldots \wedge LM\beta_n \supset \gamma \in emb(D,W).$$

For every $i$, $1 \le i \le n$, we have $M\beta_i \in ST(S)$ (since $\neg\beta_i \notin S$) and so $LM\beta_i \in ST(S)$ (by stability). Hence,

$$L\alpha \supset \gamma \in Cn_{\mathbf{S4f}}(emb(D,W) \cup \{\neg L\varphi: \varphi \notin ST(S)\}).$$

Thus, it follows that

$$Cn_{\mathbf{S4f}}(I \cup \{\neg L\varphi: \varphi \notin ST(S)\}) \subseteq Cn_{\mathbf{S4f}}(emb(D,W) \cup \{\neg L\varphi: \varphi \notin ST(S)\}).$$

Since $ST(S)$ is an **S4f**-expansion for $emb(D,W)$, we obtain that

$$ST(S) = Cn_{\mathbf{S4f}}(I \cup \{\neg L\varphi: \varphi \notin ST(S)\})$$

or, in other words, that $ST(S)$ is an **S4f**-expansion for $I$.

Consider a rule

$$\frac{\alpha}{\gamma}$$

from $D_S$. Assume that $\alpha \in S$. Then $L\alpha \in ST(S)$ and, since $I \subseteq ST(S)$, $\gamma \in ST(S)$. Hence, $\gamma \in S$, which proves that $S$ is closed under the rules from $D_S$. By the assumptions of the theorem, $S$ is closed under propositional consequence and, since $W \subseteq I \subseteq ST(S)$, $W \subseteq S$. Thus, $S$ contains $W$ and is closed under propositional consequence and the rules from $D_S$.

Consider a set $S' \subseteq S$, such that $W \subseteq S'$ and $S'$ is closed under propositional consequence and the rules from $D_S$. Since $S'$ is closed under the rules from $D_S$, $I \subseteq ST(S')$. We showed earlier that $ST(S)$ is an **S4f**-expansion of $I$. By Theorem 9.31, it follows that $S' = S$. Hence, $S$ is the least set that contains $W$ and is closed under propositional consequence and the rules from $D_S$. By Corollary 3.26, $S$ is an extension for $(D, W)$. ∎

There is an important variation of the embedding (12.2) - (12.3). Namely, for a default $d$ given by (12.1) define

$$emb^L(d) = L\alpha \wedge LM\beta_1 \wedge \ldots \wedge LM\beta_n \supset L\gamma. \tag{12.5}$$

Next, for a default theory $(D, W)$, define

$$emb^L(D, W) = \{L\varphi : \varphi \in W\} \cup \{emb^L(d) : d \in D\}. \tag{12.6}$$

The advantage of this encoding over the one given by (12.2) - (12.3) is that it naturally extends to the case of *disjunctive default logic*. Consequently, the theory of modal nonmonotonic logics applies to disjunctive default logic and, in particular, implies semantical characterizations of extensions of *disjunctive default theories*. At the end of this chapter we provide references to papers where a detailed presentation of the disjunctive default logic and its relation to modal nonmonotonic logics can be found.

For the embedding given by (12.5) - (12.6) we have the following result.

**Theorem 12.2** *Let $\mathcal{S}$ be a modal logic such that* $\mathbf{t} \subseteq \mathcal{S} \subseteq \mathbf{S4f}$ *and let $S \subseteq \mathcal{L}$ be consistent and closed under propositional consequence. Then $S$ is an extension for a default theory $(D, W)$ if and only if $ST(S)$ is an $\mathcal{S}$-expansion for $emb^L(D, W)$.*

Proof: First, assume that $S$ is an extension for a default theory $(D, W)$. By Theorem 12.1, $ST(S)$ is a t-expansion for $emb(D, W)$. That is,

$$ST(S) = Cn_{\mathbf{t}}(emb(D, W) \cup \{\neg L\varphi : \varphi \notin ST(S)\}).$$

It follows directly from the definition of stable theories that for every stable theory $T$, $emb(D, W) \subseteq T$ if and only if $emb^L(D, W) \subseteq T$ (see also Theorem 8.8). Hence,

$$ST(S) \supseteq Cn_{\mathbf{t}}(emb^L(D, W) \cup \{\neg L\varphi : \varphi \notin ST(S)\}).$$

Since $emb(D, W) \subseteq Cn_{\mathbf{t}}(emb^L(D, W))$,

$$\begin{aligned} ST(S) &= Cn_{\mathbf{t}}(emb(D, W) \cup \{\neg L\varphi : \varphi \notin ST(S)\}) \\ &\subseteq Cn_{\mathbf{t}}(emb^L(D, W) \cup \{\neg L\varphi : \varphi \notin ST(S)\}). \end{aligned}$$

Hence,

$$ST(S) = Cn_{\mathbf{t}}(emb^L(D,W) \cup \{\neg L\varphi : \varphi \notin ST(S)\}),$$

that is, $ST(S)$ is a **t**-expansion for $emb^L(D,W)$. Consequently, $ST(S)$ is an $\mathcal{S}$-expansion for $emb^L(D,W)$.

Conversely, assume that $ST(S)$ is an $\mathcal{S}$-expansion for $emb^L(D,W)$. Then $ST(S)$ is an **S4f**-expansion for $emb^L(D,W)$. Let $\varphi$ and $\varphi^L$ be formulas given by (12.2) and (12.5), respectively. Observe that

$$\vdash_{\mathbf{S4f}} \varphi \equiv (L(\alpha \wedge M\beta_1 \wedge \ldots \wedge M\beta_n) \supset \gamma).$$

(Theorem 7.17(3)). Hence (see Theorem 7.17(1)),

$$\varphi \vdash_{\mathbf{S4f}} LL(\alpha \wedge M\beta_1 \wedge \ldots \wedge M\beta_n) \supset L\gamma.$$

In addition, applying axiom 4, we obtain

$$\varphi \vdash_{\mathbf{S4f}} L(\alpha \wedge M\beta_1 \wedge \ldots \wedge M\beta_n) \supset L\gamma.$$

Consequently, again by Theorem 7.17(3),

$$\varphi \vdash_{\mathbf{S4f}} \varphi^L.$$

Hence,

$$\begin{aligned} ST(S) &= Cn_{\mathbf{S4f}}(emb^L(D,W) \cup \{\neg L\varphi : \varphi \notin ST(S)\}) \\ &\subseteq Cn_{\mathbf{S4f}}(emb(D,W) \cup \{\neg L\varphi : \varphi \notin ST(S)\}). \end{aligned}$$

Since $ST(S)$ is stable and contains $emb^L(D,W)$, it follows from Theorem 8.8 that $ST(S)$ contains $emb(D,W)$. Consequently,

$$Cn_{\mathbf{S4f}}(emb(D,W) \cup \{\neg L\varphi : \varphi \notin ST(S)\}) \subseteq ST(S).$$

Thus,

$$ST(S) = Cn_{\mathbf{S4f}}(emb(D,W) \cup \{\neg L\varphi : \varphi \notin ST(S)\}),$$

that is, $ST(S)$ is an **S4f**-expansion for $emb(D,W)$. By Theorem 12.1, $S$ is an extension for $(D,W)$. ■

## 12.2 Another semantics for default logic

Let us recall that Theorem 3.45 provides a semantic characterization of extensions for default theories. Specifically, we proved that a theory $S$ is an extension for a default theory $(D,W)$ if and only if $S$ is the collection of all formulas valid in all valuations from a set of valuations $V$, which is a fixpoint of a certain operator $\Sigma$.

In this section, a related semantic characterization of extensions will be derived as a corollary to Theorem 12.1 and the minimal-knowledge characterization of **S4f**-expansions (Corollary 9.25, Theorem 9.27 and Corollary 9.28). The approach we are going to take requires several new elements. First, based on the notion of an **S4f**-model (see Section 7.3), we define a *default model* as a pair $\langle V_1, V_2 \rangle$ of sets of valuations. It will turn out that formulas true in a set $V$ of valuations coincide with those true in the default model $\langle \emptyset, V \rangle$. Consequently, a set of valuations $V$ can be identified with a default model $\langle \emptyset, V \rangle$. In this way, the class of default models can be viewed as extending the class of sets of valuations. Among default models of the form $\langle \emptyset, V \rangle$ we will characterize those that yield extensions of $(D, W)$. This will be achieved by introducing the notion of a *minimal* default model. We will now provide the detailed description of this approach.

**Definition 12.3** A *default model* is a pair $\mathcal{V} = \langle V_1, V_2 \rangle$, where $V_1$ and $V_2$ are sets of propositional valuations of $\mathcal{L}$ and $V_2 \neq \emptyset$. A default $d$, where

$$d = \frac{\alpha : M\beta_1, \ldots, M\beta_n}{\gamma},$$

is *valid* in a default model $\mathcal{V} = \langle V_1, V_2 \rangle$ if the following conditions hold:

1. Whenever $\alpha$ is true in all valuations of $V_1 \cup V_2$ and each $\beta_i$, $1 \leq i \leq n$, is true in at least one valuation of $V_2$, then $\gamma$ is true in all valuations of $V_1 \cup V_2$.

2. Whenever $\alpha$ is true in all valuations of $V_2$ and each $\beta_i$, $1 \leq i \leq n$, is true in at least one valuation of $V_2$, then $\gamma$ is true in all valuations of $V_2$.

We write $\mathcal{V} \models d$ to denote that $d$ is valid in a default model $\mathcal{V}$.

We say that a formula $\varphi \in \mathcal{L}$ is *valid* in a default model $\mathcal{V} = \langle V_1, V_2 \rangle$ if $\varphi$ is valid in every valuation from $V_1 \cup V_2$. We write $\mathcal{V} \models \varphi$ to denote that $\varphi$ is valid in $\mathcal{V}$.

We say that a default theory $(D, W)$ is *valid* in a default model $\mathcal{V}$ if all defaults in $D$ and all formulas in $W$ are valid in $\mathcal{V}$. We write $\mathcal{V} \models (D, W)$ if $(D, W)$ is valid in $\mathcal{V}$.

Finally, we define $Th_d(\mathcal{V})$ to consist of all defaults valid in $\mathcal{V}$, $Th_f(\mathcal{V})$ to consist of all formulas valid in $\mathcal{V}$ and we denote $Th(\mathcal{V}) = (Th_d(\mathcal{V}), Th_f(\mathcal{V}))$. ▲

Observe that for a set $V$ of valuations we have $Th(V) = Th_f(\langle \emptyset, V \rangle)$.

**Theorem 12.4** *Let $V$ be a set of valuations and $(D, W)$ be a default theory. Then, $\langle\emptyset, V\rangle \models (D, W)$ if and only if $W \subseteq Th(V)$ and $Th(V)$ is closed under the defaults from $D$.* ▲

Observe that if we select from the class of all default models $\langle\emptyset, V\rangle$ of $(D, W)$ those for which $V$ is inclusion-maximal then the corresponding theories $Th_f(\langle\emptyset, V\rangle)$ form the collection of minimal sets for $(D, W)$ (see Section 3.10). Hence, to characterize extensions for $(D, W)$, a different approach is needed. As mentioned above, guided by the theory of the nonmonotonic logic **S4f**, we introduce the notion of a *minimal default model.*

**Definition 12.5** Let $(D, W)$ be a default theory. A default model $\langle\emptyset, V\rangle$ is *minimal* for $(D, W)$ if $\langle\emptyset, V\rangle \models (D, W)$ and:

**(MIN$_{dl}$)** For every set of valuations $V'$, if $\langle V', V\rangle \models (D, W)$ then $V' \subseteq V$. ▲

Our next result shows that the notion of minimality of default models given in Definition 12.5 yields a subclass of the class of minimal sets for a default theory.

**Theorem 12.6** *Let $(D, W)$ be a default theory and let $\langle\emptyset, V\rangle$ be a default model minimal for $(D, W)$. Let $V'$ be a set of valuations such that $\langle\emptyset, V'\rangle \models (D, W)$. Then, if $V \subseteq V'$, then $V = V'$. In particular, $Th(V)$ is a minimal set for $(D, W)$.*

Proof: Consider the default model $\mathcal{V}' = \langle V' \setminus V, V\rangle$. It is easy to check that $\mathcal{V}' \models (D, W)$. By minimality of the default model $\langle\emptyset, V\rangle$, $V' \setminus V \subseteq V$. Hence, $V' \subseteq V$ and, consequently, $V = V'$. ■

Below we will prove a stronger result. Namely we will show that the theories $Th_f(\langle\emptyset, V\rangle)(= Th(V))$, where $V$ is a set of valuations such that $\langle\emptyset, V\rangle$ is a minimal default model of $(D, W)$, are precisely extensions of $(D, W)$.

With each default model $\mathcal{V} = \langle V_1, V_2\rangle$ one can naturally associate an **S4f**-model as follows. For $i = 1, 2$ define $M_i = \{(v, i) : v \in V_i\}$. Next, put $\mathcal{V}_{\mathbf{S4f}} = \langle M_1, M_2, U\rangle$, where $U((v, i)) = v$ for every $(v, i) \in M_i$, $i = 1, 2$. Let us also recall that the embedding *emb* was defined in Section 12.1 by equation (12.2). We now have the following lemma.

**Lemma 12.7** *Let $\mathcal{V} = \langle V_1, V_2\rangle$ be a default model.*

1. *For every formula $\varphi \in \mathcal{L}$, $\mathcal{V} \models \varphi$ if and only if $\mathcal{V}_{\mathbf{S4f}} \models \varphi$.*

2. *For every default $d$, $\mathcal{V} \models d$ if and only if $\mathcal{V}_{\mathbf{S4f}} \models emb(d)$.* ▲

The next theorem contains a semantic minimal-model characterization of extensions for default theories.

**Theorem 12.8** *Let $(D, W)$ be a default theory and let $S \subseteq \mathcal{L}$ be consistent. Theory $S$ is an extension of $(D, W)$ if and only if $S = Th_f(\mathcal{V})$, for a default model $\mathcal{V} = \langle \emptyset, V\rangle$ minimal for $(D, W)$.*

Proof: Assume first that $S$ is an extension for $(D, W)$. Denote $T = ST(S)$. By Theorem 12.1, $T$ is an **S4f**-expansion for $emb(D, W)$. Let $\mathcal{M}_T = \langle M, U\rangle$ be the canonical model for $T$. Consider the set of valuations $V = \{U(a) : a \in M\}$. Let $\mathcal{V} = \langle \emptyset, V\rangle$. Clearly, $S = Th_f(\mathcal{V})$. To complete the proof of the necessity part of the assertion, we need to show that $\langle \emptyset, V\rangle$ is minimal for $(D, W)$. Since $emb(D, W) \subseteq T$, $\mathcal{M}_T \models emb(D, W)$. Using Lemma 12.7, it is now straightforward to verify that $\langle \emptyset, V\rangle \models (D, W)$.

Let $V'$ be any set of valuations. Define $\mathcal{V}' = \langle V', V\rangle$. Assume that $\mathcal{V}' \models (D, W)$. Then, by Lemma 12.7, $\mathcal{V}'_{\mathbf{S4f}} \models emb(D, W)$. Since $T$ is an **S4f**-expansion for $emb(D, W)$, it follows by Corollary 9.25 that $\mathcal{M}_T$ is strongly $\mathcal{C}_{\mathbf{S4f}}$-minimal for $emb(D, W)$ (let us recall that by $\mathcal{C}_{\mathbf{S4f}}$ we denote the class of all **S4f**-models). By Theorem 9.27 it follows that $V' \subseteq V$. Thus, the condition ($\mathrm{MIN}_{dl}$) holds and hence, $\langle \emptyset, V\rangle$ is minimal for $(D, W)$.

Conversely, assume that a default model $\mathcal{V} = \langle \emptyset, V\rangle$ is minimal for $(D, W)$ and that $S = Th_f(\mathcal{V})$. Let $V'$ be the set of all valuations which satisfy $S$. We have that $\langle \emptyset, V\rangle \models (D, W)$. Consequently, $W \subseteq S$ and $S$ is closed under the defaults from $D$ (Theorem 12.4). Since $S = Th(V')$, again by Theorem 12.4 it follows that $\langle \emptyset, V'\rangle \models (D, W)$. Since $V \subseteq V'$, by Theorem 12.6, $V = V'$. Hence, $V$ is the set of all valuations satisfying $S$. Consider an **S5**-model $\mathcal{M} = \langle M, U\rangle$, where $M = V$ and $U(v) = v$, for every $v \in M$. As $V$ is the set of all valuations satisfying $S$, $\mathcal{M}$ is the canonical model for $ST(S)$. Since $\mathcal{V} \models (D, W)$, $\mathcal{M} \models emb(D, W)$.

We will now show that $\mathcal{M}$ is a strongly $\mathcal{C}_{\mathbf{S4f}}$-minimal model for the theory $emb(D, W)$. To this end, consider an **S5**-model $\mathcal{M}' = \langle M', U'\rangle$, where $M' \cap M = \emptyset$. Assume that $\mathcal{M}' \odot \mathcal{M} \models emb(D, W)$. Let $V' = \{U'(a) : a \in M'\}$. Define $\mathcal{V}' = \langle V', V\rangle$. Clearly, $\mathcal{M}' \odot \mathcal{M} = \mathcal{V}'_{\mathbf{S4f}}$. By Lemma 12.7, $\mathcal{V}' \models (D, W)$. Since $\mathcal{V}$ satisfies the condition ($\mathrm{MIN}_{dl}$), $V' \subseteq V$. Thus, $\mathcal{M}$ is strongly $\mathcal{C}_{\mathbf{S4f}}$-

minimal for $emb(D, W)$ and, by Corollary 9.25, $ST(S)$ is an **S4f**-expansion for $emb(D, W)$. Hence, by Theorem 12.1, $S$ is an extension for $(D, W)$. ∎

The minimal-model characterization of default extensions (Theorem 12.8) is closely related to the characterization given in Theorem 3.45. Let us recall that for a set $V$ of valuations, we define $\Sigma^{D,W}(V)$ to be the largest set $V'$ of valuations such that:

1. $W \subseteq Th(V')$.
2. For every default $d \in D$, if $p(d) \in Th(V')$ and if for every $\beta \in J(d)$ we have $\neg\beta \notin Th(V)$, then $c(d) \in Th(V')$.

Let us assume that $S$ is consistent. Let us recall that for each $V$, $\Sigma^{D,W}(V)$ is well-defined and, by Theorem 3.45, $S$ is an extension for $(D, W)$ if and only if $S = Th(V)$, for some set $V$ of valuations such that $V = \Sigma^{D,W}(V)$.

The following three properties are straightforward consequences of the definitions of the operator $\Sigma^{D,W}$, and of the notions of default model and satisfiability by a default model. They formalize the connections between the fixpoint and minimal-model descriptions of extensions. Since we restrict here to consistent theories, in the properties listed below we assume that $V$ is nonempty.

**(P1)** For every set of valuations $V'$, if $\langle V', V\rangle \models (D, W)$, then $V' \cup V \subseteq \Sigma^{D,W}(V)$.

**(P2)** If $\langle \emptyset, V\rangle \models (D, W)$, then $\langle \Sigma^{D,W}(V), V\rangle \models (D, W)$.

**(P3)** If $V = \Sigma^{D,W}(V)$, then $\langle \emptyset, V\rangle \models (D, W)$.

Using these properties it is now easy to show that $\langle \emptyset, V\rangle$ is a minimal default model for $(D, W)$ if and only if $V = \Sigma^{D,W}(V)$. Assume that $\langle \emptyset, V\rangle$ is a minimal default model for $(D, W)$. Then, by (P1) (with $V' = \emptyset$) and (P2), $V = \Sigma^{D,W}(V)$. Conversely, if $V = \Sigma^{D,W}(V)$ then, by (P1) and (P3), $\langle \emptyset, V\rangle$ is a minimal default model for $(D, W)$.

Hence, simple derivations of one characterization from the other are possible. In particular, it is possible to obtain Theorem 12.8 without any reference to modal logics.

## 12.3 Interpreting logic programs in modal nonmonotonic logics

In Chapter 6 we showed that logic programming with stable model semantics can be regarded as a special case of default logic. Consequently, the results obtained in the two previous sections apply to logic programming, as well. But, since the syntax of logic programs is more restricted than the syntax of default theories, for the case of logic programs stronger results can be proved. In this section, we will formulate and prove results analogous to Theorems 12.1 and 12.2, thus providing modal characterizations of stable model semantics for logic programs.

We will study in this chapter stable and supported models of logic programs as well as stable answer sets of logic programs with classical negation. Let us recall that in order to define these notions we first consider a propositional grounding *ground*$(P)$ of a logic program $P$. Next, we interpret *ground*$(P)$ as a default theory $(dl(P), \emptyset)$. Then we define stable and supported models as sets of ground atomic formulas of extensions and weak extensions, respectively, of the theory $(dl(P), \emptyset)$. Similarly, stable answer sets of logic programs with classical negation are defined as sets of ground literals of extensions of $(dl(P), \emptyset)$.

Hence, from the point of view of stable and supported models, and stable answer sets, there is no difference between logic programs and their propositional groundings. Using this observation to simplify our discussion, in this chapter we will restrict our attention to the programs built of ground atomic formulas (or ground literals in the case of programs with classical negation). In other words, by a logic program we mean a program built of clauses of the form

$$c \leftarrow a_1, \ldots, a_m, \mathbf{not}(b_1), \ldots, \mathbf{not}(b_n), \tag{12.7}$$

where all $a_i$, $b_i$ and $c$ are ground atomic formulas from the Herbrand base $B_H$. By a logic program with classical negation we mean a program built of clauses of the form (12.7), where all $a_i$, $b_i$ and $c$ are ground literals from $Lit_H$. From now on we will say *atom* for a ground atomic formula from $B_H$ and *literal* for a ground literal from $Lit_H$.

Let us recall that in Chapter 6 for every literal $a$ we introduced the following notation:

$$\overline{a} = \begin{cases} \neg a & \text{if } a \text{ is an atom,} \\ c & \text{if } a = \neg c, \text{ where } c \text{ is an atom.} \end{cases}$$

Let $C$ be a clause of the form (12.7). We will encode $C$ by the modal formula

$$emb_{lp}(C) = L(a_1 \wedge \ldots \wedge a_m) \wedge LM\bar{b}_1 \wedge \ldots \wedge LM\bar{b}_n \supset c. \tag{12.8}$$

For a program $P$, define

$$emb_{lp}(P) = \{emb_{lp}(C): C \in P\}. \tag{12.9}$$

Under the encoding given by (12.8) and (12.9), $\mathcal{S}$-expansions, for any of the nonmonotonic modal logics $\mathcal{S}$, where $\mathbf{N} \subseteq \mathcal{S} \subseteq \mathbf{Sw5}$, can be regarded as a generalization of stable model semantics of logic programs. Namely, we have the following result.

**Theorem 12.9** *Let $\mathcal{S}$ be a modal logic such that $\mathbf{N} \subseteq \mathcal{S} \subseteq \mathbf{Sw5}$ and let $S$ be a consistent set of literals from $Lit_H$. Then $S$ is a stable answer set for a CN-program $P$ if and only if $ST(S)$ is an $\mathcal{S}$-expansion for $emb_{lp}(P)$.*

Proof: Assume first that $S$ is a stable answer set for a logic program $P$. By Definition 6.37, $Cn(S)$ is an extension for the default theory $(dl(P), \emptyset)$. Hence, by Theorem 12.1, $ST(S)$ is an $\mathbf{N}$-expansion for the modal theory $emb(dl(P), \emptyset)$. Since $emb(dl(P), \emptyset) = emb_{lp}(P)$, and since $\mathbf{N} \subseteq \mathcal{S}$, it follows that $ST(S)$ is an $\mathcal{S}$-expansion for $emb_{lp}(P)$.

Conversely, assume that $ST(S)$ is an $\mathcal{S}$-expansion for $emb_{lp}(P)$. Then $ST(S)$ is an $\mathbf{Sw5}$-expansion for $emb_{lp}(P)$. As in the proof of Theorem 12.1, it follows that $S$ is closed under the program $P^S$.

Consider any set $S'$ of literals such that $S' \subseteq S$ and $S'$ is closed under $P^S$. Reasoning as in the proof of Theorem 12.1 one can show that $ST(S)$ is an $\mathbf{Sw5}$-expansion of $emb_{lp}(P^S)$. Since $S'$ is closed under $P^S$, $emb_{lp}(P^S) \subseteq ST(S')$. By Theorem 9.32, $S = S'$.

Thus $S$ is the least set of literals closed under the rules in $P^S$. Consequently, by Theorem 6.44, $S$ is a stable answer set for $P$. ■

Since stable answer sets generalize stable models, as a direct corollary to Theorem 12.9 we obtain the following result.

**Corollary 12.10** *Let $\mathcal{S}$ be a modal logic such that $\mathbf{N} \subseteq \mathcal{S} \subseteq \mathbf{Sw5}$ and let $S$ be a set of atoms from $B_H$. Then $S$ is a stable model for a logic program $P$ if and only if $ST(S)$ is an $\mathcal{S}$-expansion for $emb_{lp}(P)$.* ▲

As in Section 12.1, another encoding of program clauses as modal formulas can be used. For a clause $C$ of the form (12.7) define

$$emb^L_{lp}(C) = L(a_1 \wedge \ldots \wedge a_m) \wedge LM\bar{b}_1 \wedge \ldots \wedge LM\bar{b}_n \supset Lc. \tag{12.10}$$

For a program $P$, define

$$emb_{lp}^{L}(P) = \{emb_{lp}^{L}(C): C \in P\}. \qquad (12.11)$$

**Theorem 12.11** *Let $\mathcal{S}$ be a modal logic such that* $\mathbf{t} \subseteq \mathcal{S} \subseteq \mathbf{Sw5}$ *and let $S$ be a consistent set of literals from $Lit_H$. Then $S$ is a stable answer set for a CN-program $P$ if and only if $ST(S)$ is an $\mathcal{S}$-expansion for $emb_{lp}^{L}(P)$.*

Proof: First, assume that $S$ is a stable answer set for $P$. By Theorem 12.9, $ST(S)$ is a $\mathbf{t}$-expansion for $emb_{lp}(P)$. As in the proof of Theorem 12.2, it follows that $ST(S)$ is a $\mathbf{t}$-expansion for $emb_{lp}^{L}(P)$ and hence an $\mathcal{S}$-expansion for $emb_{lp}^{L}(P)$.

Conversely, assume that $ST(S)$ is an $\mathcal{S}$-expansion for $emb_{lp}^{L}(P)$. Then $ST(S)$ is an **Sw5**-expansion for $emb_{lp}^{L}(P)$. Again, reasoning exactly as in the proof of Theorem 12.2, we show that $ST(S)$ is an **Sw5**-expansion for $emb_{lp}(P)$. Hence, by Theorem 12.9, $S$ is a stable answer set of $P$. ■

**Corollary 12.12** *Let $\mathcal{S}$ be a modal logic such that* $\mathbf{t} \subseteq \mathcal{S} \subseteq \mathbf{Sw5}$ *and let $S$ be a set of atoms from $B_H$. Then $S$ is a stable model for a logic program $P$ if and only if $ST(S)$ is an $\mathcal{S}$-expansion for $emb_{lp}^{L}(P)$.* ▲

Combining the interpretation $emb_{lp}$ with the embedding of the nonmonotonic modal logic **Sw5** into the nonmonotonic modal logic **kd45** (autoepistemic logic) one can construct a direct embedding of CN-programs in the nonmonotonic **kd45**. See comments at the end of this chapter for references.

## 12.4 More characterizations of stable answer sets

Theorem 12.9 implies two characterizations of stable answer sets of CN-programs.

The first approach is obtained as a specialization of the minimal-model characterization of **Sw5**-expansions to the case of CN-programs in a way similar to that in which the minimal-model characterization of **S4f**-expansions was adapted to the case of the default logic in Section 12.2.

Our approach is the same as in Section 12.2. In order to describe stable answer sets for a CN-program $P$ by means of sets of valuations, we first introduce the notion of a *program model* for $P$. Then, we define the notion of a minimal program model and show that sets of literals true in minimal program models for $P$ are precisely stable answer sets for $P$.

**Definition 12.13** A *program model*, or simply a *p-model*, is a pair $\mathcal{V} = \langle v, V \rangle$, where $v$ is a propositional valuation of $\mathcal{L}$ and $V$ is a set of propositional valuations of $\mathcal{L}_H$ (the language of propositional calculus with the set of atoms $B_H$). Note that no assumptions are made concerning the membership of $v$ in $V$. A clause $C$, where

$$C = c \leftarrow a_1, \ldots, a_m, \mathbf{not}(b_1), \ldots, \mathbf{not}(b_n),$$

is *valid* in a p-model $\mathcal{V} = \langle v, V \rangle$ if the following conditions hold:

1. Whenever all $a_i$, $1 \leq i \leq m$, are valid in all valuations of $\{v\} \cup V$ and every $b_i$, $1 \leq i \leq n$, is not valid in at least one valuation from $V$, then $c$ is valid in all valuations of $\{v\} \cup V$.
2. Whenever all $a_i$, $1 \leq i \leq m$, are valid in all valuations from $V$ and every $b_i$, $1 \leq i \leq n$, is not valid in at least one valuation from $V$, then $c$ is valid in all valuations of $V$.

We write $\mathcal{V} \models C$ to denote that $C$ is valid in a p-model $\mathcal{V}$.

We say that a literal $l$ is *valid* in a p-model $\mathcal{V}$ if $l$ is true in every valuation from $\{v\} \cup V$.

We say that a CN-program $P$ is *valid* in a p-model $\mathcal{V}$ if all clauses in $P$ are valid in $\mathcal{V}$. We write $\mathcal{V} \models P$ in such case. ▲

We will define next what it means for a p-model to be minimal for a program $P$.

**Definition 12.14** Let $P$ be a CN-program. A p-model $\mathcal{V} = \langle v_0, V \rangle$, is *minimal for P* if $v_0 \in V$, $\mathcal{V} \models P$ and:

($\mathrm{MIN}_{lp}$) For every valuation $v$, if $\langle v, V \rangle \models P$ then $v \in V$. ▲

Similarly as in Section 12.2, with a p-model $\mathcal{V} = \langle v, V \rangle$ we will associate an **Sw5**-model $\mathcal{V}_{\mathbf{Sw5}} = \langle (v, 1), V \times \{2\}, U \rangle$, where $U((w, i)) = w$, for $w \in \{v\} \cup V$ and $i = 1, 2$.

**Lemma 12.15** *Let* $\mathcal{V} = \langle v, V \rangle$ *be a p-model. For every clause* $C$, $\mathcal{V} \models C$ *if and only if* $\mathcal{V}_{\mathbf{Sw5}} \models emb_{lp}(C)$. ▲

We now formulate and prove a result providing a characterization of stable answer sets in terms of minimal sets of valuations.

**Theorem 12.16** *Let $P$ be a CN-program and let $S$ be a consistent set of literals. The set $S$ is a stable answer set of $P$ if and only if $S$ is the set of all literals valid in a p-model $\langle v_0, V\rangle$, minimal for $P$.*

Proof: Assume first that $S$ is a stable answer set for $P$. Denote $T = ST(S)$. By Theorem 12.9, $T$ is an **Sw5**-expansion for $emb_{lp}(P)$. Let $\mathcal{M}_T = \langle M, U\rangle$ be the canonical model for $T$. Consider the set of valuations $V = \{U(m): m \in M\}$ and let $v_0$ be any element of $V$. Clearly, $S$ is exactly the set of all literals valid in $\langle v_0, V\rangle$. Thus, to complete the proof of the necessity part of the assertion we need to show that $\langle v_0, V\rangle$ is minimal for $P$.

Since $emb_{lp}(P) \subseteq T$, $\mathcal{M}_T \models emb_{lp}(P)$. Hence, $\langle v_0, V\rangle \models P$. Consider a valuation $w$. Define $\mathcal{V} = \langle w, V\rangle$ and assume that $\mathcal{V} \models P$. Then, by Lemma 12.15, $\mathcal{V}_{\mathbf{Sw5}} \models emb_{lp}(P)$. By Corollary 9.25, $\mathcal{M}_T$ is strongly $\mathcal{C}_{\mathbf{Sw5}}$-minimal for $emb_{lp}(P)$. By Theorem 9.27, $w \in V$. Hence, $\langle v_0, V\rangle$ satisfies the condition (MIN$_{lp}$) and so $\langle v_0, V\rangle$ is minimal for $P$.

Conversely, assume that a p-model $\mathcal{V} = \langle v_0, V\rangle$ is minimal for $P$ and that $S$ is the set of all literals valid in $\mathcal{V}$. Similarly as in the proof of Theorem 12.8 one can prove that $V$ consists of all valuations which satisfy $S$. Consider an **S5**-model $\mathcal{M} = \langle M, U\rangle$, where $M = V$ and $U(v) = v$, for every $v \in M$. It follows from our remark that $\mathcal{M}$ is the canonical model for $ST(S)$. Since $\mathcal{V} \models P$, $\mathcal{M} \models emb_{lp}(P)$. We will now show that $\mathcal{M}$ is strongly $\mathcal{C}_{\mathbf{Sw5}}$-minimal for $emb_{lp}(P)$. To this end, consider a valuation $w$ and a world $a \notin V$. Define the **Sw5**-model $\mathcal{M}' = \langle a, M, U'\rangle$, where $U'(a) = w$ and $U'|_M = U$. Then $\mathcal{M}' = \mathcal{V}'_{\mathbf{Sw5}}$, where $\mathcal{V}'$ stands for the p-model $\langle w, V\rangle$. Assume that $\mathcal{M}' \models emb_{lp}(P)$. By Lemma 12.15, $\langle w, V\rangle \models P$. Since $\mathcal{V}$ satisfies the condition (MIN$_{lp}$), $w \in V$. Thus, by Corollary 9.25 and Theorem 9.27, $Th(\mathcal{M}) = ST(S)$ is an **Sw5**-expansion for $emb_{lp}(P)$. Hence, by Theorem 12.9, $S$ is a stable answer set for $P$. ■

We now move on to the second characterization of stable answer sets. For the previous characterization, as well as for the definition of stable answer sets given in Chapter 6, it is essential that clauses are treated as nonstandard inference rules rather than formulas of the propositional logic. In our present approach we will replace a CN-program $P$ by a standard propositional theory $P'$ and describe stable answer sets of $P$ in terms of certain propositional valuations in which $P'$ is valid. We will closely follow the approach taken in Section 10.3, where characterizations of nonmonotonic logic **kd45** (that is, autoepistemic logic) and nonmonotonic logic **Sw5** are presented in terms of so-called *context semantics*.

In our argument we will use the following simple corollary to Theorems 10.24 and 12.9.

**Theorem 12.17** *Let $S$ be a consistent set of literals from $\mathcal{L}$. Then $S$ is a stable answer set of a CN-program $P$ if and only if $ST(S)$ satisfies*

$$ST(S) = Cn(emb_{lp}(P) \cup \{\varphi \equiv L\varphi : \varphi \in ST(S)\} \cup \{\neg L\varphi : \varphi \notin ST(S)\}).$$

▲

Our goal is to simplify this characterization of stable answer sets by removing all explicit references to modality. To this end, we need to extend the language. For each literal $a \in Lit_H$ we add two new atoms: $a'$ (intuitively representing our knowledge of $a$) and $a''$ (intuitively representing our knowledge that $\neg a$ (or, equivalently, $\overline{a}$) is possible). We denote the extended language (which is generated by $B_H$ and all the new atoms) by $\mathcal{L}'$.

For a CN-clause $C$, of the form

$$c \leftarrow a_1, \ldots, a_m, \mathbf{not}(b_1), \ldots, \mathbf{not}(b_n),$$

by $C'$ denote the propositional formula

$$a_1' \wedge \ldots \wedge a_m' \wedge b_1'' \wedge \ldots \wedge b_n'' \supset c.$$

Define also

$$P' = \{C' : C \in P\}.$$

Note that $C'$ might be regarded as a propositional variant of (12.8).

When reasoning from $P$, we cannot treat all the atoms in $\mathcal{L}'$ as independent. Consequently, some valuations of $\mathcal{L}'$ cannot be used when verifying the validity of formulas. To define a class of "acceptable" valuations, we reason as follows. Since $a'$ represents the fact that $a$ is known, any valuation such that $v(a) \neq v(a')$ is counterintuitive. Thus, we restrict our attention to only those valuations $v$ of $\mathcal{L}'$ that satisfy the condition:

**(1)** $v(a) = v(a')$ or, equivalently, $v(a \equiv a') = \mathbf{t}$.

Furthermore, an atom $a''$, where $a \in Lit_H$, means that we know that $\neg a$ is possible. As in many other places in the book, we make that precise by selecting a set $S$ of literals from $Lit_H$ and defining *possibility* with respect to $S$. Namely, we say that $\neg a$ is possible with respect to $S$ if $a \notin S$. This notion of possibility implies another restriction on a valuation $v$:

**(2)** For every literal $a$ not in $S$, $v(a'') = \mathbf{t}$.

For a set of literals $S$, define the set $V_S$ of valuations to consist of those valuations $v$ of $\mathcal{L}'$ which satisfy the conditions (1) and (2). We say that a CN-program $P$ *S-entails* a literal $a$ if for every valuation $v$ in $V_S$ such that $v$ is a model of $P$, $v(a) = \mathbf{t}$. We write $P \models_S a$ in such case. As a corollary from Theorem 12.17, we obtain the following characterization of stable answer sets of logic programs.

**Theorem 12.18** *Let $P$ be a CN-program and let $S$ be a consistent set of literals. The set $S$ is a stable answer set of $P$ if and only if*

$$S = \{a \in Lit_H : P' \models_S a\}.$$

▲

## 12.5 Weak extensions, stable expansions and supported models

In Section 12.1 we introduced an interpretation of defaults by modal formulas and proved that under this interpretation default extensions correspond precisely to **S4f**-expansions. Specifically, we proved that under the embedding *emb* which assigns to a default $d$ of the form

$$\frac{\alpha : M\beta_1, \ldots, M\beta_n}{\gamma}$$

a modal formula $emb(d)$ of the form

$$L\alpha \wedge LM\beta_1 \wedge \ldots \wedge LM\beta_n \supset \gamma,$$

there is a one-to-one correspondence between consistent extensions of a default theory and **S4f**-expansions of its modal image. We will now prove similar results for weak extensions and for minimal sets closed under defaults. Our results relate weak extensions to autoepistemic expansions (or, equivalently, **kd45**-expansions) and sets closed under defaults to stable theories. An interesting aspect of the present theory is that we are also able to obtain inverse interpretability results. That is, not only are weak extensions of a default theory representable as **kd45**-expansions of an embedding, but also conversely, for any theory $I$ in the language $\mathcal{L}_L$, regardless of its $L$-depth, we are able to find a default theory $(D, W)$ such that the objective parts of **kd45**-expansions of $I$ are precisely the weak extensions of $(D, W)$. A similar result holds for minimal sets.

We start with an interpretability result for weak extensions. Our results are based on Theorem 11.5. This theorem provides a characterization of **kd45**-expansions of theories consisting of *epistemic clauses*, that is, formulas of the form

$$L\alpha_1 \wedge \ldots \wedge L\alpha_k \wedge \neg L\beta_1 \wedge \ldots \wedge \neg L\beta_m \supset \gamma,$$

where $\gamma \in \mathcal{L}$. The formula $\gamma$ is called the *head* of the clause. For a theory $I$ consisting of epistemic clauses by $H(I)$ we denote the set of the heads of the clauses from $I$. Let us note that every formula $\gamma \in \mathcal{L}$ is an epistemic clause (with empty antecedent).

Let $I \subseteq \mathcal{L}_L$ be a theory consisting of epistemic clauses and let $T \subseteq \mathcal{L}_L$ be stable. Recall that a formula $\gamma \in H(I)$ has an *I-support* in $T$ (Definition 11.3) if for some clause

$$L\alpha_1 \wedge \ldots \wedge L\alpha_k \wedge \neg L\beta_1 \wedge \ldots \wedge \neg L\beta_m \supset \gamma$$

from $I$, $\alpha_i \in T$, $1 \leq i \leq k$, and $\beta_i \notin T$, $1 \leq i \leq m$. In particular, every $\gamma \in I \cap \mathcal{L}$ has an $I$-support in every stable theory.

We define

$$S_T(I) = \{\varphi \in H(I) : \varphi \text{ has an } I\text{-support in } T\}.$$

It follows that, in particular, every formula $\gamma \in I \cap \mathcal{L}$ belongs to $S_T(I)$.

One of the characterizations given by Theorem 11.5 states that a consistent theory $T$ is a **kd45**-expansion for a theory $I$ consisting of epistemic clauses if and only if $T = ST(S_T(I))$. We will use this result to prove the following interpretability result for weak extensions.

**Theorem 12.19** *Let $(D, W)$ be a default theory. Let $\mathcal{S}$ be any modal logic such that $\mathbf{5} \subseteq \mathcal{S} \subseteq \mathbf{kd45}$. A consistent and closed under propositional consequence theory $S \subseteq \mathcal{L}$ is a weak extension for $(D, W)$ if and only if $ST(S)$ is an $\mathcal{S}$-expansion of $emb(D, W)$.*

Proof: It is enough to prove the result for the logic $\mathcal{S} = \mathbf{kd45}$, since all nonmonotonic logics $\mathcal{S}$, where $\mathbf{5} \subseteq \mathcal{S} \subseteq \mathbf{kd45}$, coincide.

Denote $T = ST(S)$. We will first prove that

$$W \cup c(GD(D, S)) = S_T(emb(D, W)). \tag{12.12}$$

Clearly, since $W \subseteq \mathcal{L} \cap emb(D, W)$, $W \subseteq S_T(emb(D, W))$. Consider an element $\gamma \in c(GD(D, S))$. Then there is a default in $D$ of the form

$$\frac{\alpha : M\beta_1, \ldots, M\beta_n}{\gamma} \tag{12.13}$$

such that $\alpha \in S$ and $\neg\beta_i \notin S$, $1 \leq i \leq n$. Then the formula

$$L\alpha \wedge LM\beta_1 \wedge \ldots \wedge LM\beta_n \supset \gamma \tag{12.14}$$

is in $emb(D,W)$ and, by stability of $T = ST(S)$, $\alpha \in T$ and $M\beta_i \in T$, $1 \leq i \leq n$. Hence, $\gamma$ has an $emb(D,W)$-support in $T$. That is, $\gamma \in S_T(emb(D,W))$.

To prove the converse inclusion we proceed similarly. Consider a formula $\gamma \in S_T(emb(D,W))$. Then, there are two possibilities: (1) $\gamma \in W$, or (2) there is a clause $C$ in $emb(D,W)$,

$$C = (L\alpha \wedge LM\beta_1 \wedge \ldots \wedge LM\beta_n \supset \gamma),$$

such that $\alpha \in T$ and $M\beta_i \in T$, $1 \leq i \leq n$. In the first case, clearly, $\gamma \in W \cup c(GD(D,S))$. In the second case, the default

$$d = \frac{\alpha : M\beta_1, \ldots, M\beta_n}{\gamma}$$

belongs to $D$. By stability and consistency of $T$ it follows that $\alpha \in T$ and $\neg\beta_i \notin T$, $1 \leq i \leq n$. Since $\alpha$ and all $\beta_i$ are in $\mathcal{L}$ and since $T \cap \mathcal{L} = S$ (recall that $S$ is closed under propositional consequence), $\alpha \in S$ and $\neg\beta_i \notin S$, $1 \leq i \leq n$. Hence, $d \in GD(D,S)$ and $\gamma \in c(GD(D,S))$.

By Theorem 3.80, a theory $S$ is a weak extension for $(D,W)$ if and only if $S = Cn(W \cup c(GD(D,S)))$. Since $S$ is consistent, it follows from (12.12) that $S = Cn(W \cup c(GD(D,S)))$ if and only if $S = Cn(S_T(emb(D,W)))$. Finally, since $T = ST(S)$, this last condition is equivalent to

$$T = ST(S_T(emb(D,W))).$$

By Theorem 11.5, a consistent theory $T$ is a **kd45**-expansion for a theory $emb(D,W)$ (all its formulas are epistemic clauses) if and only if $T = ST(S_T(emb(D,W)))$. Hence the assertion follows. ■

We will consider next the "converse" interpretability problem. Given a theory $I \subseteq \mathcal{L}_L$, our goal is to construct a default theory $(D,W)$ so that the modal-free parts of expansions of $I$ are precisely the weak extensions of $(D,W)$. In our considerations the key role is played by the class of epistemic clauses of the form

$$L\alpha_1 \wedge \ldots \wedge L\alpha_k \wedge \neg L\beta_1 \wedge \ldots \wedge \neg L\beta_m \supset \gamma, \tag{12.15}$$

where all $\alpha_i$, all $\beta_i$ and $\gamma$ are modal-free formulas. We will call such clauses *simple epistemic clauses.*

For a simple epistemic clause $\varphi$ of the form (12.15) define the formula $\varphi'$ as follows:

$$\varphi' = (L(\alpha_1 \wedge \ldots \wedge \alpha_k) \wedge L\neg L\beta_1 \wedge \ldots \wedge L\neg L\beta_m \supset \gamma).$$

For a theory $I$, consisting of simple epistemic clauses, define

$$I' = \{\varphi' : \varphi \in I\}. \tag{12.16}$$

It is easy to see, using the properties of stability, that for every stable theory $T$ a formula $\gamma \in \mathcal{L}$ has an $I$-support in $T$ if and only if $\gamma$ has an $I'$-support in $T$. Hence,

$$S_T(I) = S_T(I').$$

By Theorem 11.5 it follows that $I$ and $I'$ have exactly the same **kd45**-expansions. For a simple epistemic clause $\varphi$ of the form (12.15) let us define a default $emb_{dl}(\varphi)$ by

$$emb_{dl}(\varphi) = \frac{\alpha_1 \wedge \ldots \wedge \alpha_k : M\neg\beta_1, \ldots, M\neg\beta_m}{\gamma}.$$

For a theory $I$, consisting of simple epistemic clauses, define

$$emb_{dl}(I) = \{emb_{dl}(\varphi) : \varphi \in I\}.$$

The following result is a simple consequence of Theorem 12.19 and our comments above.

**Theorem 12.20** *Let $I \subseteq \mathcal{L}_L$ consist of simple epistemic clauses. Then a theory $T$ is a* **kd45***-expansion for $I$ if and only if $T \cap \mathcal{L}$ is a consistent weak extension for the default theory $(emb_{dl}(I), \emptyset)$.* ▲

In view of Theorem 12.20, to construct a promised embedding which provides an exact correspondence between **kd45**-expansions of an *arbitrary* modal theory and weak extensions of some default theory, one more step is needed. That is we need to present a method of transforming a modal theory $I$ into a theory $I^s$, consisting of simple epistemic clauses, so that $I$ and $I^s$ have the same **kd45**-expansions. One such method is implicitly provided by the following result.

**Theorem 12.21** *For every theory $I \subseteq \mathcal{L}_L$ there is a theory $I^s$ consisting of simple epistemic clauses such that $Cn_{\mathbf{kd45}}(I) = Cn_{\mathbf{kd45}}(I^s)$. Consequently, $I$, and $I^s$ have the same* **kd45***-expansions.*

Proof: Observe the following properties of logics **kd45** and **K**:

**(i)** $\vdash_{\mathbf{kd45}} L(\varphi \vee L\psi) \equiv (L\varphi \vee L\psi)$.

**(ii)** $\vdash_{\mathbf{kd45}} L(\varphi \vee \neg L\psi) \equiv (L\varphi \vee \neg L\psi)$.

**(iii)** $\vdash_{\mathbf{K}} L(\varphi \wedge \psi) \equiv (L\varphi \wedge L\psi)$.

**(iv)** $\vdash_{\mathbf{K}} (\varphi \equiv \psi) \supset (L\varphi \equiv L\psi)$.

All these properties can easily be proved semantically.

Using these properties one can now recursively reduce the $L$-depth of every formula in $I$ without changing the set of **kd45**-consequences. Once a theory with $L$-depth of at most one is obtained, using propositional calculus one can construct a propositionally equivalent theory consisting of simple epistemic clauses only (Theorem 7.3). ■

Combining Theorems 12.20 and 12.21 yields the following result.

**Theorem 12.22** *For every modal theory $I \subseteq \mathcal{L}_L$, a theory $T \subseteq \mathcal{L}_L$ is a* **kd45**-*expansion of $I$ if and only if $T \cap \mathcal{L}$ is a consistent weak extension of the default theory $(emb_{dl}(I^s), \emptyset)$.* ▲

So far, we have seen that the embedding *emb* relates consistent extensions and $\mathcal{S}$-expansions, where $\mathcal{S}$ is any modal logic between **N** and **S4f**, and consistent weak extensions and $\mathcal{S}$-expansions, where $\mathcal{S}$ is any modal logic between **5** and **kd45**. Next, we will show that the same embedding *emb* establishes a correspondence between sets (theories) closed under defaults and stable theories. Under the assumption of consistency this result associates sets closed under defaults with **S5**-expansions. We have the following theorem.

**Theorem 12.23** *Let $S \subseteq \mathcal{L}$ be closed under propositional consequence and let $D$ be a set of defaults. The theory $S$ is closed under the defaults from $D$ if and only if $\{emb(d): d \in D\} \subseteq ST(S)$.*

Proof: The assertion is trivially true if $S$ is inconsistent. Hence, assume that $S$ is consistent. Consider a default

$$d = \frac{\alpha: M\beta_1, \ldots, M\beta_n}{\gamma}.$$

The theory $S$ is closed under $d$ if and only if at least one of the following conditions holds:

**(a)** $\alpha \notin S$.

**(b)** $\neg\beta_i \in S$, for some $i$, $1 \leq i \leq n$.

**(c)** $\gamma \in S$.

Since $S$ is consistent and closed under propositional provability, it follows from the properties of stable theories that these conditions are respectively equivalent to the conditions:

**(a′)** $\neg L\alpha \in ST(S)$.

**(b′)** $\neg LM\beta_i \in ST(S)$, for some $i$, $1 \leq i \leq n$.

**(c′)** $\gamma \in ST(S)$.

Next, observe that the formula

$$emb(d) = (L\alpha \wedge LM\beta_1 \wedge \ldots \wedge LM\beta_n \supset \gamma)$$

is in $ST(S)$ precisely when at least one of the conditions (a′) - (c′) holds. Hence, the theory $S$ is closed under a default $d$ if and only if $emb(d) \in ST(S)$. Consequently, the assertion follows. ■

Let us recall that a theory $S$ is a minimal set for a default theory $(D, W)$ if $S$ is inclusion-minimal among all theories containing $W$, closed under propositional provability and closed under the defaults in $D$. To identify those stable theories which correspond to minimal sets we need one more notion. Let $T_1$ and $T_2$ be stable theories. We say that $T_1 \trianglelefteq T_2$ if $T_1 \cap \mathcal{L} \subseteq T_2 \cap \mathcal{L}$ (see Definition 11.33).

Theorem 12.23 implies the following result.

**Theorem 12.24** *Let $(D, W)$ be a default theory. Let $S \subseteq \mathcal{L}$ be closed under propositional provability. Then, $S$ is a minimal set for $(D, W)$ if and only if $ST(S)$ is a $\trianglelefteq$-minimal stable theory containing $emb(D, W)$.* ▲

The converse interpretability result for stable theories and $\trianglelefteq$-minimal stable theories also holds. Indeed, since $Cn_{\mathbf{kd45}}(I) = Cn_{\mathbf{kd45}}(I^s)$, we have $Cn_{\mathbf{S5}}(I) = Cn_{\mathbf{S5}}(I^s)$, where $I^s$ is the theory constructed in Theorem 12.21. Therefore the theories $I$ and $I^s$ are contained in exactly the same stable theories. Hence, applying Theorem 12.24 we get the following result.

**Theorem 12.25** *Let $I \subseteq \mathcal{L}_L$ and let $T \subseteq \mathcal{L}_L$ be stable. The theory $T$ contains $I$ if and only if the modal-free part of $T$, $T \cap \mathcal{L}$, is closed under the defaults in $emb_{dl}(I^s)$. The theory $T$ is a $\trianglelefteq$-minimal theory containing $I$ if and only if the modal-free part of $T$, $T \cap \mathcal{L}$, is a minimal set for the default theory $(emb_{dl}(I^s), \emptyset)$.* ▲

Let us mention here that interpretability results for weak extensions and sets closed under defaults can also be obtained for a different translation of defaults into modal formulas. For a default $d$ of the form

$$\frac{\alpha : M\beta_1, \ldots, M\beta_n}{\gamma}$$

define

$$emb_K(d) = (L\alpha \wedge \neg L\neg\beta_1 \wedge \ldots \wedge \neg L\neg\beta_n \supset \gamma).$$

Next, for a default theory $(D, W)$ define

$$emb_K(D, W) = W \cup \{emb_K(d): d \in D\}.$$

Since for every modal logic $\mathcal{S}$ such that $\mathbf{5} \subseteq \mathcal{S} \subseteq \mathbf{kd45}$ the theories $emb(D, W)$ and $emb_K(D, W)$ have the same $\mathcal{S}$-expansions and are contained in exactly the same stable theories, the assertions of Theorems 12.19, 12.23 and 12.24 hold when $emb$ is replaced by $emb_K$.

We conclude this section with a corollary to Theorem 12.19 relating supported models and $\mathcal{S}$-expansions, where $\mathcal{S}$ is a modal logic between **5** and **kd45**.

**Theorem 12.26** *Let $P$ be a logic program and let $emb_{lp}(P)$ be the modal interpretation of $P$ introduced in Section 12.3. Let $\mathcal{S}$ be any modal logic such that $\mathbf{5} \subseteq \mathcal{S} \subseteq \mathbf{kd45}$. Then, for any $M \subseteq B_H$, $M$ is a supported model of $P$ if and only if $ST(M)$ is an $\mathcal{S}$-expansion of $emb_{lp}(P)$. Moreover, every $\mathcal{S}$-expansion of $emb_{lp}(P)$ is of the form $ST(M)$, where $M$ is a supported model of $P$.*

Proof: First, recall that $M$ is a supported model of $P$ if and only if $Cn(M)$ is a weak extension of $(dl(P), \emptyset)$. Then observe that $emb(dl(P), \emptyset) = emb_{lp}(P)$ (recall that we assume that $P$ is built of ground atomic formulas). Hence, $M$ is a supported model of $P$ if and only if $ST(Cn(M))$ is an $\mathcal{S}$-expansion for $emb_{lp}(P)$ (Theorem 12.19). Since $ST(M) = ST(Cn(M))$, the assertion follows. ■

This result also remains true if instead of default clauses simple epistemic clauses are used in the translation.

## 12.6 Interpreting logic programs in autoepistemic logic

In this section we present an embedding of logic programs into modal logic language under which stable models of a program correspond to **kd45**-expansions of the modal image of the program. We have already obtained one result of this type where logic programming with stable models was embedded into nonmonotonic logic **Sw5**. Our present result is interesting for historical reasons. The embedding that we study below and its properties led Gelfond and Lifschitz to their discovery of stable models.

**Definition 12.27** *For a ground logic program clause*

$$C = (c \leftarrow a_1, \ldots, a_m, \mathbf{not}(b_1), \ldots, \mathbf{not}(b_n))$$

*put*

$$emb_G(C) = (\neg Lb_1 \wedge \ldots \wedge \neg Lb_n \supset (a_1 \wedge \ldots \wedge a_m \supset c)).$$

*For a logic program $P$ define $emb_G(P) = \{emb_G(C) : C \in ground(P)\}$.* ▲

**Theorem 12.28** *A subset $M$ of the Herbrand base $B_H$ is a stable model of a logic program $P$ if and only if there is a stable expansion (that is, a* **kd45**-*expansion) $T$ of $emb_G(P)$ such that $M = T \cap At$.*

Proof: We will first state and prove a useful auxiliary fact. Let $T$ be a stable theory and let $M$ be the set of atoms contained in $T$. Then, for any logic program $P$ we have

$$S_T(emb_G(P)) = pc(P^M). \tag{12.17}$$

(See Chapter 6 for the definitions of the classical logic interpretation $pc(P)$ of a logic program $P$, and of the reduct $P^M$ of a program $P$ with respect to a set of atoms $M$).

Indeed, let us consider the following sequence of statements:

1. A formula $a_1 \wedge \ldots \wedge a_m \supset c$ belongs to $pc(P^M)$.
2. A clause $c \leftarrow a_1, \ldots, a_m$, belongs to the reduct $P^M$.
3. There is a clause

   $$C = (c \leftarrow a_1, \ldots, a_m, \mathbf{not}(b_1), \ldots, \mathbf{not}(b_n))$$

   in $ground(P)$ such that $b_i \notin M$ for every $i$, $1 \leq i \leq n$.
4. There is a formula

   $$emb_G(C) = (\neg Lb_1 \wedge \ldots \wedge \neg Lb_n \supset (a_1 \wedge \ldots \wedge a_m \supset c))$$

   in $emb_G(P)$ such that $b_i \notin T$ for every $i$, $1 \leq i \leq n$.
5. The formula $a_1 \wedge \ldots \wedge a_m \supset c$ has an $emb_G(P)$-support in $T$, that is, $a_1 \wedge \ldots \wedge a_m \supset c$ is in $S_T(emb_G(P))$.

The statements (1) and (2) are equivalent by the definition of the mapping $pc$. The equivalence of (2) and (3) follows from the definition of the reduct $P^M$. Since each $b_i$ is an atom and $M$ is the set of all atoms contained in $T$, the equivalence of the statements (3) and (4) follows. Finally, the definition of $S_T(emb_G(P))$ implies the equivalence of (4) and (5). Consequently, (1) and (5) are equivalent and (12.17) follows.

Assume now that $M$ is a stable model for $P$. Define $T = ST(pc(P^M))$. Clearly, the theory $Cn(pc(P^M))$ is the modal-free part of $T$. Since $M$ is a stable model of $P$, it follows that $M$ is the least model of $pc(P^M)$ (Theorem 6.22). Hence, $M$ is precisely the set of all atoms true in every model of $pc(P^M)$. Consequently, $M$ is the set of all atoms of $T$. From (12.17) it follows that $T = ST(S_T(emb_G(P)))$. Observe that $S_T(emb_G(P))$ consists of program Horn clauses. Hence, $T$ is consistent. By Theorem 11.5, $T$ is a **kd45**-expansion of $emb_G(P)$.

Conversely, assume that $T$ is a **kd45**-expansion of $emb_G(P)$ and that $M$ is the set of atoms of $T$. Then it follows by Theorem 11.5 and by (12.17) that $T = ST(pc(P^M))$. Since $M$ is the set of atoms of $T$ and the modal-free part of $T$ is $Cn(pc(P^M))$, it follows that $M$ is the set of all atoms contained in $Cn(pc(P^M))$. Then $M$ is the least model of $P^M$. Consequently, $M$ is a stable model of $P$. ■

Let us comment on Theorem 12.28. First, notice that in contrast with Theorem 12.9, its assertion fails for CN-programs.

**Example 12.29** Consider the CN-program of Example 6.38, that is, the CN-program $P = \{\neg b \leftarrow, b \leftarrow \neg a\}$. As we have already seen, $P$ has a unique stable answer set $\{\neg b\}$. On the other hand, $emb_G(P) = \{\neg b, \neg a \supset b\}$ has the unique autoepistemic expansion $ST(\{\neg b, a\})$. Hence, the stable answer set of $P$ is *not* the set of literals of an autoepistemic expansion of $emb_G(P)$. ▲

Secondly, Theorem 12.28 tells us only that sets of atoms of autoepistemic expansions are stable models. Autoepistemic expansions may and usually will contain formulas not implied by stable models.

**Example 12.30** Let $P = \{a \leftarrow b\}$. Then $\emptyset$ is the unique stable model of $P$. We also have that $emb_G(P) = \{b \supset a\}$. The theory $ST(\{b \supset a\})$ is the unique autoepistemic expansion of $emb_G(P)$. Its modal-free part equals $Cn(\{b \supset a\})$ and is bigger than $Cn(\emptyset)$. ▲

## 12.7 Truth maintenance and nonmonotonic systems

Default logic and logic programming can be further generalized by abstracting the nature of the objects that are processed by defaults and logic program clauses during reasoning. Default logic deals with formulas of the language of propositional calculus and logic programming deals with atoms. The idea behind the generalization we now briefly study is to remove any references to logic. We will consider an *arbitrary* universe of objects and introduce "default-like" rules to manipulate them. In this way, we will obtain what is often called a *truth maintenance system* or a *nonmonotonic rule system*. We use the first of these two terms in this book.

**Definition 12.31** Let $U$ be a set. A triple $\langle A, B, p\rangle$, where $A, B$ are finite subsets of $U$ and $p \in U$, is called a *rule*. A rule is commonly written as $\langle A \mid B \rightarrow p\rangle$. A rule is *monotonic* if $B = \emptyset$, and *nonmonotonic* otherwise. A pair $\langle U, N\rangle$, where $U$ is a set and $N$ is a collection of rules for $U$ is called a *truth maintenance system* or a TMS. ▲

An informal interpretation of a rule $\langle A \mid B \rightarrow p\rangle$ is: "if all the elements of $A$ are IN and all the elements of B are OUT then derive $p$". The precise meaning of the notions of being IN and OUT is provided by the notion of an $S$-proof, similar to that used in default logic.

**Definition 12.32**

1. Let $\langle U, N\rangle$ be a TMS and let $S$ and $I$ be two subsets of $U$. An *$S$-derivation* or an *$S$-proof* of an element $a \in U$ from $I$ is any finite sequence $a_1, \ldots, a_n$ such that $a_n = a$ and, for every $i \leq n$, $a_i \in I$ or there exists a rule $\langle A \mid B \rightarrow a_i\rangle$ in $N$ such that:
   (a) $A \subseteq \{a_j : j < i\} \cup I$ (all elements of $A$ are IN).
   (b) $B \cap S = \emptyset$ (all elements of $B$ are OUT).

   (Notice that for $i = 1$, $\{a_j : j < i\} = \emptyset$).
2. The set of all elements in $U$ which have an $S$-proof from $I$ is denoted by $C^{U,N,S}(I)$.
3. A set $S \subseteq U$ is called an *extension* of a TMS $\langle U, N\rangle$ if $S = C^{U,N,S}(\emptyset)$. A set $S \subseteq U$ is called a *weak extension* of $\langle U, N\rangle$ if $S = C^{U,N,S}(S)$. Further, $S$ is called *closed under rules of* $N$ if for every rule $\langle A \mid B \rightarrow p\rangle \in N$, whenever $A \subseteq S$, and $B \cap S = \emptyset$ then $p \in S$. A set $S$ is called a *minimal set for* $\langle U, N\rangle$ if $S$ is an inclusion-minimal set closed under the rules of $N$. ▲

Analogies between the operator $C^{U,N,S}(\cdot)$ and the operator $Cn^{D,S}(\cdot)$ in default logic are evident. In the next result we gather some properties of the operator $C^{U,N,S}(\cdot)$. They can be derived in the same fashion as their counterparts for the operator $Cn^{D,S}(\cdot)$. We leave their proofs to the reader.

**Theorem 12.33** *Let $\langle U, N\rangle$ be a TMS and let $S$ and $I$ be subsets of $U$. Then:*

1. $I \subseteq C^{U,N,S}(I)$.
2. $C^{U,N,S}(I)$ *is closed under all monotonic rules in* $N$.
3. $C^{U,N,S}(C^{U,N,S}(I)) = C^{U,N,S}(I)$. ▲

Next, we present several properties of extensions, weak extensions and sets closed under the rules. Again the proofs mimic the corresponding proofs in default logic and are omitted.

**Theorem 12.34** *Let $\langle U, N\rangle$ be a TMS.*

1. *If $S$ is an extension of $\langle U, N\rangle$ then $S$ is a weak extension of $\langle U, N\rangle$.*
2. *If $S_1$ and $S_2$ are extensions of $\langle U, N\rangle$ and $S_1 \subseteq S_2$ then $S_1 = S_2$.*
3. *If $S$ is a weak extension of $\langle U, N\rangle$ then $S$ is closed under the rules of $N$. The same holds for extensions of $\langle U, N\rangle$.*
4. *Every extension of $\langle U, N\rangle$ is a minimal set for $\langle U, N\rangle$.* ▲

It is easy to see that default logic can be represented as a TMS. One simply has to take for $U$ the language $\mathcal{L}$ of propositional calculus. Next, each default

$$\frac{\alpha : M\beta_1, \ldots, M\beta_n}{\gamma}$$

has to be represented by the rule

$$\langle\{\alpha\} \mid \{\neg\beta_1, \ldots, \neg\beta_n\} \rightarrow \gamma\rangle.$$

In addition, every tautology $\varphi$ has to be encoded as a rule $\langle\emptyset \mid \emptyset \rightarrow \varphi\rangle$ and each instance

$$\frac{\varphi, \varphi \supset \psi}{\psi}$$

of modus ponens as the rule

$$\langle\{\varphi, \varphi \supset \psi\} \mid \emptyset \rightarrow \psi\rangle.$$

Under the translation described above, the (weak) extensions of a default theory are precisely the (weak) extensions of the corresponding TMS.

An even simpler encoding exists for logic programs. The Herbrand base $B_H$ now plays the role of $U$. Given a ground logic program $P$, each of its clauses

$$c \leftarrow a_1, \ldots, a_m, \mathbf{not}(b_1), \ldots, \mathbf{not}(b_n)$$

has to be encoded as the rule

$$\langle \{a_1, \ldots, a_m\} \mid \{b_1, \ldots, b_n\} \rightarrow c \rangle.$$

Then the stable and supported models of a logic program are precisely the extensions and weak extensions of the corresponding TMS.

We will give two examples from the field of classical mathematics where extensions of suitable TMSs describe some well-known objects.

**Example 12.35** Let $\mathcal{B} = \langle B, +, \cdot, -, 0, 1 \rangle$ be a boolean algebra. Define the following set of rules $N_{\mathcal{B}}$:

1. $\langle \emptyset \mid \emptyset \rightarrow 1 \rangle$.
2. $\langle \emptyset \mid \{x\} \rightarrow -x \rangle$, for all $x \in B$.
3. $\langle \{x\} \mid \emptyset \rightarrow y \}$, for all $x, y \in B$ such that $x \cdot y = x$.
4. $\langle \{x, y\} \mid \emptyset \rightarrow x \cdot y \}$, for all $x, y \in B$.

We have the following result. A subset $F \subseteq B$ is a proper ultrafilter in $\langle B, +, \cdot, -, 0, 1 \rangle$ if and only if $F$ is an extension of $\langle B, N_{\mathcal{B}} \rangle$. ▲

**Example 12.36** (Marriage problem). Let $S = \langle B, G, K \rangle$ be a society, that is, $B$ is a set, called the set of boys, $G$ is a set, called the set of girls, and $K \subseteq B \times G$. Moreover, assume that for every $b \in B$ there are only finitely many $g \in G$ such that $(b, g) \in K$. Then there exists a TMS $\langle K, N \rangle$ such that the extensions of $\langle K, N \rangle$ are precisely the marriages for $S$. ▲

We have seen that every logic program can be regarded as a certain TMS. It is easy to see that the converse is also true. Every TMS can be encoded as a suitably chosen propositional logic program. We outline this construction below.

**Definition 12.37** Let $\langle U, N \rangle$ be a TMS. We consider the language of propositional calculus with the set of propositional variables $U$. Given a rule

$R = \langle A \mid B \rightarrow p \rangle$, where $A = \{a_1, \ldots, a_n\}$ and $B = \{b_1, \ldots, b_m\}$, assign to $R$ the clause:

$$emb_{tms}(R) = (p \leftarrow a_1, \ldots, a_n, \mathbf{not}(b_1), \ldots, \mathbf{not}(b_m)).$$

Finally, we define $emb_{tms}(U, N) = \{emb_{tms}(R) : R \in N\}$. ▲

We then have the following result:

**Theorem 12.38** *Let $\langle U, N \rangle$ be a TMS. Let $P = emb_{tms}(U, N)$ be a logic program constructed as above. Then for every $M \subseteq U$:*

1. *$M$ is an extension of $\langle U, N \rangle$ if and only if $M$ is a stable model of $P$.*
2. *$M$ is a weak extension of $\langle U, N \rangle$ if and only if $M$ is a supported model of $P$.*
3. *$M$ is a minimal set for $\langle U, N \rangle$ if and only if $M$ is a minimal Herbrand model of $P$.* ▲

We have seen that logic programming can be embedded into the nonmonotonic logic **Sw5**. Theorem 12.38 shows that, in fact, a large class of nonmonotonic systems, including default logic, logic programming and truth maintenance systems, can be embedded into nonmonotonic modal logic **Sw5**. We have seen earlier in this chapter that without modifying the language, default logic can be embedded into a weaker logic, the nonmonotonic logic **S4f**. It is especially interesting to note that at the price of regarding the formulas of $\mathcal{L}$ as the elements of $U$, and adding new rules which encode the propositional calculus, default logic can be represented by the nonmonotonic logic **Sw5**.

## 12.8 Comments and remarks

An expectation that nonmonotonic reasoning formalism such as default logic can be modeled faithfully within a nonmonotonic modal logic is present already in the earliest papers on the subject. The notation used by Reiter [Rei80] for defaults is perhaps the best witness to that. However, it was not obvious which modal nonmonotonic logic to use. In particular, it was not clear whether the scheme of McDermott and Doyle, [McD82] can be employed. The influential paper of Moore [Moo85] initially sidetracked efforts to establish the connection by promoting autoepistemic logic as the main modal nonmonotonic formalism. As a result, there have been several attempts to embed default logic into autoepistemic logic, the most important of them by Konolige [Kon88, Kon89]. Konolige used the translation $emb_K$. As we have seen, this translation works well in the case of weak extensions and relates them to autoepistemic expansions but is not appropriate for the case of extensions. Some extralogical machinery (the notion of reduct) is necessary to make the translation $emb_K$ work in this case, too. In [MT90] the source of the problems was identified. Default logic treats prerequisites and justifications as two different types of premises, and this distinction disappears if the embedding $emb_K$ is used. Consequently, the translation that assigns to a default rule $\frac{\alpha:M\beta_1,\ldots,M\beta_k}{\gamma}$ the formula $L\alpha \wedge \neg LL\neg\beta_1 \wedge \ldots \wedge \neg LL\neg\beta_k \supset \gamma$ was proposed in [MT90]. It was also shown there that this translation allows us to properly model default logic in the nonmonotonic logic **N**. But that translation is rather unintuitive and works only with the logic **N** and some other, rather obscure, modal subnormal logics. In a subsequent development, the translation *emb*, studied in this chapter, was introduced by Truszczyński [Tru91a, Tru91b].

The combination of interpreting logic programming in default logic and interpreting default logic within modal language results in a similar embedding $emb_{lp}$ of logic programming into modal nonmonotonic logics. This interpretation not only models stable semantics for logic programs in nonmonotonic **S4f** but also works correctly with a stronger (and semantically simpler) nonmonotonic logic **Sw5**.

We have also mentioned two related translations $emb^L$ and $emb^L_{lp}$. They allow us to embed disjunctive default logic (introduced by Gelfond, Lifschitz, Przymusinska and Truszczyński [GLPT91]) into nonmonotonic modal logic **S4f** and disjunctive databases (introduced in [GL91]) into nonmonotonic modal logic **Sw5**.

The relationship between supported models and stable expansions was investigated Marek and Subrahmanian [MS89].

The translation $emb_G$ relating stable models and stable expansions (**kd45**-expansions) was introduced by Gelfond [Gel87] and was later rediscovered by Elkan [Elk90] in the context of truth maintenance systems.

A common generalization of default logic with extensions, logic programming with stable semantics, and modal nonmonotonic logic with expansions is known in the literature as truth maintenance system (also called nonmonotonic rule system). The construction in a rudimentary form is due to Doyle ([Doy79]). A precise, formal, definition of nonmonotonic rule system is given by Marek, Nerode and Remmel [MNR90] (see also Reinfrank, Dressler and Brewka [RDB89]). The representation by means of extensions of the solutions to "marriage problem" and of the ultrafilters in boolean algebras was introduced by Marek, Nerode and Remmel [MNR90]. Many other structures in classical combinatorics and algebra are representable by means of extensions of suitably chosen nonmonotonic rule system (see Marek, Nerode and Remmel [MNR92c] and Mikitiuk and Truszczyński [MT92]). An abstract representation of the nonmonotonic reasoning by means of nonmonotonic rule systems provides means for a natural analysis of complexity of default extensions and stable models of logic programs (Marek, Nerode, and Remmel [MNR92b]).

Let us notice, though, that *finite* default theories and finite theories in modal language are representable by *infinite* nonmonotonic rule systems. The reason is that, in the abstract setting of nonmonotonic rule systems we need to write the rules of logic explicitly. Therefore, as far as default theories are concerned, the role of nonmonotonic rule systems as means of analysis of the complexity of the existence problem for extensions is limited. For infinite default theories, infinite propositional logic programs, and predicate logic programs with function symbols, nonmonotonic rule systems provide a useful tool in the complexity analysis (Marek, Nerode and Remmel [MNR92a]).

# 13. Complexity of some forms of nonmonotonic reasoning

In this chapter we investigate the complexity of some forms of nonmonotonic reasoning. We assume familiarity with basic notions of the computational complexity theory and the recursion theory. Definitions of some of these concepts are provided in the first section of this chapter. For the undefined concepts the reader is referred to Garey and Johnson [GJ79] and Rogers [Rog87].

In this chapter we present results on the complexity of default and modal nonmonotonic reasoning. Specifically, we study the complexity of determining the existence of an extension or expansion, and the complexity of the membership in all (some) extensions or expansions. We also discuss the complexity of similar problems for stable models of finite and infinite propositional logic programs.

## 13.1 Basic notions of the complexity theory

We assume that the reader is familiar with such concepts as deterministic and nondeterministic Turing machines, and Turing machines with oracles.

Let $\Sigma$ be a fixed finite alphabet. By a *decision problem* we mean any subset $P$ of the set $\Sigma^*$ of all finite strings over $\Sigma$. The strings belonging to a problem $P$ are called its YES *instances*, and all other strings its NO *instances*.

In what follows we need the concept of a Turing machine with an oracle for a decision problem $Q$. Speaking informally, such an oracle Turing machine, let us call it $T$, works in the same way as a standard Turing machine except that in the course of computation, from time to time, it will build an input data, say $I$, for the problem $Q$ and then it will consult its oracle to determine whether $I$ is a YES or NO instance of $Q$. Subsequent computation depends on the answer provided by the oracle. Each call to the oracle is regarded as a single step in the computation of the Turing machine $T$ and under this convention the time complexity of a Turing machine with an oracle is defined in the standard way. In particular, the time complexity of a nondeterministic Turing machine $T$ with an oracle is the following function $comp_T$, defined on positive integers:

$$comp_T(n) = \max\{\{1\} \cup \{at(s) \colon s \in \Sigma^*,\ |s| = n \text{ and } T \text{ accepts } s\}\},$$

where $at(s)$ is the *accept time* for $s$ — the minimum length of a computation that accepts $s$. If $comp_T(n) = O(p(n))$, for some polynomial $p(n)$, we say that $T$ is a *polynomial-time* nondeterministic Turing machine with an oracle.

Sets of decision problems are called classes. Two important complexity classes are $P$ and $NP$. The class $P$ consists of those decision problems which can be solved by a polynomial-time deterministic Turing machine. The decision problems that can be solved by a polynomial-time nondeterministic Turing machine form the class $NP$.

Let $\mathcal{C}$ be a class of decision problems. By co-$\mathcal{C}$ we mean the class consisting of the complements of the problems in $\mathcal{C}$. The class $P^{\mathcal{C}}$ consists of the problems solvable by a polynomial-time deterministic Turing machine with an oracle for a problem from $\mathcal{C}$. Similarly, the class $NP^{\mathcal{C}}$ is the class of problems solvable by a polynomial-time nondeterministic Turing machine with an oracle for a problem in $\mathcal{C}$. The *polynomial hierarchy* of complexity classes is defined as follows:

$$\Delta_0^P = \Sigma_0^P = \Pi_0^P = P,$$

$$\Delta_{k+1}^P = P^{\Sigma_k^P}, \quad \Sigma_{k+1}^P = NP^{\Sigma_k^P} \quad \text{and} \quad \Pi_{k+1}^P = \text{co-}\Sigma_{k+1}^P,$$

for $k \geq 0$. The classes $\Sigma_k^P$, $\Pi_k^P$, and $\Delta_{k+1}^P$ are often called the $k^{th}$ level of the polynomial hierarchy.

We say that a problem $A$ is *complete* for a class $\mathcal{C}$ if $A \in \mathcal{C}$ and if for every problem $B$ in the class $\mathcal{C}$ there exists a polynomial transformation of $B$ to $A$. That is, there exists a function $g$ and a polynomial $f$ such that a string $w$ belongs to $B$ if and only if the string $g(w)$ belongs to $A$ and the cost of computing $g(w)$ is bounded from above by $f(|w|)$.

The satisfiability problem for propositional calculus is an archetypical $NP$-complete problem. For the higher classes of the polynomial hierarchy the problems related to the satisfiability of boolean expressions in the algebra of truth values serve as complete problems. In our case we will mostly be interested in the second level of the polynomial hierarchy. By $QBF(2, \exists)$ we denote the problem of deciding whether a quantified boolean formula

$$(\exists p_1) \ldots (\exists p_n)(\forall q_1) \ldots (\forall q_m)\psi,$$

is true in the algebra of truth values ($p_1, \ldots, p_n, q_1, \ldots, q_m$ are distinct boolean variables and $\psi$ is a boolean expression, not containing quantifiers, built of $p_1, \ldots, p_n, q_1, \ldots, q_m$). It is well known ([MS72, GJ79]) that $QBF(2, \exists)$ is $\Sigma_2^P$-complete.

## 13.2 Complexity of default reasoning

In this section we will investigate the computational complexity of some problems associated with default logic. We will prove that default reasoning is located at the second level of the polynomial hierarchy. In particular, we will prove that the problem of deciding whether a finite default theory $(D, W)$ possesses an extension is $\Sigma_2^P$-complete.

**Theorem 13.1** (Existence problem) *The problem of deciding whether a finite default theory $(D, W)$ possesses at least one extension is $\Sigma_2^P$-complete.*

Proof: We need to prove that:

1. Our problem belongs to $\Sigma_2^P$.
2. $QBF(2, \exists)$ is polynomially reducible to our problem.

(1) We will show that there is a polynomial-time nondeterministic Turing machine $T$ with an oracle for the satisfiability problem (given a finite set of formulas in the language of propositional calculus, determine if it is consistent) that decides the existence of extensions.

Consider a finite default theory $(D, W)$. In what follows, extensions will be represented by subsets of the set of consequents of all the defaults in $D$. Our polynomial-time nondeterministic Turing machine $T$ with an oracle for the satisfiability problem will first (nondeterministically) guess a subset $Z$ of the set of the consequents of the defaults in $D$. To complete this part of the proof of the theorem, all that we really have to do is to show that there is a polynomial-time (assuming that each call to the oracle is counted as one step) deterministic method to check whether $S = Cn(W \cup Z)$ is an extension. However, this is quite easy using the methods developed in Chapter 3. Given a set of defaults $D$, it takes $m$ calls to the oracle, where $m$ is the total number of justifications in the defaults in $D$, to compute the reduct $D' = D_S$ (a default $d = \frac{\varphi : M\psi_1, \ldots, M\psi_k}{\gamma}$ contributes the rule $\frac{\varphi}{\gamma}$ to the reduct if and only if for every $i$, $1 \leq i \leq k$, $\{\psi_i\} \cup W \cup Z$ is consistent). Next, it takes at most $|D'|^2$ calls to the oracle to construct the set $B = B^{D'} \uparrow \omega(W)$ (see the algorithm given in Figure 3.1). Finally, we need to check whether

$$Cn(B) = Cn(W \cup Z).$$

That is, we need to check whether $\Phi \equiv \Psi$ is a tautology, where $\Phi$ is the conjunction of the formulas in $B$ and $\Psi$ is the conjunction of the formulas in $W \cup Z$. Clearly, it suffices to decide if $\neg(\Phi \equiv \Psi)$ is consistent, which can be accomplished by just one additional call to the oracle. Summarizing, whether $Cn(W \cup Z)$ is an extension or not can be decided by means of $O(n^2)$ calls to the oracle, where $n$ is the size of the input default theory. It follows that the problem of existence of an extension is in $\Sigma_2^P$.

(2) We will now prove that the existence of an extension problem is complete for the class $\Sigma_2^P$. To this end, we polynomially reduce to it the problem $QBF(2,\exists)$. Let $\varphi = \exists p_1 \ldots \exists p_n \forall q_1 \ldots \forall q_m \psi$ be an instance of $QBF(2,\exists)$. In particular, $\psi$ does not contain quantifiers. We will construct a default theory $(D,W)$ such that $(D,W)$ possesses an extension if and only if $\varphi$ is satisfied in the algebra of truth values. We define $W_\varphi = \emptyset$ and put

$$D_\varphi = \left\{ \frac{:Mp_1}{p_1}, \frac{:M\neg p_1}{\neg p_1}, \ldots, \frac{:Mp_n}{p_n}, \frac{:M\neg p_n}{\neg p_n}, \frac{:M\neg\psi}{\bot} \right\}.$$

It is quite clear that the construction of the default theory $(D_\varphi, W_\varphi)$ can be accomplished in time polynomial (in fact linear) in the size of $\varphi$.

We will prove that $\varphi$ is satisfied if and only if $(D_\varphi, W_\varphi)$ possesses an extension. Assume first that $\varphi$ is satisfied in the algebra of truth values. Then there is a valuation $v$ of $\{p_1, \ldots, p_n\}$ such that for every extension $V$ of $v$ to the set of variables $\{p_1, \ldots, p_n, q_1, \ldots, q_m\}$, the formula $\psi$ is satisfied by $V$.

Consider the theory

$$S = Cn(\{p : v(p) = \mathbf{t}\} \cup \{\neg p : v(p) = \mathbf{f}\}).$$

We claim that $S$ is an extension of $(D_\varphi, W_\varphi)$. Indeed, observe that since $\psi$ is satisfied by *every* $V$ extending $v$, $\neg\psi$ is not consistent with $S$. Therefore the last default in $D_\varphi$ does not contribute to the reduct $(D_\varphi)_S$. In addition, for each pair of defaults $\frac{:Mp_j}{p_j}, \frac{:M\neg p_j}{\neg p_j}$, exactly one of them is applicable with respect to $S$. Hence,

$$(D_\varphi)_S = \left\{ \frac{}{\epsilon_i p_i} : 1 \le i \le n \right\},$$

where $\epsilon_i p_i = p_i$ if $p_i \in S$ and $\epsilon_i p_i = \neg p_i$, otherwise. Clearly,

$$B^{(D_\varphi)_S}\!\uparrow\!\omega(W_\varphi) = B^{(D_\varphi)_S}\!\uparrow\!1(W_\varphi) = \{p : v(p) = \mathbf{t}\} \cup \{\neg p : v(p) = \mathbf{f}\}.$$

Hence, $S$ is an extension for $(D_\varphi, W_\varphi)$.

Conversely, assume now that $(D_\varphi, W_\varphi)$ possesses an extension, say $S$. This extension is generated by the consequents of some of the defaults in $D_\varphi$. Since $W_\varphi$ is consistent and each of the defaults possesses at least one justification, $S$ is consistent (Corollary 3.60). It follows from the form of the defaults in $D_\varphi$ that $S$ is complete with respect to $\{p_1, \ldots, p_n\}$. Consequently, $S = Cn(\{\epsilon_i p_i\colon 1 \leq i \leq n\})$, where $\epsilon_i p_i = p_i$ if $p_i \in S$ and $\epsilon_i p_i = \neg p_i$, otherwise. Consider now a valuation $v$ of $\{p_1, \ldots, p_n\}$ such that $v(p_i) = \mathbf{t}$ if and only if $p_i \in S$. Clearly, $v$ and every valuation $V$ of $\{p_1, \ldots, p_n, q_1, \ldots, q_m\}$ such that $V|_{\{p_1,\ldots,p_n\}} = v$ satisfy $S$. In addition, since $S$ is a consistent extension of $(D_\varphi, W_\varphi)$, $\neg\psi$ is not consistent with $S$. In other words, $S \vdash \psi$. Consequently, for every extension $V$ of $v$ to $\{p_1, \ldots, p_n, q_1, \ldots, q_m\}$, $V$ satisfies $\psi$. Therefore, $\varphi$ is satisfied in the algebra of truth values. ■

Since all the extensions of the theory $(D_\varphi, W_\varphi)$ are consistent, the problem of deciding whether a finite default theory has a *consistent* extension is also $\Sigma_2^P$-complete. Hence, we have the following corollary.

**Corollary 13.2** *The problem of deciding whether a finite default theory $(D, W)$ possesses at least one* consistent *extension is $\Sigma_2^P$-complete.* ▲

Now, we will deal with several versions of the problem of membership of a formula in some or all extensions.

**Theorem 13.3** *The following two problems are $\Sigma_2^P$-complete:*

1. *Given a finite default theory $(D, W)$ and a formula $\varphi$, decide whether there is an extension of $(D, W)$ which contains $\varphi$.*
2. *Given a finite default theory $(D, W)$ and a formula $\varphi$, decide whether there is an extension of $(D, W)$ which* does not *contain $\varphi$.*

Proof: We will prove (2) first. As before, we need to prove two facts. First, that our problem is in the class $\Sigma_2^P$. Second, that it is complete for that class. To prove the first property, let us recall (proof of Theorem 13.1) that there is a polynomial-time nondeterministic Turing machine $T$ with an oracle for the satisfiability problem that decides, for a given $Z \subseteq D$, whether $S = Cn(W \cup c(Z))$ is an extension of $(D, W)$. One additional call to the oracle suffices to check that $\varphi \notin S$.

To prove completeness, we reduce $QBF(2, \exists)$ to problem (2). Let $\vartheta = \exists p_1 \ldots \exists p_n \forall q_1 \ldots \forall q_m \psi$ be an instance of $QBF(2, \exists)$. Consider the default theory $(D_\vartheta, W_\vartheta)$, where $W_\vartheta = \emptyset$, and

$$D_\vartheta = \left\{ \frac{: Mp_1}{p_1}, \frac{: M\neg p_1}{\neg p_1}, \ldots, \frac{: Mp_n}{p_n}, \frac{: M\neg p_n}{\neg p_n}, \frac{: M\neg\psi}{\neg\psi} \right\}.$$

The theory $(D_\vartheta, W_\vartheta)$ can be constructed in a polynomial time (in the size of $\vartheta$). Notice also that if $S$ is an extension of $(D_\vartheta, W_\vartheta)$ then $S$ is consistent.

It is straightforward to check that $\vartheta$ is satisfied in the algebra of truth values if and only if $\neg\psi$ does not belong to some extension of $(D_\vartheta, W_\vartheta)$. Hence, the proof of (2) is complete.

We will now deal with problem (1). To see that problem (1) is in $\Sigma_2^P$ we proceed as in the case of problem (2), with the only difference that the additional call to the oracle is made to decide whether $\varphi \in S$. To prove that problem (1) is $\Sigma_2^P$-complete observe that the theory constructed in the proof of completeness of problem (2) consists exclusively of normal prerequisite-free defaults. In particular, it follows that for every extension $S$ of $(D_\vartheta, W_\vartheta)$, $\neg\psi$ does not belong to $S$ if and only if $\psi$ belongs to $S$. Hence, $QBF(2, \exists)$ is polynomially reducible to the problem (1). ■

We have observed in the proof of the last result that every extension of $(D_\vartheta, W_\vartheta)$ is consistent. Hence, we have the following corollary.

**Corollary 13.4** *The following two problems are $\Sigma_2^P$-complete:*

1. *Given a finite default theory $(D, W)$ and a formula $\varphi$, decide whether there is a* consistent *extension of $(D, W)$ which contains $\varphi$.*
2. *Given a finite default theory $(D, W)$ and a formula $\varphi$, decide whether there is a* consistent *extension of $(D, W)$ which* does not *contain $\varphi$.*

▲

The default theory we have constructed in the proof of Theorem 13.4 is normal. In addition, each default is prerequisite-free (this property has, in fact, been used in the proof). Hence, we obtain the following corollary.

**Corollary 13.5** *The following two problems are $\Sigma_2^P$-complete:*

1. *Given a formula $\varphi$ and a finite normal default theory $(D, W)$, with all defaults in $D$ prerequisite-free, decide whether there is an extension of $(D, W)$ which contains $\varphi$.*
2. *Given a formula $\varphi$ and a finite normal default theory $(D, W)$, with all defaults in $D$ prerequisite-free, decide whether there is an extension of $(D, W)$ which* does not *contain $\varphi$.* ▲

Let us notice that the problem to decide whether a given formula belongs to the intersection of all extensions is the complement of the problem considered in part (2) of Theorem 13.3. Similarly, the problem to decide whether a formula does not belong to the union of all extensions is the complement of the problem considered in part (1) of Theorem 13.3. Thus, Theorem 13.3 and Corollaries 13.4 and 13.5 imply the following three results.

**Corollary 13.6** *The following two problems are $\Pi_2^P$-complete:*

1. *Given a finite default theory $(D, W)$ and a formula $\varphi$, decide whether $\varphi$ belongs to all extensions of $(D, W)$.*
2. *Given a finite default theory $(D, W)$ and a formula $\varphi$, decide whether $\varphi$ is in no extension of $(D, W)$.* ▲

**Corollary 13.7** *The following two problems are $\Pi_2^P$-complete:*

1. *Given a finite default theory $(D, W)$ and a formula $\varphi$, decide whether $\varphi$ belongs to all* consistent *extensions of $(D, W)$.*
2. *Given a finite default theory $(D, W)$ and a formula $\varphi$, decide whether $\varphi$ is in no* consistent *extension of $(D, W)$.* ▲

**Corollary 13.8** *The following two problems are $\Pi_2^P$-complete:*

1. *Given a formula $\varphi$ and a finite normal default theory $(D, W)$, with all defaults prerequisite-free, decide whether $\varphi$ belongs to all extensions of $(D, W)$.*
2. *Given a formula $\varphi$ and a finite normal default theory $(D, W)$, with all defaults prerequisite-free, decide whether $\varphi$ is in no extension of $(D, W)$.* ▲

All our results (Theorems 13.1 and 13.3, Corollaries 13.2, 13.4, 13.5, 13.6, 13.7 and 13.8) remain valid if the notion of an extension is replaced by that of a weak extension. To see this observe that Theorem 3.80 and Corollary 3.81 imply that a default theory $(D, W)$ has a weak extension if and only if there is a set of defaults $Z \subseteq D$ such that $Z = GD(D, Cn(W \cup c(Z)))$. In fact, in such a case, the theory $S = Cn(W \cup c(Z))$ is a weak extension of $(D, W)$. This result shows that problems discussed in Theorems 13.1 and 13.3, with the notion of an extension replaced by the notion of a weak extension, are in $\Sigma_2^P$. In addition, the theories used in the proofs of these theorems are prerequisite-free. For such theories the concepts of an extension and a weak extension coincide. Hence, these three problems are $\Sigma_2^P$-complete and their complements are $\Pi_2^P$-complete. We gather these results in the theorem below.

**Theorem 13.9** *The following problems are $\Sigma_2^P$-complete:*

1. *Given a finite default theory $(D, W)$, decide whether $(D, W)$ has a weak extension.*
2. *Given a finite default theory $(D, W)$, decide whether $(D, W)$ has a consistent weak extension.*
3. *Given a finite default theory $(D, W)$ and a formula $\varphi$, decide whether there is a weak extension of $(D, W)$ which contains $\varphi$ (this problem remains $\Sigma_2^P$-complete under the restriction to normal default theories with all defaults prerequisite-free).*
4. *Given a finite default theory $(D, W)$ and a formula $\varphi$, decide whether there is a* consistent *weak extension of $(D, W)$ which contains $\varphi$.*
5. *Given a finite default theory $(D, W)$ and a formula $\varphi$, decide whether there is a weak extension of $(D, W)$ which* does not *contain $\varphi$ (this problem remains $\Sigma_2^P$-complete under the restriction to normal default theories with all defaults prerequisite-free).*
6. *Given a finite default theory $(D, W)$ and a formula $\varphi$, decide whether there is a* consistent *weak extension of $(D, W)$ which* does not *contain $\varphi$.*

*The following problems are $\Pi_2^P$-complete:*

1. *Given a finite default theory $(D, W)$ and a formula $\varphi$, decide whether $\varphi$ belongs to all weak extensions of $(D, W)$ (this problem remains $\Pi_2^P$-complete under the restriction to normal default theories with all defaults prerequisite-free).*
2. *Given a finite default theory $(D, W)$ and a formula $\varphi$, decide whether $\varphi$ is in no weak extension of $(D, W)$ (this problem remains $\Pi_2^P$-complete under the restriction to normal default theories with all defaults prerequisite-free).*
3. *Given a finite default theory $(D, W)$ and a formula $\varphi$, decide whether $\varphi$ belongs to all* consistent *weak extensions of $(D, W)$.*
4. *Given a finite default theory $(D, W)$ and a formula $\varphi$, decide whether $\varphi$ is in no* consistent *weak extension of $(D, W)$.* ▲

## 13.3 Complexity of reasoning with modal nonmonotonic logics

In this section we will briefly discuss complexity issues for modal nonmonotonic logics. We will consider the following three cases:

1. $\mathcal{S}$ is contained in the modal logic **S4f** and is characterized by a cluster-closed class of Kripke models (or structures).

2. $\mathcal{S} =$ **kd45**.

3. $\mathcal{S} =$ **Sw5**.

The upper estimates on the complexity of reasoning with these logics will be derived from Theorems 9.50, 10.15 (see also Remarks 9.33, 9.56 and 10.18) in the case of logics satisfying (1), from Theorem 10.21 for **kd45** and from Theorem 10.25 for **Sw5**. In all three cases our approach is similar and we will provide a proof for only one of these cases — the first one.

We start with a result on the complexity of the problem to decide the membership of a formula $\varphi$ in a stable theory $T_I(\Phi, \Psi)$.

**Theorem 13.10** *The following problem is in the class $\Delta_2^P$:*

*Given a formula $\varphi \in \mathcal{L}_L$, a finite theory $I \subseteq \mathcal{L}_L$ and a partition $(\Phi, \Psi)$ of $MA(I)$, introspection-consistent with $I$, decide whether $\varphi \in T_I(\Phi, \Psi)$.*

Proof: We will use Theorem 9.40. It provides a method to test whether an arbitrary formula $\xi$ belongs to $T_I(\Phi, \Psi)$, where $(\Phi, \Psi)$ is a partition of $MA(I)$, introspection-consistent with $I$. Namely, $\xi \in T_I(\Phi, \Psi)$ if and only if

$$I \cup \{\neg L\varphi : \varphi \in \Phi\} \cup \{L\psi : \psi \in \Psi\} \cup \Psi \cup \\ \{L\psi : \psi \in MA(\{\xi\}) \cap T_I(\Phi, \Psi)\} \cup \{\neg L\psi : \psi \in MA(\{\xi\}) \setminus T_I(\Phi, \Psi)\} \vdash \xi.$$

Consequently, the membership of $\xi$ in $T_I(\Phi, \Psi)$ can be decided by checking for every modal atom $L\alpha$ of $\xi$, in the order of increasing $L$-depth, whether $\alpha$ belongs to $T_I(\Phi, \Psi)$ or not. Each such decision requires one call to an oracle for the propositional provability problem. Since in each case the size of the theories and formulas involved is polynomial in the total size of $I$ and $\xi$, it follows that checking whether $\xi \in T_I(\Phi, \Psi)$ takes polynomially many calls to the oracle. ∎

**Corollary 13.11** *The following problem is in $\Delta_2^P$:*

*Given a formula $\varphi \in \mathcal{L}_L$ and a finite theory $S \subseteq \mathcal{L}$, decide whether $\varphi \in ST(S)$.*

Proof: First, observe that one call to an oracle for the propositional satisfiability problem decides whether $S$ is consistent. If $S$ is inconsistent the answer to our problem is YES. Otherwise, notice that $(\emptyset, \emptyset)$ is a partition of $MA(S)$ introspection-consistent with $S$. Since $ST(S) = T_S(\emptyset, \emptyset)$, the assertion follows from Theorem 13.10. ■

**Theorem 13.12** *Let $\mathcal{S}$ be a modal logic contained in* **S4f** *and characterized by a cluster-closed class of Kripke models (or structures). In addition assume that the problem:*

($\vdash_S$) *Given a finite theory $I$ and a formula $\varphi \in \mathcal{L}_L$, decide whether $I \vdash_S \varphi$*

*is in $\Pi_k^P$, for some $k \geq 1$. The following problems are in $\Sigma_{k+1}^P$:*

1. *Given a finite theory $I$, decide whether $I$ has an $\mathcal{S}$-expansion.*
2. *Given a finite theory $I$ and a formula $\varphi \in \mathcal{L}_L$, decide whether $\varphi$ belongs to at least one $\mathcal{S}$-expansion for $I$.*
3. *Given a finite theory $I$ and a formula $\varphi \in \mathcal{L}_L$, decide whether $\varphi$ does not belong to at least one $\mathcal{S}$-expansion for $I$.*

*The following problem is in $\Pi_{k+1}^P$:*

4. *Given a finite theory $I$ and a formula $\varphi \in \mathcal{L}_L$, decide whether $\varphi$ belongs to all $\mathcal{S}$-expansions for $I$.*
5. *Given a finite theory $I$ and a formula $\varphi \in \mathcal{L}_L$, decide whether $\varphi$ belongs to no $\mathcal{S}$-expansion for $I$.*

Proof: Recall (Theorem 9.50) that for logics $\mathcal{S}$ that are contained in logic **S4f** and are characterized by a cluster-closed class of Kripke models (or structures), $\mathcal{S}$-expansions of a theory $I$ are characterized as exactly the theories of the form $T_I(\Phi, \Psi)$, for partitions $(\Phi, \Psi)$ of $MA(I)$ which are $\mathcal{S}$-perfect for $I$.

Recall also that a partition $(\Phi, \Psi)$ of $MA(I)$ is $\mathcal{S}$-perfect for $I$ if:

1. $I \cup \{\neg L\varphi : \varphi \in \Phi\} \cup \{L\psi : \psi \in \Psi\} \cup \Psi$ is propositionally consistent.
2. For each $\vartheta \in \Phi$, $I \cup \{\neg L\varphi : \varphi \in \Phi\} \cup \{L\psi : \psi \in \Psi\} \cup \Psi \not\vdash \vartheta$.
3. For each $\vartheta \in \Psi$, $I \cup \{\neg L\varphi : \varphi \in \Phi\} \vdash_{\mathcal{S}} \vartheta$.

Observe that the size of $MA(I)$ is polynomial in the size of $I$. Hence, it takes polynomially many calls to an oracle for propositional provability to check that a given partition $(\Phi, \Psi)$ satisfies conditions (1) and (2). In addition, by the same argument, it takes polynomially many calls to an oracle for provability in $\mathcal{S}$ to determine whether $(\Phi, \Psi)$ satisfies condition (3).

Next notice that all theories and formulas occurring in conditions (1) - (3) can be constructed in time bounded by a polynomial in the size of $I$. Hence, the problem of existence of an $\mathcal{S}$-expansion for a finite theory $I$ can be solved in polynomial time by a nondeterministic Turing machine which first guesses (nondeterministically) a partition $(\Phi, \Psi)$ and then checks that this partition is $\mathcal{S}$-perfect for $I$ by means of polynomially many calls to oracles for propositional provability and $\mathcal{S}$-provability problems. Since the problem of deciding propositional provability is in $\Pi_1^P$ and the problem of deciding $\mathcal{S}$-provability is in $\Pi_k^P$, it follows that problem (1) is in $\Sigma_{k+1}^P$.

In a similar way we prove the same upper estimate on the complexity of problems (2) and (3). The only difference is that after guessing a partition $(\Phi, \Psi)$ of $MA(I)$ and verifying that it is $\mathcal{S}$-perfect for $I$ (and, in particular, introspection-consistent with $I$), our Turing machine has still to verify that $\varphi \in T_I(\Phi, \Psi)$ ($\varphi \notin T_I(\Phi, \Psi)$ in the case of problem (3)). These tasks can be accomplished by means of polynomially many calls to an oracle for the propositional provability problem which follows directly from Theorem 13.10.

Hence, problems (2) and (3) are in $\Sigma_{k+1}^P$. Since problems (4) and (5) are the complements of problems (3) and (2), respectively, the theorem follows. ■

Using semantic characterizations of modal logics **N**, **t** and **S4f** one can show that for each of these logics the corresponding provability problem is in $\Pi_1^P$ (see the last section for more information). Hence, we obtain the following corollary.

**Corollary 13.13** *Let $\mathcal{S}$ be* **N**, **t** *or* **S4f**. *The following problems are in $\Sigma_2^P$:*

1. *Given a finite theory $I$, decide whether $I$ has an $\mathcal{S}$-expansion.*
2. *Given a finite theory $I$ and a formula $\varphi \in \mathcal{L}_L$, decide whether $\varphi$ belongs to at least one $\mathcal{S}$-expansion for $I$.*
3. *Given a finite theory $I$ and a formula $\varphi \in \mathcal{L}_L$, decide whether $\varphi$ does not belong to at least one $\mathcal{S}$-expansion for $I$.*

*The following problem is in $\Pi_2^P$:*

4. *Given a finite theory $I$ and a formula $\varphi \in \mathcal{L}_L$, decide whether $\varphi$ belongs to all $\mathcal{S}$-expansions for $I$.*
5. *Given a finite theory $I$ and a formula $\varphi \in \mathcal{L}_L$, decide whether $\varphi$ belongs to no $\mathcal{S}$-expansion for $I$.*

▲

A similar reasoning can be applied to modal logics **kd45** and **Sw5**. The only difference is that instead of Theorem 9.50 we use Theorems 10.21 and 10.25.

**Theorem 13.14** *Let $\mathcal{S}$ be* **kd45** *or* **Sw5**. *The following problems are in $\Sigma_2^P$:*

1. *Given a finite theory $I$, decide whether $I$ has an $\mathcal{S}$-expansion.*
2. *Given a finite theory $I$ and a formula $\varphi \in \mathcal{L}_L$, decide whether $\varphi$ belongs to at least one $\mathcal{S}$-expansion for $I$.*
3. *Given a finite theory $I$ and a formula $\varphi \in \mathcal{L}_L$, decide whether $\varphi$ does not belong to at least one $\mathcal{S}$-expansion for $I$.*

*The following problem is in $\Pi_2^P$:*

4. *Given a finite theory $I$ and a formula $\varphi \in \mathcal{L}_L$, decide whether $\varphi$ belongs to all $\mathcal{S}$-expansions for $I$.*
5. *Given a finite theory $I$ and a formula $\varphi \in \mathcal{L}_L$, decide whether $\varphi$ belongs to no $\mathcal{S}$-expansion for $I$.* ▲

We will now prove hardness results for the problems dealt with in Corollary 13.13 and Theorem 13.14. In this way we will establish the following theorem.

**Theorem 13.15** *Let $\mathcal{S}$ be* **N**, **t**, **S4f**, **kd45** *or* **Sw5**. *The following problems are $\Sigma_2^P$-complete:*

1. *Given a finite theory $I$, decide whether $I$ has an $\mathcal{S}$-expansion.*
2. *Given a finite theory $I$ and a formula $\varphi \in \mathcal{L}_L$, decide whether $\varphi$ belongs to at least one $\mathcal{S}$-expansion for $I$.*

3. *Given a finite theory $I$ and a formula $\varphi \in \mathcal{L}_L$, decide whether $\varphi$ does not belong to at least one $\mathcal{S}$-expansion for $I$.*

*The following problem is $\Pi_2^P$-complete:*

4. *Given a finite theory $I$ and a formula $\varphi \in \mathcal{L}_L$, decide whether $\varphi$ belongs to all $\mathcal{S}$-expansions for $I$.*

5. *Given a finite theory $I$ and a formula $\varphi \in \mathcal{L}_L$, decide whether $\varphi$ belongs to no $\mathcal{S}$-expansion for $I$.*

Proof: Theorem 12.19 states that a consistent and closed under propositional consequence theory $S$ is a weak extension of a default theory $(D, W)$ if and only if $ST(S)$ is a **kd45**-expansion of $emb(D, W)$. Since the theory $emb(D, W)$ can be constructed in time polynomial in the size of $(D, W)$, one of the consequences of Theorem 12.19 is that the problem of existence of a weak extension of a finite default theory is polynomially reducible to the problem of existence of a **kd45**-expansion of a finite modal theory. Hence, the complexity results on default reasoning with weak extensions given in the previous section (Theorem 13.9) imply the assertion in the case of logic $\mathcal{S} =$ **kd45**.

Recall now that, by Theorem 10.31, a theory $T$ is a **kd45**-expansion of $I$ if and only if $T$ is an **Sw5**-expansion of $I^K$. Since the size of $I^K$ is polynomial in the size of $I$, the assertion (for each of the problems) follows for $\mathcal{S} =$ **Sw5**.

A relationship between default reasoning and nonmonotonic modal logics **N**, **t** and **S4f** (Theorem 12.1) and the complexity results for default logic with extensions from the previous section (Corollaries 13.4 and 13.7) imply the assertion for each of the logics **N**, **t** and **S4f**. ■

The provability problem for modal logic **S4** is known to be PSPACE-complete. Hence (unless PSPACE collapses to the first level of the polynomial hierarchy), Theorem 13.12 does not apply to **S4**. However, let us recall that, by Theorem 11.10, for *finite* theories $I$ nonmonotonic logics **S4** and **S4f** coincide. Hence, we obtain the following result.

**Corollary 13.16** *The assertion of Theorem 13.15 holds for modal logic $\mathcal{S} =$* **S4**. ▲

This last result is particularly interesting. Assuming that PSPACE does not collapse to the second level of the polynomial hierarchy, reasoning with nonmonotonic logic **S4** is easier than provability in **S4**. No other monotonic logics whose nonmonotonic counterparts have better computational properties are known so far.

## 13.4 Complexity issues for finite logic programs

In this section we will study the complexity of the problem of existence of stable models of finite ground logic programs. We start by analyzing the case of finite ground Horn programs.

**Theorem 13.17** *Let $P$ be a finite ground Horn program. There is a linear-time algorithm for computing the least model of $P$.*

Proof: We assume that each clause of $P$ is represented by a linked list of ground atoms. The first element in the list is the head of the clause, all other elements are precisely the elements of the body of the clause (no repetitions are allowed). From this input structure one can construct (in linear-time, we leave the verification as an easy exercise):

1. For each ground atom $p$ occurring in $P$, the list $cl(p)$ of clauses containing $p$ in the body.
2. For each clause $C$ of $P$, the number $l(C)$ of elements in the body of $C$.

Next, a version of the topological sort is performed on these data structures. The effect of the algorithm, the least Herbrand model, is stored in a set $M$. The set $M$ is initialized to $\emptyset$ at the beginning of the algorithm. In addition, an auxiliary queue, say $Q$, is used. Initially, the heads of clauses with empty bodies are placed on $Q$ (they are easy to compute since the numbers $l(C)$ are available). Next, we repeat the following operations while $Q$ is nonempty (main loop): We remove the front element, say $p$, from the queue. Then we mark $p$ and include $p$ in $M$. Next, for each clause $C$ on $cl(p)$ we decrease $l(C)$ by 1. If $l(C)$ becomes 0 and if its head is unmarked, we add the head of $C$ to $Q$.

It should be clear that the running time of this algorithm is linear in the size of our input representation of $P$. To see the correctness of the algorithm, it suffices to prove that the following statement holds true after every iteration of the main loop of the algorithm:

> The union of $M$ and the atoms on $Q$ is the least Herbrand model of the program consisting of those clauses $C$ for which $l(C) = 0$.

We omit an easy inductive proof of this fact. ∎

As a corollary, we obtain the following useful result.

**Theorem 13.18**

1. *If $P$ is a finite ground Horn program then we can test whether $M \subseteq At$ is a stable model of $P$ in time linear in the size of $P$.*
2. *The problem to decide whether a finite ground logic program $P$ possesses a stable model is in the class NP.*
3. *The problem to decide whether a ground atom $p$ belongs to some stable model of a finite ground logic program $P$ is in the class NP.*
4. *The problem to decide whether a ground atom $p$ belongs to all stable models of a finite ground logic program $P$ is in the class co-NP.* ▲

We will prove *NP*-completeness of the problem of existence of a stable model of a ground logic program by reducing to it a well-known *NP*-complete problem of graph theory, namely the problem of existence of a kernel in a directed graph.

**Definition 13.19** Given a directed graph $G = \langle V, E\rangle$ ($V$ stands for the set of vertices, $E$ for the set of edges), a set $K \subseteq V$ is called a *kernel* if:

1. The complement of $K$, $V \setminus K$, is a *vertex cover* (that is, for every edge $(w, v) \in E$, $\{w, v\} \cap (V \setminus K) \neq \emptyset$).
2. For every node $w \in V \setminus K$, there exists a node $v \in K$ such that $(w, v) \in E$. ▲

It is well-known (see [GJ79]) that the problem of existence of a kernel in a finite directed graph is *NP*-complete. We will now encode directed graphs as logic programs.

**Definition 13.20** Let $G = \langle V, E\rangle$ be a directed graph. We will assign to $G$ a logic program in which the vertices of $G$ are used as atoms. Namely, by $P_G$ we denote the program consisting of all clauses of the form $v \leftarrow \mathbf{not}(w)$, where $(v, w) \in E$. ▲

Clearly, the size of the program $P_G$ is linear in the size of $G$. Kernels of a graph $G$ and stable models of the program $P_G$ are closely related.

**Lemma 13.21** *Let $G = \langle V, E\rangle$ be a directed graph. Then $K \subseteq V$ is a kernel of $G$ if and only if the set $V \setminus K$ is a stable model of $P_G$.*

Proof: Since $P_G$ has a particularly simple form, it is easy to describe its reducts. Given a set $M \subseteq V$, $P_G^M$ consists of clauses with empty bodies. Consequently, the least Herbrand model of $P_G^M$ is the set of the heads of the clauses in $P_G^M$. Therefore, $M$ is a stable model of $P_G$ if and only if $M$ is precisely the set of those atoms $v$ that satisfy the following condition:

$$\text{there is a clause } v \leftarrow \mathbf{not}(w) \text{ in } P_G \text{ such that } w \notin M,$$

or, equivalently,

$$\text{there is an edge } (v, w) \in E \text{ such that } w \notin M. \tag{13.1}$$

Assume first that $V \setminus K$ is a stable model of $P_G$. Consider an edge $(w, v) \in E$. If $v \in V \setminus K$ then $\{w, v\} \cap (V \setminus K) \neq \emptyset$. If $v \notin V \setminus K$ then, by (13.1), $w \in V \setminus K$. Hence, again $\{w, v\} \cap (V \setminus K) \neq \emptyset$. Consequently, $V \setminus K$ is a vertex cover for $G$. The second requirement of the definition of a kernel follows directly from (13.1). Hence, $K$ is a kernel of $G$.

Conversely, assume that $K$ is a kernel of $G$. If $w \in V \setminus K$, then there is a vertex $v \in K$ such that $(w, v) \in E$. Hence, (13.1) holds for $w$. On the other hand, if $w \notin V \setminus K$ and $(w, v) \in E$, then since $V \setminus K$ is a vertex cover, $v \in V \setminus K$. Hence, (13.1) fails for such $w$. Summarizing, $V \setminus K$ consists of precisely those vertices for which (13.1) holds. In other words, $V \setminus K$ is a stable model for $P_G$. ■

Thus we reduce the problem of existence of a kernel (a kernel exists if and only if the complement of a kernel exists) to the problem of existence of a stable model of a suitably chosen program. Therefore we get the following result.

**Theorem 13.22** *The problem of existence of a stable model of a finite general propositional program is NP-complete.* ▲

Similarly, we get a precise estimate of the complexity of the problem of deciding whether a given atom belongs to some (all) stable models.

**Theorem 13.23** *The problem to decide whether a given atom p belongs to some stable model of a finite ground logic program P is NP-complete. The problem if a given atom p belongs to all stable models of a finite ground logic program P is co-NP-complete.* ▲

## 13.5 Issues in complexity of models of infinite logic programs

In this section we apply classical methods of the recursion theory to establish the complexity of families of models, minimal models and stable models for infinite ground logic programs. Because the recursion theory requires an involved technical machinery that goes beyond the scope of the book, we will only touch on these issues. We will give some basic definitions here, enough to formulate results and give a flavor of the proof methods that are used. Our goal is to convince the reader that this particular area abounds in challenging and still vastly unexplored research problems.

Our point of departure is the observation that all objects discussed in this book can be coded by natural numbers. Let us assume that the set of all ground atoms in the language is enumerated as $\{p_0, p_1, \ldots\}$. We will use now a modification of the process called "Gödelization", adapted to the case of logic program clauses. First, we assign the natural number 1 to the symbol $\leftarrow$ and the natural number 2 to **not**. Next, each ground atom $p_i$ is assigned $i+3$. Under this assignment, each clause becomes a sequence of natural numbers. Finally, to each such sequence we assign a natural number. This number is regarded as an encoding of the clause.

**Example 13.24** The clause $p_0 \leftarrow p_1, \mathbf{not} p_3$ is represented by the sequence $\langle 3, 1, 4, 2, 6 \rangle$ which, in turn, is assigned the number:

$$2^3 \cdot 3^1 \cdot 5^4 \cdot 7^2 \cdot 11^6.$$

▲

Once we have encoded program clauses with natural numbers, programs can be encoded by sets of natural numbers. We define a program $P$ to be *recursive* if the set of natural numbers that codes $P$ is recursive.

We will now study issues related to existence and the complexity of objects associated with a program $P$ such as models, minimal models and stable models of $P$. Clearly, a (Herbrand) model of a program can be regarded as a collection of natural numbers that code its elements. Hence, the set of models becomes a family of sets of natural numbers, that is, a subset of $\mathcal{P}(\omega)$. In the recursion theory, there is Kleene-Mostowski hierarchy of subsets of space $\mathcal{P}(\omega)$. This hierarchy is discussed at length in Rogers [Rog87] and we refer the reader to that book for details. It should be obvious that a program $P$ (or, to be precise, its code) must enter as a parameter into the description of classes of its models. Some important classes of this hierarchy, which we will use in our discussion, are the classes $\Pi_1^{0,P}$, $\Pi_2^{0,P}$, $\Pi_1^{1,P}$ and $\Sigma_1^{1,P}$.

The proofs we will give in this section are meant to serve as a short introduction to the methods of the recursion theory useful when studying families of stable models of logic programs. They rely on the concept of a *tree*. A *tree* is a set $T$ of finite sequences of natural numbers such that $T$ is closed under the initial segments. That is, if $s \in T$ and $t$ is an initial segment of the sequence $s$ then $t \in T$. The fact that $t$ is an initial segment of $s$ is denoted by $t \sqsubseteq s$. We denote by $lh(s)$ the length of a finite sequence $s$. When we discuss subsets of $\omega$, our trees will be composed of binary sequences. Later on we will also consider trees composed of sequences of arbitrary natural numbers. The difference is that in the case of binary sequences, each sequence possesses at most two immediate successors in the tree.

Given a tree $T$, a *branch* through $T$ is a function $f$ with the domain $\omega$ and with values in $\omega$ such that all its proper initial segments $f|_n$ belong to $T$. The set of all branches through $T$ is denoted by $[T]$. Since a tree consists of finite sequences of natural numbers which, as we have seen, can be coded as natural numbers, a tree can be regarded as a set of natural numbers. Consequently, we can talk about *recursive* trees and trees *recursive in* $P$.

Let us introduce a topology, called *Cantor topology*, on $\mathcal{P}(\omega)$. The basic open neighborhoods in this topology are determined by pairs of disjoint finite sets. Given a pair $\langle A, B\rangle$ such that $A \cap B = \emptyset$, the corresponding basic open neighborhood $U_{A,B}$ is defined by $U_{A,B} = \{X : A \subseteq X, B \cap X = \emptyset\}$. Next, recall, that the set $\mathcal{P}(\omega)$ can be identified with the set of all functions from $\omega$ to $\{0,1\}$ (simply identify a subset $A \subseteq \omega$ with its characteristic function). This operation allows us to reproduce Cantor topology for the space of all functions from $\omega$ to $\{0,1\}$. In this setting, the closed sets of Cantor topology are precisely the sets of the form $[T]$, where $T$ is a binary tree.

**Theorem 13.25**

1. *The set of all models of a program $P$ is a $\Pi^{0,P}_1$ subset of $\mathcal{P}(\omega)$. In particular, if $P$ is a recursive program then the set of all models of $P$ is $\Pi^0_1$.*
2. *The set of all minimal models of a program $P$ is a $\Pi^{0,P}_2$ subset of $\mathcal{P}(\omega)$.*
3. *The set of all stable models of a program $P$ is a $\Pi^{0,P}_2$ subset of $\mathcal{P}(\omega)$.*

Proof: (1) We apply the Kuratowski-Tarski test (see [Rog87]). We need to describe the set *Mod*($P$) of models of $P$ by a formula $\Phi$ of the form $\forall x \psi$ where $\psi$ has only bounded quantifiers. It should be clear that the following formula (which can be subsequently formalized to a $\Pi^{0,P}_1$ form) will do:

"For every clause $C$ of $P$, $C$ is satisfied in $M$".

The fact that a clause $C$ is satisfied requires checking that: Whenever $M$ satisfies the body of $C$ then the head of $C$ belongs to $M$ as well. This is, of course, a bounded formula. Since we have a universal quantifier ranging over clauses in $P$ (but remember that clauses are encoded by natural numbers) the resulting formula is $\Pi_1^{0,P}$. Hence $Mod(P)$ is a $\Pi_1^{0,P}$ set of sets of natural numbers.

(2) By (1) the set of all models of $P$ is a $\Pi_1^{0,P}$ subset of $\mathcal{P}(\omega)$. We do not distinguish in our considerations between $\mathcal{P}(\omega)$ and $2^\omega$. Hence, there is a tree $T$ (that is, a set of finite sequences closed under initial segment) such that the set of all models of $P$ is precisely the set of all the branches $[T]$ through the tree $T$. This tree $T$ is recursive in $P$. Define now the *bottom* of a closed subset $\mathcal{A}$, $B(\mathcal{A})$, as the set of all inclusion-minimal elements of $\mathcal{A}$. When $\mathcal{A}$ is the set of all models of $P$, $Mod(P)$, then $B(\mathcal{A})$ is precisely the set of all minimal models of $P$. The "naive" evaluation of the definition of $B(\mathcal{A})$ yields a $\Pi_1^{1,P}$ definition. It is so because minimality of a model of $P$ is "naively" described by a universal quantifier ranging over subsets of the Herbrand base ("there is no smaller subset of the Herbrand base that is a model of $P$"). This is, of course, a very complex definition, and in order to get a simpler definition we will need an additional, roundabout construction.

We consider a partial ordering $\preceq$ on finite binary sequences defined as follows:

$$t \preceq s \text{ if and only if } lh(t) = lh(s) \text{ and for all } n < lh(s),\ t(n) \leq s(n).$$

We write $s \prec t$ if $s \preceq t$ and $s \neq t$. Thus $s \preceq t$ says that the "partial set" coded by $s$ is included in the "partial set" encoded by $t$. We write $s \parallel t$, where $s, t$ are finite or infinite binary sequences, if there exist $n_1, n_2$ such that $s(n_1) = 1$ and $t(n_1) = 0$ and $s(n_2) = 0$ and $t(n_2) = 1$.

Given a tree $T$, define a set $\mathcal{J}(T)$ of functions as follows:

$$\begin{aligned}\mathcal{J}(T) = \{f \in 2^\omega \colon f \in [T] \wedge \\ \forall k \exists n (n > k \wedge (\forall s \in T)((lh(s) = n \wedge s|_k \prec f|_k) \supset s \parallel f))\}\end{aligned}$$

This formal definition of $\mathcal{J}(T)$ will facilitate reading off the class of the family of sets $\mathcal{J}(T)$.

We will now prove the following auxiliary result:

**Lemma 13.26** *Let $\mathcal{X}$ be closed in Cantor topology. Let $\mathcal{X} = [T]$. Then $B(\mathcal{X}) = \mathcal{J}(T)$.*

Proof: If $X \in B(\mathcal{X})$, $k \in X$, then let $T_{X,k}$ be the tree consisting of initial segments of sequences $s$ satisfying the following condition:

$$s \in T \text{ and } s|_k = \chi(X)|_k \text{ and } s(k) = 0 \text{ and } s \preceq \chi(X)|_{lh(s)},$$

where $\chi(X)$ stands for the characteristic function of $X$. For all $k$, $T_{X,k}$ is finite. Hence $T_{X,k}$ is of finite height. Let $m_k$ be the height of $T_{X,k}$. Then $n = max\{m_i : i \leq k\}$ is witnessing to the fact that $X$ belongs to $\mathcal{J}(T)$.

Conversely, if $X \in \mathcal{J}(T)$, $Y \in [T]$, $Y \subseteq X$, and $k = min(X \setminus Y)$, then $\chi(Y)|_{k+1} \prec \chi(X)|_{k+1}$. Since $X \in \mathcal{J}(T)$, there is $n$ such that $\chi(Y)|_n \parallel \chi(X)|_n$, contradicting $Y \subseteq X$. ■

Now we are in a position to complete our proof of (2). As proved in part (1), the family $Mod(P)$ of models of $P$ is closed in Cantor topology, in fact it is $\Pi_1^{0,P}$. Thus there is a tree $T$, recursive in $P$, such that $Mod(P) = [T]$. We noticed that the family $\mathcal{M}$ of minimal models of $P$ is equal to $B(Mod(P))$. Using Lemma 13.26, we just need to evaluate the form of $\mathcal{J}(T)$. So let us look at the formula defining the set $\mathcal{J}(T)$. The formula $f \in [T]$ is a $\Pi_1^0$ formula. Then we have the universal quantifier followed by the existential quantifier. Finally, we have the universal quantifier $\forall s \in T$. But this quantifier ranges over a finite set (since it is restricted to sequences $s$ of length $n$) and so it can be eliminated. Therefore the formula defining $\mathcal{J}(T)$ is $\Pi_2^{0,P}$.

(3) We use the characterization of stable models by means of proof schemes. We introduced this technique in Section 6.8.

Recall that a subset $M$ of $At$ is a stable model of $P$ if and only if for every $a \in M$ there exists an $M$-proof for $a$ and for every $M$-proof $s$ the conclusion of that proof belongs to $M$. The first of these formulas is easily seen to be $\Pi_2^{0,P}$, the second one is $\Pi_1^{0,P}$. The conjunction of these two formulas is $\Pi_2^{0,P}$, and so the collection of all stable models of $P$ is a $\Pi_2^{0,P}$ subset of $\mathcal{P}(\omega)$. ■

In order to describe more recursion theory results about logic programs and classes of models we need more terminology. We will not give the proofs of the subsequent theorems, but we hope that the user will realize that the methods of the classical recursion theory can be used to analyze the issues related to the recursion-theoretic properties of models of programs.

First of all, we need to consider not only binary functions but functions taking arbitrary natural values. The set of all such functions is called the *Baire* space. As before, in this space we can classify subsets. In particular, we can talk about $\Pi_1^0$ subsets. Again, such subsets are the sets of branches of a recursive tree (except that now the tree is no more binary).

Let $\{e\}^B$ denotes the function computed by the $e$th oracle Turing machine with oracle $B$. Also, write $\{e\}^B = A$ for a set $A$ if $\{e\}^B$ is the characteristic function of $A$. For a function $f: \omega \rightarrow \omega$, define $gr(f) = \{\langle x, f(x)\rangle : x \in \omega\}$, where $\langle x, f(x)\rangle$ is the natural number coding the pair $(x, f(x))$.

We say that there is an effective, one-to-one degree-preserving correspondence between the set of stable models, we will denote it by *Stab*($P$), of a recursive program $P$ and the set of infinite paths $[T]$ through a recursive tree $T$ if there are indices $e_1$ and $e_2$ of oracle Turing machines such that:

1. For all $f$ belonging to $[T]$, $\{e_1\}^{gr(f)} \in Stab(P)$, that is, $\{e_1\}^{gr(f)}$ is a stable model for $P$.

2. For all $M$ belonging to $Stab(P)$, $\{e_2\}^M \in [T]$.

3. For all $f$ belonging to $[T]$, and for all $M$ in $Stab(P)$, $(\{e_1\}^{gr(f)} = M$ if and only if $\{e_2\}^M = f)$.

Hence we require that $e_1$ is an algorithm that, uniformly, produces stable models of $P$ out of branches of $T$. Then, $e_2$ is an algorithm for the converse mapping, and the condition (3) requires that our mapping be *one-to-one* and *onto*.

We say that a mapping is *degree-preserving*, if the Turing-degrees of the argument and the value are the same.

Notice that if $P$ is a finite predicate program then the Herbrand base of $P$ can be encoded either by a finite set (if there are no function symbols) or by $\omega$ and then *ground*($P$) is a recursive propositional program.

**Theorem 13.27**

1. *For every finite predicate program $P$ there exists a recursive tree $T$ and an effective, one-to-one, degree-preserving correspondence of Stab($P$) and $[T]$.*

2. *Conversely, for every recursive tree $T$ there is a finite predicate program $P$ and an effective, one-to-one, degree-preserving correspondence of Stab($P$) and $[T]$.* ▲

Out of many results which can be obtained as corollaries to Theorem 13.27 (after all, recursion theorists studied trees and the complexity of branches in trees for many decades), let us quote one.

**Theorem 13.28** *The set of indices of recursive programs possessing stable models is a $\Sigma^1_1$-complete subset of natural numbers.* ▲

## 13.6 Comments and remarks

Niemelä [Nie92] proved that the existence problem for stable expansions is in $\Sigma_2^P$. The completeness results, for autoepistemic reasoning and default logic (including the case of normal default logic) presented in this chapter, are due to Gottlob [Got92] (see also Niemelä [Nie92]). The complexity results for nonmonotonic modal logics **N** and **S4f** were also first given by Gottlob in [Got92]. Using the technique introduced by Ladner [Lad77] to show that **S5**-consistency problem is in *NP*, Schwarz and Truszczyński [ST93] proved that the problems of testing **N**- and **t**-consistency are in *NP*. The complexity of reasoning with nonmonotonic logic **S4** has been established by Schwarz and Truszczyński [ST92b].

The complexity of computation of the least model of a propositional Horn program is due to Dowling and Gallier [DG84]. The results on the complexity of stable semantics of finite logic programs are due to Bidoit and Froidevaux [BF91b, BF91a], and independently to the authors [MT91].

The results on the complexity of infinite logic programs quoted in Section 13.5 are due to Marek, Nerode and Remmel [MNR92c]. The estimate of the complexity of minimal models of logic programs is due to Just [Jus90]. A close relationship of the arithmetical hierarchy and perfect models of infinite logic programs describable as finite stratified predicate logic programs has been established by Apt and Blair [AB90].

# References


[Apt90] K.R. Apt. Logic programming. In J. van Leeuven, editor, *Handbook of theoretical computer science*, pages 493–574. Cambridge, MA: MIT Press, 1990

[AB90] K.R. Apt and H.A. Blair. Arithmetical classification of perfect models of stratified programs. *Fundamenta Informaticae*, 12:1–17, 1990

[ABW88] K.R. Apt, H.A. Blair, and A. Walker. Towards a theory of declarative knowledge. In J. Minker, editor, *Foundations of deductive databases and logic programming*, pages 89–142. Los Altos, CA: Morgan Kaufmann, 1988

[AvE82] K.R. Apt and M.H. van Emden. Contributions too the theory of logic programming. *Journal of the ACM*, 29:841–862, 1982

[BS91] C. Baral and V.S. Subrahmanian. Dualities between alternative semantics for logic programming and nonmonotonic reasoning. In A. Nerode, W. Marek, and V.S. Subrahmanian, editors, *Logic programming and non-monotonic reasoning*, pages 69–86. Cambridge, MA: MIT Press, 1991

[BED91] R. Ben-Eliyahu and R. Dechter. Default logic, propositional logic and constraints. In *Proceedings of AAAI-91*. Cambridge, MA: MIT Press, 1991

[Bes89] P. Besnard. *An introduction to default logic*. Berlin: Springer-Verlag, 1989

[BF91a] N. Bidoit and C. Froidevaux. General logical databases and programs: default logic semantics and stratification. *Information and Computation*, 91:15–54, 1991

[BF91b] N. Bidoit and C. Froidevaux. Negation by default and unstratifiable logic programs. *Theoretical Computer Science*, 78:85–112, 1991

[BMS91] H.A. Blair, W. Marek, and J. Schlipf. Expressiveness of locally stratified programs. Technical Report SU-CIS-91-28, University of Syracuse, 1991

[Bre91a] G. Brewka. Cumulative default logic: in defense of nonmonotonic inference rules. *Artificial Intelligence*, 50:183–205, 1991

[Bre91b] G. Brewka. *Nonmonotonic reasoning: logical foundations of commonsense*. Cambridge, UK: Cambridge University Press, 1991

[Che80] B.F. Chellas. *Modal logic, an introduction*. Cambridge, UK: Cambridge University Press, 1980

[Cla78] K.L. Clark. Negation as failure. In H. Gallaire and J. Minker, editors, *Logic and data bases*, pages 293–322. New York, NY: Plenum Press, 1978

[Ded88] R. Dedekind. *Was sind und was sollen die Zahlen*. Braunschweig: Friedrich Vieweg, 1888

[Ded48] R. Dedekind. *Essays on the theory of numbers, I. Continuity and irrational numbers, II. The nature and meaning of numbers*. La Salle, IL: The Open Publishing, 1948

[DG84] W.F. Dowling and J.H. Gallier. Linear-time algorithms for testing the satisfiability of propositional Horn formulae. *Journal of Logic Programming*, 3:267–284, 1984

[Doy79] J. Doyle. A truth maintenance system. *Artificial Intelligence*, 12:231–272, 1979

[Elk90] C. Elkan. A rational reconstruction of nonmonotonic truth maintenance systems. *Artificial Intelligence*, 43:219–234, 1990

[Eth88] D. W. Etherington. *Reasoning with incomplete information*. London: Pitman, 1988

[ER83] D. W. Etherington and R. Reiter. On inheritance hierarchies with exceptions. In *Proceedings of AAAI-83*, pages 104–108. Los Altos, CA: W. Kaufmann, 1983

[Fit83] M. C. Fitting. *Proof methods for modal and intuitionistic logics*. Dordrecht: D. Reidel, 1983

[Fit90] M. C. Fitting. *First order logic and automated theorem proving*. Berlin: Springer-Verlag, 1990

[FMT92] M. C. Fitting, W. Marek, and M. Truszczyński. The pure logic of necessitation. *Journal of Logic and Computation*, 2:349–373, 1992

[GJ79] M.R. Garey and D.S. Johnson. *Computers and intractability; a guide to the theory of NP-completeness*. New York, NY: W.H. Freeman, 1979

[Gel87] M. Gelfond. On stratified autoepistemic theories. In *Proceedings of AAAI-87*, pages 207–211. Los Altos, CA: Morgan Kaufmann, 1987

[GL88] M. Gelfond and V. Lifschitz. The stable semantics for logic programs. In R. Kowalski and K. Bowen, editors, *Proceedings of the 5th international symposium on logic programming*, pages 1070–1080. Cambridge, MA: MIT Press, 1988

[GL90] M. Gelfond and V. Lifschitz. Logic programs with classical negation. In D. Warren and P. Szeredi, editors, *Proceedings of the 7th international conference on logic programming*, pages 579–597. Cambridge, MA: MIT Press, 1990

[GL91] M. Gelfond and V. Lifschitz. Classical negation in logic programs and disjunctive databases. *New Generation Computing*, 9:365–385, 1991

[GLPT91] M. Gelfond, V. Lifschitz, H. Przymusinska, and M. Truszczyński. Disjunctive defaults. In J. Allen, R. Fikes and E. Sandewall, editors, *Proceedings of the second international conference on principles of knowledge representation and reasoning, KR '91*, San Mateo, CA: Morgan Kaufmann, 1991

[Got92] G. Gottlob. Complexity results for nonmonotonic logics. *Journal of Logic and Computation*, 2:397–425, 1992

[GC90] R. Guerreiro and M. Casanova. An alternative semantics for default logic. Preprint. The Third International Workshop on Nonmonotonic Reasoning, South Lake Tahoe, 1990

[HM85] J.Y. Halpern and Y. Moses. Towards a theory of knowledge and ignorance: preliminary report. In K. Apt, editor, *Logics and models of concurrent systems*, pages 459–476. NATO ASI Series F, Vol 13. Berlin: Springer-Verlag, 1985

[HM86] S. Hanks and D. McDermott. Default reasoning, nonmonotonic logics and frame problem. In *Proceedings of AAAI-86*, pages 328–333. Los Altos, CA: Morgan Kaufmann, 1986

[HC84] G.E. Hughes and M.J. Cresswell. *A companion to modal logic*. London: Methuen, 1984

[Imi87] T. Imielinski. Results on translating defaults to circumscription. *Artificial Intelligence*, 32:131–146, 1987

[Jec78] T. Jech. *Set theory*. New York, NY: Academic Press, 1978

[Jus90] W. Just. Personal communication. 1990

[Kam91] M. Kaminski. Embedding a default system into nonmonotonic logic. *Fundamenta Informaticae*, 14:345–354, 1991

[KS89] H.A. Kautz and B. Selman. Hard problems for simple default logics. In R.J. Brachman, H.J. Levesque and R. Reiter, editors, *Proceedings of the 1st international conference on principles of knowledge representation and reasoning, KR '89*, pages 189–197. San Mateo, CA: Morgan Kaufmann, 1989

[Kon88] K. Konolige. On the relation between default and autoepistemic logic. *Artificial Intelligence*, 35:343–382, 1988

[Kon89] K. Konolige. Errata to: On the relation between default and autoepistemic logic. *Artificial Intelligence*, 41:115, 1989

[Kri63] S. Kripke. Semantical analysis of modal logic I: normal modal propositional calculi. *Zeitschrift für mathematische Logik and Grundlagen der Mathematik*, pages 67–96, 1963

[Kri65] S. Kripke. Semantical analysis of modal logic II: non-normal modal propositional calculi. In J.W. Addison, L. Henkin, and A. Tarski, editors, *The theory of models*, pages 206–220. Amsterdam: North-Holland, 1965

[KM76] K. Kuratowski and A. Mostowski. *Set theory*. Warszawa: Polish Scientific Publishers, 1976

[Lad77] R. Ladner. The computational complexity of provability in systems of modal propositional logic. *SIAM J. Comput.*, 6:467–480, 1977

[Lev90] H. J. Levesque. All I know: a study in autoepistemic logic. *Artificial Intelligence*, 42:263–309, 1990

[Lif88] V. Lifschitz. On the declarative semantics of logic programs with negation. In J. Minker, editor, *Foundations of deductive databases and logic programming*, pages 177–192. Los Altos, CA: Morgan Kaufmann, 1988

[Lif90] V. Lifschitz. On open defaults. In J. Lloyd, editor, *Computational logic. Symposium proceedings*, pages 80–95. ESPRIT Basic Research Series. Berlin: Springer-Verlag, 1990

[Llo84] J. Lloyd. *Foundations of logic programming*. Berlin: Springer-Verlag, 1984, 2nd, extended edition, 1987

[Łuk84] W. Łukaszewicz. Considerations on default logic. In R. Reiter, editor, *Proceedings of the international workshop on non-monotonic reasoning*, pages 165–193. 1984

[Mar89] W. Marek. Stable theories in autoepistemic logic. *Fundamenta Informaticae*, 12:243–254, 1989

[MNR90] W. Marek, A. Nerode, and J.B. Remmel. Nonmonotonic rule systems I. *Annals of Mathematics and Artificial Intelligence*, 1:241–273, 1990

[MNR92a] W. Marek, A. Nerode, and J. B. Remmel. The stable models of predicate logic programs. In K. Apt, editor, *Proceedings of the international joint conference and symposium on logic programming*. Boston, MA: MIT Press, 1992

[MNR92b] W. Marek, A. Nerode, and J.B. Remmel. How complicated is the set of stable models of a recursive logic program? *Annals of Pure and Applied Logic*, 33:229–263, 1992

[MNR92c] W. Marek, A. Nerode, and J.B. Remmel. Nonmonotonic rule systems II. *Annals of Mathematics and Artificial Intelligence*, 5:229–263, 1992

[MST91] W. Marek, G.F. Shvarts, and M. Truszczyński. Modal nonmonotonic logics: ranges, characterization, computation. In J. Allen, R. Fikes and E. Sandewall, editors, *Proceedings of the 2nd international conference on principles of knowledge representation and reasoning, KR '91*, pages 395–404. San Mateo, CA: Morgan Kaufmann, 1991. An extended version of this article will appear in the Journal of the ACM

[MS89] W. Marek and V.S. Subrahmanian. The relationship between logic program semantics and non-monotonic reasoning. In G. Levi and M. Martelli, editors, *Proceedings of the 6th international conference on logic programming*, pages 600–617. Cambridge, MA: MIT Press, 1989

[MT89a] W. Marek and M. Truszczyński. Negation as failure to prove and fixed points. *Fundamenta Informaticae*, 12:255–268, 1989

[MT89b] W. Marek and M. Truszczyński. Relating autoepistemic and default logics. In R.J. Brachman, H.J. Levesque and R. Reiter, editors, *Proceedings of the 1st international conference on principles of knowledge representation and reasoning, KR '89*, pages 276–288. San Mateo, CA: Morgan Kaufmann, 1989

[MT89c] W. Marek and M. Truszczyński. Stable semantics for logic programs and default theories. In E.Lusk and R. Overbeek, editors, *Proceedings of the North American conference on logic programming*, pages 243–256. Cambridge, MA: MIT Press, 1989

[MT90] W. Marek and M. Truszczyński. Modal logic for default reasoning. *Annals of Mathematics and Artificial Intelligence*, 1:275–302, 1990

[MT91] W. Marek and M. Truszczyński. Autoepistemic logic. *Journal of the ACM*, 38:588–619, 1991

[McC80] J. McCarthy. Circumscription — a form of non-monotonic reasoning. *Artificial Intelligence*, 13:27–39, 1980

[McC90] J. McCarthy. *Formalization of common sense, papers by John McCarthy edited by V. Lifschitz.* Norwood, NJ: Ablex, 1990

[McD82] D. McDermott. Nonmonotonic logic II: nonmonotonic modal theories. *Journal of the ACM*, 29:33–57, 1982

[MD80] D. McDermott and J. Doyle. Nonmonotonic logic I. *Artificial Intelligence*, 13:41–72, 1980

[Men64] E. Mendelson. *Introduction to mathematical logic.* New York, NY: Van Nostrand Reinhold, 1964

[MS72] A.R. Meyer and L.J. Stockmeyer. The equivalence problem for regular expressions with squaring requires exponential time. In *Proceedings of the 13th annual symposium on switching and automata theory*, pages 125–129. New York, NY: IEEE Computer Society, 1972

[Moo84] R.C. Moore. Possible-world semantics for autoepistemic logic. In R. Reiter, editor, *Proceedings of the workshop on non-monotonic reasoning*, pages 344–354. 1984. Reprinted in: M. Ginsberg, editor, *Readings on nonmonotonic reasoning*, pages 137–142. Los Altos, CA: Morgan Kaufmann, 1990

[Moo85] R.C. Moore. Semantical considerations on non-monotonic logic. *Artificial Intelligence*, 25:75–94, 1985

[MT92] A. Mikitiuk and M. Truszczyński. Problems on graphs as default theories. In preparation, 1992

[Nie88] I. Niemelä. Decision procedure for autoepistemic logic. In R. Overbeek and E. Lusk, editors, *Proceedings of the 9th international conference on automated deduction*, Lecture Notes in Computer Science, Vol. 310, pages 675–684. Berlin: Springer-Verlag, 1988

[Nie92] I. Niemelä. On the decidability and complexity of autoepistemic reasoning. *Fundamenta Informaticae*, 17:117–155, 1992

[NS93] A. Nerode and R. Shore. *Logic and logic programming.* Berlin: Springer-Verlag, 1993

[Prz88] T.C. Przymusinski. On the declarative semantics of deductive databases and logic programs. In J .Minker, editor, *Foundations of deductive databases and logic programming*, pages 193–216. Los Altos, CA: Morgan Kaufmann, 1988

[Prz89] T.C. Przymusinski. An algorithm to compute circumscription. *Artificial Intelligence*, 38:49–73, 1989

[RS70] H. Rasiowa and R. Sikorski. *Mathematics of metamathematics.* Warszawa: Polish Scientific Publishers, 1970

[RDB89] M. Reinfrank, O. Dressler, and G. Brewka. On the relation between truth maintenance and non-monotonic logics. In *Proceedings of IJCAI-89*, pages 1206–1212. San Mateo, CA: Morgan Kaufmann, 1989

[Rei78] R. Reiter. On closed world data bases. In H. Gallaire and J. Minker, editors, *Logic and data bases*, pages 55–76. New York, NY: Plenum Press, 1978

[Rei80] R. Reiter. A logic for default reasoning. *Artificial Intelligence*, 13:81–132, 1980

[Rei87] R. Reiter. Nonmonotonic reasoning. *Ann. Rev. Comput. Sci.*, 2:147–186, 1987

[Rog87] H. Rogers. *Theory of recursive functions and effective computability.* Cambridge, MA: MIT Press, 1987

[Sch87] J. Schlipf. Decidability and definability with circumscription. *Annals of Pure and Applied Logic*, 35:173–191, 1987

[Sch91a] G.F. Schwarz. Autoepistemic logic of knowledge. In A. Nerode, W. Marek, and V.S. Subrahmanian, editors, *Logic programming and nonmonotonic reasoning*, pages 260–274. Cambridge, MA: MIT Press, 1991

[Sch91b] G.F. Schwarz. Bounding introspection in nonmonotonic logics. In B. Nebel, C. Rich and W. Swartout, editors, *Proceedings of the 3rd international conference on principles of knowledge representation and reasoning, KR '92*. San Mateo, CA: Morgan Kaufmann, 1992

[Sch92] G.F. Schwarz. Minimal model semantics for nonmonotonic modal logics. In *Proceedings of LICS-92*. Washington, DC: IEEE Computer Society Press, 1992

[ST92a] G.F. Schwarz and M. Truszczyński. Modal logic **S4F** and the minimal knowledge paradigm. In *Proceedings of TARK 1992*. San Mateo, CA: Morgan Kaufmann, 1992

[ST92b] G.F. Schwarz and M. Truszczyński. Nonmonotonic reasoning is sometimes simpler. In *Proceedings of the 3rd Kurt Gödel Colloquium*. Berlin: Springer-Verlag, 1993

[ST93] G.F. Schwarz and M. Truszczyński. Subnormal modal logics for knowledge representation. In *Proceedings of AAAI-93*, Menlo Park, CA: AAAI Press, 1993

[Seg71] K. Segerberg. *An essay in classical modal logic*. Uppsala University, Filosofiska Studier, 13, 1971

[Shv90] G.F. Shvarts. Autoepistemic modal logics. In R. Parikh, editor, *Proceedings of TARK 1990*, pages 97–109. San Mateo, CA: Morgan Kaufmann, 1990

[Sta80] R.C. Stalnaker. A note on nonmonotonic modal logic. Unpublished manuscript, 1980

[Sti90] J. Stillman. It is not my default: the complexity of membership problem for restricted propositional default logics. In *Proceedings of AAAI-90*. Los Altos, CA: Morgan Kaufmann, 1990

[TK90] M. Tiomkin and M. Kaminski. Nonmonotonic default modal logics. In R. Parikh, editor, *Proceedings of TARK 1990*, pages 73–84. San Mateo, CA: Morgan Kaufmann, 1990

[Tru91a] M. Truszczyński. Embedding default logic into modal nonmonotonic logics. In A. Nerode, W. Marek, and V.S. Subrahmanian, editors, *Logic programming and non-monotonic reasoning*, pages 151–165. Cambridge, MA: MIT Press, 1991

[Tru91b] M. Truszczyński. Modal interpretations of default logic. In *Proceedings of IJCAI-91*, pages 393–398. San Mateo, CA: Morgan Kaufmann, 1991

[Tru91c] M. Truszczyński. Modal nonmonotonic logic with restricted application of the negation as failure to prove rule. *Fundamenta Informaticae*, 14:355–366, 1991

[Tru92] M. Truszczyński. A generalization of Kripke semantics to the case of logics without axiom K. Manuscript, 1992

[vEK76] M.H. van Emden and R.A. Kowalski. The semantics of predicate logic as a programming language. *Journal of the ACM*, 23(4):733–742, 1976

[YBB92] F. Yang, H. Blair, and A. Brown. Programming in default logic. Technical Report SU-CIS-93-26, Syracuse University, June, 1993.

[Zha90] A. Zhang. Personal communication. 1990

# Index

## A



## B



## C

## O



## P



## R

Springer Series
*Artificial Intelligence*

N. J. Nilsson: Principles of Artificial Intelligence. XV, 476 pages, 139 figs., 1982

J. H. Siekmann, G.Wrightson (Eds.): Automation of Reasoning 1. Classical Papers on Computational Logic 1957–1966. XXII, 525 pages, 1983

J. H. Siekmann, G. Wrightson (Eds.): Automation of Reasoning 2. Classical Papers on Computational Logic 1967–1970. XXII, 638 pages, 1983

L. Bolc (Ed.): The Design of Interpreters, Compilers, and Editors for Augmented Transition Networks. XI, 214 pages, 72 figs., 1983

M. M. Botvinnik: Computers in Chess. Solving Inexact Search Problems. With contributions by A. I. Reznitsky, B. M. Stilman, M. A. Tsfasman, A. D.Yudin. Translated from the Russian by A. A. Brown. XIV, 158 pages, 48 figs., 1984

L. Bolc (Ed.): Natural Language Communication with Pictorial Information Systems. VII, 327 pages, 67 figs., 1984

R. S. Michalski, J. G. Carbonell, T. M. Mitchell (Eds.): Machine Learning. An Artificial Intelligence Approach. XI, 572 pages, 1984

C. Blume, W. Jakob: Programming Languages for Industrial Robots. XIII, 376 pages, 145 figs., 1986

J. W. Lloyd: Foundations of Logic Programming. Second, extended edition. XII, 212 pages, 1987

L. Bolc (Ed.): Computational Models of Learning. IX, 208 pages, 34 figs., 1987

L. Bolc (Ed.): Natural Language Parsing Systems. XVIII, 367 pages, 151 figs., 1987

N. Cercone, G. McCalla (Eds.): The Knowledge Frontier. Essays in the Representation of Knowledge. XXXV, 512 pages, 93 figs., 1987

G. Rayna: REDUCE. Software for Algebraic Computation. IX, 329 pages, 1987

D. D. McDonald, L. Bolc (Eds.): Natural Language Generation Systems. XI, 389 pages, 84 figs., 1988

L. Bolc, M. J. Coombs (Eds.): Expert System Applications. IX, 471 pages, 84 figs., 1988

Springer Series
*Artificial Intelligence*

---

C.-H. Tzeng: A Theory of Heuristic Information in Game-Tree Search. X, 107 pages, 22 figs., 1988

H. Coelho, J. C. Cotta: Prolog by Example. How to Learn, Teach and Use It. X, 382 pages, 68 figs., 1988

L. Kanal, V. Kumar (Eds.): Search in Artificial Intelligence. X, 482 pages, 67 figs., 1988

H. Abramson, V. Dahl: Logic Grammars. XIV, 234 pages, 40 figs., 1989

R. Hausser: Computation of Language. An Essay on Syntax, Semantics, and Pragmatics in Natural Man-Machine Communication. XVI, 425 pages, 1989

P. Besnard: An Introduction to Default Logic. XI, 201 pages, 1989

A. Kobsa, W. Wahlster (Eds.): User Models in Dialog Systems. XI, 471 pages, 113 figs., 1989

B. D'Ambrosio: Qualitative Process Theory Using Linguistic Variables. X, 156 pages, 22 figs., 1989

V. Kumar, P. S. Gopalakrishnan, L. N. Kanal (Eds.) Parallel Algorithms for Machine Intelligence and Vision. XI, 433 pages, 148 figs., 1990

Y. Peng, J. A. Reggia: Abductive Inference Models for Diagnostic Problem-Solving. XII, 284 pages, 25 figs., 1990

A. Bundy (Ed.): Catalogue of Artificial Intelligence Techniques. Third, revised edition. XV, 179 pages, 1990

D. Navinchandra: Exploration and Innovation in Design. XI, 196 pages, 51 figs., 1991

R. Kruse, E. Schwecke, J. Heinsohn: Uncertainty and Vagueness in Knowledge Based Systems. Numerical Methods. XI, 491 pages, 59 figs., 1991

Z. Michalewicz: Genetic Algorithms + Data Structures = Evolution Programs. XVII, 250 pages, 48 figs., 1992

V. W. Marek, M. Truszczyński: Nonmonotonic Logic. Context-Dependent Reasoning. XIII, 417 pages, 14 figs., 1993

MIX
Papier aus verantwortungsvollen Quellen
Paper from responsible sources
FSC® C105338

If you have any concerns about our products,
you can contact us on
**ProductSafety@springernature.com**

In case Publisher is established outside the EU,
the EU authorized representative is:
**Springer Nature Customer Service Center GmbH**
**Europaplatz 3, 69115 Heidelberg, Germany**

Printed by Libri Plureos GmbH
in Hamburg, Germany